P9-DNU-125

Genetic Manipulation of Woody Plants

BASIC LIFE SCIENCES

Alexander Hollaender, Founding Editor

Recent volumes in the series:

Volume 32 TISSUE CULTURE IN FORESTRY AND AGRICULTURE
Edited by Randolph R. Henke, Karen W. Hughes, Milton J. Constantin, and Alexander Hollaender

Volume 33 ASSESSMENT OF RISK FROM LOW-LEVEL EXPOSURE TO RADIATION AND CHEMICALS: A Critical Overview
Edited by Avril D. Woodhead, Claire J. Shellabarger, Virginia Pond, and Alexander Hollaender

Volume 34 BASIC AND APPLIED MUTAGENESIS: With Special Reference to Agricultural Chemicals in Developing Countries
Edited by Amir Muhammed and R. C. von Borstel

Volume 35 MOLECULAR BIOLOGY OF AGING
Edited by Avril D. Woodhead, Anthony D. Blackett, and Alexander Hollaender

Volume 36 ANEUPLOIDY: Etiology and Mechanisms
Edited by Vicki L. Dellarco, Peter E. Voytek, and Alexander Hollaender

Volume 37 GENETIC ENGINEERING OF ANIMALS: An Agricultural Perspective
Edited by J. Warren Evans and Alexander Hollaender

Volume 38 MECHANISMS OF DNA DAMAGE AND REPAIR: Implications for Carcinogenesis and Risk Assessment
Edited by Michael G. Simic, Lawrence Grossman, and Arthur C. Upton

Volume 39 ANTIMUTAGENESIS AND ANTICARCINOGENESIS MECHANISMS
Edited by Delbert M. Shankel, Philip E. Hartman, Tsuneo Kada, and Alexander Hollaender

Volume 40 EXTRACHROMOSOMAL ELEMENTS IN LOWER EUKARYOTES
Edited by Reed B. Wickner, Alan Hinnebusch, Alan M. Lambowitz, I. C. Gunsalus, and Alexander Hollaender

Volume 41 TAILORING GENES FOR CROP IMPROVEMENT: An Agricultural Perspective
Edited by George Bruening, John Harada, Tsune Kosuge, and Alexander Hollaender

Volume 42 EVOLUTION OF LONGEVITY IN ANIMALS: A Comparative Approach
Edited by Avril D. Woodhead and Keith H. Thompson

Volume 43 PHENOTYPIC VARIATION IN POPULATIONS: Relevance to Risk Assessment
Edited by Avril D. Woodhead, Michael A Bender, and Robin C. Leonard

Volume 44 GENETIC MANIPULATION OF WOODY PLANTS
Edited by James W. Hanover and Daniel E. Keathley

Genetic Manipulation of Woody Plants

Edited by

James W. Hanover and
Daniel E. Keathley
Michigan State University
East Lansing, Michigan

Technical Editors

Claire M. Wilson and
Gregory Kuny
Council for Research Planning in Biological Sciences, Inc.
Washington, D.C.

PLENUM PRESS • NEW YORK AND LONDON

Library of Congress Cataloging in Publication Data

Conference on Genetic Manipulation of Woody Plants (1987: Michigan State University)
Genetic manipulation of woody plants / edited by James W. Hanover and Daniel E. Keathley; technical editors, Claire M. Wilson and Gregory Kuny.
p cm.—(Basic life sciences; v. 44)
"Proceedings of a Conference on Genetic Manipulation of Woody Plants, held June 21–25, 1987, at the Kellogg Center, Michigan State University, East Lansing, Michigan
Bibliography: p.
Includes index.
ISBN 0-306-42815-6
1. Woody plants—Genetic engineering—Congresses. 2. Woody plants—Propagation—In vitro—Congresses. I. Hanover, James W. II. Keathley, Daniel E. III. Title. IV. Series.
SB123.57.C65 1987
634.9′56—dc19 87-34298
CIP

This book was copyedited and entirely retyped by the staff of the Council for Research Planning in Biological Sciences, Inc., located on the premises of Associated Universities, Inc., of which the Council is a guest.

The opinions expressed herein reflect the views of the authors, and mention of any trade names or commercial products does not necessarily constitute endorsement by the funding sources.

Proceedings of a conference on Genetic Manipulation of Woody Plants, held June 21–25, 1987, at the Kellogg Center, Michigan State University, East Lansing, Michigan

Printed in the United States of America

PREFACE

This Volume contains the papers presented by twenty-eight invited speakers at the symposium entitled, "Genetic Manipulation of Woody Plants," held at Michigan State University, East Lansing, Michigan, from June 21-25, 1987. Also included are abstracts of contributed poster papers presented during the meeting.

That the molecular biology of woody plants is a rapidly expanding field is attested to by the large attendance and high level of enthusiasm generated at the conference. Leading scientists from throughout the world discussed challenging problems and presented new insights into the development of in vitro culture systems, techniques for DNA analysis and manipulation, gene vector systems, and experimental systems that will lead to a clearer understanding of gene expression and regulation for woody plant species. The presence at the conference of both invited speakers and other scientists who work with nonwoody plant species also added depth to the discussions and applicability of the information presented at the conference.

The editors want to commend the speakers for their well-organized and informative talks, and feel particularly indebted to the late Dr. Alexander Hollaender and others on the planning committee who assisted in the selection of the invited speakers. The committee consisted of David Burger (University of California, Davis), Don J. Durzan (University of California, Davis), Bruce Haissig (U.S. Department of Agriculture Forest Service), Stanley Krugman (U.S. Department of Agriculture Forest Service), Ralph Mott (North Carolina State University), Otto Schwarz (University of Tennessee, Knoxville), and Roger Timmis (Weyerhaeuser Company).

We would also like to thank the following sponsors for their generous financial help in making the conference possible: Argonne National Laboratory, Argonne Universities Association, Michigan State University Agricultural Experiment Station, Michigan State University Office of Research and Graduate Studies, Michigan State University Provost, Michigan Biotechnology Institute, the U.S. Department of Agriculture Forest Service Experiment Stations (Northeastern, North Central, and Pacific Southwest), U.S. Department of Energy, and the U.S. Department of Agriculture Cooperative State Research Service.

Preparation and coordination of on-site details were handled by Mary Brown of the Forestry Department, Myrtle F. Jones of the Kellogg Conference Center, and Raymond O. Miller of the Michigan State Cooperative Tree Improvement Program. Their tireless efforts were responsible for the smooth conduct of the symposium.

Finally, special appreciation goes to Claire M. Wilson and Gregory Kuny of the Council for Research Planning in Biological Sciences, Inc., for their key role in manuscript preparation and formatting for these proceedings.

James W. Hanover
Daniel E. Keathley

CONTENTS

KEYNOTE ADDRESS

TISSUE CULTURE SYSTEMS

BIOTECHNOLOGY APPLIED TO THE IMPROVEMENT OF UNDERGROUND SYSTEMS OF WOODY PLANTS

John G. Torrey

Harvard Forest
Harvard University
Petersham, Massachusetts 01366

ABSTRACT

The optimal functioning of the root system of a plant and its most favorable association with the soil and the microorganisms of the rhizosphere are crucial to the growth and development of the whole plant. A variety of technical approaches directed toward improving the functioning of the underground systems of woody plants are available. Genetic manipulations, including root transformations using microbial vectors, may result in more effective root systems and better understanding of root development. Chemical treatments, including mineral nutrient additions, hormonal modifications, and soil amendments, result in improved plant development. The root system provides sites for complex and subtle microbial interactions. Optimization of host-microorganism associations managed for maximal productivity is a major mode for biotechnical improvement of plants.

INTRODUCTION

Plant biotechnology encompasses the application of many diverse technical achievements directed toward the improvement of plants for use by man. This research effort in its many forms and ramifications will have a profound impact on all fields of plant production. The magnitude of the effect and the timing of the application will almost certainly be different in different fields of agriculture. Forestry, one of the most conservative fields of plant husbandry, bounded by constraints imposed by long-term investments in land and capital, and by longevity of the crop, will be one of the slowest to apply novel methods to improving plant production.

The biotechnological options open to foresters directed toward the improvement of forest tree productivity are not significantly different from those available to agriculturalists concerned with most other plant crops, namely, genetic improvement, better cultural methods, and more complete utilization of the biomass. What is significantly different is the time-span over which the investment of time, land, and cultivation occurs. The long-term scale in forestry leads to a conservative approach to management and to all its associated parameters: limited diversity in plantations (one works with the time-tested crop), careful attention to input costs, use of well-proven methodologies, and cautious exploration of new methods or techniques.

Where, one would like to know, does forestry stand in relation to the agricultural revolutions and advancing technologies developing so rapidly in plant research at the present time? Having recently taken the occasion to look at the history and present state of plant biotechnology applied to herbaceous plant species (67), I have found it intriguing to explore to what extent recent advances are being, and can be expected to be, applied to forest trees. For the purposes of this exploration I have chosen to limit my review to biotechnology applied to the improvement of underground systems of woody plants--that is, the root systems and their biotic and edaphic associations. I will examine briefly the following topics: genetic improvement, including genetic manipulations combined with the use of plant organ and tissue culture methods; mineral nutrition; and the improvement of plant-soil microorganism associations leading to optimum symbioses. The last topic involves the examination of inorganic nutrient uptake facilitated by mycorrhizal relationships, and biological nitrogen fixation including Rhizobium and Frankia symbioses.

ROOT FORM AND STRUCTURE

Roots have genetically determined form; they are conservative organs with such basic similarities in meristem organization and branching patterns that they do not offer good taxonomic characters. Yet root systems in higher plants are remarkably diverse depending upon whether the seedling radicle forms a vigorous tap root with lesser secondary branchings, or the lateral roots replace the main axis root and form an elaborately branched fibrous root system. The ramification and distribution of branched root systems may provide an extensive underground surface area. This is the case, for example, in the herbaceous rye plant, described by Dittmer, which at four months he calculated to possess a root system comprised of nearly 14 million branch roots estimated to be 623 km in combined length with a surface area of 639 m^2, including about 14 billion root hairs (21). Root systems of trees are much larger than those of herbaceous plants and often more complex. According to calculations by Lyford (38), a mature red oak tree possesses of the order of 500 million live root tips. Increasing evidence of turnover in these fine root tips (43) suggests that root form is in a continuous state of change that must influence its function in the whole plant.

Root systems are notoriously difficult to measure and describe, so that, despite heroic efforts to obtain reliable and accurate estimates of root form and structure (7), usually only seedling roots of forest trees are carefully observed and their development and final form only surmised or very inadequately sampled.

What can one do to manipulate root form to benefit the plant and its productivity? Several approaches come to mind. First is the very practical matter of starting out seedlings with well-developed root systems. Much attention has been paid in the forestry industry to the design and handling of nursery plantations to optimize seedling root form and vigor. In bare root seedling outplantings, care is directed toward soil mixes favoring lateral root branching, field practices of root pruning are applied, and optimum conditions for transplanting are monitored. In containerized seedling production, containers are designed to favor root branch development, and vertical downward growth with air pruning. I keep up on one point of view by reading the "Rootrainers Newsletter" of Spencer-Lemaire (Alberta, Canada), just one of many publications on methods where attention is focused on seedling root form (cf. also Ref. 66).

A more basic approach and one worth exploring is to raise the question of whether one can select genetic traits for optimum root form that can lead to best seedling establishment in the field and optimum adult root form for given sites. These questions have been raised for agronomic crops such as soybean, corn, and tomato, as illustrated by the studies of Zobel (83,84). Have these questions been pursued in forest trees? I know of few studies focused on the genetic control of root systems of forest trees through breeding directed toward improving seedling or adult root form. There exists a literature in the horticultural field in which grafting plays a major role in propagation. Here, a root stock favorable for disease resistance, winter hardiness, or other characters is selected as the root system onto which to graft choice selected scions. In general, these methods have not been and may not be applicable to forest trees.

ROOT TRANSFORMATION AND GENETIC MANIPULATIONS

Recently new approaches to the problem of manipulating root form have become available through studies on the nature of the "hairy root disease." The soil bacterium *Agrobacterium rhizogenes*, closely related to *Agrobacterium tumefaciens*, carries a plasmid that induces abundant adventitious root formation when it infects shoot or root tissues of a wide variety of plants. The genetic information for root initiation is conveyed on a piece of DNA of the plasmid, called the T-DNA (transferred DNA), which, upon infection of host cells, is transferred to the host chromosomal DNA and is there stably integrated (15,64). In this transfer the evidence is that the Ri (root-inducing) plasmid acts genetically in the same way that the Ti (tumor-inducing) plasmid of *A. tumefaciens* is known to act (18). In the latter case, transformed tissues carry genetic information that leads to tumor formation (19); in the former case, the DNA codes for, among other characteristics, the biochemical events leading to lateral root initiation. The Ri T-DNA also codes for opine synthesis in a way analogous to the Ti-plasmid.

Tepfer (61,62) has pointed out that the Ri T-DNA transformation need not be deleterious to the plant but may encode beneficial information bearing on root formation that could prove useful in improving plant establishment that depends on effective root systems. His experiments, performed first with tobacco, carrot, and *Convolvulus* roots, have been extended to a number of plants, including woody species such as apple (63). The host range of *A. rhizogenes* is broad (20) and can be expected to be applicable widely.

Let us look more closely at what we know about the Ri-plasmid and its root-inducing properties. Cultured callus tissues, root or stem, can be readily infected by *A. rhizogenes* and induced to form large numbers of adventitious roots in two to three weeks. Root tips excised and grown in culture can be freed of bacterial contaminants using antibiotics. Transformed excised roots grow rapidly in relatively simple media such as modified Murashige and Skoog medium, showing faster rates of elongation and much more frequent lateral root formation. Chemical analysis demonstrates that these roots form opines. Extraction of host chromosomal DNA from the cultured roots, digested and hybridized using Southern blot procedures against Ri-plasmid T-DNA, showed that some of the plasmid DNA had been transferred into the host DNA, that is, that the roots were transformed. A higher frequency of formation of lateral roots and the production of opines confirm the transformation. Such cultured roots grow much faster than normal roots and can be sustained in culture indefinitely with much greater ease.

Callus tissue can be induced in turn from excised root segments using supplementary callus-inducing hormones [2,4-dichlorophenoxyacetic acid (2,4-D) and kinetin] from which cell suspensions develop in liquid culture. In tobacco and in carrot, whole plant regeneration from callus or cell suspensions, possibly via embryoids, is possible. Such regenerated plants show high levels of opine, and roots are more highly branched. In the herbaceous species studied (61), leaves on shoots showed some wrinkling, shoots showed reduced apical dominance, and in carrot, plants became annuals rather than biennials, and showed slight reduction in seed production. All of these characters segregate together. Of greatest interest to us is the fact that the host plant produces a more elaborate root system that might function to benefit the plant.

It seems clear that the T-DNA derived from the Ri-plasmid contains, along with other genes, a gene cluster regulating root initiation. We know little in detail about that group of genes but it is now accessible to analysis, understanding, and utilization. As has been summarized by Torrey (68), it seems reasonable that the gene cluster will code for the synthesis of indoleacetic acid, which is known to be involved in lateral root initiation in many plants. Also, it is possible that genes controlling cytokinin biosynthesis are part of the Ri-plasmid T-DNA complex. We lack specific information on these genes in the Ri-plasmid but can expect that unraveling these events at the genetic level will provide tools for the elegant and subtle manipulation of root form so important to seedling establishment and whole plant survival.

The applicability of the Ri T-DNA transformation to woody species has already begun to be explored by Tepfer and his colleagues (63). Following the lead of Moore et al. (47), who claimed that apple trees infected with *A. rhizogenes* withstood drought better than uninfected trees, Tepfer et al. (63) developed a procedure for improving the rooting of cuttings of apple trees by transforming the root system with the Ri-plasmid T-DNA. Thus, they claim to have produced genetic grafts with normal aerial plant parts attached to transformed, more effective root systems containing introduced foreign genes.

Since lateral root formation in seedling or mature growing roots is a normal morphogenetic event, it is clear that there exists within the host genome a gene cluster homologous to the T-DNA of the Ri-plasmid from *A. rhizogenes*, and that the transformation by the Ri-plasmid increases the dosage of these genes or amplifies their expression in some way. Dissection of this phenotype into its genetic components should provide considerable insight into the normal morphogenetic events controlling root development.

GENE CONTROL OF LATERAL ROOT FORMATION

Another promising approach to our better understanding of the basic events leading to root initiation and branching should be mentioned. This approach involves efforts to identify and characterize the genes controlling root meristem formation by the identification of distinctive proteins associated with specific morphogenetic events in root meristem initiation, and then tracking them back through the mRNA that elicited them to the cDNA bearing the genes controlling the event.

This approach has been used with some success in pursuing the "nodulins" in Rhizobium-induced nodule formation in legumes. In these studies, recently reviewed by Verma and Nadler (76), efforts have been made to identify and isolate host genes involved in the formation of root nodules in response to Rhizobium infection. The experimental approach involved immunological methods to identify organ-specific proteins and DNA-RNA hybridization techniques to locate and identify the cDNA ultimately responsible for specific phenotypic expressions. A large number of nodule-specific gene products (proteins termed nodulins) have been identified. Nodule-specific proteins such as leghemoglobin, a product of the symbiotic interaction of Rhizobium and legume host genomes, have been well characterized as well as enzymes associated with the dinitrogen-reducing system uniquely associated with root nodules. According to Verma et al. (75), perhaps 100 different genes can be estimated to be involved in root nodule formation and maintenance based on DNA-RNA hybridization experiments.

Bisseling et al. (6) have emphasized the events in root development elicited by Rhizobium infection that center on host tissue differentiation events. These are cytodifferentiation and morphogenetic events that, while unique to root nodule development, bear close relationship to the events leading to lateral root initiation and development. The method of in vitro translation of extracted root nodule RNA followed by two-dimensional gel electrophoresis of the translation products served as a sensitive and efficient method for revealing mRNAs associated with developmental stages in nodule formation.

Of the more than 500 polypeptides identified by this method, most were present in both roots and root nodules. Only 21 polypeptides were products of nodulin mRNA. These included leghemoglobin nodulins and others of varying molecular weight that changed with time of nodule development. Molecular weights varied from 68,000 down to 21,000 among the major nodulins characterized. Nodulin N-40, with a molecular weight of 40,000, appeared eight days after nodule initiation and before leghemoglobin nodulins were apparent. This early nodulin has continued to be of intense interest, probably representing early events in nodule initiation. Studies directed to determine the locus of the genes in the Rhizobium that elicit the nodulin genes in pea roots show that chromosomal genes of Rhizobium rather than plasmid DNA are involved in eliciting the nodulin gene expression in the host plants.

This same approach that has begun to be used so successfully in analyzing the specific events in nodulation in legumes, gives promise for the study of developmental events in normal root development. Using techniques similar to those outlined above for the analysis of nodule-specific gene expression, Christianson and Warnick (16) have attempted to study the developmental events in the process of root initiation from cultured leaf explants of Convolvulus. Instead of Rhizobium initiation of the morphogenetic event of nodule formation, they have elicited root initiation in cultured leaf discs with root-inducing concentrations of indolebutyric acid (IBA), in a process analogous to root induction on stem cuttings. Roots are initiated and develop in a predictable time-sequence over a period of three weeks. They have discovered that the process can be interfered with by chemical treatments which are effective at precise times in the initiation process. For example, tri-iodobenzoic acid (TIBA), a reagent believed to block auxin transport, applied at 5 mg/l in the root-inducing medium, completely prevents root initiation if applied continuously. It acts, however, only at a specific state in root formation which is between day 7 and day 12 in their three-week treatment period. Another and different block can be placed on auxin-induced root development by treatment of the explants with high levels (20 g/l) of galactose for one day only at day 11. The authors believe they are interfering with the sequential developmental program of root initiation.

Using this experimental system, Warnick and Christianson (77) have attempted to apply in vitro translation of mRNAs extracted from different stages in the root initiation process, hoping to characterize the proteins associated with these developmental events. Their studies showed that a novel pattern of mRNA occurrence develops after root-induction in leaf explants is initiated (two to four days). A shift in the pattern of mRNA formation occurs at the time the root primordia are "determined" (six days), and this pattern is essentially identical with mRNA preparations from roots. With their experimental system, these authors are in the position to document at the molecular level some of the developmental events involved in root formation and, in addition, chemical manipulations that may foster improved root formation.

CHEMICAL TREATMENTS AFFECTING ROOT FORM AND FUNCTION

Historically, the 1930s stand out as the decade for progress in roots and root biotechnology. It had been postulated earlier by Loeb (36) that root formation in plants was controlled by a special root-forming substance or hormone, formed in the buds or leaves of green plants. In rooting of woody stem cuttings, van der Lek contributed much to our understanding of the correlations involved in root initiation (cited by Ref. 65 and 78). With the characterization of the plant hormone indole-3-acetic acid (IAA) as the auxin active in root initiation on stems (78), it became possible to induce root formation by chemical treatment in many woody species. One of my oldest lantern slides is an excellent photograph of woody stem cuttings of Ilex, one set untreated, the other dipped in IBA and then maintained in rooting medium [taken from Zimmerman and Hitchcock (81)]!

The old idea of a root-forming substance, the "rhizocaline" of Bouillenne and Went (cf. Ref. 78), took on a real and practical meaning as stable analogs of IAA were synthesized and tested on a range of herbaceous and woody species (82). Thus, IBA and naphthaleneacetic acid (NAA) became established in a commercial way which has persisted up to the present--root initiation on demand by application of a chemical off the shelf. How much more successful can one be in biotechnology?

CULTURE OF EXCISED ROOTS

The 1930s also marked the first success with the isolation and continuous cultivation in vitro of excised roots (79). This work was preceded by a flurry of activity both in Europe and in the United States that led to our clear understanding of the heterotrophic and autotrophic nature of roots, concepts from which could be extrapolated an understanding of roots and root systems in whole plants. Excised roots of tomato could be grown by regular subculture every few weeks of root segments with lateral root meristems that formed successive new roots. The nutrient requirements of several species of herbaceous roots were relatively simple: macro- and micronutrient mineral elements, a carbon and energy source such as sucrose, and a few growth factors including the B vitamins thiamin, pyridoxine, and nicotinic acid. Subtle differences among species tested suggested that the roots of species or even of cultivars possessed different genetic capacities to synthesize these vitamins. For example, isolated pea roots and radish roots required thiamin and nicotinic acid in culture, while flax roots required only thiamin and tomato roots required all three B vitamins (10). Otherwise, excised roots were able to synthesize the organic constituents necessary for continued growth and development. The dependence of the root apex on the shoot for its organic nutrition was not unexpected, and sucrose is the most common translocated form of carbohydrate in higher plants. That the B vitamins were also photosynthetic products upon which the root depended must have been a surprise. For most species, the excised root was able to synthesize all of its requirement for hormones, including IAA, gibberellic acid, cytokinins, abscisic acid, and ethylene, in whatever amounts that were required. Occasional reports of excised root cultures being stimulated by adding very low amounts of auxin and cytokinin (56) suggest that species may differ in their requirements (i.e., they may differ in their capacity to synthesize the essential organic growth factors necessary for their optimum growth in vitro).

Even so, excised root cultures of many species of plants have failed despite considerable effort. Particularly prominent on the list of hard or impossible-to-grow plants in excised root culture are species of woody plants. In 1942, Bonner (9) reported his efforts to grow excised roots of woody species in culture. He reported limited success with excised roots of *Sterculia diversifolia*, *Bauhinia purpurea*, *Wisteria sinensis*, and *Acacia melanoxylon* grown in culture media with known mineral nutrients, the B vitamins, sucrose, and other complex addenda.

Goforth and Torrey (27) reported the successful culture of excised roots of the woody shrub *Comptonia peregrina* by providing the B vitamins, thiamin, pyridoxine and nicotinic acid, at trace amounts (0.1-0.5 ppm), and, in addition, near nutrient level amounts (10-100 ppm) of the sugar alcohol myoinositol. These roots in culture grew reasonably well through a number of subcultures and formed secondary thickening at their bases. Also, they were able, at least in early passages, to initiate endogenous buds--the characteristic mode of vegetative propagation in this species. Hormone additions to the culture medium had no beneficial effect.

Roots of woody species are difficult to grow in culture, and because of the difficulties they have been neglected. Goforth and Torrey (27) summarized the very limited literature. As listed by them, limited success has been reported for *Acacia melanoxylon*, *Acer rubrum*, *Comptonia peregrina*, *Picea abies*, *Pinus* spp., and *Robinia pseudoacacia*. In our laboratories we have tried a good number of other species in the hope that we could embark on studies with them in relation to vegetative propagation, secondary growth, or root nodulation. Our success has been very limited (unpublishable!). Over the years we have tried to grow excised roots in culture, under a range of conditions and media, of the following woody species: Acer rubrum, Allocasuarina decaisneana, Alnus rubra, Casuarina cunninghamiana, Casuarina glauca, Myrica gale, and Quercus rubra, among others. All were provided with various combinations of vitamins, hormones, sugar alcohols, sugars, and complex factors like yeast extract without much success. We were successful with random individual roots of the Allocasuarina species, whose seeds produce a large seedling radicle that exhibits vigorous and rapid seedling root development.

In the culture of excised woody roots we seem to be missing some component, presumably a photosynthetic product, necessary for continuous activity of the root meristems when isolated from the plant. Progress in our understanding of this problem would open up the use of cultured roots for experimentation among woody species. It is possible that transfer of the T-DNA from the Ri-plasmid of A. rhizogenes would transfer the necessary genes to invigorate excised root culture of woody species, as it does with a number of herbaceous species. To date, I have found no successful published experiments in this direction with woody plants, although increasing numbers of reports are published on this technique as a device to improve root culture in vitro for biosynthesis of important secondary product formation [e.g., excised transformed and cultured roots of Atropa belladonna (29); excised "hairy root" cultures of Hyoscyamus muticus, producing tropane alkaloids (23); and excised transformed roots of Ambrosia, Bidens, Rudbeckia, and Tagetes, all in the family Asteraceae, which synthesize in the roots sesquiterpene lactones and polyacetylenes, compounds active against bacteria, fungi, and nematode pathogens (23)].

Excised root cultures have proved useful in the study of a number of other developmental phenomena in herbaceous plants. Although roots of most herbaceous plants undergo limited secondary thickening and tend not to become woody, the process of cambial activation and vascular cambium activity was studied in excised roots of radish by Loomis and Torrey (37,70). Using basal feeding of excised roots in culture, they showed that hormones entering from the shoot end of the root provide the stimulation to vascular cambium activation and sustained meristematic activity to produce secondary diameter growth in roots. Auxins, cytokinins, and myoinositol were all implicated in this stimulation. Since roots vary in their inherent tendency to undergo secondary thickening, one would expect the hormonal supply necessary for cambial activation in the root to be under genetic control. Thus far, no experimental system has allowed the characterization and possible manipulation of the genes for cambial activation. Thickened root systems are characteristic of tree species, and management of this character could be a useful tool in programming the most effective root systems either for support or for storage.

Excised root cultures have also been used for the study of endogenous bud formation, for example, in the herbaceous species Convolvulus arvensis (11,12). Endogenous bud formation, like lateral root formation, is under hormonal control, involving both auxins and cytokinins. Although we do not yet know of a genetic system, possibly involving bacterial vectors, that regulates bud initiation in roots, it is reasonable to make a search for such vectors. Bacteria, such as Corynebacterium faciens, that are known to produce witches' brooms, i.e., multiple shoot formation, in above-ground plant parts can be expected to occur, which may affect roots similarly. These microorganisms may provide genetic tools for experimental manipulation of shoot formation by roots, either to exploit in plant propagation or to help to understand and manipulate bud formation for purposes of inhibition or control.

Root nodulation by Rhizobium has been studied in an in vitro system (55), and infection by vesicular-arbuscular mycorrhizal organisms has been achieved in cultured roots of clover (48). No success has been reported in establishing this type of mutualism in excised roots of woody plants.

IMPROVED MINERAL NUTRITION

We have known since the early 1940s in considerable detail the mineral elemental requirements for plant development. Of the more than 100 elements in the earth's crust, only 16 are essential, namely, C, H, O, P, K, N, S, Ca, Fe, Mg, Mo, B, Cu, Mn, Zn, and Cl, with possibly another one or two trace elements to be added. The first three of these serve as major inputs from the air and water (CO_2, H_2O, O_2) which, in the presence of the energetic input from the sun, are incorporated as the major carbohydrate building blocks of the plant. All the other mineral elements, some used in substantial quantities, come from the soil. The basis for the usual mineral fertilizer applications are the macronutrient elements "NPK," applied in various ratios in applied mineral fertilizer to many agricultural, horticultural, or home-garden crops. The availability of these elements frequently sets limits on the productivity of crops, including trees.

For perhaps decades, foresters have made use of this recently acquired knowledge, using high-technology methods such as helicopters to spread mineral fertilizers including nitrogen in the form of urea and phosphate and potash as salts and perhaps added trace elements. Specific tree crops have been shown to respond by growing faster and larger. Detailed studies of forest tree responses to mineral fertilizers are still not very extensive. Only recently has Eucalyptus been shown to respond to fertilizer treatment, so this is still a relatively new technology for foresters. Taking 1940 as the date we had in hand explicit information concerning the beneficial role of mineral fertilizers in plant growth, one can see that many stands of trees were established well before such information was fully available, and 40 years for experimentation is not a long time for forest tree production.

Even so, the use of mineral fertilizers on such a large scale (by the hundreds of hectares) is both expensive and not without risk. Mineral fertilizers are produced by methods involving costly fossil-fuel-consuming technologies which reduce long-term energy resources and add considerably to the production costs. Run-off of applied minerals may be wasteful and even hazardous, leading to contamination of potable water supplies and eutrophication of standing bodies of water such as lakes and ponds. So alternate methods of meeting the mineral nutrient needs of forest trees, as in agriculture, are actively sought.

In the following section, I want to review briefly the developing knowledge and technologies which center on improved mineral nutrition of forest trees by the use of improved associations between tree roots and soil microorganisms, leading to symbiotic systems that facilitate the uptake and utilization of mineral elements, in particular, phosphate (PO_4^{3-}) and nitrogen from the dinitrogen (N_2) of the atmosphere.

MINERAL NUTRIENT AVAILABILITY FROM SYMBIOTIC SYSTEMS

In recent years, agriculturalists and foresters have become increasingly interested in biological methods of increasing mineral nutrient availability for plant growth and development through establishing symbiotic relationships among soil microorganisms and the host plants. The first major group discussed below concerns plant host-soil microorganisms involving fungus associations with roots called mycorrhizae ("fungus roots"). Mycorrhizae are symbiotic associations between certain soil fungi and plant roots. The roots are infected by the fungus, typically in a nonpathogenic relationship. The fungus usually obtains essential nutrients from the host plant, especially carbohydrate; the host, in turn, benefits from increased uptake of minerals from the soil, especially phosphate ion. A comprehensive and readable review of mycorrhizal symbioses has been written by Harley and Smith (28). Some of the detailed methodology used in the study of mycorrhizae has been assembled in useful form by Schenck (57).

Ectotrophic Mycorrhizae

This type of association is found exclusively in trees and shrubs. The fungal mycelium forms an extensive mantle or sheath around the infected root, often producing swollen and shortened root tips that have characteristic form and color. The fungus mycelium grows between the root cortex cells, forming a network called the "Hartig net" but not entering the cells. Mycelial strands extend out into the soil. Soil nutrients in ionic form, both cationic and anionic, are absorbed from the soil and pass through the fungal sheath and then into the roots of the host plant. Vascular tissue serves to carry the nutrients in aqueous solution to all the parts of the plant.

The common forest trees infected by ectotrophic mycorrhizal fungi include many gymnosperms, especially the conifers, such as pines, spruces, larches, and firs. Many hardwood angiosperms are also infected by ectomycorrhizae, including oak, beech, birch, and eucalyptus. According to Meyer (45), about 3% of the flowering plants are infected by ectotrophic mycorrhizal fungi, and all are woody plants.

The fungi may belong to several of the major fungal groups which include basidiomycetes (toadstool or mushroom-forming fungi) and ascomycetes. Many of these fungi are relatively easy to isolate from their natural occurrences and grow in pure culture on simple or complex nutrient media. Production of the fruiting bodies in the laboratory proves to be difficult. In the northern temperate zone, some of the frequently occurring basidiomycetes that form mycorrhizae include the following genera: *Boletus*, *Russula*, *Amanita*, *Pisolithus*, *Lactarius*, and *Cenococcum*.

Endotrophic Mycorrhizae

Far more common are the fungal-root associations termed endotrophic mycorrhizae. The hosts for these fungal symbionts are very diverse, ranging from the mosses, through the ferns, to the gymnosperms and angiosperms. The hosts are both herbaceous and woody, including important agricultural species and forest trees, and the associations range from the tropics to the arctic and antarctic. Within the endomycorrhizal fungi there are three distictive types of associations, namely, ericoid, orchid, and vesicular-arbuscular mycorrhizae. For forestry, this last group is most important.

Endotrophic mycorrhizal fungi actually penetrate the roots, invade cortical cells, and within these cells, differentiate elaborate mycelial branching. The external appearance of the root is typically less modified than in ectotrophic mycorrhizae, and one must fix and clear the root and stain the fungal mycelia within the root in order to determine the occurrence of these endomycorrhizae (53). There are a number of well-studied fungi belonging to this group, which is diverse and may include members of the basidiomycetes, the phycomycetes, and the fungi imperfecti.

The vast majority of plants infected by endotrophic mycorrhizal fungi belong to the group termed vesicular-arbuscular mycorrhizae, usually referred to as VA mycorrhizae or VAM. In the VAM, the fungal associates are aseptate or irregularly septate forms of phycomycetes. Within the cortical cells of the infected root, they form elaborately branched terminal filaments called arbuscules. They also develop enlarged terminal endings called vesicles. The two most common genera are Glomus and Gigaspora. None of the VAM fungi has been grown in pure culture independent of the host.

As with the ectotrophic mycorrhizal fungi, the VAM hyphae extend out into the soil well past the zone that is accessible even to the longest root hairs formed at the epidermal surface of the root. Cations and anions in the soil solution or released from soil particles in these zones are absorbed in the fungal hyphae and translocated within the hyphae into the root cortical cells, where they become available to the living cells of the host root. Phosphate may also be stored in the fungal hyphae as polyphosphate granules. The fungi derive carbon and energy sources from the host root tissues, usually as carbohydrate.

The evidence is now conclusive that both types of mycorrhizal fungi increase the absorptive surfaces available to the roots by their extensions into the soil and that they facilitate the uptake and transport of mineral nutrients, especially the phosphate anion (PO_4^{3-}) but also sulfate (SO_4^{2-}) and the major cations (K^+, Ca^{2+}, and Mg^{2+}). In soils in which these mineral elements are in limiting supply, the presence of mycorrhizae results in strongly beneficial effects to plant growth that are especially evident in phosphorus-deficient soils. There is no evidence that mycorrhizae can fix atmospheric nitrogen, but there is experimental evidence that in some host plant groups (e.g., the Ericales), mycorrhizae increase the uptake of nitrogenous compounds which in turn benefit the growth of the host plants (59).

There are claims that mycorrhizae increase water uptake and benefit plants by improved water availability (26,44). This subject is a matter of current debate. At the present time, the demonstrated beneficial effects of mycorrhizal associations, both ecto- and endo-, result primarily from increased inorganic nutrient uptake, especially phosphate.

In perennial plant nurseries where plants are started from seed or cuttings in sterilized soil or synthetic soil mixes, the need for introducing appropriate fungal organisms to allow the development of mycorrhizal associations is particularly evident. The beneficial responses resulting from inoculation with mycorrhizal fungus strains are well documented in Citrus (31), apple (54), and the conifers such as Pinus (41,42).

Also, when exotic trees are introduced into new sites, associated fungal organisms may need to be introduced by soil inoculation. Thus in New

Zealand, South Africa, and Australia, establishment of man-made forests using containerized plants depends on the appropriate introduction of mycorrhizal fungi.

In recent years, the problems of selecting the proper fungus to introduce for forest plantations have been complicated by the recognition that, in fact, no single fungus occupies the root system of a given plant. In birch, for example, young seedling trees can be shown to have their roots infected predominantly by one species of ectomycorrhizal fungus growing in the field. One can show that over successive years (22), there is a succession of fungal organisms that occupy the growing roots. In mature stands of trees in the forest, quite different populations of fungi occupy the growing roots than infect young seedling roots. Much more needs to be learned about the succession of root-infecting fungi and how they benefit their hosts.

The inability to successfully grow endomycorrhizal fungi in culture has been a great impediment to research on VAM associations. Using methods developed by Gerdemann and his students, one can screen out the large (80 μm or greater diameter) spores of Glomus or Gigaspora from the soil around root systems infected with these fungi. But despite much effort and study, such spores may be difficult to germinate in sterile culture and the germinating filaments fail to grow. At present, research on VAM associations depends upon propagating the root-infecting VA fungus in tne roots of growing plants, the so-called pot-culture technique, and harvesting infected roots and reinoculating the soil with root pieces containing the endophyte.

In natural forests, mycorrhizal associations are commonplace. In man-made ecosystems, artificial inoculation with mycorrhizal fungi is of increasing importance in agriculture and in forestry. Inoculation with ectomycorrhizal fungi is common practice in Europe (35,48) and in North America (41). Millions of seedlings grown for forest plantation are inoculated with mycorrhizal fungi at low cost. As has been pointed out by Abelson (1), further improvement of mycorrhizal associations will come from additional conventional selection procedures of both host and microsymbiont and from development and application of genetic engineering techniques.

Progress in the improvement of VAM associations in forest tree plantations is slowed by the difficulty in producing inoculum (25). With the unavailability of axenic pure cultures of VAM fungi, commercial inoculum is produced by large-scale modifications of the pot-culture technique. Inoculum has been prepared in a variety of rooting media which include infected root pieces, spores, and hyphae. Methods of large-scale application are also being considered. Use of inoculation techniques under appropriate conditions will assure improved plant growth, drought resistance, increased pH tolerance, and in some cases increased resistance to plant pathogens. These potential gains continue to attract intense research efforts.

Rhizobium-Legume Symbioses of Interest in Forestry

The importance of symbiotic nitrogen fixation by legumes whose root systems are nodulated by the soil bacterium Rhizobium has long been recognized in agriculture. Culture of Rhizobium for the production of inoculum for assuring nodulation of legume crops such as soybeans, common beans, alfalfa, clover, and other important agricultural crops in the legume family has been common practice for many years. Even the home gardener has access to inoculum through *Rhizobium* prepared and packaged in a peat mix, appropriate for species-specific infection and nodulation.

Leguminous tree species also benefit from *Rhizobium* infection and nodule formation. Foresters searching for trees suitable for a wide range of sites and climates have turned increasingly in recent years to leguminous trees, especially for use in tropical and subtropical climates where native legumes are most widespread. But even in temperate regions, searches have been made for more effective multipurpose trees in the legume family. This interest has been expressed by the formation of such groups as the Nitrogen-Fixing Tree Association, based in Hawaii but international in its scope.

Like herbaceous annual legume crops, seedling legume trees may be deliberately associated by inoculation with appropriate strains of *Rhizobium*, which infect the seedling root system, either by root hair infection such as is seen and well studied in white clover (*Trifolium*), or by direct epidermal penetration and cortical infection such as is seen in *Stylosanthes* and other subtropical herbaceous legumes (69). Although limited research has been performed on the establishment of root nodules by *Rhizobium* on leguminous trees, many such associations are known and have been described (2). Recent efforts have been made to improve the selection of highly infective and effective strains of *Rhizobium* for inoculation of leguminous trees, especially for reforestation in the tropics.

Some of the important tropical or subtropical trees under active study, with support from a number of international funding agencies, include the following genera (13): *Acacia*, *Albizia*, *Calliandria*, *Erythrina*, *Gliricidia*, *Leucaena*, *Prosopis*, *Samanea*, and *Sesbania*. Most research on trees is with wild, genetically undefined populations, and research on the infective and effective (dinitrogen-fixing symbioses) strains of *Rhizobium* is still in its infancy, centering on collections of native strains from diverse soils and sites which require testing at the most basic level. Such research has been fostered by the establishment of international networks in the research for germplasm, the exchange of seed sources, and the collection, testing, and exchange of *Rhizobium* strains by such agencies as the Board on Science and Technology for International Development, National Academy of Sciences (BOSTID), the Food and Agriculture Organization of the United Nations (FAO), the International Council for Research in Agroforestry (ICRAF), the International Development Research Centre of Canada (IDRC), Nitrogen Fixation for Tropical Agricultural Legumes (NiFTAL), the United States Agency for International Development (AID), and other organizations.

The stimulus for improving biological nitrogen fixation in forest trees has come from the plight of less-developed countries around the world faced with impossibly high costs of chemical fertilizers and urgent needs for extensive reforestation. Nitrogen-fixing trees often establish easily in nutrient-poor soils and difficult sites, giving them some advantage in competing with other species by virtue of the use of dinitrogen from the atmosphere. The leguminous trees have multiple uses which also encourage their establishment--uses as fuelwood, timber and/or pulpwood, fodder, food for man, shelter belts, and soil stabilizers and improvers.

During the past decade, remarkable strides have been made in unraveling the molecular genetics of the Rhizobium-legume symbiosis (cf. Ref. 74 and 75). Rhizobial genes have been characterized for dinitrogen fixation (nif), nodulation (nod), and many other elements of the gene complex involved in symbiosis. Such information may lead to improved bacterial strains for more effective associations beneficial to agriculture. To date, almost none of this information has been applied to leguminous forest trees.

Dissection of the nod gene cluster has led recently to novel approaches to facilitating the Rhizobium-legume association. One of the earliest steps in the infection process is that of recognition, i.e., the bacteria in the soil sensing the presence of an appropriate host. Peters et al. (52) have shown in the case of alfalfa (Melilotus) that the root system exudes a chemical signal in the form of a flavone, luteolin, which induces the transcription of nodulation genes in Rhizobium, leading to nodule initiation. Already it is clear that luteolin is only one of a family of flavones that serve as chemical stimuli for soil bacteria. Supplementing the immediate environment of a seedling root with chemical applications of such a signal may be expected to enhance the effectiveness of the symbiosis (30). Closely related compounds may be inhibitory to these early recognition steps. These discoveries may open a new approach to field manipulation of this symbiotic relationship.

Frankia-Actinorhizal Symbiosis

In the tropics and subtropics, woody legumes represent the major group of importance in biological fixation of dinitrogen. In temperate regions, both north and south, the major woody species involved in symbiotic nitrogen fixation are the actinorhizal plants--a diverse group of woody dicotyledonous plants nodulated by the filamentous bacterium Frankia of the Actinomycetales. In excess of 200 species in 24 genera distributed among eight families are encompassed by this broad group (8) including such familiar trees as the alders (Alnus spp.), autumn olive and Russian olive (Elaeagnus spp.), and a large number of shrubs. An actinorhizal group important in the tropics and subtropics includes members of the Casuarinaceae, i.e., Casuarina, Allocasuarina, and Gymnostoma, which taken together comprise about 80 species. Only in the last decade has it become possible to envisage technological manipulation of symbiotic associations in these plants. With the isolation and growth in pure culture of a Frankia strain from the root nodules of Comptonia peregrina (14), methods became available which have made possible the isolation and culture of a large number of Frankia strains. Studies on the cultivation and use of Frankia as inoculum for selected plantations have been published, and research in practical applications continues. In a special issue of Plant and Soil (Vol. 87, No. 1, 1985), resulting from a Frankia meeting at Laval University in Quebec, Canada, several papers were devoted to culture of Frankia, methods of inoculation, and seedling establishment for forestation. This issue represents one of the more or less biennial publications resulting from Frankia meetings held over the last decade that were devoted to the biology of Frankia and actinorhizal plants.

Practical Importance of Actinorhizal Plants

Interplanting of dinitrogen-fixing alders with black cottonwood has been demonstrated to benefit the crop plant (17); other successful beneficial uses of actinorhizal plants in combination with nonfixing trees include black walnut and autumn olive (24) and succession planting of red alder followed by Douglas fir (3). Some of the economic considerations in forest plantation management in relation to nitrogen-fixing plants have been discussed recently by Turvey and Smethurst (73) and by Tarrant (60). In these applications, the dinitrogen-fixing trees provide the nitrogen for the growth not only of the actinorhizal plant but also of the companion crop through cycling of fixed nitrogen via leaf litter decomposition.

In the tropical and subtropical countries of the world, fast-growing, dinitrogen-fixing actinorhizal plants serve multiple purposes. Many of them are harvested for fuel wood, some for timber or poles; others are planted for soil stabilization, wind breaks, or soil improvement. *Casuarina* species have been spread by man from their native habitat in Australia to many places in the tropical world. Their utilization and management was the topic of a recent special volume (46). As with the woody legumes in the tropics, these trees can establish quickly on sites unfavorable for most vegetation, and can grow quickly and serve many purposes, not least of which is the amelioration of the site and the conservation and improvement of the soil.

The microbial symbiont in the root nodules of actinorhizal plants is the filamentous bacterium *Frankia* of the Actinomycetales. Although only a single genus, *Frankia*, is known, many different strains have been isolated from different host plants and grown in pure culture. There exist several distinctive groups of *Frankia* with respect to the hosts that they nodulate (4). Cross-inoculation groups which are somewhat reminiscent of the situation in *Rhizobium* are described in *Frankia* with varying degrees of host-microbial strain specificity. For example, *Frankia* strains isolated from root nodules of *Alnus rubra* will nodulate other alder species but show no capacity to infect *Casuarina*. *Frankia* isolates from *Casuarina* nodulate seedlings of a number of *Casuarina* species tested and also some species of *Allocasuarina* in the same family Casuarinaceae but fail to nodulate many other actinorhizal genera (80). In some cases, pure cultured strains of *Frankia* will nodulate the host from which they were isolated but fail to fix dinitrogen, i.e., they are ineffective [e.g., *Elaeagnus umbellata* (5)].

Because of this specificity and the likelihood that only specific host-microbial strain interactions can result in optimum symbioses, it is desirable or necessary to inoculate seedlings with selected *Frankia* strains to achieve plant establishment in much the same way that the practice of inoculation has proved to be valuable with *Rhizobium*-legume plantations. To date, relatively little attention has been paid to the production of *Frankia* inoculum for plantation, although methods have been described (32) and some large-scale inoculations effected (51). Commercial production of *Frankia* inoculum for forestation with such trees as *Casuarina* in the developing countries of the world, although technologically feasible, remains for the future.

Genetic research on *Frankia* is still in its infancy. Because of its slow growth in culture and its filamentous nature, the microbial genetic tools used with *Rhizobium* do not apply directly to *Frankia*. The closest parallel model is the research on *Streptomyces*, another filamentous genus of the Actinomycetales. Plasmids have been described in *Frankia*, and protoplast formation from *Frankia* hyphae is possible. DNA hybridization experiments point out the closeness of *nif* and *nod* genes in *Frankia* to such bacterial species as *Rhizobium meliloti*. A review on the genetics of *Frankia* has recently been published (50).

Extending Dinitrogen Fixation to New Host Plants

Great interest and optimism continues in the minds of molecular geneticists concerning the possibilities of extending the dinitrogen-fixing capacities of the known symbiotic systems, such as those involving Rhizobium and Frankia, to other host plants, including, perhaps, some of the world's primary food crops or timber trees. Early optimism over transferring the nif genes to nonleguminous or nonactinorhizal genera has been sobered by our increasing understanding of the complexity of the genetically controlled host-microbial environment within which the nif complex operates. The Rhizobium-legume association may be the most highly evolved and complex. In the Frankia-actinorhizal host association, the microbial component can actually fix dinitrogen under aerobic conditions in isolation, and simply finds in symbiosis a congenial source of substrate and environment. Whether favorable induced associations can be generated by genetic manipulation remains to be seen.

In recent years excitement concerning symbiotic nitrogen fixation in forest trees was markedly increased by the reports initiated by Trinick in 1973 (71), that the tropical tree Parasponia (first erroneously reported as Trema) in the family Ulmaceae was effectively nodulated by a strain of Rhizobium that also effectively nodulated some leguminous herbaceous species in the cowpea group. This report, which subsequently has been confirmed and extended (72), provides the first case of Rhizobium producing symbiotic dinitrogen-fixing nodules outside the family Leguminosae. The association is unique in a number of ways. The infection itself is anomalous, involving multiple root hair development in seedling roots but invasion via epidermal intercellular penetration with infection occurring in the inner cortical cells (33). Furthermore, nodules that develop are more like actinorhizal nodules than rhizobial nodules, involving stimulation of lateral root initiation and subsequent infection (34,72), with the rhizobial endosymbiont held within the infection threads throughout the effective life of the nodule (58).

Clearly, the major excitement centers on the question of host susceptibility. If Rhizobium strains can infect a host so far removed from its normal host range, can we learn enough from this exception to open up other host plants to associations with Rhizobium or with Frankia which would allow us to transfer the dinitrogen-fixing capacities of these microorganisms to other important food or fiber crop plants? Some genetic research on these questions has begun (39,40) but much work remains.

REFERENCES

1. Abelson, P.H. (1985) Plant fungal symbiosis. Science 229:617.
2. Allen, O.N., and E.K. Allen (1981) The Leguminosae, University of Wisconsin Press, Madison, Wisconsin.
3. Atkinson, W.A., B.T. Bormann, and D.S. De Bell (1979) Crop rotation of Douglas fir and red alder: A preliminary biological and economic assessment. Bot. Gaz. 140(Suppl.):S102-S107.
4. Baker, D.D. (1987) Relationships among pure cultured strains of Frankia based on host specificity. Physiol. Plant. 70:245-248..
5. Baker, D., W. Newcomb, and J.G. Torrey (1980) Characterization of an ineffective actinorhizal microsymbiont, Frankia sp. EμI1 (Actinomycetales). Can. J. Microbiol. 26:1072-1089.

6. Bisseling, T., H. Franssen, F. Govers, T. Gloudemans, J. Louwerse, M. Moreman, J.-P. Nap, and A. van Kammen (1985) Nodulin gene expression in *Pisus sativum*. In *Nitrogen Fixation Research Progress*, H.J. Evans, P.J. Bottomley, and W.E. Newton, eds. Martinus Nijhoff Publishers, Dordrecht, The Netherlands, pp. 53-59.
7. Böhm, W. (1979) *Methods of Studying Root Systems*, Springer-Verlag, Berlin.
8. Bond, G. (1983) Taxonomy and distribution of non-legume nitrogen-fixing systems. In *Biological Nitrogen Fixation in Forest Ecosystems: Foundations and Applications*, J.C. Gordon and C.T. Wheeler, eds. Martinus Nijhoff/Dr. W. Junk Publishers, The Hague, pp 55-87.
9. Bonner, J. (1942) Culture of isolated roots of *Acacia melanoxylon*. *Bull. Torrey Bot. Club* 69:130-133.
10. Bonner, J., and P.S. Devirian (1939) Growth factor requirements of four species of isolated roots. *Am. J. Bot.* 26:661-665.
11. Bonnett, Jr., H.T., and J.G. Torrey (1965) Chemical control of organ formation in root segments of *Convolvulus* cultured in vitro. *Plant Physiol.* 40:1228-1236.
12. Bonnett, Jr., H.T., and J.G. Torrey (1966) Comparative anatomy of endogenous bud and lateral root formation in *Convolvulus arvensis* roots cultured in vitro. *Am. J. Bot.* 53:496-507.
13. Brewbaker, J.L., J. Halliday, and J. Lyman (1983) Economically important nitrogen fixing tree species. *Nitrogen Fixing Tree Research Reports* 1:35-40.
14. Callaham, D., P. Del Tredici, and J.G. Torrey (1978) Isolation and cultivation *in vitro* of the actinomycete causing root nodulation in *Comptonia*. *Science* 199:899-902.
15. Chilton, M.-D., D.A. Tepfer, A. Petit, C. David, F. Casse-Delbart, and J. Tempe (1982) *Agrobacterium rhizogenes* inserts T-DNA into the genomes of the host plant root cells. *Nature* 295:432-434.
16. Christianson, M.L., and D.A. Warwick (1987) Transient biochemical sensitivities during rhizogenesis *in vitro*. *Planta* (submitted for publication).
17. De Bell, D.S., and M.A. Radwan (1979) Growth and nitrogen relations of coppiced black cottonwood and red alder in pure and mixed plantings. *Boz. Gaz.* 140(Suppl.):S97-S101.
18. De Block, M., L. Herrera-Estrella, M. Van Montagu, J. Schell, and P. Zambryski (1984) Expression of foreign genes in regenerated plants and their progeny. *EMBO J.* 3:1681-1689.
19. De Cleene, M., and J. De Ley (1976) The host range of crown gall. *Bot. Rev.* 42:389-466.
20. De Cleene, M., and J. De Ley (1981) The host range of infectious hairy-root. *Bot. Rev.* 47:147-194.
21. Dittmer, H.J. (1937) A quantitative study of the roots and root hairs of a winter rye plant (*Secale cereale*). *Am. J. Bot.* 24:417-420.
22. Fast, F.T., P.A. Mason, J. Wilson, K. Ingleby, R.C. Munro, L.V. Fleming, and J.W. Deacon (1985) "Epidemiology of sheathing (ecto-) mycorrizhas in unsterile soils: A case study of *Betula pendula*. *Proc. Royal Soc. Edinburgh* 85B:299-315.
23. Flores, H.E., M.W. Hoy, and J.J. Pickard (1987) Secondary metabolites from root cultures. *Trends in Biotechnology* (in press).
24. Funk, D.T., R.C. Schlesinger, and F. Ponder, Jr. (1979) Autumn-olive as a nurse plant for black walnut. *Bot. Gaz.* 140(Suppl.):S110-S114.
25. Gianinazzi, S., and V. Gianinazzi-Pearson (1986) Progress and headaches in endomycorrhiza biotechnology. *Symbiosis* 2:139-149.

26. Gianinazzi-Pearson, V., and S. Gianinazzi (1983) The physiology of vesicular-arbuscular mycorrhizal roots. Plant and Soil 71:197-209.
27. Goforth, P.L., and J.G. Torrey (1977) The development of isolated roots of Comptonia peregrina (Myricaceae) in culture. Am. J. Bot. 64:476-482.
28. Harley, J.L., and S.E. Smith (1983) Mycorrhizal Symbiosis, Academic Press, Inc., London.
29. Kamada, H., O. Okamura, M. Satake, H. Harada, and K. Shimomura (1986) Alkaloid production by hairy root cultures of Atropa belladonna. Plant Cell Reports 5:239-242.
30. Kapulnik, Y., C.M. Joseph, and D.A. Phillips (1987) Flavone limitations to root nodulation and symbiotic nitrogen fixation in alfalfa. Plant Physiol. (in press).
31. Kleinschmidt, G.D., and J.W. Gerdemann (1972) Stunting of citrus seedlings in fumigated soils in relation to the absence of endomychorrhizas. Phytopathology 62:1447-1453.
32. Lalonde, M., and H.E. Calvert (1979) Production of Frankia hyphae and spores as an infective inoculant for Alnus species. In Symbiotic Nitrogen Fixation in the Management of Temperate Forests, J.C. Gordon, C.T. Wheeler, and D.A. Perry, eds. Forest Research Laboratory, Oregon State University, Corvallis, Oregon, pp. 95-110.
33. Lancelle, S.A., and J.G. Torrey (1984) Early development of Rhizobium-induced root nodules of Parasponia rigida. I. Infection and early nodule initiation. Protoplasma 123:26-37.
34. Lancelle, S.A., and J.G. Torrey (1985) Early development of Rhizobium-induced root nodules of Parasponia rigida. II. Nodule morphogenesis and symbiotic development. Can. J. Bot. 63:25-35.
35. Le Tacon, F., G. Jung, J. Mugnier, P. Michelot, and C. Maujprim (1984) Efficiency in a forest nursery of an ectomycorrhizal fungus inoculum produced in a fermentor and entrapped in polymeric gels. Can. J. Bot. 63:1664-1668.
36. Loeb, J. (1917) Influence of the leaf upon root formation and geotropic curvature in the stem of Bryophyllum calycinum and the possibility of a hormone theory of these processes. Bot. Gaz. 63:25-50.
37. Loomis, R.S., and J.G. Torrey (1964) Chemical control of vascular cambium initiation in isolated radish roots. Proc. Natl. Acad. Sci., USA 52:3-11.
38. Lyford, W.H. (1975) Rhizography of non-woody roots of trees in the forest floor. In The Development and Function of Roots, J.G. Torrey and D.T. Clarkson, eds. Academic Press, Inc., London, pp. 179-196.
39. Marvel, D.J., J.G. Torrey, and F.M. Ausubel (1987) Rhizobium symbiotic genes required for nodulation of legume and non-legume hosts. Proc. Natl. Acad. Sci., USA 84:1319-1323.
40. Marvel, D.J., G. Kuldau, A. Hirsch, E. Richards, J.G. Torrey, and F.M. Ausubel (1985) Conservation of nodulation genes between Rhizobium meliloti and a slow-growing Rhizobium strain which nodulates a non-legume host. Proc. Natl. Acad. Sci., USA 82:5841-5848.
41. Marx, D.H., and D.S. Kenney (1982) Production of ectomycorrhizal fungus inoculum. In Methods and Principles of Mycorrhizal Research, N.C. Schenck, ed. American Phytopathology Society, St. Paul, Minnesota, pp. 131-146.
42. Marx, D.H., J.L. Ruehle, D.S. Kenney, D.E. Cordell, J.W. Riffle, R.J. Molina, W.H. Pawuk, S. Navratil, R.W. Tinus, and O.C. Goodwin (1982) Commercial vegetative inoculum of Pisolithus tinctorius and inoculation techniques for development of ectomycorrhizae on container-grown tree seedlings. For. Sci. 28:373-400.

43. McClaugherty, C.A., J.D. Aber, and J.A. Melillo (1982) The role of fine roots in the organic matter and nitrogen budgets of two forested ecosystems. Ecology 63:1481-1490.
44. Mexal, J., and C.P.P. Reid (1973) The growth of selected mycorrhizal fungi in response to induced water stress. Can. J. Bot. 51: 1579-1588.
45. Meyer, F.H. (1973) Distribution of ectomycorrhizae in native and man-made forests. In Ectomycorrhizae, G.C. Marks and T.T. Kozlowski, eds. Academic Press, Inc., New York, pp. 23-31.
46. Midgley, S.J., J.W. Turnbull, and R.D. Johnston (1983) Casuarina Ecology, Management and Utilization, CSIRO, Melbourne, Australia, 286 pp.
47. Moore, L., G. Warren, and G. Strobel (1979) Involvement of a plasmid in the hairy root disease of plants caused by Agrobacterium rhizogenes. Plasmid 2:617-626.
48. Moser, M., and K. Hanselwandter (1983) Ecophysiology of mycorrhizal symbiosis. Encycl. Plant Physiol. N.S. 12C:391-421.
49. Mosse, B., and C. Hepper (1975) Vesicular-arbuscular mycorrhizal infections in root organ cultures. Physiol. Plant Pathol. 5:215-223.
50. Normand, P., and M. Lalonde (1986) The genetics of actinorhizal Frankia: A review. Plant and Soil 90:429-453.
51. Perinet, P., J.G. Brouillette, J.A. Fortin, and M. Lalonde (1985) Large scale inoculation of actinorhizal plants with Frankia. Plant and Soil 87:175-183.
52. Peters, K., J.W. Frost, and S.R. Long (1986) A plant flavone, luteolin, induces expression of Rhizobium meliloti nodulation genes. Science 233:977-980.
53. Phillips, J.M., and D.S. Hayman (1970) Improved procedure for clearing roots and staining parasitic and vesicular-arbuscular mycorrhizal fungi for rapid assessment of infection. Trans. British Mycol. Soc. 55:158-161.
54. Plenchette, C., V. Furlan, and J.A. Fortin (1981) Growth stimulation of apple trees in unsterilized soil under field conditions with VA mycorrhiza inoculation. Can. J. Bot. 59:2003-2008.
55. Raggio, M., N. Raggio, and J.G. Torrey (1957) The nodulation of isolated leguminous roots. Am. J. Bot. 44:325-334.
56. Robbins, W.J., and A. Hervey (1971) Cytokinin and growth of excised roots of Bryophyllum calycinium. Proc. Natl. Acad. Sci., USA 68:347-348.
57. Schenck, N.C., ed. (1982) Methods and Principles of Mycorrhizal Research, American Phytopathology Society, St. Paul, Minnesota.
58. Smith, C.A., R.C. Skvirsky, and A.M. Hirsch (1986) Histochemical evidence for the presence of a suberinlike compound in Rhizobium-induced nodules of the nonlegume Parasponia rigida. Can. J. Bot. 64:1474-1483.
59. Stribley, D.P., and D.J. Read (1975) Some nutritional aspects of the biology of ericaceous mycorrhizas. In Endomycorrhizas, F.E. Sanders, B. Mosse, and P.B. Tinker, eds. Academic Press, Inc., London, pp. 195-207.
60. Tarrant, R.F. (1983) Nitrogen fixation in North American forestry: Research and application. In Biological Nitrogen Fixation in Forest Ecosystems: Foundations and Applications, J.C. Gordon and C.T. Wheeler, eds. Martinus Nijhoff/Dr. W. Junk Publishers, The Hague, pp. 261-277.

61. Tepfer, D. (1983) The biology of genetic transformation of higher plants by Agrobacterium rhizogenes. In Molecular Genetics of the Bacteria-Plant Interaction, S. Pühler, ed. Springer-Verlag, New York, pp. 248-258.
62. Tepfer, D. (1984) Transformation of several species of higher plants by Agrobacterium rhizogenes: Sexual transmission of the transformed genotype and phenotype. Cell 35:959-967.
63. Tepfer, D., A. Yacoub, C. Lambert, A. Goldmann, C. Rosenberg, J. Denaire, G. Jung, and J. Slightom (1986) Applications of genetic transformation by Ri T-DNA from Agrobacterium rhizogenes in plant biotechnology. Symbiosis 2:9 (Abstr.).
64. Tepfer, M., and F. Casse-Delbart (1987) Agrobacterium rhizogenes as a vector for transforming higher plants. Microbiol. Sciences 4:24-28.
65. Thimann, K.V. (1977) Hormone Action in the Whole Life of Plants, University of Massachusetts Press, Amherst, Massachusetts.
66. Tinus, R.W., and S.E. McDonald (1979) How to Grow Tree Seedlings in Containers in Greenhouses, Rocky Mountain Forest and Range Experimental Station General Technical Report RM-60, Forest Service, U.S. Department of Agriculture.
67. Torrey, J.G. (1985) The development of plant biotechnology. Am. Scientist 73:354-363.
68. Torrey, J.G. (1986) Endogenous and exogenous influences on the regulation of lateral root formation. In New Root Formation in Plants and Cuttings, M.B. Jackson, ed. Martinus Nijhoff Publishers, Dordrecht, The Netherlands, pp. 31-66.
69. Torrey, J.G. (1988) Cellular interactions between host and endosymbiont in dinitrogen-fixing root nodules of woody plants. In Cell to Cell Signals in Plant, Animal and Microbial Symbiosis, D. Smith, P. Bonfante, and V. Gianinazzi-Pearson, eds. Springer-Verlag, Vienna (in press).
70. Torrey, J.G., and R.S. Loomis (1967) Auxin-cytokinin control of secondary vascular tissue formation in isolated roots of Raphanus. Am. J. Bot. 54:1098-1106.
71. Trinick, M.J. (1973) Symbiosis between Rhizobium and the nonlegume, Trema aspera. Nature (London) 244:459-460.
72. Trinick, M.J. (1979) Structure of nitrogen-fixing nodules formed by Rhizobium on roots of Parasponia andersonii Planch. Can. J. Microbiol. 25:565-578.
73. Turvey, N.D., and P.J. Smethurst (1983) Nitrogen fixing plants in forest plantation management. In Biological Nitrogen Fixation in Forest Ecosystems: Foundations and Applications, J.C. Gordon and C.T. Wheeler, eds. Martinus Nijhoff/Dr. W. Junk Publishers, The Hague, pp. 233-259.
74. Verma, D.P.S. (1982) Host-Rhizobium interactions during symbiotic nitrogen fixation. In The Molecular Biology of Plant Development, H. Smith and D. Grierson, eds. Blackwell Science Publishers, Oxford, pp. 437-466.
75. Verma, D.P.S., and T. Hohn, eds. (1984) Genes Involved in Microbe-Plant Interactions, Springer-Verlag, New York.
76. Verma, D.P.S., and K. Nadler (1984) Legume-Rhizobium-symbiosis: Host's point of view. In Genes Involved in Microbe-Plant Interactions, D.P.S. Verma and T. Hohn, eds. Springer-Verlag, New York, pp. 57-93.
77. Warwick, P.A., and M.L. Christianson (1987) Messenger RNA populations during rhizogenesis in vitro. Planta (submitted for publication).

78. Went, F.W., and K.V. Thimann (1937) Phytohormones, The Macmillan Company, New York.
79. White, P.R. (1934) Potentially unlimited growth of excised tomato root tips in liquid medium. Plant Physiol. 9:585-600.
80. Zhang, Z., and J.G. Torrey (1985) Biological and cultural characteristics of the effective Frankia strain HFPCcI3 (Actinomycetales) from Casuarina cunninghamia (Casuarinaceae). Ann. Bot. 56:367-378.
81. Zimmerman, P.W., and A.E. Hitchcock (1929) Vegetative propagation of holly. Am. J. Bot. 16:556-570.
82. Zimmerman, P.W., and F. Wilcoxon (1935) Several chemical growth substances which cause initiation of roots and other responses in plants. Contrib. Boyce Thompson Inst. 7:209-229.
83. Zobel, R.W. (1975) The genetics of root development. In The Development and Function of Roots, J.G. Torrey and D.T. Clarkson, eds. Academic Press, Inc., London, pp. 261-275.
84. Zobel, R.W. (1986) Rhizogenetics (root genetics) of vegetable crops. Hort. Sci. 21:956-959.

TISSUE CULTURE SYSTEMS

GENE TRANSFER IN FOREST TREES

M.R. Ahuja

Federal Research Centre for Forestry and Forest Products
Institute of Forest Genetics and Forest Tree Breeding
D-2070 Grosshansdorf, Federal Republic of Germany

ABSTRACT

Genes can be transferred in plants in several different ways. The conventional method involving hybridization is effective but takes a number of years, even in the herbaceous crops. Because of long generation cycles, this method would be impractical in forest trees. Recent advances in biotechnology have made it possible to attempt gene transfer in the forest tree species involving vector systems. In vivo and in vitro genetic transformation will be reviewed in forest tree species. Some potential "gene" candidates for transfer in forest trees will be discussed. However, the isolation and cloning of commercially important genes will still present substantial problems. For this reason, it might be worthwhile to consider the transfer of genes from other organisms to forest trees. But we should be cautious in such endeavors.

INTRODUCTION

Gene transfer by hybridization, that is, crosses involving any two genetically different individuals, has played an important role in the evolution and speciation of plants and animals. Hybridization may occur at the intraspecific, interspecific, or intergeneric levels. A number of agricultural crops, for example, wheat, potato, cotton, sugarcane, coffee, and tobacco, are polyploids (74), which are thought to have arisen by hybridization followed by genomic doubling/rearrangements in nature. There is adequate evidence to suggest the actual parentage in wheat [*Triticum aestivum* (56)], cotton [*Gossypium hirsutum* (44)], and tobacco [*Nicotiana tabacum* (36,37)]. Triticale, derived by hybridization between wheat and rye, is the first man-made polyploid crop. Among forest tree species polyploidy is uncommon. A few exceptions include coastal redwood (*Sequoia sempervirens*) and naturally occurring triploids in European aspen [*Populus tremula* (45)] and quaking aspen [*Populus tremuloides* (17)]. The triploid aspens found in the diploid stands were initially identified on the basis of leaf size and unusually long fibers, and later confirmed by chromosome counts. Because of longer fibers and greater volume in the triploid aspens as compared to their diploid counterparts, there has been interest in polyploidy as an avenue for aspen improvement (31). Triploid aspens have been produced artificially following crosses between tetraploid and diploid aspens both in Europe and the United States for testing their potential for biomass production. Since triploids are usually infertile, tissue culture technology has been employed for clonal propagation of triploid aspen (5,6,12). Allopolyploids derived by hybridization between *Populus tremula* and *P. tremuloides* (53) have not been successfully exploited.

HYBRIDIZATION

Backcross Method

For improvement of plants involving specific genes, the breeder employs a backcrossing program. This is often combined with mutation selection for the introduction of new sets of genes. By employing backcrossing programs, it is possible to transfer one or a few genes from one species to another. Essentially this hybridization procedure, which may involve several generations, selects for a desirable gene(s) from one species on the genetic background of the second. Although the process of gene transfer by this method is slow and time consuming, it is effective and transferred genes are generally stably transmitted through the seed. Because of long generation cycles, hybridization as an approach for gene transfer in forest tree species seems to have rather limited possibilities.

PROTOPLAST FUSION

Gene transfer by hybridization is limited to those plant species that are sexually compatible. Protoplast fusion, on the other hand, offers prospects for gene transfer in sexually incompatible species. Indeed, protoplasts of sexually incompatible plant species have been fused and in some combinations somatic hybrids have been regenerated. Partial chromosomal losses of either parental species have been commonly observed in the somatic hybrids. For example, random chromosomal losses have been reported in somatic hybrids between potato (*Solanum tuberosum*) and tomato (*Lycopersicon esculentum*) (57,58), *Arabidopsis thaliana* and *Brassica campestris* (34,40), and *Datura innoxia* and *Atropa belladonna* (49). However, most of these somatic hybrids were sterile. On the other hand, in the somatic hybrid cell line of *Nicotiana glauca* and *Glycine max*, there was preferential loss of *N. glauca* chromosomes (47). In another intergeneric somatic hybrid involving *Aegopodium podograria* and *Daucus carota*, all the chromosomes of *A. podograria* were apparently lost. However, molecular hybridization studies suggested integration of small *A. podograria* chromosome segments in the carrot genome (30).

In addition to random or preferential loss of chromosomes, segregation may also occur in cytoplasmic organelles, as plastids and mitochondria. Although chloroplasts generally undergo random sorting out resulting in the survival of a single chloroplast type in the somatic hybrid cell (75), multiple plastid types may survive in the tissues of somatic hybrids (35). The fate of mitochondria in the somatic hybrids is less clear. Sequence alteration and multiple mitochondrial types have been detected in a somatic hybrid involving *N. tabacum* and *N. debneyi* (13). Restoration or loss of male sterility presumably depends on the mitochondrial type in the somatic hybrid.

Protoplast fusion as an avenue for gene transfer has limited possibilities in plant species. This is because there is extensive variability and instability in the somatic hybrids produced so far. Recovery of desirable genotypes following fusion from the mixture of homokaryons and heterokaryons in culture may be yet another major difficulty. Further, in most economically important plants, such as cereals, fruit trees, and forest trees, it is still difficult to regenerate viable plants from the protoplasts (4,7,9,24,55). Plant protoplasts have also been fused with animal cells (interkingdom fusions) to monitor the expression of plant and animal genes in the heterokaryon (4).

AGROBACTERIUM-MEDIATED GENE TRANSFER

Host Range

Agrobacterium tumefaciens is a soil bacterium which causes crown gall disease. Forest tree species exhibit differential sensitivity to *A. tumefaciens*. Initial reports indicated that among 112 forest tree species that were inoculated with *A. tumefaciens*, 65 species developed tumors of various sizes (25). Included in the immune species were *Fagus sylvatica* and species of the genera *Quercus*, *Pinus*, *Picea*, and *Larix*. In recent years, tumor formation potential of forest tree species following infection with different strains of *A. tumefaciens* has been reexamined. Successful but variable infection, as evidenced by tumor formation (Tab. 1), has been observed in *Pinus taeda* (69,70), *Abies nordmanniana*, and *Picea abies* (21), and in *Betula verrucosa* and *B. mandschurica* (79). We have also investigated in vivo response of several forest tree species to nopaline (C58) and octopine (B6S3) strains of *A. tumefaciens*. In general, tree species responded more positively to the nopaline than the octopine strain of crown gall bacteria. In one set of experiments, we inoculated two- to three-year-old plants of conifer species with *A. tumefaciens* strains. Each

Tab. 1. *Agrobacterium tumefaciens*-mediated tumor formation in some forest tree species.

Species	Strain	Age of plant inoculated	Percentage tumors	Reference
Abies nordmanniana	C58	Seedling	65	21
Betula verrucosa	B6/T37	Seedling	23	79
Betula mandschurica	B6/T37	Seedling	23	79
Fagus sylvatica	B6/T37	Seedling	17	79
Larix decidua	C58	2 to 3 years	0	This study
Larix decidua	B6S3	2 to 3 years	0	This study
Picea abies	C58	Seedling	13	21
Picea abies	C58	2 to 3 years	15	This study
Picea abies	B6S3	2 to 3 years	10	This study
Pinus sylvestris	C58	2 to 3 years	10	This study
Pinus sylvestris	B6S3	2 to 3 years	0	This study
Pinus taeda	U3	Seedling	17	69

plant was infected at two to four sites. Our results also show variable tumor formation in Pinus sylvestris (Fig. 1) and Picea abies (Fig. 2), and practically no tumor formation in Larix decidua (Tab. 1).

A number of European aspen (Populus tremula) and hybrid aspen (Populus tremula x P. tremuloides) clones were also investigated for their infectivity with C58 and B6S3 strains of A. tumefaciens. Inoculations were made on plantlets maintained in vivo (Fig. 3) and in vitro (Fig. 4), and on one- to two-year-old plants. These experiments showed that aspen clones developed tumors following infection with the nopaline strain (C58), but generally responded poorly to the octopine (B6S3) strain. More recently, we have been experimenting with a slightly different method of inoculation. After wounding, indoleacetic acid (IAA) is applied to the wound, followed by inoculation with bacteria. The wound site is then covered with an adhesive tape. By using this approach, we observed tumors on all six clones infected with C58, and five of six clones infected with B6S3. However, C58 (nopaline) tumors were much larger than B6S3 (octopine) tumors. With this approach tumors develop earlier, as compared to plain infection, and can be scored in aspen seven to ten days following infection. We are now applying this technique to conifers for reevaluation of crown gall formation. It should be mentioned though that tumor formation is not necessarily the only criterion for monitoring transformation.

Biology of Agrobacterium

The tumors caused by A. tumefaciens are mediated by the Ti-plasmid (tumor-inducing). Experimental evidence suggests that only part of the Ti-plasmid is transferred to the host plant cell. The transferred DNA (T-DNA) is integrated in the nuclear genome and stably maintained in the descendants of the transformed cells (20,50). The physiology and molecular biology of crown gall have been extensively reviewed (59,67,68). Here only salient features relevant to gene transfer will be discussed. The

Fig. 1. Tumor induced on a two- to three-year-old Pinus sylvestris plant following a single inoculation with the nopaline (C58) strain of Agrobacterium tumefaciens.

Ti-region of the Ti-plasmid is approximately 23 kilobases (kb) in size and contains genes for the initiation and maintenance of the tumorigenic state (67). The so-called onc genes, located in the T-region, code for or regulate the enzymes involved in the synthesis of phytohormones, auxins, and cytokinins. In addition to the onc genes, the T-region also contains genes which code for the enzymes that synthesize specific compounds called opines. These compounds are utilized by the bacteria as their sole source of carbon and nitrogen. Crown gall tumors are classified on the basis of opines present in them: nopaline, octopine, or agropine tumors. The opines are amino acid derivatives that plant cells are unable to metabolize. The T-DNA controls the biosynthetic machinery of the plant cells and leads to the development of unorganized tumors or teratomas, which are tumors with partially organized shoots [cf. genetic tumors in Nicotiana (1,2)]. The transformed cells produce high levels of phytohormones essential for the maintenance of the tumorous state (14).

Genetic studies indicate that ends or borders of the T-region, consisting of imperfect 25-base pair (bp) direct repeats (59,67), are essential for the integration of the T-region in the plant genome. The presence of right border repeats seems to be essential for T-DNA integration in the plant cell genome (71,77), whereas the left border repeats are apparently not required for the integration process (46).

Fig. 2. Tumors induced on a two- to three-year-old Picea abies plant following single inoculations with the octopine (B6S3) strain of Agrobacterium tumefaciens.

Fig. 3. Tumor induced on a one-year-old Populus tremula plant following a single inoculation with the nopaline (C58) strain of Agrobacterium tumefaciens.

For the transfer of the T-region from the Ti-plasmid to the plant cell, another part of the Ti-plasmid, the virulence (vir) region, and the bacterial chromosomal virulence (chv) locus are essential (29,72). The activation of the vir and chv loci is mediated by specific phenolic compounds present in the exudates of the wound cells. One such compound, acetosyringone (4-acetyl-2,6-dimethoxyphenol), has been identified in tobacco (73). Recent experiments indicate that vir induction by acetosyringone results in the production of S1 nuclease-sensitive sites at the T-DNA borders (73). The T-DNA borders are nicked after vir gene activation (78), probably representing an early step in the generation of single-stranded T-DNA and the transfer process (73).

Since the T-region of the Ti-plasmid seems to be integrated in the plant cell genome, this system has been extensively employed as an experimental vector for gene transfer in plants. The Ti-plasmid can be used as a gene vector for the transfer of DNA sequences up to 50 kb long (67). Any sequences inserted between the direct repeats or "ends" of the T-DNA are co-transferred and integrated in the host cell genome following transformation (67,68). This means that the onc genes could be entirely removed (82) and new DNA sequences introduced in the T-region for gene transfer via Agrobacterium (39,41,48). By employing disarmed Ti-plasmids, gene transfer and recovery of normal plants have been demonstrated in a variety of different plant species (32,39,41,42,51,54,62,68).

The T-DNA contains a number of onc genes that are linked, and these are involved in the growth hormone autonomy of the tumor cells. At least six transcripts involved in the growth autonomy are derived from the "common" or "core" segment of the T-DNA, which was identical in nopaline and octopine tumors (67). Each gene on the T-DNA has its own signal for transcription in the plant cell. Several of these transcripts act by "turning on" the biosynthetic systems involved in continuous synthesis of phytohormones, thereby suppressing organ development, that is, shoot or root

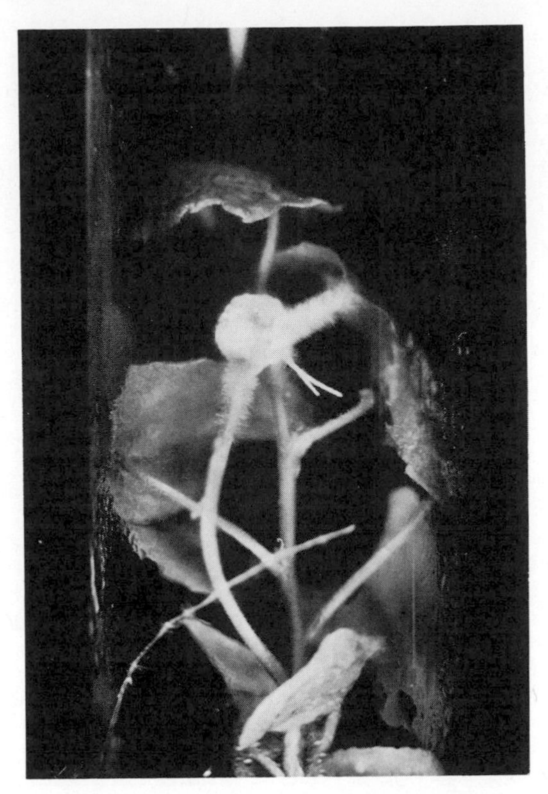

Fig. 4. Tumor induced on a Populus tremula plantlet in culture following a single inoculation with the octopine (B6S3) strain of Agrobacterium tumefaciens. Note roots emerging from the tumor.

formation is inhibited. Transcripts 1 and 2 specifically suppress shoot formation; they are presumably involved in conditioning high levels of auxin. Deletion of transcripts 1 and 2 leads to shoot differentiation (67). Transcript 4 prevents root formation; therefore it is presumably involved in the production of high levels of cytokinins. Removal of transcript 4 leads to root formation (67). Transcripts 5, 6a, and 6b are somehow involved in accentuating the tumor phenotype (67). Regions of DNA in the T-DNA involved in the suppression of shoot or root formation have been called shi (shoot-inhibiting) and roi (root-inhibiting) loci (27) or tms (tumor morphology shoot) and tmr (tumor morphology root) loci (33). Detailed analysis of the onc genes would certainly lead to a better understanding of plant development, in particular the genetics of organ development. We are currently employing strains of A. tumefaciens harboring mutant plasmids to investigate growth and development in forest tree species.

By employing recombinant DNA technology, cloning vectors have been constructed that ideally contain a wide host range promoter region, a coding region of a dominant selectable marker, and a terminator region. Chimeric genes have been constructed that utilize a broad spectrum promoter region of nopaline synthase (nos), that seems to be more eukaryotic than prokaryotic and is functional in a broad range of dicotyledonous plants in its recognition signal, a kanamycin-resistant gene, neomycin phosphotransferase (NPTII), and a polyadenylation signal of an octopine synthase gene (ocs) (39). Such chimeric genes are fused and inserted into disarmed Ti-plasmids. One such construction which we have employed in our laboratory is the A. tumefaciens strain called C58 CI pGV3850::1103 (39). Similar gene constructions have been employed for incorporating kanamycin resistance into tobacco, tomato, and Petunia (41,42).

Transformation in Forest Trees

We have carried out in vitro transformation studies in European aspen (Populus tremula), hybrid aspen (Populus tremula x P. tremuloides), birch (Betula pendula), and redwood (Sequoia sempervirens) by employing several different strains of A. tumefaciens. However, here only preliminary results following inoculations with an oncogenic nopaline strain (C58) and disarmed C58CI pGV3850::1103 carrying a kanamycin-resistant gene will be presented. Most of our work is on the aspen system. Tumors induced following infection with C58 on aspen plants were excised and cultured on a hormone-free low-salt medium (5,8), supplemented with 500 mg/l Claforan (Hoechst). It was difficult to maintain infection-free tumor tissues in the cultures. However, tumors from two aspen plants have been successfully maintained on a hormone-free medium, indicating their hormone autonomy. One tumor was derived by inoculation of a dormant bud. All the cells in the bud were not transformed, as there were normal-looking shoots that developed along with the tumor from the inoculated bud. Obviously all the cells in the target area are not transformed, and tumors resulting from such events may consist of a heterogenous population of cells.

For in vitro transformation, leaf explants and microshoots from stabilized shoot cultures were co-cultured with bacteria overnight or for shorter durations. After the co-culture, the tissues were blotted dry and cultured on medium containing 500 mg/l Claforan, and 60 to 200 mg/l kanamycin sulfate. Kanamycin sensitivity of normal aspen, birch, and redwood was tested beforehand by culturing untransformed tissues in media containing a wide range of kanamycin concentrations (10, 20, 40, 50, 60, 80, 100, 200, 500, 1000 mg/l). Inhibition of growth and differentiation and bleaching of leaves were observed from upward of 60 mg/l kanamycin in aspen and birch. Dosages higher than 200 mg/l kanamycin were toxic. On the other hand, growth and differentiation of redwood seemed to be inhibited by the lowest concentrations of kanamycin (10 to 20 mg/l).

Following inoculation with pGV3850::1103, kanamycin-resistant tissues were screened. Kanamycin-resistant shoots have not been recovered in Sequoia so far. In one set of experiments, six of 40 birch microshoots and eight of 26 aspen microshoots survived on a 100 mg/l kanamycin selection medium. These microshoots were transferred after four weeks again to 100 mg/l kanamycin medium for continued kanamycin tolerance. In another set of experiments, two of 17 birch microshoots and one to two of 14 aspen microshoots survived on a 200 mg/l kanamycin medium. The surviving microshoots were transferred after four weeks to 100 mg/l kanamycin for further selection. In another aspen clone, 42 leaf explants from in vitro cultures were inoculated with pGV3850::1103, and three of 42 leaf explants are still green on a 60 mg/l kanamycin medium. Leaf discs from one- to two-year-old aspen plants from several different clones are being screened for kanamycin resistance following co-culture with the bacteria. Further studies are under way in birch and sequoia for monitoring genetic transformation events. Molecular studies will follow after initial screening.

More recently, a mutant aroA gene from Salmonella typhimurium that confers tolerance to herbicide glyphosate and a dominant selectable kanamycin marker have been co-transferred and expressed in tobacco (22) and Populus (32). The genetically engineered Populus plants are undergoing glyphosate tolerance tests at the U.S. Department of Agriculture Forest Service, Forestry Science Laboratory, Rhinelander, Wisconsin (B. Haissig, pers. comm.).

Although most genetic transformation studies have employed A. tumefaciens, another species involved in the hairy root disease A. rhizogenes (carrying Ri-plasmid) has been explored for its potential in gene transfer. The Ri-plasmid contains regions homologous to the shi locus of the Ti-plasmid of A. tumefaciens, but does not seem to contain the roi locus (43). Following inoculation of the hypocotyl and shoot apex of larch (Larix decidua) with A. rhizogenes, blue needles, multiple buds, and adventitious roots were observed (28).

INTRODUCTION OF FOREIGN GENES INTO PROTOPLASTS

In order to introduce foreign genes into protoplasts one needs: (a) totipotent protoplasts, and (b) a delivery system to protect DNA from nuclease digestion. Several delivery systems have been explored. In one system, bacterial (Escherichia coli) spheroplasts are used to mediate gene transfer via a DNA vector into Nicotiana protoplasts. Gene transfer occurs by fusion of the DNA-spheroplast complex with the protoplasts (38). Another delivery system involves the use of liposomes, which are small artificial lipid vesicles. Nucleic acid encapsulated in liposomes is highly tolerant to degradation by nucleases. Uptake of liposome-encapsulated plasmid DNA by plant protoplasts has been described (18,60).

More recently, direct gene transfer to plants has been described (65). The procedure requires a gene under the control of a plant expression signal, calf thymus carrier DNA, and protoplasts (65). By employing a selectable marker (kanamycin resistance), stable transformation has been described in Nicotiana (65), Lolium multiflorum (66), Triticum monococcum (52), and Oryza sativa (76). Since members of the Graminaceae and monocots in general are not susceptible to Agrobacterium-mediated transformation, a direct gene transfer method seems to have potential for gene transfer in this group. However, protoplast regeneration still remains a limiting factor in Graminaceous plants, although a recent report on regeneration of plantlets from rice protoplasts (80) is encouraging.

Microinjection has also been discussed for DNA transfer into protoplasts or cells. In a recent study, random portions of plasmid DNA were found to be integrated following microinjection into the tobacco genome (23), indicating that T-DNA borders were not used. This is in contrast to the situation in Agrobacterium-mediated gene transfer in which the border sequences are essential for integration (59). Microinjection of "protected" DNA would also be limited to those species whose regeneration from protoplasts is possible. If microinjection could become an efficient method for gene transfer in cells or tissues, this might offer unique opportunities for cereals and forest tree species in which protoplast regeneration remains a limiting factor.

DIRECT INJECTION OF DNA INTO PLANTS

Recently, DNA has been directly injected into plants. De la Pena et al. (26) injected plasmid carrying the kanamycin-resistant gene (aminoglycoside phosphotransferase II gene [APH(3')II] into young floral shoots of about 100 rye (Secale cereale) plants and demonstrated that a small number of the progeny (seven of 3,000 seedlings) had acquired kanamycin resistance. Of these seven, only two were positive for the APH(3')II activity, while DNA from one of five kanamycin-resistant plants apparently did not exhibit the APH(3')II activity, but escaped the selection pressure of kanamycin (26). In a repeat experiment, 37 rye plants were injected with DNA carrying the kanamycin-resistance gene. Three of 1,000 seedlings were kanamycin-resistant, and one of these three showed APH(3')II activity (26). The efficiency of transformation seems to be rather low (0.07 to 0.1%) in these experiments, but this could be influenced by several factors, including genotype, number of available competent cells for genetic transformation, copy number of the introduced gene, and subsequent DNA rearrangements.

CAULIFLOWER MOSAIC VIRUS AS VECTOR

Cauliflower mosaic virus (CaMV) is a double-stranded DNA virus which has been investigated for its potential as a vector for foreign gene transfer in plants. The DNA of the virus is not integrated in the host genome, but replicates independently. The virus spreads in the host after local infection. At the present time there are several limitations to the CaMV transfer system, including: (a) limited host range, (b) limited space available for inserting foreign gene sequences in the virus, and (c) lack of transmission of CaMV through the germ cells to the next generation (16).

GENETICS OF TRANSFORMANTS

At least three important questions regarding the fate of a foreign gene (DNA) in the plant cell may be asked: (a) is the transferred DNA integrated?, (b) is it expressed?, and (c) is it transmitted to the sexual offspring? Previous studies (15,81) indicated that it is not very likely that the tumor trait (T-DNA) is transmitted across the meiosis barrier. However, later investigations with partially deleted T-DNA have revealed that it was stably passed on through meiosis (both male and female gametes) and that the marker gene (octopine synthase) on T-DNA was transmitted as a dominant trait (63).

In a Populus hybrid, leaf explants from stabilized shoot cultures were co-cultured with: (a) an oncogenic Agrobacterium tumefaciens nopaline strain C58 harboring a binary vector containing three chimeric genes, of which two code for kanamycin resistance (NPTII') and the third (aroA) codes for herbicide (glyphosate) tolerance; and (b) the same construction as above but in a nononcogenic strain of A. tumefaciens (32). Selection for transformants was carried out on a medium containing 60 mg/l kanamycin. Shoots were not recovered in the selection medium following cocultivation with the nononcogenic strain. However, two types of shoots developed following transformation with the oncogenic strain. Phenotypes of type A shoots were normal, while those of type B were abnormal and teratomatous (32). In type A shoots, 38 of 40 showed NPTII' activity, but none of the 40 shoots exhibited nopaline synthase activity. On the other hand, in the type B shoots, 18 of 20 exhibited NPTII' activity and all 20 showed nopaline synthase production (32). In a different Populus hybrid (Populus trichorpa x P. deltoides), molecular evidence for transformation of the cells by an Agrobacterium strain has been obtained, but there has been no recovery of transformed plants (64).

It would appear that the genotypes of the host and the Agrobacterium strain or the vector system seem to play a very important role in genetic transformation. A foreign gene(s) may or may not be expressed in the plant. Unexpected changes in the expression of a foreign gene may be due to loss of a gene, or to loss of gene expression during development. Variability recovered in the somatic tissues in vitro that seems to be associated with genetic transformation and gene transfer may be termed "somatoclonal variation," as opposed to the general term "somaclonal variation." The alteration of the latter term indicates the possible effect of a transformation/gene transfer event.

PROSPECTS AND LIMITATIONS

A number of different approaches have been employed for the transfer of foreign genes into plants. Hybridization is effective and will always be required for checking the transmission/inheritance of a foreign gene.

Chimeric genes have been constructed and transferred to plants. The chimeric genes, carrying selectable markers, have been expressed in a number of plant species. Some progress has been made with Agrobacterium-mediated gene transfer in forest tree species. The gene for herbicide tolerance has been transferred and expressed in Populus. Other genes will follow. But what other commercially important genes do we have in our repertoire that can be transferred to forest tree species? Since recombinant DNA technologies work best with a single gene or a block of closely linked genes, we might be limited to single gene-controlled traits. Most of the commercially important traits, such as yield, vigor, growth, height, and biomass, are controlled by multiple genes which are not well understood at the molecular level. Resistance to diseases and pests seems to be governed by major genes. In this respect, a good candidate for transfer might be the blister rust-resistance gene in pines. Blister rust resistance has been detected in sugar pine, but how does one identify and clone this gene? Very little is known about the structure of the genome. Other traits that might be considered for transfer are drought resistance, salt resistance, and cold hardiness.

Initially, genetic engineering in forestry will rely on genes obtained from other organisms. Although availability of in vitro regenerating systems seems essential for gene transfer work at the present time, recent experiments with direct injection of DNA in plants (26) might suggest an entirely new approach. For example, plasmid DNA with a dominant selectable marker could be injected in young male cones/flowers before the onset of meiosis, and the pollen from such injected cones/flowers could be used for pollination. Any resulting transformants could be selected in the first seed generation in vivo.

In spite of the excitement in the genetic engineering of plants, we should continue to search for new variants by other methods. We might ask, do we really need to introduce foreign genes in forest tree species?

Have we already effectively utilized the enormous existing genetic variability in forest tree species? Have we adequately explored the potential of somaclonal variation/mutation selection (10) for forest tree improvement? We do not have satisfactory answers to these questions. But the search for new variants in vitro might be worthwhile. Cells could be exposed to a chemical (for example, herbicide) in vitro and selection made for cells tolerant/resistant to the chemical. Tobacco plants resistant to herbicide chlorosulfuron have been isolated by challenging cells to the herbicide in vitro (19).

There are certain genuine concerns in the application of genetic engineering to forest trees. Forest trees have long generation cycles, and it is possible that some of the genetic changes associated with a foreign gene transfer might occur after a long "latent" period. Foreign genes may be immediately expressed, or they may remain inactive for a long time. If T-DNA-associated genes can be as easily integrated in the host genome, they could be as easily lost from the host genome. The foreign gene sequences or T-DNA may cause position effect and/or rearrangements, thus resulting in genetic changes. Some of these genetic changes might be detected during the vegetative phase, while others might be detected after the mature state (i.e., the flowering stage). Certain genetic changes may be involved in turning on the cellular tumor genes or oncogenes presumably present in all living organisms (3,11), including forest trees. The onc genes might be expressed after the flowering stage, as in the genetic tumor system in Nicotiana (1,2), to cause growth abnormalities or tumors. And that might take ten to 50 years or more after the initial foreign gene transfer. Therefore, we should be cautious in introducing foreign genes into forest tree species. That does not imply that this type of research should be curtailed, but rather that we should intensify research on the molecular basis of cell differentiation for a better understanding of gene action.

ACKNOWLEDGEMENTS

The author thanks Drs. G.H. Melchior and H.J. Muhs for discussions, Dr. Jeff Schell, Max-Planck-Institut für Züchtungsforschung, Cologne, Federal Republic of Germany, for the generous gift of Agrobacterium strains, and Mrs. R. Amenda and Miss A. Schellhorn for technical assistance.

REFERENCES

1. Ahuja, M.R. (1965) Genetic control of tumor formation in higher plants. Quart. Rev. Biol. 40:329-340.
2. Ahuja, M.R. (1968) An hypothesis and evidence concerning the genetic components controlling tumor formation in Nicotiana. Mol. Gen. Genet. 103:176-184.
3. Ahuja, M.R. (1979) On the nature of genetic change as an underlying cause for the origin of neoplasms. In Antiviral Mechanisms in the Control of Neoplasia, P. Chandra, ed. Plenum Press, New York, pp. 17-37.
4. Ahuja, M.R. (1982) Isolation, culture and fusion of protoplasts: Problems and prospects. Silvae Genet. 31:66-77.
5. Ahuja, M.R. (1983) Somatic cell genetics and rapid clonal propagation of aspen. Silvae Genet. 32:131-135.

6. Ahuja, M.R. (1984) A commercially feasible micropropagation method for aspen. Silvae Genet. 33:174-176.
7. Ahuja, M.R. (1984) Protoplast research in woody plants. Silvae Genet. 33:32-37.
8. Ahuja, M.R. (1986) Aspen. In Handbook of Plant Cell Culture, D.A. Evans, W.R. Sharp, and P.J. Ammirato, eds. Macmillan Publishing Company, New York, pp. 626-651.
9. Ahuja, M.R. (1986) Perspectives in plant biotechnology. Curr. Sci. 55:217-224.
10. Ahuja, M.R. (1987) Somaclonal variation. In Cell and Tissue Culture in Forestry, Vol. 1, J.M. Bonga and D.J. Durzan, eds. Martinus Nijhoff Publishers, Dordrecht, The Netherlands, pp. 272-285.
11. Ahuja, M.R., and F. Anders (1977) Cancer as a problem of gene regulation. In Recent Advances in Cancer Research: Cell Biology, Molecular Biology, and Tumor Virology, R.C. Gallo, ed. CRC Press, Inc., Cleveland, Ohio, pp. 103-117.
12. Barocka, K.H., M. Baus, E. Lontke, and F. Sievert (1985) Tissue culture as a tool for in vitro mass propagation of aspen. Z. Pflanzenzuchtg. 94:340-343.
13. Belliard, G., G. Vedel, and G. Pelletier (1979) Mitochondrial recombination in cytoplasmic hybrids of Nicotiana tabacum by protoplast fusion. Nature 281:401-403.
14. Braun, A.C. (1958) A physiological basis of autonomous growth of crown gall tumor cell. Proc. Natl. Acad. Sci., USA 44:344-349.
15. Braun, A.C., and H.N. Wood (1976) Supression of the neoplastic state with the acquisition of specialized functions in cells, tissues and organs of crown gall teratomas of tobacco. Proc. Natl. Acad. Sci., USA 73:496-500.
16. Brisson, N.J., J.R. Paszkowski, B. Penswick, I. Gronenhorn, I. Potrykus, and T. Hohn (1984) Expression of a bacterial gene in plants using a viral vector. Nature 310:511-514.
17. Buijtenen, J.P. van, P.N. Joranson, and D.W. Einspahr (1957) Naturally occurring triploid quaking aspen in the United States. Proc. Soc. Am. For. 1957:62-64.
18. Caboche, M., and A. Deshayes (1984) Utilization de loposome pour la transformation de protoplasts de mesophylle de tabac par plasmide recombinant de E. coli leur conferant la resistance a la Kanamycine. C.R. Acad. Sci. (Paris) 229:663-666.
19. Chaleleff, R.S., and T.B. Ray (1984) Herbicide resistant mutants from tobacco cell cultures. Science 223:1148-1151.
20. Chilton, M.D., M.H. Drummond, D.J. Merlo, D. Sciaky, A.L. Montoya, M.P. Gordon, and E.W. Nester (1977) Stable incorporation of plasmid DNA into higher plants: The molecular basis of crown gall tumorigenesis. Cell 11:263-271.
21. Clapham, D.H., and I. Ekberg (1986) Induction of tumors by various stains of Agrobacterium tumefaciens on Abies nordmanniana and Picea abies. Scand. J. For. Res. 1:435-437.
22. Comai, L., D. Facciotti, W.R. Hiatt, G. Thompson, R.E. Rose, and D.M. Stalker (1985) Expression in plants of a mutant aro A gene from Salmonella typhimurium confers tolerance to glyphosate. Nature 317:741-744.
23. Crossway, A., H. Hauptli, C.M. Houck, J.M. Irvine, J.V. Oakes, and L.A. Perani (1986) Micromanipulation techniques in plant manipulation. Biotechniques 4:320-334.
24. David, A. (1987) Conifer protoplasts. In Cell and Tissue Culture in Forestry, Vol. 2, J.M. Bonga and D.J. Durzan, eds. Martinus Nijhoff Publishers, Dordrecht, The Netherlands, pp. 2-15.

25. De Cleene, M., and L. De Ley (1976) The host range of crown gall. Bot. Rev. 42:389-466.
26. De la Pena, A., H. Lörz, and J. Schell (1987) Transgenic rye plants obtained by injecting DNA into young floral tillers. Nature 325:174-176.
27. Depicker, A., M. van Montagu, and J. Schell (1983) Plant cell transformation by Agrobacterium plasmids. In Genetic Engineering of Plants: An Agricultural Perspective, T. Kosuge, C.P. Meredith, and A. Hollaender, eds. Plenum Press, New York, p. 499.
28. Diner, A.M., and D.F. Karnosky (1987) Tissue culture technology to forest pathology and pest control. In Cell and Tissue Culture in Forestry, J.M. Bonga and D.J. Durzan, eds. Martinus Nijhoff Publishers, Dordrecht, The Netherlands, pp. 351-373.
29. Douglas, C.J., R.J. Staneloni, R.A. Rubin, and E.W. Nester (1985) Identification and genetic analysis of an Agrobacterium tumefaciens chromosomal virulence region. J. Bacteriol. 161:850-860.
30. Dudits, D., G. Hadlaczky, G. Bajszar, C. Koncz, G. Lazar, and G. Horvarth (1979) Plant regeneration from intergeneric cell hybrids. Plant Sci. Lett. 15:101-112.
31. Einspahr, D.W., and L.L. Winton (1977) Genetics of Quaking Aspen, Forest Service, U.S. Department of Agriculture, U.S. Government Printing Office, Washington, D.C., pp. 1-23.
32. Fillatti, J., J. Sellmer, B. McCown, B. Haissig, and L. Comai (1987) Agrobacterium mediated transformation and regeneration of Populus. Mol. Gen. Genet. 206:192-199.
33. Garfinkel, D.J., R.B. Simpson, L.W. Ream, F.F. White, M.P. Gordon, and E.W. Nester (1981) Genetic analysis of crown gall: Fine structure map of the T-DNA by site directed mutagenesis. Cell 27: 143-153.
34. Gleba, Y.Y., and F. Hoffmann (1980) Arabidobrassica: A novel plant obtained by protoplast fusion. Planta (Berl.) 149:112-117.
35. Glimelius, K., K. Chen, and H. Bonnet (1981) Somatic hybridization in Nicotiana: Segregation of organellar traits among hybrids and cybrid plants. Planta (Berl.) 153:504-510.
36. Goodspeed, T.H., and R.E. Clausen (1928) Interspecific hybridization in Nicotiana. VIII. The sylvestris-tomentosa-tabacum triangle and its bearing on the origin of Tabacum. Univ. California Publ. Bot. II: 245-256.
37. Greenleaf, W. (1941) Sterile and fertile amphidiploids: Their possible relation to the origin of Nicotiana tabacum. Genetics 26:301-324.
38. Hain, R., H.H. Steinbiss, and J. Schell (1984) Fusion of Agrobacterium and E. coli spheroplasts with Nicotiana tabacum protoplasts--Direct gene transfer from microorganisms to higher plants. Plant Cell Reports 3:60-64.
39. Hain, R., P. Stabel, A.P. Czernilofsky, H.H. Steinbiss, L. Herrera-Estrella, and J. Schell (1985) Uptake, integration, expression and genetic transmission of a selectable chimaeric gene by plant protoplasts. Mol. Gen. Genet. 199:161-168.
40. Hoffmann, F., and T. Adachi (1981) Arabidobrassica: Chromosomal recombination and morphogenesis in asymmetric intergeneric hybrid cells. Planta (Berl.) 153:586-593.
41. Horsch, R.B., R.T. Fraley, S.G. Rogers, P.R. Sanders, A. Lloyd, and N. Hoffmann (1984) Inheritance of functional foreign genes in plants. Science 223:496-498.
42. Horsch, R.B., J.E. Fry, N.L. Hoffmann, D. Eichholtz, S.G. Rogers, and R.T. Fraley (1985) A simple and general method for transferring genes into plants. Science 227:1229-1281.

43. Huffman, G.A., F.A. White, M.P. Gordon, and E.W. Nester (1984) Hairy-root inducing plasmid: Physical map and homology to tumor-inducing plasmids. J. Bacteriol. 157:269-276.
44. Hutchinson, J.B., and S.G. Stephens (1947) The Evolution of Gossypium, Oxford University Press, London, 160 pp.
45. Johnsson, H. (1940) Cytological studies of diploid and triploid Populus tremula and of crosses between them. Hereditas 26:321-352.
46. Joos, H., D. Inze, A. Caplan, M. Sormann, M. van Montagu, and J. Schell (1983) Genetic analysis of T-DNA transcripts in nopaline crown galls. Cell 32:1057-1067.
47. Kao, K.N. (1977) Chromosomal behaviour in somatic hybrids of soybean-Nicotiana glauca. Mol. Gen. Genet. 150:225-230.
48. Klee, H.J., M.F. Yanofsky, and E.W. Nester (1985) Vectors for transformation of higher plants. Bio/Technology 3:637-642.
49. Krumbiegel, G., and O. Schieder (1979) Selection of somatic hybrids after fusion of protoplasts from Datura innoxia Mill and Atropa belladonna L. Planta (Berl.) 145:371-375.
50. Lemmers, M., M. de Benekeleer, M. Holsters, P. Zambryski, A. Depicker, J.P. Hernalsteens, M. von Montagu, and J. Schell (1980) Genetic identifications of functions of TL-DNA transcripts in octopine crown gall. EMBO J. 1:147-152.
51. Lloyd, A.M., A.R. Barnason, S.G. Rogers, M.C. Byrne, R.T. Fraley, and R.B. Horsch (1986) Transformation of Agrabidopsis thaliana with Agrobacterium tumefaciens. Science 234:464-466.
52. Lörz, H., B. Baker, and J. Schell (1985) Gene transfer to cereal cells mediated by protoplast transformation. Mol. Gen. Genet. 199: 178-182.
53. Mattila, R.E. (1961) On the production of tetraploid hybrid aspen by colchicine treatment. Hereditas 47:631-640.
54. McCormick, S., J. Niedermeyer, J. Fry, A. Barnason, R. Horsch, and R. Fraley (1986) Leaf disc transformation of cultivated tomato (Lycopersicon esculentum) using Agrobacterium tumefaciens. Plant Cell Reports 5:81-84.
55. McCown, B.H., and J.A. Russell (1987) Protoplast culture of hard woods. In Cell and Tissue Culture in Forestry, J.M. Bonga and D.J. Durzan, eds. Martinus Nijhoff Publishers, Dordrecht, The Netherlands, pp. 16-30.
56. McFadden, E.S., and E.R. Sears (1946) The origin of Triticum spelta and its free-threshing hexaploid relatives. J. Hered. 37:81-89.
57. Melchers, G. (1980) Protoplast fusion, mechanisms and consequences for potato breeding and production of potatoes + tomatoes. In Advances in Protoplast Research, L. Ferency and G.L. Farkas, eds. Pergamon Press, Oxford, pp. 283-286.
58. Melchers, G., M.D. Sacristan, and A.A. Holder (1978) Somatic hybrid plants of potato and tomato regenerated from fused protoplasts. Carlsberg Res. Comm. 43:203-218.
59. Nester, E.W., M.P. Gordon, R.M. Amasino, and M.F. Yanofsky (1984) Crown gall: A molecular and physiological analysis. Ann. Rev. Plant Physiol. 35:387-413.
60. Ohgawara, T., H. Uchimiya, and H. Harada (1983) Uptake of liposome-encapsulated plasmid DNA by plant protoplasts and molecular fate of foreign DNA. Protoplasma 116:145-148.
61. Ooms, G., P.J.J. Hooykaas, G. Moolenaar, and R.A. Schilpevoort (1981) Crown gall plant tumors of abnormal morphology, induced by Agrobacterium tumefaciens carrying mutated octopine Ti-plasmids; Analysis of T-DNA functions. Gene 14:33-50.

62. Ooms, G., M.M. Burrell, A. Karp, and J. Hille (1987) Genetic transformation in two potato cultivars with T-DNA from disarmed Agrobacterium. Theor. Appl. Genet. 73:744-750.
63. Otten, L., H. de Greve, J.P. Hernalsteens, M. van Montagu, O. Schieder, J. Straub, and J. Schell (1981) Mendelian transmission of genes introduced into plants by the Ti-plasmid of Agrobacterium tumefaciens. Mol. Gen. Genet. 183:209-213.
64. Parson, T.J., V.P. Sinkar, R.F. Stettler, E.W. Nester, and M.P. Gordon (1986) Transformation of poplar by Agrobacterium tumefaciens. Bio/Technology 4:533-536.
65. Potrykus, I., R.D. Shillito, M. Saul, and J. Paszkowski (1985) Direct gene transfer-state of the art and future potential. Plant Mol. Biol. Reporter 3:117-128.
66. Potrykus, I., M. Saul, I. Petruska, J. Paszkowski, and R.D. Shillito (1985) Direct gene transfer to cells of monocots. Mol. Gen. Genet. 199:183-188.
67. Schell, J., M. van Montagu, L. Willmitzer, J. Leemans, R. Deblaere, H. Joos, D. Inze, A. Wostemeyer, L. Otten, and P. Zambryski (1984) Transfer of foreign genes to plants and its use to study developmental processes. In Cell Fusion, Gene Transfer and Transformation, R.F. Beers and E.G. Bassett, eds. Raven Press, New York, pp. 113-128.
68. Schilperoort, R.A. (1986) Integration, expression and stable transmission through seeds of foreigns in plants. In Genetic Manipulation in Plant Breeding, W. Horn, C.J. Jensen, W. Odenbach, and O. Schieder, eds. Walter de Gruyter, Berlin, pp. 837-858.
69. Sederoff, R., A. Stomp, W.C. Chilton, and L.W. Moore (1986) Gene transfer into loblolly pine by Agrobacterium tumefaciens. Bio/Technology 4:647-649.
70. Sederoff, R., A. Stomp, B. Gwynn, E. Ford, C. Loopstra, P. Hodgkiss, and W.S. Chilton (1987) Application of recombinant DNA techniques to pines: A molecular approach to genetic engineering in forestry. In Cell and Tissue Culture in Forestry, J.M. Bonga and D.J. Durzan, eds. Martinus Nijhoff Publishers, Dordrecht, The Netherlands, pp. 314-329.
71. Shaw, C.H., M.D. Watson, G.H. Carter, and C.H. Shaw (1984) The right and copy of the nopaline Ti-plasmid 25bp repeat is required for tumor formation. Nucl. Acids Res. 12:6031-6041.
72. Stachel, S.C., and E.W. Nester (1986) The genetic and transcriptional organization of the vir region of the A6 Ti plasmid of Agrobacterium tumefaciens. EMBO J. 5:1445-1454.
73. Stachel, S.C., B. Timmermann, and P. Zambryski (1985) Generation of single stranded T-DNA molecules during the initial stages of T-DNA transfer from Agrobacterium tumefaciens. Nature 322:706-712.
74. Stebbins, G.L. (1950) Variation and Evolution in Plants, Columbia University Press, New York.
75. Uchimiya, H. (1982) Somatic hybridization between male sterile Nicotiana tabacum and N. glutinosa through protoplast fusion. Theor. Appl. Genet. 61:69-72.
76. Uchimiya, H., T. Fushimi, H. Hashimoto, H. Harada, K. Syono, and Y. Sugawara (1986) Expression of foreign gene in callus derived from DNA-treated protoplasts of rice (Oryza sativa L). Mol. Gen. Genet. 204:204-207.
77. Wang, K., L. Herrera-Estrella, M. van Montagu, and P. Zambryski (1984) Right 25-bp terminus sequence of the nopaline T-DNA is essential for and determines direction of DNA transfer from Agrobacterium to the plant genome. Cell 38:455-462.

78. Wang, K., S.E. Stachel, B. Timmermann, M. van Montagu, and P. Zambryski (1987) Site specific nick in the T-DNA border sequence as a result of Agrobacterium vir gene expression. Science 235:587-591.
79. Wicker, M. (1987) Tumors. In Cell and Tissue Culture in Forestry, Vol. 2, J.M. Bonga and D.J. Durzan, eds. Martinus Nijhoff Publishers, Dordrecht, The Netherlands, pp. 374-389.
80. Yamada, Y., Y. Zhi-Qi, and T. Ding-Tai (1986) Plant regeneration from protoplast-derived callus of rice (Oryza sativa L.). Plant Cell Reports 5:85-88.
81. Yang, F.M., and R.B. Simpson (1981) Revertant seedlings from the crown gall tumors retain a portion of the bacterial Ti-plasmid sequences. Proc. Natl. Acad. Sci., USA 78:4151-4155.
82. Zambryski, P., H. Joos, C. Gentello, J. Leemans, M. van Montagu, and J. Schell (1983) Ti plasmid vector for the introduction of DNA into plant cells without alteration of their normal regeneration capacity. EMBO J. 2:2143-2150.

IN VITRO MANIPULATION OF SLASH PINE (PINUS ELLIOTTII)

M.S. Lesney, J.D. Johnson, Theresa Korhnak, and M.W. McCaffery

Department of Forestry
University of Florida
Gainesville, Florida 32611

ABSTRACT

Slash pine is being developed as a model system for the study of genetic and biochemical manipulation in southern conifers. Slash pine is of tremendous economic importance to the southeastern United States, and has been the subject of intensive tree improvement efforts. The species is amenable to standard cotyledonary micropropagation techniques and to protoplast isolation and culture (as yet only to first-division stage). Suspension cultures have been established and have a doubling time of four to six days. Cultures are green (contain chlorophyll) and possess RuBPCarboxylase (as determined by electrophoresis and immunoassay). Cell wall-bound and cytoplasmic peroxidases have been isolated and partially characterized for guaiacol oxidase, IAA oxidase, and syringaldazine oxidase activities. Studies on photosynthesis and heat shock response in culture are underway.

INTRODUCTION

Importance of Slash Pine

Slash pine (*Pinus elliottii* Engelm.) is one of the major tree species in the southern United States and provides such forest products as pulp, poles, lumber, and gum naval stores. In 1980, slash pine plantations amounted to 11.5 billion cubic feet of growing stock. The species grows in the Coastal Plain from South Carolina to Florida and west to eastern Louisiana (19). In Florida nearly 50% of the land area (15.7 million acres) is forested. Thirty-four percent of this forested land (5.34 million acres) is occupied by slash pine, which makes it the major single tree species of interest to the forestry industry in this state (19). It is widely grown in countries other than the United States, including Australia, New Zealand, Colombia, Argentina, and South Africa (8).

History of Slash Pine Research

Research on slash pine genetics and breeding began in Florida in the early 1950s with the United States Forest Service and is currently continued by the Cooperative Forest Genetics Research Program at the University of Florida. Advanced generation selection based on intensive progeny testing is currently underway for several characteristics, including growth, form, and fusiform rust resistance. A broad and defined genetic base exists (as good as for any of the southern pines) and is available for research use in both clone banks and family seed sources. Early on, tissue culture approaches such as callus culture techniques (10) and cotyledonary micropropagation (19) were applied to slash pine, and much work on the biology of fusiform rust interactions in slash pine has been pursued (11). Currently, the Forest Biotechnology Program at the University of Florida is trying to integrate and extend this research into the realm of the molecular biology and genetics of stress metabolism.

Stress Metabolism in Woody Plants

A coherent view of the role of stress metabolism and the wound response in plants is currently forming (9,12,25). Wounds, which trigger the stress metabolism of the plant, may be induced by physical (including temperature), chemical (herbicides and pollutants), or biological (viruses, bacteria, fungi, or insects) agents in the environment. Certain processes of stress metabolism (specifically, linked changes in ethylene biosynthesis, peroxidase production, and cell wall lignification) have been seen to occur universally in higher plants and are involved intimately in all facets of stress resistance, natural or induced (9,25). The ethylene-peroxidase-lignification system is also of critical importance to the normal life of the plant, especially for a woody plant species such as pine, with its profound seasonal changes in dormancy, cell growth, phenolic production, and lignification, even independent of the disease/stress associations in which the system participates. These processes are physiologically complex and difficult to study in whole plant systems, especially in woody species. Without elucidation of the biochemical and genetic bases for stress metabolism in pine, it is unlikely that further genetic or biotechnological manipulation in these areas can reasonably be attempted. Thus, the analysis of such processes and the development and application of methodologies to study them comprise the major thrust of biotechnology research in our program.

METHODOLOGIES AND RESULTS

Tissue Culture

Cotyledonary micropropagation. Currently, the main consistently viable method of micropropagation in conifers involves the use of adventitious buds initiated from cotyledons and induced to subsequent growth and shoot development followed by rooting and transfer to soil. Such techniques have worked well for numerous gymnosperm genera, including *Thuja*, *Cupressus*, and *Picea* (23). In *Pinus*, particular success has been seen for *Pinus pinaster* and *Pinus sylvestris* among others (23). Of the southern pines, *Pinus taeda* (loblolly pine) has had the most success reported. Several field studies are already providing interesting data on the behavior of tissue-culture-derived plantlets under normal planting conditions as compared to seedlings of the same genotypic source (1). Work was undertaken to further adapt this system to slash pine in our laboratory so that in the future we would be able to take advantage of clonal techniques for the production of uniform genetic material for physiological testing.

Seeds of fusiform rust-resistant and fusiform rust-susceptible clones of P. elliottii (clone numbers 70-56 and 245-55, respectively) were obtained from the Cooperative Forest Genetics Research Program at the University of Florida. Seeds were surface-sterilized, dried and nicked at the micropylar end, and placed in 3% hydrogen peroxide for two days at 31°C. Seeds were then placed in 1% hydrogen peroxide for four to five days, depending on the rate of germination. The seed coats were then removed by hand and the gametophyte tissue surface-sterilized for 15 min in 15% household bleach and a small amount of detergent (used as a wetting agent), followed by two rinses with sterile distilled water. Gametophytes were then slit gently and the embryos removed. Cotyledons were cut from the embryos and placed separately in horizontal orientation on GD_1 medium (17) modified by the addition of 10 mg/l benzylaminopurine (BAP) and 0.01 mg/l naphthaleneacetic acid (NAA) (17).

Cotyledons on the modified GD_1 medium were cultured for six to eight weeks under continuous light at 22°C until they became swollen, shiny, and dark green in appearance. Cotyledons were transferred to fresh, modified GD_1 medium every three to four weeks while developing. Subsequently cotyledons were plated on GD 1/2 medium (one-half GD_1 salts, no hormones) for six to eight weeks under the above incubation conditions until shoots became differentiated from the bumpy epidermal layers (Fig. 1). When shoots reached 2 to 5 mm in height, they were excised

Fig. 1. Developmental sequence for Pinus elliottii from shoot induction on cotyledons (top left) to excised shoot (top right) to rooted shoot (bottom left) to plant in soil (bottom right).

from the cotyledon and placed on GD 1/2 medium under the same incubation conditions. Shoots were continuously subcultured to fresh GD 1/2 medium each month, each time receiving a fresh cut near the base of the stem. Shoots reaching 1.5 cm were cut basally and transferred to rooting medium containing GD 1/2 with 0.5 mg/l NAA and 0.1 mg/l BAP. When the stem base became swollen (five to ten days), the shoots were transferred back to GD 1/2 without hormones. No cut was made at the base of the stem for this transfer.

Approximately 90% of shoots transferred to the rooting step rooted within one week to one month after treatment. Very often, long taproot-like single roots formed with little secondary root development. Such roots could be pruned at the tips and the plants put back in hormone-free medium so that lateral root development was triggered, making for better subsequent survival in soil.

Although plants could be transferred to the soil when roots were 3 mm or longer, best results were obtained when roots were allowed to reach 1 cm or more and lateral roots were naturally produced or induced via pruning before transfer to soil (Fig. 1).

As has been seen in other *Pinus* species, apparent mature characteristics occur in the plantlets derived from tissue culture as compared to the phenology of normal seedlings: (a) needles are encased in fascicles and are longer and thicker then seedling needles, and (b) the growth habit of the plantlet seems more compact, initially almost a rosette before stem elongation and growth occur.

Projected efficiencies of the system as operable at present for slash pine in our laboratory are from 20 to 40 clonal plantlets into soil from each seed, taking a minimum of five months. Work on loblolly pine in other laboratories (1,17) suggests that with additional fine-tuning of the system, there is the potential for much greater yields (up to hundreds of shoots).

Callus and suspension cultures. Callus was obtained either from cambial explants removed from mature trees in the University of Florida clone bank using a 3/4-inch bore cork borer (clones: 311-56, 196-57, 78-57, and 61-56), or from stem segments of an individual two-month-old *P. elliottii* seedling of female parent clone 52-56. Tissue was plated on solid Murashige and Skoog (MS) medium containing 2 mg/l 2,4-dichlorophenoxyacetic acid (2,4-D) to produce callus. Callus was removed from plant tissues and subcultured two to three times at monthly intervals. Callus was then transferred to liquid Litvay (LM) medium (22) to form suspensions. Initial liquid subcultures were at one- to two-week intervals and involved settling the cells and removing old medium to replace with fresh medium. After several such transfers, cells were placed on a four- to six-day subculturing regime using 1:1 dilutions.

Protoplast isolation and culture. Protoplast methodology is only recently becoming common in the gymnosperms. Needle protoplasts have been isolated and cultured back to callus stage for a number of pine species (6). Suspension culture protoplasts are ideal material for genetic and physiological manipulation studies, especially due to the inherent genetic and physiological uniformity possible and the ability to obtain large quantities of protoplasts reproducibly. Suspension culture protoplasts also make ideal partners for protoplast fusion due to their differing morphology and ease of manipulation compared to most mesophyll protoplast systems.

Loblolly pine suspension culture protoplasts have been reported upon (22). We have attempted to adapt such methodology to slash pine suspension cultures with some limited success. The following reports on our current level of progress.

Slash pine suspension cultures are maintained as reported above. Two to three days post-subculture, 10 to 12 ml of cells are settled, washed once in fresh LM medium, incubated in osmoticum for one-half hr [CPW salts (13) with 9% added glucose], and then placed in two volumes of an enzyme solution containing 2% pectinase (Sigma Chemical Company, St. Louis, Missouri) and 1% Cellulysin (Calbiochem Company, La Jolla, California) in osmoticum. Incubation is for 2 to 3 hr at 60 rpm in a flask at 23°C. Protoplasts are centrifuged and enzyme removed. The pellet is resuspended in CPW osmoticum containing 6% glucose, and the solution is layered onto a CPW 12% sucrose cushion. The gradient is then centrifuged and the resulting interface band of protoplasts is harvested and placed in CPW 6% glucose. These protoplasts are then concentrated by centrifugation and transferred to various LM-based media. Consistent yields of 0.5-1.0 x 10^6 protoplasts per ml of initial settled cell volume are obtained. Protoplast viability is not as high as desirable after incubation in medium: 62% viability after three days in LM plus 0.35% casein hydrolysate being the best treatment so far. After one week, viable protoplasts have nearly all regenerated cell walls, and in most of the media massive budding and "near divisions" occur (a "near division" can be described as a protoplast that achieves the typical figure eight appearance of a dividing protoplast but never achieves true division or the formation of an intervening cell wall).

Occasional apparently authentic first divisions are seen in the best media, but these are rare (<1%) and as yet not consistent. Since subsequent second divisions and return to callus have not been noted, the system requires much more extensive analysis before it is perfected. There is a strong beneficial effect of added casein hydrolysate to viability, such that protoplasts die within two days without this amendment. This leads us to suspect that variations in nitrogen source will be significant. Preliminary attempts to screen specific amino acids show that glutamic acid can mimic part of the casein hydrolysate effect. We have tested the media and enzymes used in the loblolly system and have found that these provide no added response and are somewhat less successful than the above enzyme combination and medium. Addition of 2 mg/l BAP or NAA to the LM plus CH medium has not stimulated division above that found in the unamended medium.

Further hormone combinations as well as alterations of the nitrogen source are being tried. More work on the initial isolation technique (enzymes) and protoplast purification technique may increase initial viability and improve subsequent medium response. During protoplast isolation, enzyme digestion reveals the presence of tracheary elements formed in the suspension culture: these appear heavily walled and uniformly pitted as would be expected for gymnosperm xylem cells. Since these cultures have been in liquid suspension for nearly two years, the chance of lingering initial explant material being present is quite minimal, and these cells most probably represent a limited ability for in vitro differentiation which may be quite beneficial in the study of lignification and metabolic control in a woody plant system.

Stress Physiology

Peroxidases: guaiacol oxidase and IAA oxidase activities. Peroxidase isozymes were purified as follows (3,14): 100 g frozen suspension culture cells were homogenized in 200 ml 0.01 M KPO_4 buffer (pH 8.0) in a blender. The homogenate was centrifuged and the supernatant designated as "buffer fraction." The pellet was washed and rehomogenized in 1% Triton-X and the centrifugation repeated. The resulting supernatant was designated the "Triton fraction." The remaining pellet was washed six times with water to wash the cell walls free of contaminating cytoplasm and cytoplasmic peroxidases. The resulting pellet was washed with 1 M NaCl and centrifuged. The resulting supernatant was designated as "NaCl fraction." The pellet was then washed with $NaHCO_3$ followed by water washing. The resultant pellet was air-dried overnight and incubated in 1% Cellulysin in 0.05 M Na-citrate buffer (pH 5.2) for 8 hr. Cell wall material was then pelleted and the supernatant designated "cellulase fraction." Concentrated fractions were passed first through an anion-exchange column (DEAE cellulose). Bound (negatively charged) peroxidases were eluted using 1 M NaCl. Unbound peroxidases were passed through a cation exchange column to bind positively charged peroxidases, which were then eluted as above (14).

Guaiacol peroxidase activity was assayed at 470 nm as in Castillo and Greppin (5). Indoleacetic acid-oxidase activity was assayed at 247 nm by following increased absorbance as generated by IAA breakdown as described elsewhere (24). Specific activity was calculated against protein using Coumassie staining for protein determination.

Table 1 shows the specific activities determined for each of the fractions. As in the model for peroxidase activity proposed by Gaspar et al. (9), there was more significant IAA-oxidase activity in the positively charged fractions, except for those covalently bound to the cell walls.

Figure 2 shows the comparative peroxidase patterns for buffer, NaCl, and cellulased fractions, respectively, for five different slash pine clone suspension cultures derived from cambial explants. Procedures, weights, and dilutions were identical for each clone so that comparisons can be made within fractions, although not between fractions. It can be seen that for each fraction there is variation in either or both isozyme banding pattern and intensities, and that positively charged peroxidases predominate in the NaCl fraction. (This is to be expected for the ionically bound peroxidases due to the net negative charge of the cell walls.) This variation has promising potential both for clonal identification purposes and, more importantly, for physiological and genetic studies on the role of specific peroxidases in stress metabolism and normal developmental processes.

Ethylene and 1-aminocyclopropane-1-carboxylic acid. Changes in ethylene production in many plant species has been tied to stress metabolism, including water stress (21), wounding, and disease resistance. Studies have indicated that the production of 1-aminocyclopropane-1-carboxylic acid (ACC) and/or ACC synthase is the step involved in triggering stress and wound ethylene production (21,26). In loblolly pine seedlings, ACC levels and ethylene production were shown to be related to water stress (21). It is of obvious benefit to be able to follow ethylene production in cell suspension culture as both a monitor and possible modulator of stress metabolism, especially in light of the peroxidase/ethylene model presented by Gaspar and co-workers described above (9).

Tab. 1. Peroxidase activity from suspension cultures.

Extract		Guaiacol S.A.*	IAA S.A.
Buffer	(-)	623	0.54
	(+)	367	42.1
Triton	(-)	527	0.27
	(+)	158	17.8
NaCl	(-)	2,167	8.9
	(+)	24,563	29.5
Cellulysin	(-)	35	0.35
	(+)	1.3	0.16

*Specific activity ($mg^{-1} \cdot min^{-1}$).

Note: Peroxidases were purified and separated according to charge using ion exchange chromatography. Designated fractions were analyzed spectrophotometrically for enzyme reactivity using guaiacol or IAA. Protein concentration for specific activity was determined using Coumassie Blue staining.

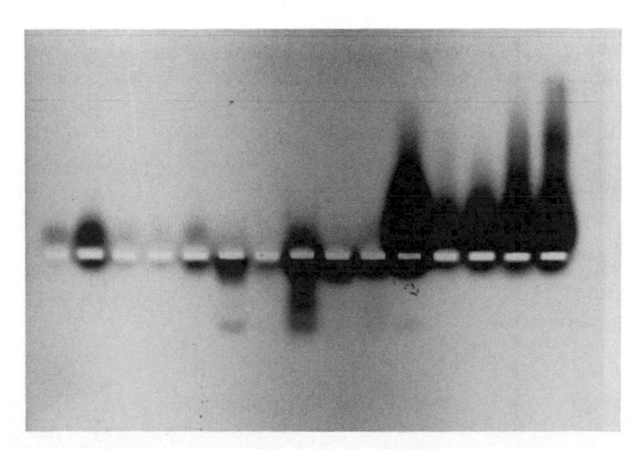

Fig. 2. Peroxidase isozyme patterns. Extracts from 15 g each of five slash pine clones (52-56, 61-56, 78-57, 196-57, and 311-56, respectively) were extracted with: (a) 0.01 M K-phosphate buffer, pH 8.0 (wells 1-5); (b) 1 M NaCl (wells 6-10); and (c) cellulase treatment of cell walls (wells 11-15). The extracts were then concentrated and samples electrophoresed on a 1% agarose gel. Guaiacol staining for detection of peroxidase activity was used.

Preliminary work in our laboratory has found ethylene to be produced in easily detectable amounts in slash pine suspension cultures, but high variability was seen, most probably due to the difficulty of measuring a water soluble gas under normal cell bath conditions. The purification and quantifying of ACC, however, is more uniform because it is not a gas, and once purified can be assayed unequivocally using gas chromatography to monitor the production of ethylene in vitro under controlled conditions in the absence of cells. Under normal cellular conditions, ACC in slash pine suspensions is detectable at levels comparable within an order of magnitude (400 nmoles/g) on a fresh weight basis to ACC in dry weight of needles (92 nmoles/g) and roots (122 nmoles/g) of loblolly pine (21). The production of ACC under elevated temperature conditions is discussed in the following section.

High-Temperature Stress

One of the most pressing problems of decline in productivity due to abiotic stress in the world today is the linked stresses caused by high temperature and drought (15). Even without the predicted global warming trend, the day-to-day necessity of crop and woody plants for dealing with supraoptimal temperatures and suboptimal water creates a critical metabolic barrier to food and fiber production throughout the world. And yet, biological mechanisms exist for dealing with high temperature stress. Nearly ubiquitous mechanisms for providing apparent protection against nonpermissive temperatures exist in organisms ranging from the lowest bacteria to the most highly evolved plants and animals (4). Some theorize that it is the modulation and the elaborated differences in this extraordinarily widespread defense system that provide much of the ability of some plants and animals to be far more heat-tolerant than others (4). Thus, the heat shock and high-temperature response has become a model system with tremendous potential applied benefits for the study of stress regulation and the molecular biology and genetic control of the biological response to exogenous stress. It has led to phenomenal progress in gene regulation studies in the more conventional crop plants and lends itself well to the utilization of current gene transfer and analysis techniques.

Strangely, little has been done in the examination of high-temperature response in woody perennials, where the ability of the individual to adapt to long-term environmental change from season to season and from year to year is of even greater pertinence compared to annuals. In gymnosperms, temperatures well into normal heat shock ranges have been detected routinely (needles and twigs 9° to 11.8°C above air temperature and stems having temperatures 12.2° above needles), with cambial temperatures of 55°C being recorded on the southwest side of a spruce tree in the open at air temperatures of 37°C (15). Young pine seedlings are, in fact, frequently killed at soil level due to overheating of the soil surface (15). It must be considered that, since heat shock leads to a decrease in normal protein synthesis and delay in cell growth ability, any thermotolerance effect that can mitigate or decrease the effect of high temperatures could have a significant effect on growth, even in those cases where survival is not at issue.

For these reasons we have begun looking at the behavior of the slash pine cell suspensions under high-temperature stress conditions. Preliminary experiments have examined the behavior of generalized protein synthesis and ACC production under high-temperature conditions. Slash pine suspension culture cells were placed under stationary conditions at room temperature (23°C) and in incubators at 35°, 40°, and 45°C for 1 hr, at which time [^{3}H]leucine (50 μCi/ml) was added and the incubation continued for 3 hr at these temperatures to monitor differences in radioactive incorporation into proteins. Tissue was washed and then homogenized in buffer containing detergent and mercaptoethanol following methodology similar to Lopato and Gleba (16). The homogenate was centrifuged and the supernatant examined for radioactive incorporation. Table 2 shows that radioactive incorporation as determined by liquid scintillation counting followed an increasing pattern from 23° to 40°C, decreasing abruptly by 45°C. Preliminary examination of ACC production under similar conditions after 1 and 4 hr of the various temperature treatments showed a similar increase from 23° to 40°C with a less precipitous drop by 45°C as assayed using ethylene evolution monitored by gas chromatography (Tab. 3).

These early results show the potential value of the use of such techniques and slash pine suspension cultures for controlled analysis of stress metabolism as it occurs at the cellular level. Much further corroborative work is needed and specific proteins must be examined using fluorographic techniques to determine if authentic heat shock responses are occurring, and if the apparent changes in the behavior of ACC are truly high temperature-related. It must also be determined if these changes have any regulatory role in stress metabolism or are simply the result of shut-down of the ACC → ethylene pathway under high-temperature conditions.

Molecular Biology

Ribulose bisphosphate carboxylase (Rubisco). We are developing semiphotoautotrophic suspension cultures by the use of continuous light and the addition of CO_2 for use in photosynthetic studies. Cultures have been observed to be bright green and to possess at least the large subunit of ribulose bisphosphate carboxylase (as demonstrated by electrophoretic comparison of cell culture protein patterns with purified Rubisco from spinach (Fig. 3), and as conformed by co-reactivity of pine cell suspension extracts with antiserum made to the spinach holoenzyme following standard methodology. The presence of chlorophyll in the suspensions was confirmed spectrophotometrically.

Tab. 2. Incorporation of [^{3}H]leucine in slash pine cells.

Temperature	DPM/μg protein
23°C	255.4 ± 57
35°C	613.0 ± 106
40°C	746.2 ± 63
45°C	209.1 ± 71

Note: Cells were incubated at the designated temperatures for 1 hr at which time [^{3}H]-leucine (50 μCi/ml) was added and incubation continued for 3 hr. Tissue was washed, homogenized, pelleted, and the supernatant assayed for radioactive incorporation into soluble protein using scintillation counting.

Tab. 3. Production of 1-aminocyclopropane-1-carboxylic acid (ACC) (high-temperature stress).

Temperature		ACC* (nmoles/g fresh weight)
23°C	(1 hr)	410
	(4 hr)	950
35°C	(1 hr)	820
	(4 hr)	4,590
40°C	(1 hr)	2,610
	(4 hr)	6,410
45°C	(1 hr)	1,760
	(4 hr)	2,210

*1-Aminocyclopropane-1-carboxylic acid.

Note: Cells were incubated at the designated temperature and duration, and ACC concentration was measured for 0.2-g samples. ACC concentration was determined via conversion to ethylene followed by gas chromatographic analysis.

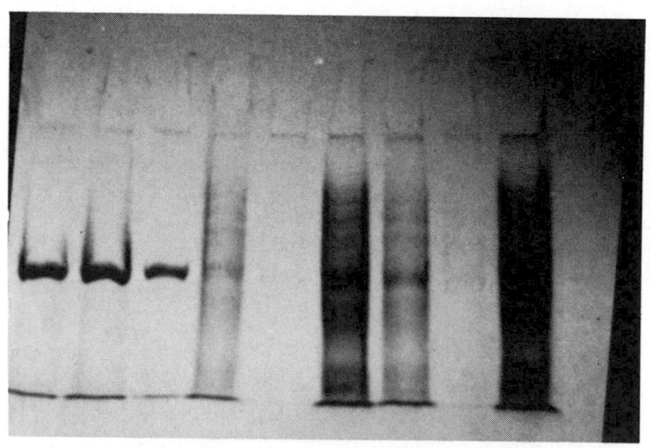

Fig. 3. Apparent ribulose bisphosphate carboxylase (Rubisco) from slash pine suspensions. Lanes 1-3: purified spinach Rubisco (Sigma Chemical Company, St. Louis, Missouri); lanes 5 and 8: buffer extract of slash pine suspension culture cells; lanes 6 and 9: 3% Tween extract of suspension culture cells; lanes 4 and 7: 3% Tween re-extraction of pellet remaining from original Tween extraction (see lanes 6 and 9). Gel is 10% polyacrylamide run under denaturing conditions with sodium dodecyl sulfate.

Isolation of Polysomal RNA

We are currently beginning isolation of total and polysomal RNA from the greened slash pine suspension cultures. In the process, we are comparing several methodologies, including use of guanidine isothiocyanate (7) and magnesium sulfate precipitation (18). Figure 4 shows apparent ribosomal RNA banding patterns from slash pine suspension cultures using the two techniques. RNA will be bulked and eventually put through oligo-dT column purification to obtain purified mRNA. Since total RNA yields are fairly low and quite variable in our initial experiments (100 to 1,000 μg per 40 g fresh weight of cells), further work is necessary.

CONCLUSIONS

The thrust of our research is to use biotechnological approaches to study the biochemistry, physiology, and genetics of stress metabolism in slash pine as a model system for the southern pines and pines in general. To this end we are developing and adapting methodologies as described above to enable us to analyze the various parameters of stress metabolism. Slash pine was chosen both for its economic importance and for its long history of rust reaction and genetic studies that have made available defined clonal and familial plant and seed sources from the Cooperative Forest Genetics Research Program and the U.S. Forest Service. It is our interest to be able to isolate and understand specific genes and gene processes important in stress metabolism in order to have the resource base to ultimately utilize developing gene transfer and analysis techniques to greatest advantage in the improvement of slash pine and other forest species. Additionally, we hope that the developing models of stress metabolism--including ethylene, IAA-oxidase, and peroxidases--will prove of more universal import to plant systems in general, as has been speculated (9).

Fig. 4. Agarose electrophoresis of RNA. Lanes 1-3 show RNA pattern from the guanidine isothiocyanate purification procedure used on slash pine suspension cells. Lanes 4-6 show marker rRNA bands from Escherichia coli. Lanes 7-10 show RNA from slash pine cells purified via magnesium sulfate precipitation. The intense bands are probably rRNA; the minor bands are most likely breakdown products.

REFERENCES

1. Amerson, H.V., L.J. Frampton, Jr., S.E. McKeand, R.L. Mott, and R.J. Weir (1985) Loblolly pine tissue culture: Laboratory, greenhouse and field studies. In Tissue Culture in Forestry and Agriculture, R.R. Henke, K.W. Hughes, M.J. Constantin, and A. Hollaender, eds. Plenum Press, New York, pp. 271-287.
2. Bircham, R., H.E. Sommer, and C.L. Brown (1980) Comparison of plastids and pigments of Pinus palustris Mill. and Pinus elliottii Engelm. in callus tissue culture. Z. Pflanzenphysiol. 102:101-107.
3. Bireka, H., and A. Miller (1974) Cell wall and protoplast isoperoxidases in relation to injury, indole-acetic acid and ethylene effects. Plant Physiol. 53:569-574.
4. Burke, J.J., J.L. Hatfield, R.R. Klein, and J.E. Mullet (1985) Accumulation of heat shock proteins in field-grown cotton. Plant Physiol. 78:394-398.
5. Castillo, F.J., and H. Greppin (1986) Balance between anionic and cationic extracellular peroxidase activity in Sedum album leaves after ozone exposure. Analysis by high performance liquid chromatography. Physiol. Plant. 68:201-208.
6. David, H., E. Jarlet, and A. David (1984) Effects of nitrogen source, calcium concentration and osmotic stress on protoplasts and protoplast-derived cell cultures of Pinus pinaster cotyledons. Physiol. Plant. 61:477-482.
7. Davis, L.G., M.D. Dibner, and J.F. Battey (1986) Basic Methods in Molecular Biology, Elsevier, New York, 388 pp.
8. Dorman, K.W. (1976) The Genetics and Breeding of Southern Pines, Agriculture Handbook No. 471, U.S. Department of Agriculture Forest Service, Washington, D.C., 407 pp.
9. Gaspar, T., C. Penel, F.J. Castillo, and H. Greppin (1985) A two-step control of basic and acidic peroxidases and its significance for growth and development. Physiol. Plant. 64:418-423.
10. Hall, R.H., P.S. Baur, and C.H. Walkinshaw (1972) Variability in oxygen consumption and cell morphology in slash pine tissue cultures. For. Sci. 18:298-307.
11. Hare, R.C. (1972) Physiology and biochemistry of resistance to pine rusts. In Biology of Rust Resistance in Forest Trees, U.S. Department of Agriculture Forest Service Miscellaneous Publication No. 1221, Washington, D.C.
12. Hoque, E. (1982) Biochemical aspects of stress physiology of plants and some considerations of defense mechanisms in conifers. Eur. J. For. Pathol. 12:280-296.
13. Lesney, M.S., P.W. Callow, and K.C. Sink (1986) A technique for bulk production of cytoplasts and miniprotoplasts from suspension culture-derived protoplasts. Plant Cell Reports 5:115-118.
14. Lesney, M.S., J.D. Johnson, and Theresa Korhnak (1988) Partial purification of peroxidase isozymes and associated IAA-oxidase activity from suspension-cultured cells of Pinus elliottii (Slash Pine). Physiol. Plant. (submitted for publication).
15. Levitt, J. (1980) Responses of plants to environmental stresses. In Chilling, Freezing and High Temperature Stress, Vol. 1, Academic Press, Inc., New York, 489 pp.
16. Lopato, S.V., and Y.Y. Gleba (1985) Heat shock proteins from cell cultures of higher plants and their somatic hybrids. Plant Cell Reports 4:19-22.

17. Mott, R.L., and H.V. Amerson (1982) A Tissue Culture Process for the Clonal Production of Loblolly Pine Plantlets, North Carolina ARS Technical Bulletin No. 271.
18. Palmiter, R.D. (1974) Magnesium precipitation of ribonucleoprotein complexes. Expedient techniques for the isolation of undegraded polysomes and messenger ribonucleic acid. Biochemistry 13:3606-3615.
19. Scheffield, Knight, and McClure (1983) The managed slash pine ecosystem. School of Forestry Research Communication, University of Florida, Gainesville, Florida.
20. Sommer, H.E., C.L. Brown, and P. Kormanik (1975) Differentiation of plantlets in longleaf pine (*Pinus palustris* Mill.) tissue cultured in vitro. Bot. Gaz. 136:196-200.
21. Stumpff, N.J., and J.D. Johnson (1987) Ethylene production by loblolly pine seedlings associated with water stress. Physiol. Plant. 69:167-172.
22. Teasdale, R.D., and E. Rugini (1983) Preparation of viable protoplasts from suspension cultured loblolly pine (*Pinus taeda*) cells and subsequent regeneration to callus. Plant Cell Tissue Organ Culture 3:253-260.
23. Thorpe, T.A., and S. Biondi (1983) Conifers. In Handbook of Plant Cell Culture, Vol 2, W.R. Sharp, D. Evans, P. Ammirato, and Y. Yamada, eds. Macmillan Publishing Company, New York, pp. 435-470.
24. Ueng, P.P., and J.M. Daly (1985) Comparison of indole-3-acetic acid oxidation in peroxidases from rust-infected resistant wheat leaves. Plant Cell Physiol. 26:77-87.
25. Van Loon, L.C. (1984) Regulation of pathogenesis and symptom expression in diseased plants by ethylene. In Ethylene: Biochemical, Physiological and Applied Aspects, Martinus Nijhoff/Dr. W. Junk Publishers, The Hague, pp. 171-180.
26. Yang, S.F., and H.K. Pratt (1978) The physiology of ethylene in wounded plant tissues. In Biochemistry of Wounded Plant Tissues, Gunter Kahl, ed. Walter D. Gruyter, Berlin, pp. 595-622.

POTENTIAL APPLICATION OF HAPLOID CULTURES OF TREE SPECIES

J.M. Bonga, P. von Aderkas, and D. James

Canadian Forestry Service Maritimes
P.O. Box 4000
Fredericton, New Brunswick, Canada E3B 5P7

ABSTRACT

Haploid cultures are useful: (a) to unmask cryptic recessive genetic information; (b) to produce completely homozygous diploid plants in the shortest possible time; (c) to establish gene maps; and (d) to produce somatic hybrids through protoplast fusion. During each of the last three summers we have obtained embryogenic haploid callus from megagametophytes of *Larix decidua*. Key factors in establishing these cultures are collection date, and 2,4-D and glutamine in the medium. Cultures have remained embryogenic through more than 50 subcultures over three years. The callus is initiated by repetitive, strongly polar divisions. Groups of small cells eventually form embryos. Embryos are also formed from multinucleate cells in a fashion somewhat similar to zygotic embryogenesis.

INTRODUCTION

To obtain haploid and doubled-haploid plants has long been a cherished goal for plant breeders (71). The primary reason for this is the desire to (a) avoid some of the complications encountered in traditional breeding, and (b) replicate the successes in genetic manipulation achieved with haploid microorganisms. The main problem with diploid plants and those of higher ploidy levels is that genetic investigations are complicated by such phenomena as dominance and segregation (72,118), and that many useful normal or mutant genes remain undetected (71,72,76,107). Therefore, major research efforts to obtain haploid cell cultures or plants are warranted.

GENERAL FUNCTIONS OF HAPLOID (OR POLYHAPLOID) PLANTS, CELLS, OR PROTOPLASTS IN CROP IMPROVEMENT

Use of Haploid Plants

Haploid or doubled-haploid plants are useful to:

(a) Unmask cryptic, recessive genetic variation.

(b) Produce homozygous diploid lines for controlled hybridization to capture, in a reproducible manner, such genetically controlled features as heterosis and disease resistance. The production of relatively homozygous diploids of herbaceous species has been achieved through several generations of inbreeding. For tree species, with their long life cycle and their often high degree of heterozygosity and load of deleterious recessives, inbreeding is impractical (70). This problem is exacerbated in dioecious tree species, like poplar, because in these a degree of inbreeding can be obtained only through sib-mating, never by selfing (48).

(c) Speed up selection in a variety of breeding schemes (24, 41,54), including the recurrent selection schemes commonly used for diploid cross-fertilizing species (23).

(d) Study the inheritance of qualitative traits (23,24).

(e) Establish linkage maps (24).

Use of Polyhaploid Plants

In some tetraploid tree species, reducing the number of chromosomes in half (polyhaploid) may make these species sexually compatible with diploid species within the genus (56,104). This has been attempted with *Ulmus americana* (69).

Use of Haploid Cells or Protoplasts

The recent development of techniques to obtain haploid cells or protoplasts through androgenesis or gynogenesis in vitro has opened new avenues to:

(a) Detect and select cells or protoplasts with useful genes (82,122). Such genes are mostly either natural recessives, mutated genes, or somaclonal variants. Selection will be predominately for biochemical (79) rather than morphological traits, e.g., for resistance to a specific herbicide or phytotoxin, for salt tolerance, or for such physical traits as tolerance to low temperature or drought, if these are expressed at the cellular level.

(b) Establish gene maps or find genetic markers (isozymes) for a better understanding of the genetics of the plant (67). With regard to isozyme analysis, conifers are uniquely suited. By analyzing both the haploid female gametophyte and its embryo, the male as well as the female genetic component of the embryo can be determined (18). In vitro culture of the female gametophyte could provide sufficient tissue for analysis in those cases where the gametophytes are too small for that purpose.

(c) Carry out karyological studies. Not only are karyological studies easier when one has to deal only with half the genome, but some haploid tissues show details of chromosome structure that are difficult to discern in diploid tissues (89).

(d) Produce somatic hybrids to bypass sexual barriers.

(e) Act as acceptors of foreign DNA prior to plantlet regeneration. Gene transfer is thus obtained much faster than is possible by sexual means.

(f) Avoid chimeras. Mutagenic treatment of multicellular plants generally forms chimeras. These are not always useful in breeding because the mutant trait often disappears in sexual reproduction. Therefore, regeneration of mutant plants from single mutant cells in vitro is preferred (82).

(g) Manipulate the cytoplasm. In most angiosperm plants cytoplasmic inheritance is maternal. However, androgenesis produces haploid plants with paternal cytoplasm. This is of interest in studies of the effect of cytoplasmic factors in genetic mechanisms (1).

(h) Make use of typical, haploid-associated metabolic pathways (109). For example, pollen of several tree species have an amino acid profile that is different from that of their diploid (leaf) tissues (123). Some metabolic characteristics of pollen remain expressed in haploid callus derived from these (120). Such haploid tissues could be used to study, or to exploit commercially, a number of specific metabolic pathways (79,129). A few genes are expressed in the haploid stage only; the products formed by these genes could be of interest (110).

(i) Remove deleterious recessive genes from trees. In the case of deleterious recessives expressing themselves at the cellular level, haploid callus relatively free of such genes would have a higher survival value, i.e., calli with many deleterious recessives would be rogued out (33).

METHODS OF OBTAINING HAPLOID PLANTS

There are five general methods of obtaining haploid plants (46): (i) find naturally occurring haploid plants, (ii) induce parthenogenetic growth from gametes in diploid plants, (iii) eliminate chromosomes by chemical or physical treatments, or, in a few cases, by culture of interspecific hybrid embryos, i.e., of embryos with an unstable genome, (iv) use a haploid initiation gene, and (v) use androgenesis or gynogenesis in vitro.

Occurrence of Natural Haploids

Natural haploids of higher plants, including conifers (29,107) and hardwood trees (125), arise infrequently, mostly through parthenogenetic development of the male or the female nucleus in the embryo sac (4,59, 107). Haploid embryos are sometimes found in polyembryonic seeds (4,15, 59,65,107). Embryos from polyembryonic seeds and reverse embryos of

Picea abies are occasionally haploid (52) or mosaic-aneuploid (21), with chromosome numbers varying from 1n to 2n. Most haploid embryos are weak and, even with careful handling, grow poorly (98).

Among trees only a few well-developed haploid specimens of natural origin are known. One of these is a mature tree of Thuja plicata. The good vigor of this tree, and the fact that this species shows a high degree of self-fertility, suggest that it is devoid of lethal and sublethal recessives (101). This tree is not entirely genetically stable, since it has produced diploid (presumably doubled-haploid) sports (101). Most conifer species are highly heterozygous and often carry a heavy load of lethal recessives (19,38,68). Consequently, reasonably vigorous natural haploids of these species are unlikely ever to be found (101). The same is probably the case with most hardwood species.

Parthenogenetic Development of Gametes

Parthenogenetic development, i.e., embryo development from a male or a female gamete, is generally achieved by either pollinating with pollen partially inactivated by chemicals [chloral hydrate, Toluidine blue (125), colchicine, nitrous oxide (97)], heat or radiation (118,125), or by delayed pollination or distant crossing (97,118). Pollen that have received a sublethal dose of radiation sometimes will transmit single genes or larger sections of the parental genome to the parthenogenic egg (87). Furthermore, the parthenogenic embryo is sometimes homozygous diploid rather than haploid (87).

The most frequently practiced of the above methods is distant crossing. Formation of haploid embryos in such crosses generally appears to depend on cytoplasmic factors (118). Remote crossing of male and female gametes occasionally leads to semigamy, i.e., fusion of the cytoplasm but not of the nuclei of the male and female gametes. This can result in chimeric plants with some sectors of haploid tissue being of paternal and others of maternal nuclear origin (121).

The above techniques have been highly successful with some herbaceous crops, e.g., cereals. However, these techniques have worked with only a few hardwood species, most notably with Populus, and have been ineffective with conifers (125).

Chromosome Elimination

Sometimes interspecific hybridization will result in a genome imbalance in the young embryo, causing a gradual elimination of chromosomes as the embryo grows (10,24,47,54,82). In some crosses, this will eventually lead to an increasing degree of haploidy in the maturing embryo. Interspecific crosses often have to be followed by excision and in vitro culture of the young haploid embryo because of failure of the endosperm to develop (24, 57).

In some cases a reduction in chromosome number of somatic cells can be induced by chemicals or cold or heat treatment (64,82). Occasionally this process takes place in a meiosis-like manner (57,127). With 3-fluorophenylalanine, ploidy has been reduced gradually in somatic cells from tetraploid to haploid (81,112); chloramphenicol has been used to obtain haploid cells from diploid cells (112,127).

Chromosome elimination sometimes results in ploidy reduction, not in the whole plant but only in some of its tissues. For example, somatic reduction, to the haploid and lower-than-haploid aneuploid levels, occurs naturally in some of the cells of the apogeotropic root of cycads (111), and occurs naturally, or is chemically induced, in tissues of some other plants (51). Cells with highly reduced ploidies (with micronuclei) are potentially useful for genetic mapping and engineering (99).

Haploid Initiation Genes

Mutant genes have been found that favor the formation and survival of haploids. Examples of such genes are the *ig* gene in corn and the *hap* gene in barley (23,24).

In Vitro Androgenesis or Gynogenesis

All of the above-mentioned methods are limited in application. The main limitation is that only a few work well, and only for a few species, e.g., the *bulbosum* method with barley (24). A technique that probably will find wider application is regeneration of haploid plants from haploid cells cultured in vitro.

In vitro androgenesis or gynogenesis has resulted in haploid plantlets now being available for several herbaceous and some tree crops. Most of the success has been achieved through anther or pollen culture (71). Lately, however, gynogenesis, i.e., development of unfertilized eggs into haploid plants in cultures of unpollinated ovaries or ovules (126), has become an alternative method for some species.

The most important advantage of in vitro techniques is that selection can be carried out in large populations of cells or protoplasts in suspension, at least for traits expressing themselves at the cellular level. Even though not approaching the adaptability and population density of microbial culture systems, such selection in vitro is far more effective than selection among haploid individuals obtained by the methods described in the previous sections. Androgenesis allows selection within large populations of cells that are genetically diverse, a diversity created by prior meiotic segregation and crossing over. Gynogenesis has the disadvantage that for practical reasons we can culture only a limited number of female gametophytes and thus only a limited number of genotypes. Consequently, in gynogenesis we have to wait for genetic variation to arise naturally (i.e., somaclonally) in culture, or we have to induce it experimentally before selection at the cellular level among a great variety of genotypes can take place.

Working with large, genetically diverse cell populations may make it possible to rogue out some of the lethal recessives (at least the ones expressing themselves early in the life cycle, i.e., in the cellular stage rather than after organ formation) that are so prominent in many tree species. After sufficient selection of cultures with fewer lethal recessives, prior to chromosome doubling, it may be possible to regenerate doubled-haploid plants vigorous enough to reach sexual maturity. These plants could then be used for controlled hybridization (33). However, removal of recessive lethals and semilethals may carry a price, because some of these may be beneficial in heterozygous conditions (106,107).

In androgenesis the regenerated plants inherit the paternal cytoplasm. In gymnosperms this presumably will not create problems, because in these the natural (sexual) cytoplasmic inheritance is mostly paternal (78). However, in angiosperms, which generally have a maternal mode of cytoplasmic inheritance, a paternal cytoplasm in androgenetically produced plants could create problems. In addition, since all pollen in one anther or microsporophyll have different genotypes, embryos arising from these pollen will be genetically different from each other. If embryogenesis or organogenesis is preceded by callus formation, the resulting plants could be chimeric if the callus is composed of a mixture of cells originating from more than one pollen.

Gymnosperm androgenesis. LaRue (see Ref. 117) probably was the first to obtain cell masses from gymnosperm pollen in vitro. He cultured pollen of several gymnosperms, but obtained callus masses only in cultures of *Taxus*. These calli were composed mostly of elongated cells and did not enter organogenesis. Since then, callus has been produced from pollen of a few other gymnosperms (12,15,83,95). However, most of these calli probably have arisen from several pollen within microsporophylls rather than from single pollen (95). Consequently, such calli would not be composed of genetically identical cells but of a mosaic of genetically different cells.

In vitro culture of microspores has greatly facilitated the study of spermatogenesis, which is often difficult to study with fresh field material. For example, the ultrastructural aspects of the release of the spermatogenic cell from pollen tubes of *Taxus* (95) and *Juniperus* (30) were studied more effectively with cultured material. Extensive electron-microscopy studies of pollen ontogeny have also been carried out with pollen of *Juniperus communis* developing in vitro (31).

For most species the microspores in vitro formed callus composed of typical callus cells, i.e., isodiametric (more or less) cells (12,95). The microspores of *Taxus* initially form masses of pollen tube-like cells (95,117) which subsequently divide in polar fashion, thus forming small cells (117). These divisions are similar to strongly polar divisions in callus developing from megagametophyte explants of *Larix decidua* (see below). Callus derived from *Ginkgo biloba* pollen was albino and only had a limited capability to differentiate (102,116). Organogenesis and embryogenesis are rare and generally incomplete in callus of pollen origin. Microspore callus of *Picea abies* produced small shoot initials and short roots (102). Small structures, somewhat resembling the early stages of cleavage proembryos, were noted in microsporophyll cultures of *Pinus resinosa* (11); proembryo-like structures also occurred in pollen cultures of *Zamia floridana* (66).

In comparison with cells of callus derived from megagametophytes (see below), those derived from microspores often display a variety of ploidies (95). For example, in callus obtained in microsporophyll cultures of *Pinus resinosa*, some cells contained far fewer than the haploid number of chromosomes. Other cells each possessed several micronuclei (each containing a much lower than the haploid number of chromosomes), and some cells had lagging chromosomes. Diploid cells occurred among the haploid ones. Most of these diploid cells probably originated from fusion of nuclei in binucleate cells, although one cannot exclude the possibility that some may have arisen from diploid tapetal and connective tissue cells or from unreduced pollen, i.e., from cells with the same diploid heterozygous genotype as the parent (11,12). If in such cultures these heterozygous diploid cells would be more vigorous than the haploid and homozygous diploid cells, they would eventually take over the culture.

Large coenocytes arise in vitro in pollen of some gymnosperm species. For example, pollen of *Torreya nucifera* produced coenocytes with up to 28 nuclei (119), those of *Pinus resinosa* with up to 16 nuclei (12), those of *Ginkgo biloba* with up to 50 nuclei (116), and those of *Taxus cuspidata* with 18 nuclei (66). Bi- or multinucleate cells are rare in microspore callus of all species. For example, callus of *Thuja orientalis* was solely composed of uninucleate cells (92).

In some tobacco anther cultures, callus arose from naturally abnormal pollen ["S pollen" (49)]; in others, it developed from normal pollen (45). For gymnosperms, it is not known whether callus arises from pollen that are naturally aberrant or from normal ones. In microsporophyll cultures of *Pinus resinosa*, some callus develops from immature pollen in which the formation of a normal linear four-celled gametophyte is interrupted and a cell with two equal, or sometimes more, nuclei is formed instead (12). Similarly, in *Torreya nucifera* microsporophyll cultures, callus forms from multinucleate cells. In some cultures the numbers of abnormal pollen are very high. For example, in *Torreya nucifera* pollen in vitro, only 2% to 6% developed normally (119). There are other causes of abnormal pollen development than in vitro culture. In *Pinus resinosa* pollen, abnormal development was induced by cold storage of microstrobili prior to culture of the pollen (11). Many of the observed abnormalities probably can be induced by a wide variety of nonspecific stress factors.

Gymnosperm gynogenesis. For so far unexplained reasons, the megagametophytes of gymnosperms have been far more responsive in vitro than the microspores. Callus derived from pollen demonstrated little tendency to organogenesis or embryogenesis, and often showed abnormal ploidies; callus from megagametophytes, on the other hand, was often highly organo- or embryogenic and, in general, stably haploid or diploid (95). In addition, megagametophyte cultures have the distinct advantage that each explant contains only one genotype, all cells of the megagametophyte having arisen from one megaspore.

Experimentally induced organogenesis in megagametophytes is not new. As early as 1888, Duchartre recorded the formation of roots on megagametophytes of *Cycas* (see reviews in Ref. 28 and 83). Since then, root and shoot formation was induced in megagametophytes, or their callus, of several gymnosperm species (12,15,83,95,102). Morphogenesis or organogenesis was sometimes obtained on very simple nutrient media. For example, LaRue (66) obtained roots and shoots on *Zamia floridana* megagametophytes on a simple mineral and sucrose nutrient, while some cultures stayed alive for years on just water. Megagametophytes of *Pinus nigra* var. *austriaca* showed a high degree of internal morphogenesis (formation of meristems, tracheids, and unusual cell types) on just buffer with indoleacetic acid (91). Clearly, the megagametophytes contain sufficient food reserves to maintain growth for a long time.

The morphogenetic capacity of megagametophyte tissues in vitro is often determined by the stage of development of the gametophyte. For example, megagametophytes of *Zamia* excised about two months before fertilization were the ones most morphogenetically responsive. Those excised three months before fertilization or shortly after fertilization were nonresponsive (66). Megagametophytes of Picea abies were most responsive if excised about one month after pollination. This is about the time when the fluid megagametophyte begins to harden and the first lipid droplets and starch grains appear (50). A distinct interaction between stage of development and morphogenetic capacity also appears in megagametophytes of Pinus nigra (91) and Larix decidua (see below) in vitro. These changes in morphogenetic potential are not unexpected. During gametophyte and subsequent early embryo development, a sequence of distinct developmental changes occur. Most noticeable among these changes are the initiation of archegonia, fertilization, proembryo development, formation of a corrosion cavity, and formation of meristems in the embryos. Each of these initiations presumably requires its own specific ultrastructural and nutritional conditions, and morphogenetic stimuli such as specific combinations of hormones, enzymes, and mRNAs. These changes may have a strong effect on the morphogenetic capacity of the gametophytic tissues in vitro and thus may explain the noted importance of collection date.

There are only a few reports dealing with chemical changes in the gametophyte during its development, most of the literature being limited to changes occurring during and after germination. Carbohydrates, lipids, amino acids, and enzymes change as the gametophyte develops (9,34,35,40, 55,60,61). Unidentified growth promoters appear temporarily in immature megagametophytes of Ginkgo biloba (108) and Pseudotsuga menziesii (74), and gibberellic acid-like substances have been found in immature Ginkgo and pine seed (8,63). However, most of these chemical changes probably are only of minor consequence in determining the morphogenetic capacity of the megagametophyte. More relevant may be the observation that for a two-week period after fertilization, a changed pattern of RNA distribution is present in the ovules of Pinus strobus (62). Another relevant factor could be the changes in the ultrastructure that occur during gametophyte and embryo development (34).

Most of the calli obtained from megagametophytes were haploid; some were diploid or polyploid (95,102). Some of the nonhaploid tissue may have arisen from multinucleate cells in the natural megagametophyte in which the nuclei fused (15,95). For example, diploid and polyploid cells that resulted from nuclear fusion occur naturally in megagametophytes of Ginkgo biloba (5) and Taxus baccata (131). Ball (7) has reported that megagametophytes of Sequoia sempervirens contain not only haploid but also diploid, polyploid, and aneuploid cells. He has regenerated diploid plantlets from callus, presumably from the small number of diploid cells contained in these megagametophytes. During the first four years after out-planting of these plantlets in the field, their growth rates were about the same as those of seedlings but only half of that of plantlets regenerated from stem pieces in vitro.

Gametophytic embryogenesis occurred in callus from megagametophytes of Zamia integrifolia (83,84), Ceratozamia mexicana (27), and Larix decidua (see below) in vitro.* This contrasts with the nonresponse in microsporophyll or pollen cultures where advanced embryogenesis has never been observed.

---

*The common term "somatic embryogenesis" is not appropriate here because we are dealing with gametophytic (i.e., nonsomatic) tissues.

Embryogenesis in haploid callus of megagametophytes of *Larix decidua*. During each of the last three summers we have obtained callus from immature megagametophytes of *Larix decidua*. Subcultures of this callus produced adventitious embryos in high numbers, with several of these embryos growing into plantlets large enough to be planted in soil (77). The most notable features of these cultures are that (a) embryogenic callus arises only from megagametophytes collected over a four-week period, starting at about the time of fertilization, and (b) 2,4-dichlorophenoxyacetic acid (2,4-D) is essential during the callus initiation phase, and glutamine is essential during the initiation and subsequent maintenance phase. The callus is fast-growing and slimy white in appearance if frequently (every two to four weeks) subcultured. If not subcultured for several months, the cultures turn dark (sometimes black) and growth slows down. However, such cultures have a remarkable capacity for recovery if returned to a regime of frequent subculture. Several clones of the first cultures, now about three years old and having passed through about 50 subcultures, are still growing rapidly and are still producing embryos at an undiminished rate. It has been difficult to obtain good chromosome counts from cells of these embryos. However, reliable counts were obtained for about 20 cells, and these counts showed that the cells were haploid.

The first cells that arise from the megagametophytes are very long; in fact, they are sometimes long enough to be discernible by the naked eye. These cells divide in a strongly polar (unequal) manner, i.e., they produce a small cell at their tip. These small cells divide and form microcalli of loosely packed small cells. Many of the small cells of the microcallus form long cells again, which in turn will divide polarly, initiating new microcalli of small cells. This sequence of cell formation often continues through several subcultures. Eventually the microcalli become more densely packed and form proembryos with long suspensor-like protrusions.

A similar sequence of alternating long and short cells has also been described for pollen cultures of *Taxus* (117), and may have occurred in megagametophyte cultures of *Picea abies* (50). This pronounced and repetitive polarity may be unique for gymnosperm tissues, because it has not been described for haploid or diploid angiosperm embryogenic tissues. Carrot cultures show polarity initially during embryogenesis (100), but the process is far less extreme than in our haploid *Larix decidua* cultures, and is nonrepetitive.

There is a second pathway of embryogenesis in our haploid *Larix decidua* cultures. Some of the large cells go through two nuclear divisions without formation of phragmoplasts and new walls. Thus, a four-nucleate coenocyte is formed. These four nuclei then migrate to one pole of the large cell. New cell walls form and separate the four nuclei. The four cells thus formed develop directly into proembryos. A similar development has been noted in somatic embryogenesis in cultures of diploid cells of *Pinus taeda* (42).

As was pointed out earlier, several authors, the first one being Duchartre in 1888, have noted the appearance of root-like structures in their megagametophyte cultures, without further describing these. It is conceivable that, at least in some of the cases where roots were reported, these in fact were proembryos that did not develop into embryos and thus went unrecognized.

Angiosperm androgenesis. Angiosperms, unlike the gymnosperms, do not possess a large haploid megagametophyte that could be used as explant in haploid tissue culture. Therefore, most efforts to obtain haploid plants of angiosperms in vitro have been carried out with anthers or pollen as explants. Recent reviews (13,20) list only a few tree species for which haploid plantlets have been obtained by androgenesis. Several cases of angiosperm androgenesis not listed in these reviews are the following. Callus formation, without subsequent plantlet development, has been reported for Betula pendula (102), Peltophorum pterocarpum, Albizzia lebbeck (26), and Ulmus americana (94). Haploid plantlets have been obtained from anthers of Populus spp. (20,48,58). Embryogenesis, but not plantlet formation, was reported for Cassia fistula, Jacaranda acutifolia, and Poinciana regia (6); embryogenesis followed by plantlet formation was reported for Carica papaya (115) and Vitis vinifera (75). Considerable success has been achieved in China with anther cultures of Hevea brasiliensis and Populus berolinensis (20). These cultures produced plantlets that grew well enough to be planted in the field for testing. Most of the plantlets were mixoploids, with cells ranging in ploidy from 1n to 2n and the ratio of diploid over other ploidies increasing over the years as the plantlets grew into trees. The trees with the highest percentage of diploid (presumably homozygous diploid) cells in the shoot apices grew the best. Clones derived from a tree obtained from Hevea brasiliensis anthers in vitro increased more in girth than clones of the parent tree. This success with Hevea brasiliensis may have been due to this species probably being tetraploid (88), i.e., plantlets from pollen would still be heterozygous and thus would not suffer from the effects of deleterious recessives.

Angiosperm gynogenesis. For some angiosperm herbaceous species, unpollinated ovules have been used as a source of haploid tissue (126). Haploid plantlets have been produced from unpollinated ovules of only one tree species, Robinia pseudoacacia (124).

DISCUSSION

For a few gymnosperm and angiosperm species the techniques of andro- and gynogenesis have improved to the point where plantlets produced by these methods have reached the field testing stage. However, in spite of this success, considerable short- and long-term difficulties are anticipated for the haploid in vitro approach for most species. These will be discussed first before we deal with some of the more promising aspects of haploid culture of tree species.

Problems

The two major objectives of most research with haploid cells in vitro are the production of homozygous diploid plants for controlled hybridization, and somatic hybridization by protoplast fusion. However, both these objectives present obstacles that may be difficult to overcome.

Whereas most herbaceous crops have been bred and inbred for decades or centuries, most tree crops are composed of natural populations or populations that have undergone little controlled breeding. Consequently, most trees are highly heterozygous and contain a heavy load of lethal or semilethal recessives (68,70). Therefore, the haploid as well as the homozygous diploid plants could be of low vigor and may be incapable either of reaching the sexual stage or of being fertile when reaching that stage. It is of interest to note that androgenesis or gynogenesis has been most successful with polyploid tree species, i.e., Hevea brasiliensis among the angiosperms and Sequoia sempervirens among the gymnosperms (see above).

Not all tree species are highly heterozygous. Among the usually extremely heterozygous conifers, Pinus resinosa is relatively homozygous (38,70). Parallel to observations made with herbaceous crop plants, where androgenesis is often the easiest to obtain with relatively homozygous lines (37), one may expect that it will be easier to obtain vigorous haploid plantlets from gametophyte cultures of such tree species as Pinus resinosa than of more heterozygous species. One advantage of androgenesis over gynogenesis is that we are dealing with a large number of gametophytes that have undergone recombination and random segregation in meiosis. This means that we can select among a large number of genotypes, some of which, depending on the distribution of the deleterious genes on the parental chromosomes, may be relatively free of deleterious recessives, at least in some species. On the other hand, the efficiency of androgenesis is often severely limited by only a low percentage of pollen being embryogenic (85).

In many cases where haploid cells have been cultured in vitro, the ploidy level has not been stable, with the callus or the regenerated plants gradually becoming predominantly diploid (see above). Most of the diploid cells appear to arise through nuclear fusion or endoreduplication and thus are presumably homozygous. Most haploid and aneuploid cells eventually disappear through competition. This, of course, is an ideal situation if the purpose is to produce homozygous diploids. However, such diploidization does not always occur. In particular, among the conifers the cultures often remain haploid for years through a large number of subcultures (see above). Diploidization in these by the use of colchicine and other mitotoxins may be difficult (10,39), and even if successful may lead to complications, e.g., mutations and chimeras (53). For some herbaceous species homozygous diploids have been obtained by the use of irradiated pollen, i.e., by an in vivo technique (86,87). If such a technique could be made to work for tree species, it could eventually be a more effective method of obtaining homozygous diploids than androgenesis or gynogenesis in vitro, particularly when chromosome doubling is a problem.

For species for which pedigree breeding is practical, such breeding can, depending on heritability and population size, be more effective in crop improvement than the use of homozygous diploids obtained from haploids by in vitro or other techniques (24). After each successive generation of breeding, a selection for new, valuable meiotic recombinants can be carried out, although the opportunity to do this diminishes in each successive cycle. Homozygotic plants produced from doubled-gametes have had only one cycle of recombination (i.e., the cycle that resulted in the formation of the gametes) (85). However, breeding, although providing more generations of recombination, does not always produce more recombinants (16). Breeding has the advantage that, while the homozygosity level for desired characteristics is improved in successive generations of breeding, a sufficient degree of heterozygosity is generally maintained to retain sufficient vigor for the plants to survive. Anther-derived doubled-haploids are often (14), although not always (16), weaker than the selfed progenies of the parental anther source. Homozygous lines produced by selfing also have the advantage that they are adapted to the environmental conditions of the area where they are produced. Doubled-haploids lack this advantage, but may be better suited when selecting for different climatic conditions (46). To make up for the loss of valuable recombinants in the doubled-haploid method, it has been suggested either to obtain a maximum number of recombinants by crossings and then to produce haploids from these crossings (22,37,103), or to produce various doubled-haploids first, then hybridize these to obtain the recombinants, and then produce doubled-haploids again (96). A disadvantage of the doubled-haploid method is that by selecting against deleterious recessives during the production of the doubled-haploid, we may lose valuable combinations of these recessives with other alleles (106,107). On the other hand, doubled-haploid lines have the advantage of exhibiting only additive and (additive x additive) variance, whereas early selfed generations contain a degree of nonselectable, nonadditive variance (46). Furthermore, breeding may result in gametic selection, by pollen competition, during fertilization (16). Such competition would be absent in doubled-haploid production by androgenesis and, consequently, genotypes different from those arising in breeding may become available.

Use of Haploid Cultures of Tree Species in Future Research

From the foregoing it appears that for most tree species, the use of haploid cultures as a means of obtaining homozygous diploids for controlled hybridization may be limited. However, there are other aspects of haploid culture that could be of value in future research:

(a) One of the major functions of haploid culture is the unmasking of natural, recessive genetic variation. For example, our haploid *Larix decidua* plantlets (see above), even though remaining small, could be used for selection of traits expressed early in the life cycle of the plant. Presumably this could include selection for cold tolerance and biochemical traits, and resistance to some diseases and herbicides. Because of the heavy load of lethal recessives in most tree species, it probably would, in general, not be possible to produce sexually viable doubled-haploid plants from haploid cultures of these species. To introduce the genes selected in the haploid cultures of these species into the breeding population, one could fuse protoplasts from the selected haploid cells with protoplasts with a different haploid genotype and regenerate heterozygous diploid plants from the fusion product. Through breeding of the plants thus produced, progeny in which the recessive trait would appear in homozygous diploid form could be acquired (46). Ideally, protoplasts should be fused in which the desired recessive trait is present in both of the different haploid genotypes. Subsequent breeding would not be necessary in that case. For field testing of plants produced by the fused protoplasts, one can visualize a scenario similar to that used for the testing of sexually produced hybrids that are mass-propagated in vitro [e.g., *Pinus radiata* (2)]. Part of each of the selected haploid cultures would be preserved in cold storage while the fusion products are being field tested. After field testing, cultures of the best combiners could be retrieved from cold storage and used for hybridization by protoplast fusion on a massive scale. This scenario would provide controlled hybridization while bypassing the problems encountered when using doubled-haploids or inbred lines for controlled hybridization. A scenario as described above could also be used for haploid protoplasts that are somaclonal variants or that have been genetically modified by mutagen treatment or genetic engineering.

(b) For genetic engineering it is important to pinpoint genes on the chromosomes first (80). Modern gene mapping techniques, such as mapping by in situ hybridization (36,128), may be easier to carry out with haploid than with diploid cells. This would particularly be the case for species with numerous chromosomes or with chromosomes that are long and difficult to disperse. Andro- or gynogenesis could supply haploid cells in sufficient numbers for this type of work. One feature of interest is that some haploid cultures produce a great number of cells with micronuclei (see the "Gymnosperm androgenesis" section above). Micronuclei have been useful in the mapping of the human genome and could similarly be of use in establishing maps for tree species genomes (99).

(c) As was pointed out earlier, Pandey (86,87) has used sublethally irradiated pollen to introduce paternal DNA fragments into the maternal genome. One might consider using a similar approach with haploid protoplasts, i.e., one could fuse protoplasts from sublethally irradiated pollen, or sublethally irradiated protoplasts derived from haploid callus, with nonirradiated haploid protoplasts. Pandey has argued that DNA fragments thus obtained may be of better quality than DNA fragments obtained with restriction enzymes. However, a similar approach with protoplasts may not work, because by irradiating pollen one irradiates nuclei at the prometaphase stage, which, according to Pandey, is significant. Irradiated haploid protoplasts have been used to transfer organelles without simultaneous nuclear transfer (73), and sublethally irradiated ones have been used for mutant production (105). Irradiated diploid protoplasts have been employed for limited chromosome or gene transfer (43).

(d) Androgenesis, gynogenesis, and genetic manipulation of haploid cells would be more amenable to experimental control if development of sexual structures and subsequent meiosis could be induced in vitro (17,130). With tree species some advances have been made in that direction. For example, male strobili with sporogenic tissue were induced in vegetative shoot tips of _Thuja plicata_ (25), and well developed male and female cones were induced in vegetative shoot tips of _Sequoia sempervirens_ (114) in vitro. Male and female flower buds were initiated in seedlings derived from embryos of _Phoenix dactylifera_ cultured in vitro (3). Plantlets developed from meristems of _Manihot esculenta_ produced male and female flowers in vitro, with the male flowers going through meiosis (113). In some species, gametophytes produced in vitro are more embryogenic than those produced in vivo (114). Cone or flower formation and fertilization in vitro could considerably shorten the breeding cycle of trees (114).

(e) Haploid protoplasts have been prepared from gymnosperm pollen (32) and megagametophytes (44), and from angiosperm tree pollen (93), but none of these have divided and formed plantlets. Since this inability to regenerate plantlets from haploid protoplasts is a major stumbling block in genetic engineering in tree species, one should consider alternatives that do not require regeneration from protoplasts. It has been suggested by Ledig (67) that with conifers, genetic modification by non-in vitro means could possibly be achieved by injecting DNA into the megagametophyte during the free nuclear state. Presumably this DNA would be incorporated readily into all nuclei. After formation of cell walls, all cells of the megagametophyte would have transformed nuclei, including the one that would become the egg cell. After fertilization, the zygote, and consequently all cells of the ensuing embryo, would contain the injected DNA. Injection of DNA into immature flowers of the intact plant is gaining acceptance as a means of transforming plants of species that cannot be infected by _Agrobacterium_ or that are difficult to regenerate from protoplasts or single cells (90).

CONCLUSION

Considerable progress has been made with haploid tissue cultures of tree species. For a few species this progress has reached the stage where progeny derived from pollen or megagametophytes are now being field tested. However, as has been pointed out in this review, for most tree species we can expect to encounter major problems with haploids and doubled-haploids that will have to be solved before in vitro culture of haploid tissues will find general application in genetic improvement programs of tree species.

ACKNOWLEDGEMENT

We wish to thank Dr. T.M. Choo, Agriculture Canada, Charlottetown, Prince Edward Island, Canada, for his thorough review of the manuscript.

REFERENCES

1. Ahuja, M.R. (1982) Isolation, culture, and fusion of protoplasts: Problems and prospects. _Silvae Genet._ 31:66-77.
2. Aitken-Christie, J., and A.P. Singh (1987) Cold storage of tissue cultures. In _Cell and Tissue Culture in Forestry_, Vol. 2, J.M. Bonga and D.J. Durzan, eds. Martinus Nijhoff Publishers, Dordrecht, The Netherlands, pp. 285-304.
3. Ammar, S., A. Benbadis, and B.K. Tripathi (1987) Floral induction in date palm seedlings (_Phoenix dactylifera_ var. _Deglet Nour_) cultured _in vitro_. _Can. J. Bot._ 65:137-142.

4. Asker, S. (1980) Gametophytic apomixis: Elements and genetic regulation. Hereditas 93:277-293.
5. Avanzi, S., and P.G. Cionini (1971) A DNA cytophotometric investigation on the development of the female gametophyte of Ginkgo biloba. Caryologia 24:105-116.
6. Bajaj, Y.P.S., and M.S. Dhanju (1983) Pollen embryogenesis in three ornamental trees--Cassia fistula, Jacaranda acutifolia, and Poinciana regia. J. Tree Sci. 2:16-19.
7. Ball, E.A. (1987) Tissue culture multiplication of Sequoia. In Cell and Tissue Culture in Forestry, Vol. 3, J.M. Bonga and D.J. Durzan, eds. Martinus Nijhoff Publishers, Dordrecht, The Netherlands, pp. 146-158.
8. Bannerjee, S.N. (1968) Changes in the amounts of gibberellin-like and cytokinin-like substances in developing seeds of Ginkgo biloba L. Bot. Mag. Tokyo 81:67-73.
9. Bannerjee, S.N., and N.W. Radforth (1966) Changes in the pool of free amino acids in the gametophyte of Ginkgo biloba Linn. at different stages of embryonic differentiation. Indian Agric. 10:33-39.
10. Bennett, M.D. (1981) Nuclear instability and its manipulation in plant breeding. Phil. Trans. Royal Soc. London B 292:475-485.
11. Bonga, J.M. (1974) In vitro culture of microsporophylls and megagametophyte tissue of Pinus. In Vitro 9:270-277.
12. Bonga, J.M. (1981) Haploid culture and cytology of conifers. In Colloque International sur la Culture "In Vitro" des Essences Forestieres, IUFRO Section S2 01 5, AFOCEL, 77370 Nangis, France, pp. 283-293.
13. Bonga, J.M. (1987) Tree tissue culture applications. In Advances in Cell Culture, Vol. 5, K. Maramorosch, ed. Academic Press, Inc., Orlando, Florida, pp. 209-239.
14. Brown, J.S., and E.A. Wernsman (1982) Nature of reduced productivity of anther-derived dihaploid lines of flue-cured tobacco. Crop Sci. 22:1-5.
15. Brunkener, L. (1974) A Review of Methods for the Production of Haploids in Seed Plants, Royal College of Forestry, Department of Forestry and Genetics, Stockholm, Sweden, Research Notes 13.
16. Charmet, G., and G. Branlard (1985) A comparison of androgenetic doubled-haploid, and single seed descent lines in Triticale. Theor. Appl. Genet. 71:193-200.
17. Chase, S.S. (1983) Plant cell culture technology in relation to plant breeding. In Plant Cell Culture in Crop Improvement, S.K. Sen and K.L. Giles, eds. Plenum Press, New York, pp. 1-8.
18. Cheliak, W.M., T. Skroppa, and J.A. Pitel (1987) Genetics of the polycross: 1. Experimental results from Norway spruce. Theor. Appl. Genet. 73:321-329.
19. Cheliak, W.M., K. Morgan, B.P. Dancik, C. Strobeck, and F.C.H. Yeh (1984) Segregation of allozymes in megagametophytes of viable seed from a natural population of jack pine, Pinus banksiana Lamb. Theor. Appl. Genet. 69:145-151.
20. Chen, Z. (1987) Induction of androgenesis in hardwood trees. In Cell and Tissue Culture in Forestry, Vol. 2, J.M. Bonga and D.J. Durzan, eds. Martinus Nijhoff Publishers, Dordrecht, The Netherlands, pp. 247-268.
21. Ching, K., and M. Simak (1971) Competition among embryos in polyembryonic seeds of Pinus silvestris L. and Picea abies (L.) Karst. Royal College of Forestry, Department of Forestry and Genetics, Stockholm, Sweden, Research Notes 30.

22. Choo, T.M. (1981) Doubled haploids for studying the inheritance of quantitative characters. Genetics 99:525-540.
23. Choo, T.M., and L.W. Kannenberg (1978) The efficiency of using doubled haploids in a recurrent selection program in a diploid, cross-fertilized species. Can. J. Genet. Cytol. 20:505-511.
24. Choo, T.M., E. Reinbergs, and K.J. Kasha (1985) Use of haploids in breeding barley. Plant Breed. Rev. 3:219-252.
25. Coleman, W.K., and T.A. Thorpe (1978) In vitro culture of western red cedar (Thuja plicata). II. Induction of male strobili from vegetative shoot tips. Can. J. Bot. 56:557-564.
26. De, D.N., and P.V.L. Rao (1983) Androgenetic haploid callus of tropical leguminous trees. In Plant Cell Culture in Crop Improvement, S.K. Sen and K.L. Giles, eds. Plenum Press, New York, pp. 469-474.
27. De Luca, P., A. Moretti, and S. Sabato (1979) Regeneration in megagametophytes of Cycads. Giorn. Bot. Ital. 113:129-143.
28. Dogra, P.D. (1966) Observations on Abies pindrow with a discussion on the question of occurrence of apomixis in gymnosperms. Silvae Genet. 15:11-20.
29. Dogra, P.D. (1983) Reproductive biology of conifers and its application in forestry and forest genetics. Phytomorphology 33:142-156.
30. Duhoux, E. (1973) Développement du gamétophyte mâle et gamétogénèse in vitro chez le Juniperus communis (Cupressacees). Premiers résultats. C.R. Acad. Sci. (Paris) 277:2665-2668.
31. Duhoux, E. (1980) Le développement cellulaire du tube pollinique du Juniperus communis L. (Cupressacees) cultivé in vitro. Rev. Cytol. Biol. Végét. Bot. 3:95-145.
32. Duhoux, E. (1980) Protoplast isolation of gymnosperm pollen. Z. Pflanzenphysiol. 99:207-214.
33. Durzan, D.J. (1980) Progress and promise in forest genetics. In Paper Science and Technology, the Cutting Edge, Proceedings of the 50th Anniversary Conference, Appleton, Wisconsin, 1978, Institute of Paper Chemistry, pp. 31-60.
34. Engels, P.M. (1981) Storage products and tissue interaction in the ovule of Pinus sylvestris (L.). Acta Soc. Bot. Poloniae 50:339-344.
35. Favre-Duchartre, M. (1958) Ginkgo, an oviparous plant. Phytomorphology 8:377-390.
36. Flavell, R.B. (1982) Recognition and modification of crop plant genotypes using techniques of molecular biology. In Plant Improvement and Somatic Cell Genetics, I.K. Vasil, W.R. Scowcroft, and K.J. Frey, eds. Academic Press, Inc., New York, pp. 277-291.
37. Fossard, R.A. de (1974) Summation: Methods for producing haploids. In Haploids in Higher Plants: Advances and Potential, K.J. Kasha, ed. University of Guelph, Ontario, Canada, pp. 145-150.
38. Fowler, D.P., and R.W. Morris (1977) Genetic diversity in red pine: Evidence for low genic heterozygosity. Can. J. For. Res. 7:343-347.
39. Fragata, M. (1970) A hypothesis concerning the use of colchicine as a polyploidy inducer. Experientia 26:104-106.
40. Franssen-Verheijen, M.A.W., and M.T.M. Willemse (1982) Histochemical study of ovular development in Pinus sylvestris. Phytomorphology 32:345-363.
41. Griffing, B. (1975) Efficiency changes due to use of doubled-haploids in recurrent selection methods. Theor. Appl. Genet. 46:367-386.
42. Gupta, P.K., and D.J. Durzan (1987) Biotechnology of somatic polyembryogenesis and plantlet regeneration in Loblolly pine. Bio/Technology 5:147-151.

43. Gupta, P.P., O. Schieder, and M. Gupta (1984) Intergeneric nuclear gene transfer between somatically and sexually incompatible plants through asymmetric protoplast fusion. Mol. Gen. Genet. 197:30-35.
44. Hakman, I., and S. von Arnold (1986) Isolation and DNA analysis of protoplasts from developing female gametophytes of Picea abies (Norway spruce). Can. J. Bot. 64:108-112.
45. Harada, H., M. Kyo, and J. Imamura (1986) Induction of embryogenesis and regulation of the developmental pathway in immature pollen of Nicotiana species. In Current Topics in Developmental Biology. Vol. 20. Commitment and Instability in Cell Differentiation, A.A. Moscona and A. Monroy, eds. Academic Press, Inc., Orlando, Florida, pp. 397-408.
46. Hermsen, J.G.T., and M.S. Ramanna (1981) Haploidy and plant breeding. Phil. Trans. Royal Soc. London B 292:499-507.
47. Ho, K.M., and K.J. Kasha (1975) Genetic control of chromosome elimination during haploid formation in barley. Genetics 81:263-275.
48. Ho, R.H., and Y. Raj (1985) Haploid plant production through anther culture in poplars. For. Ecol. Manag. 13:133-142.
49. Horner, M., and H.E. Street (1978) Pollen dimorphism--Origin and significance in pollen plant formation by anther culture. Ann. Bot. 42:763-771.
50. Huhtinen, O. (1976) In vitro culture of haploid tissue of trees. In XVI IUFRO World Congress, Division II, Forest Plants and Forest Protection, Norwegian Forest Research Institute, N-1432 AS-NHL, Norway, pp. 28-30.
51. Huskins, C.L. (1952) Nuclear reproduction. In International Review of Cytology, Vol. 1, G.H. Bourne and J.F. Danielli, eds. Academic Press, Inc., New York, pp. 9-25.
52. Illies, Z.M. (1964) Auftreten haploider Keimlinge bei Picea abies. Naturwissenschaften 51:442.
53. Jensen, C.J. (1974) Chromosome doubling techniques in haploids. In Haploids in Higher Plants: Advances and Potential, K.J. Kasha, ed. University of Guelph, Ontario, Canada, pp. 153-190.
54. Jensen, C.J. (1977) Monoploid production by chromosome elimination. In Applied and Fundamental Aspects of Plant Cell, Tissue and Organ Culture, J. Reinert and Y.P.S. Bajaj, eds. Springer-Verlag, Berlin, pp. 299-330.
55. Johnson, M.A., J.A. Carlson, J.H. Conkey, and T.L. Noland (1987) Biochemical changes associated with zygotic pine embryo development. J. Exp. Bot. 38:518-524.
56. Karnosky, D.F., and R.A. Mickler (1984) Propagation and preservation of elms via tissue culture systems. In Seedling Physiology and Reforestation Success, M.L. Duryea and G.N. Brown, eds. Martinus Nijhoff Publishers, Dordrecht, The Netherlands, pp. 29-36.
57. Kasha, K.J. (1974) Haploids from somatic cells. In Haploids in Higher Plants: Advances and Potential, K.J. Kasha, ed. University of Guelph, Ontario, Canada, pp. 67-87.
58. Kim, J.H., H.K. Moon, and J.I. Park (1986) Haploid plantlet induction through anther culture of Populus maximowiczii. Res. Rep. Inst. For. Genet. Korea 22:116-121.
59. Kimber, G., and R. Riley (1963) Haploid angiosperms. Bot. Rev. 29:480-531.
60. Konar, R.N. (1958) A qualitative survey of the free amino acids and sugars in the developing female gametophyte and embryo of Pinus roxburghii Sar. Phytomorphology 8:168-173.

61. Konar, R.N. (1958) A quantitative survey of some nitrogenous substances and fats in the developing embryos and gametophytes of Pinus roxburghii Sar. Phytomorphology 8:174-176.
62. Kriebel, H.B., and F.W. Whitmore (1985) The application of molecular biology to mass cell cloning of conifers. In Canadian Tree Improvement Association, Proceedings of the 20th Meeting, Part 2, Quebec City, Canada, pp. 92-93.
63. Krugman, S.L. (1967) A gibberellin-like substance in immature pine seed. For. Sci. 13:29-37.
64. Lacadena, J.R. (1974) Spontaneous and induced parthenogenesis and androgenesis. In Haploids in Higher Plants: Advances and Potential, K.J. Kasha, ed. University of Guelph, Ontario, Canada, pp. 13-32.
65. Lakshmanan, K.K., and K.B. Ambegaokar (1984) Polyembryony. In Embryology of Angiosperms, B.M. Johri, ed. Springer-Verlag, Berlin, pp. 445-474.
66. LaRue, C.D. (1954) Studies on growth and regeneration in gametophytes and sporophytes of gymnosperms. Brookhaven Symp. Biol. 6:187-208.
67. Ledig, F.T. (1985) Genetic transformation in forest trees. For. Chron. 61:454-458.
68. Ledig, F.T., and M.T. Conkle (1983) Gene diversity and genetic structure in a narrow endemic, Torrey pine (Pinus torreyana Parry ex Carr.). Evolution 37:79-85.
69. Lester, D.T. (1970) An attempt to induce polyhaploidy in American elm. For. Sci. 16:137-138.
70. Libby, W.J., R.F. Stettler, and F.W. Seitz (1969) Forest genetics and forest-tree breeding. Ann. Rev. Genet. 3:469-494.
71. Maheshwari, S.C., A.K. Tyagi, and K. Malhotra (1980) Induction of haploidy from pollen grains in angiosperms. Theor. Appl. Genet. 58:193-206.
72. Maheshwari, S.C., A. Rashid, and A.K. Tyagi (1982) Haploids from pollen grains--Retrospect and prospect. Am. J. Bot. 69:865-879.
73. Maliga, P., L. Menczel, V. Sidorov, L. Marton, A. Cseplo, P. Medgyesy, T.M. Dung, G. Lazar, and F. Nagy (1982) Cell culture mutants and their uses. In Plant Improvement and Somatic Cell Genetics, I.K. Vasil, W.R. Scowcroft, and K.J. Frey, eds. Academic Press, Inc., New York, pp. 221-237.
74. Mapes, M.O., and J.B. Zaerr (1981) The effect of the female gametophyte on the growth of cultured Douglas-fir embryos. Ann. Bot. 48:577-582.
75. Mauro, M.Cl., C. Nef, and J. Fallot (1986) Stimulation of somatic embryogenesis and plant regeneration from anther culture of Vitis vinifera cv. Cabernet-Sauvignon. Plant Cell Reports 5:377-380.
76. Mulcahy, D.L., and G.B. Mulcahy (1987) The effects of pollen competition. Am. Scientist 75:44-50.
77. Nagmani, R., and J.M. Bonga (1985) Embryogenesis in subcultured callus of Larix decidua. Can. J. For. Res. 15:1088-1091.
78. Neale, D.B., N.C. Wheeler, and R.W. Allard (1986) Paternal inheritance of chloroplast DNA in Douglas-fir. Can. J. For. Res. 16:1152-1154.
79. Nelson, Jr., O.E., and B. Burr (1973) Biochemical genetics of higher plants. Ann. Rev. Plant Physiol. 24:493-518.
80. Nilan, R.A. (1981) Induced gene and chromosome mutants. Phil. Trans. Royal Soc. London B 292:457-466.
81. Nitzsche, W. (1973) Mitotische Chromosomenreduktion in höheren Pflanzen durch 3-Fluor-phenylalanin. Naturwissenschaften 60:390.

82. Nitzsche, W., and G. Wenzel (1977) Haploids in plant breeding. In Advances in Plant Breeding (Suppl. 8 to J. Plant Breed.), W. Horn and G. Robbelen, eds. Verlag Paul Parey, Berlin, 101 pp.
83. Norstog, K. (1982) Experimental embryology of gymnosperms. In Experimental Embryology of Vascular Plants, B.M. Johri, ed. Springer-Verlag, Berlin, pp. 25-51.
84. Norstog, K., and E. Rhamstine (1967) Isolation and culture of haploid and diploid cycad tissues. Phytomorphology 17:374-381.
85. Orton, T.J. (1985) Model systems for biotechnological applications in vegetable crop improvements. For. Chron. 61:429-435.
86. Pandey, K.K. (1983) Evidence for gene transfer by the use of sub-lethally irradiated pollen in Zea mays and theory of occurrence by chromosome repair through somatic recombination and gene conversion. Mol. Gen. Genet. 191:358-365.
87. Pandey, K.K. (1986) Gene transfer through the use of sublethally irradiated pollen: The theory of chromosome repair and possible implication of DNA repair enzymes. Heredity 57:37-46.
88. Paranjothy, K. (1987) Hevea tissue culture. In Cell and Tissue Culture in Forestry, Vol. 3, J.M. Bonga and D.J. Durzan, eds. Martinus Nijhoff Publishers, Dordrecht, The Netherlands, pp. 326-337.
89. Pederick, L.A. (1967) The structure and identification of the chromosomes of Pinus radiata D. Don. Silvae Genet. 16:69-77.
90. Pena, A. de la, H. Lorz, and J. Schell (1987) Transgenic rye plants obtained by injecting DNA into young floral tillers. Nature 325:274-276.
91. Radforth, N.W., and J.M. Bonga (1960) Differentiation induced as season advances in the embryo-gametophyte complex of Pinus nigra var. austriaca, using indoleacetic acid. Nature 185:332.
92. Rao, N.M., and A.R. Mehta (1969) Callus tissue from the pollen of Thuja orientalis L. Indian J. Exp. Biol. 7:132-133.
93. Redenbaugh, M.K., R.D. Westfall, and D.F. Karnosky (1980) Protoplast isolation from Ulmus americana L. pollen mother cells, tetrads, and microspores. Can. J. For. Res. 10:284-289.
94. Redenbaugh, M.K., R.D. Westfall, and D.F. Karnosky (1981) Dihaploid callus production from Ulmus americana anthers. Bot. Gaz. 142:19-26.
95. Rohr, R. (1987) Haploids (gymnosperms). In Cell and Tissue Culture in Forestry, Vol. 2, J.M. Bonga and D.J. Durzan, eds. Martinus Nijhoff Publishers, Dordrecht, The Netherlands, pp. 230-246.
96. Rongxuan, W., and H. Han (1985) Genetic manipulation--A new approach to crop improvement. Newslett. Intl. Assoc. Plant Tissue Cult. No. 45, pp. 24-27.
97. Rowe, P.R. (1974) Methods of producing haploids: Parthenogenesis following interspecific hybridization. In Haploids in Higher Plants: Advances and Potential, K.J. Kasha, ed. University of Guelph, Ontario, Canada, pp. 43-52.
98. Sarkar, K.R. (1974) Genetic selection techniques for production of haploids in higher plants. In Haploids in Higher Plants: Advances and Potential, K.J. Kasha, ed. University of Guelph, Ontario, Canada, pp. 33-41.
99. Schlarbaum, S.E. (1987) Cytogenetic manipulations in forest trees through tissue culture. In Cell and Tissue Culture in Forestry, Vol. 1, J.M. Bonga and D.J. Durzan, eds. Martinus Nijhoff Publishers, Dordrecht, The Netherlands, pp. 330-352.
100. Sharp, W.R., M.R. Sondahl, L.S. Caldas, and S.B. Maraffa (1980) The physiology of in vitro asexual embryogenesis. In Horticultural

Reviews, Vol. 2, J. Janick, ed. Avi Publishing Company, Inc., Westport, Connecticut, pp. 268-310.
101. Simak, M., A. Gustafsson, and W. Rautenberg (1974) Meiosis and pollen formation in haploid Thuja plicata gracilis Oud. Hereditas 76:227-238.
102. Simola, L.K. (1987) Structure of cell organelles and cell wall in tissue cultures of trees. In Cell and Tissue Culture in Forestry, Vol. 1, J.M. Bonga and D.J. Durzan, eds. Martinus Nijhoff Publishers, Dordrecht, The Netherlands, pp. 389-418.
103. Snape, J.W., and E. Simpson (1981) The genetical expectations of doubled haploid lines derived from different filial generations. Theor. Appl. Genet. 60:123-128.
104. Sondahl, M.R., and W.R. Sharp (1979) Research in Coffea spp. and applications of tissue culture methods. In Plant Cell and Tissue Culture, Principles and Applications, W.R. Sharp, P.O. Larsen, E.F. Paddock, and V. Raghaven, eds. Ohio State University Press, Columbus, Ohio, pp. 527-584.
105. Steffen, A., and O. Schieder (1984) Biochemical and genetical characterization of nitrate reductase deficient mutants of Petunia. Plant Cell Reports 3:134-137.
106. Steinhauer, A. (1981) Haploid tissue culture for hybrid breeding in forestry: Some fundamental reflections on experiences of the last years. In Colloque International sur la Culture "In Vitro" des Essences Forestieres, IUFRO Section S2 01 5, AFOCEL, 77370 Nangis, France, pp. 295-306.
107. Stettler, R.F. (1976) Haploidy and forest-tree breeding. In XVI IUFRO World Congress, Division II, Forest Plants and Forest Protection, Norwegian Forest Research Institute, N-1432 AS-NLH, Norway, pp. 260-266.
108. Steward, F.C., and S.M. Caplin (1952) Investigations on growth and metabolism of plant cells. IV. Evidence on the role of the coconut-milk factor in development. Ann. Bot. N.S. 16:491-504.
109. Steward, F.C., and D.J. Durzan (1965) Metabolism of nitrogenous compounds. In Plant Physiology: A Treatise, Vol. IV, F.C. Steward, ed. Academic Press, Inc., New York, pp. 379-686.
110. Stinson, J.R., A.J. Eisenberg, R.P. Willing, M.E. Pe, D.D. Hanson, and J.P. Mascarenhas (1987) Genes expressed in the male gametophyte of flowering plants and their isolation. Plant Physiol. 83:442-447.
111. Storey, W.B. (1968) Somatic reduction in Cycads. Science 159:648-650.
112. Sybenga, J. (1983) Genetic manipulation in plant breeding: Somatic versus generative. Theor. Appl. Genet. 66:179-201.
113. Tang, A.F., M. Cappadocia, and D. Byrne (1983) In vitro flowering in cassave (Manihot esculenta Crantz). Plant Cell Tissue Organ Culture 2:199-206.
114. Tran Thanh Van, K., D. Yilmaz-Lentz, and T.H. Trinh (1987) In vitro control of morphogenesis in conifers. In Cell and Tissue Culture in Forestry, Vol. 2, J.M. Bonga and D.J. Durzan, eds. Martinus Nijhoff Publishers, Dordrecht, The Netherlands, pp. 168-182.
115. Tsay, H.S., and C.Y. Su (1985) Anther culture of papaya (Carica papaya L.). Plant Cell Reports 4:28-30.
116. Tulecke, W. (1957) The pollen of Ginkgo biloba: In vitro culture and tissue formation. Am. J. Bot. 44:602-608.
117. Tulecke, W. (1959) The pollen cultures of C.D. LaRue: A tissue from the pollen of Taxus. Bull. Torrey Bot. Club 86:283-289.

118. Tulecke, W. (1965) Haploidy versus diploidy in the reproduction of cell type. In Reproduction: Molecular, Subcellular, and Cellular, M. Locke, ed. Academic Press, Inc., New York, pp. 217-241.
119. Tulecke, W., and N. Sehgal (1963) Cell proliferation from the pollen of Torreya nucifera. Contr. Boyce Thompson Inst. 22:153-163.
120. Tulecke, W., L.H. Weinstein, A. Rutner, and H.J. Laurencot, Jr. (1962) Biochemical and physiological studies of tissue cultures and the plant parts from which they were derived. II. Ginkgo biloba L. Contr. Boyce Thompson Inst. Plant Res. 21:291-301.
121. Turcotte, E.L., and C.V. Feaster (1974) Methods of producing haploids: Semigametic production of cotton haploids. In Haploids in Higher Plants: Advances and Potential, K.J. Kasha, ed. University of Guelph, Ontario, Canada, pp. 53-64.
122. Vasil, I.K., and C. Nitsch (1975) Experimental production of pollen haploids and their uses. Z. Pflanzenphysiol. 76:191-212.
123. Virtanen, A.I., and S. Kari (1955) Free amino acids in pollen. Acta Chem. Scand. 9:1548-1551.
124. Wang, Q.Z., Z.X. Wang, and X.H. Zhang (1982) Initial success in the culture of monoploid plants from the unpollinated ovary of Robinia pseudoacacia. For. Abstr. 43:845.
125. Winton, L.L., and R.F. Stettler (1974) Utilization of haploidy in tree breeding. In Haploids in Higher Plants: Advances and Potential, K.J. Kasha, ed. University of Guelph, Ontario, Canada, pp. 259-273.
126. Yang, H.Y., and C. Zhou (1982) In vitro induction of haploid plants from unpollinated ovaries and ovules. Theor. Appl. Genet. 63:97-104.
127. Yoshida, H., and H. Yamaguchi (1973) Arrangement and association of somatic chromosomes induced by chloramphenicol in barley. Chromosoma (Berlin) 43:399-407.
128. Zabel, B.U., S.L. Naylor, A.Y. Sakaguchi, G.I. Bell, and T.B. Show (1983) High-resolution chromosomal localization of human genes for amylase, proopiomelanocortin, somatostatin, and a DNA fragment (D3S1) by in situ hybridization. Proc. Natl. Acad. Sci., USA 80:6932-6936.
129. Zenk, M.H. (1974) Haploids in physiological and biochemical research. In Haploids in Higher Plants: Advances and Potential, K.J. Kasha, ed. University of Guelph, Ontario, Canada, pp. 339-353.
130. Zenkteler, M.A. (1984) In vitro pollination and fertilization. In Cell Culture and Somatic Cell Genetics of Plants. Vol. 1. Laboratory Procedures and their Applications, I.K. Vasil, ed. Academic Press, Inc., Orlando, Florida, pp. 269-275.
131. Zenkteler, M.A., and I. Guzowska (1970) Cytological studies on the regenerating mature female gametophyte of Taxus baccata L. and mature endosperm of Tilia platyphyllos Scop. in in vitro culture. Acta Soc. Bot. Poloniae 39:161-173.

BIOCHEMICAL AND ANATOMICAL STUDIES OF BIRCH (*BETULA PENDULA* ROTH) BUDS EXPOSED TO DIFFERENT CLIMATIC CONDITIONS IN RELATION TO GROWTH IN VITRO

Margareta Welander

Department of Horticultural Science
The Swedish University of Agricultural Sciences
S-230 53 Alnarp, Sweden

ABSTRACT

In vitro growth of birch (*Betula pendula* Roth) buds taken from 10-month-old seedlings or adult material was studied. The seedlings were exposed to either 24 hr of an irradiance of 40 W/m^2 at 18°C, or 8 hr of an irradiance of 9 W/m^2 at 15°C. Buds from these plants were taken from upper branches and from the upper and lower parts of the stem. Buds from adult material were taken from dormant twigs and at two occasions after flushing. Considerable differences in in vitro growth were obtained due to bud position, stock plant treatment, and growth phase. Correlation between differences in in vitro growth and anatomical/biochemical changes will be discussed.

INTRODUCTION

It is well-recognized that several woody species are still recalcitrant to establishment in vitro. This is especially true for trees in adult growth phase. The transition from juvenile to adult growth phase often occurs gradually, and within the same tree some portions may retain juvenile or transient characters for many years (4). For example, epicormic shoots, formed at the trunk of many tree species, show juvenile morphology. These shoots are much easier to establish in vitro compared to mature shoots (3,14). Furthermore, the position of the explant on the parent plant influences the morphogenetic capacity with respect to growth in vitro. This has been shown for both woody species (11) and herbaceous species (10,35). Another important factor is the physiological condition of the explant, which in turn is determined by the environment to which the parent plant is subjected. For woody species, with their episodic growth, the time of excision is often a key factor in the success of establishing cultures in vitro (5,13,20,27,34). Although it is well known that the condition of the initial explant is very important, very few attempts have been made to relate the biochemical state to in vitro growth and development.

The aim of this chapter is to investigate the anatomical and biochemical state of the initial explant as influenced by bud position, stock plant treatment, and juvenile and adult growth phases, and the possible correlation with growth in vitro. Since phenolic exudation is a serious problem in tissue culture of many woody species (9), it was of interest to investigate whether these exudates were always inhibitory. The object was also to examine whether the levels of endogenous phenolic compounds were affected by the stock plant treatment and bud position.

Birch was chosen as a model plant for two main reasons. First, basic studies have been undertaken regarding (a) photoperiodically induced dormancy and the possible role of growth inhibitors and promoters (16,19); (b) partitioning in carbohydrates during bud break (12); and (c) growth of callus in relation to nitrogen metabolism (32). Micropropagation studies have also been undertaken (7,17,30). These studies in sum constituted a good basis for the present investigation. Second, there is a demand for micropropagated birch plants. Due to a great diversity within the birch family, several varieties have an ornamental value. A special variety of _Betula pendula_ Roth named curly-birch (var. _carelia_ Merklin) has a very strong and decorative wood which commands a high price. It is also more interesting to include birch in a breeding program today because of its positive influence on soil pH and its high fuel value.

MATERIALS AND METHODS

Plant Material and Climatic Conditions

The plant material consisted of juvenile and adult trees of birch (_Betula pendula_ Roth). The adult trees, 20 years old, originated from a selected 40-year-old plus tree grafted onto seedlings. The trees were grown in a plantation in the south of Sweden. Half-sib family seeds from the plantation were harvested in August 1984. The seeds were sown in boxes in April 1985, and placed in the greenhouse under natural day length at 21°C. After 50 days, the seedlings were planted in pots and placed at 18°C. After five months, 50 potted plants (approximately 70 cm in height) were transferred to a controlled climate chamber at 18°C. The irradiance of 40 W/m^2 was supplied by cool-white fluorescent lamps and incandescent lamps for 24 hr [long day (LD)]. After 84 days, 25 plants were placed in a climate chamber at 15°C with an irradiance of 9 W/m^2 for 8 hr per day [short day (SD)]. After an additional 67 days, explants from the LD- and SD-treated trees were used in tissue culture. Shoot tips from LD-treated trees and buds from the middle branches and the upper and lower stem of both LD- and SD-treated trees were used in the experiments (Fig. 1). Twigs from the adult trees were collected in February and stored at 2°C until use. Buds were excised from dormant twigs and after flushing when the female catkins had developed (flush I). Shoot tips were excised when the vegetative buds had developed (flush II) (Fig. 2). Twenty buds from each position of the LD- and SD-treated trees and 50 buds from the adult material excised on different occasions were used for in vitro culture. The following year the entire procedure was repeated.

Culture Conditions

Stem or branch segments, 2 to 3 cm long, were surface-sterilized in 70% ethanol for 1 min followed by 15 min in 7% calcium hypochlorite

Fig. 1. Bud positions on LD- and SD-treated trees. (A) Elongated terminal and lateral shoot tips. (B) Buds from middle branches. (C) Buds from upper stem (0-30 cm). (D) Buds from lower stem (0-30 cm).

containing 0.1% Tween 20. Segments were then rinsed three times with sterile distilled water. In order to avoid browning of the tissue and the culture medium, the segments were placed in a solution of 0.1 mM L-cystein and the buds excised under a dissecting microscope. The bud scales were removed and the buds were placed in a fresh L-cystein solution for 24 hr or in sterile distilled water for 3 hr. Dormant buds from adult material were more difficult to sterilize. They required an additional sterilization in 7% calcium hypochlorite for 3 min after the bud scales were removed.

Fig. 2. Different types of bud development from adult material after flush II. (1) Female catkin; (2) elongated shoot tip; (3) short shoot.

Bud development was initiated on a medium containing macroelements according to Chu et al. (8), microelements and vitamins based on Murashige and Skoog's (MS) medium (23), 10 mM Fe-EDTA, 2% sucrose, 4.4 μM 6-benzylaminopurine (BAP), 0.005 μM α-naphthaleneacetic acid (NAA), and 0.6% Difco Bacto agar. The pH was adjusted to 5.5. Buds from adult material in dormant stage and flush I secreted a high amount of phenolic compounds and needed to be transferred after two days to new induction medium. The buds were cultured in 25 mm x 150 mm test tubes with 10 ml medium and sealed with plastic caps. After three weeks buds were transferred to fresh induction medium and cultured for an additional three weeks.

Both axillary and adventitious shoots were formed from the initial explant. Shoots of different origin were kept separated during shoot multiplication. For shoot proliferation, the WPM medium (21) was used with 2.2 μM BAP, 0.005 μM NAA, 2% sucrose, and 0.6% agar at pH 5.5. The multiplication phase took place in glass jars containing 40 ml of medium and sealed with plastic caps. Shoots were subcultured after three to four weeks. The rooting medium was composed of WPM macroelements diluted to one fifth, microelements and vitamins at full strength, 2% sucrose, 0.5 μM indolebutyric acid (IBA), 0.6% agar, and pH 5.5. All media were autoclaved at 120°C for 15 min. The cultures were maintained at a 16-hr photoperiod with 12 W/m^2 (4,000 lux) from cool-white fluorescent lamps at 23°C. The root induction phase was also performed under the same conditions, since preliminary experiments showed no differences between the light and dark treatments. The tissue culture experiments were all repeated.

Sampling for Analysis

Buds for analyses were collected concurrently with buds for tissue culture. The protective bud scales were removed and buds were immediately frozen at -70°C before being freeze-dried. For determination of carbohydrates, each sample contained 10 to 25 buds (15 to 30 mg fresh weight); for determination of total nitrogen, each sample contained 50 buds (100 to 170 mg fresh weight). For cytokinin analyses, 20 buds were immediately soaked in 80% methanol and frozen at -70°C. Four parallel assays including two separate extractions were performed for each sample.

Carbohydrate Determination

The samples were ground in glass tubes with a glass rod and suspended in 2 ml distilled water plus 0.05 g/g fresh weight polyvinylpyrrolidone (molecular weight 24,000). The glass tubes were sealed with screw caps and the soluble carbohydrates extracted at 60°C for 3 hr in a shaking water bath. The homogenate was filtered and the filtrate assayed enzymatically for glucose and sucrose according to the Boehringer Mannheim Manual (2). Samples assayed for starch were first extracted with 40% ethanol to remove free glucose. Starch was then extracted from the tissue with HCl/dimethylsulfoxide, hydrolized with amyloglucosidase, and the released glucose enzymatically determined according to the above method.

Nitrogen Determination

Total nitrogen was determined by the Kjeldahl method in a block digestor (Tecator AB, Sollentuna, Sweden) with copper as catalyst. Prior to digestion, the nitrate was reduced with salicylic acid (24).

Cytokinin Determination

Extraction and purification. Lyophilized buds were extracted for 24 hr at 0°C in 80% methanol, filtered through paper, washed, and the residue extracted once more in methanol. The pooled filtrates were purified by removing phenolic substances on a polyvinylpyrrolidone column equilibrated in 80% methanol. The sample was run through and the column washed with 80% and 100% methanol to recover the cytokinins. The samples were reduced in volume to the water phase under reduced pressure at 40°C. Further purification was achieved on Bondelute C18 reversed-phase cartridges in phosphate-buffered saline (pH 7.4). After sample application, the columns were washed with buffer and distilled water followed by elution of cytokinins with 40% acetonitrile in water. The eluates were dried on a Vortex evaporator and dissolved in a small volume of high-pressure liquid chromatography (HPLC) solvent.

Cytokinins were separated by HPLC under the following conditions: phenylsilica column (250 mm x 4.6 mm) eluted with a gradient of acetonitrile in 20 mM triethylamine-formic acid buffer, pH 5.6, at 1 ml/min. Fractions were collected every minute, evaporated and dissolved in phosphate buffer. The overall recovery was estimated to be 75%.

Immunological analysis of zeatin and zeatin riboside. The HPLC fractions were analyzed for immunoreactivity using an enzyme-linked immunosorbent assay (ELISA) based on polyclonal antibodies against 9-ribosyl-zeatin. Typical assays gave standard curves, linear after logit transformation, that were in the range 0.02 to 2.0 pmol, with $r>0.98$ and a midpoint at around 0.25 pmol. The detection limit is 2 ng/g fresh weight.

Estimation of Phenolic Compounds

Extraction and separation by HPLC. Buds (50 mg) from the B-position on LD- or SD-treated trees were extracted with 3 ml distilled water in an ultra-turrax. The extract was filtered in a filter funnel, evaporated to dryness in vacuo at 40°C, and dissolved in 1.0 ml methanol. The residue in the funnel was extracted twice with 50% ethanol, evaporated, and dissolved in 1.0 ml methanol. Before separation by HPLC, 0.3 mg 2,4-dichlorphenol was added as an internal standard. The sample (10 µl) was injected onto a C18 column (Chrompack CP® sphere C18, flow rate 1.5 ml/min). The solvents used were 0.01 M phosphate buffer (pH 2.8) and methanol in a gradient ranging from 10 to 55% over 30 min, then to 90% over 2 min. The peaks were detected at 280 nm.

Extraction and separation by thin-layer chromatography. Buds (20 mg) from either the B, C, or D position of LD- and SD-treated trees were extracted in 5 ml 80% ethanol in an ultra-turrax. The mixture was filtered and rinsed with 4.0 ml 80% ethanol and 2 ml 96% ethanol. The filtrate was evaporated to dryness and dissolved in 1.0 ml 80% ethanol and 0.5 ml 96% ethanol. The solution was filtered by a millipore filter (0.45 µm) and an amount corresponding to 0.73 mg bud material was subjected to thin-layer chromatography (TLC) [Merck (Darmstadt, Federal Republic of Germany) Silicagel; solvent system: chloroform:methanol:H_2O, 8:3:2]. The spots were located by spraying with diazotized sulphanilic acid plus 10% sodium carbamate, followed by 50% sulfuric acid.

Extraction and test on growth in vitro. Buds of adult material (166 g) were extracted with 500 ml 80% ethanol in an ultra-turrax. The ethanol was evaporated in vacuo at 40°C. When 220 ml of water suspension remained, it was extracted three times with 220 ml of chloroform followed by three times with 220 ml ethyl acetate. The organic phases were dried over anhydrous sodium sulfate and evaporated below 40°C. The water phase was freeze-dried. The amounts of the different extracts were as follows: chloroform, 32.27 g; ethyl acetate, 2.73 g; and water extract, 11.24 g. These extracts were tested on growth and development of buds in vitro.

The water and ethyl acetate extracts were diluted with 2 ml 70% ethanol and the chloroform phase was diluted with 2 ml chloroform. Amounts of 20, 100, and 200 µl corresponding to 1, 5, and 10 buds were applied to sterile filter paper discs (0.5 mm diameter). The discs were placed on the surface of the medium in test tubes containing induction medium. Buds of adult material from flush II were established on induction medium for one week before being transferred to test tubes containing the extracts. Five buds per concentration and extract were used in the experiment. The control buds included discs of filter paper with and without pure solvents.

Anatomical Studies

Buds from different origins were fixed, embedded, and sectioned according to Welander (37). Sections were stained in Delafield's hemotoxylin and counterstained in Eosin Y to identify meristematic centers. Carbohydrates were stained with Periodic acid-Schiff's (PAS) reaction.

RESULTS

In Vitro Culture and Anatomy

Seedlings. Results from the study of in vitro phenolic exudation and development of buds at different positions from LD- and SD-treated trees are shown in Tab. 1. Despite procedures to avoid browning of tissue and culture medium, buds from the basal part of the stem (D) continued to exude phenolic substances. D buds from LD-treated trees produced most of these exudates. However, this exudation was not inhibitory to the growth and survival of the explants, since D buds exhibited the best development in vitro. Figure 3 shows that the physiological state of the stock plant strongly influenced growth and development of buds. Generally, buds from SD-treated plants were much easier to establish. The most pronounced differences were obtained for buds on branches (B) and the upper stem (C) from the different environments. Between 50 and 60% of the B and C buds from SD-treated trees developed into shoots, whereas the percentages for corresponding buds from LD-treated trees were less than 20%. Sectioning of buds at the B position from LD and SD showed that there were also differences in the number of leaves within the bud. More leaves were present in buds from SD-treated trees (Fig. 4). In buds from LD-treated trees, there was a gradient of morphogenetic capacity with respect to growth in vitro due to bud position which was not observed in SD-treated trees (Fig. 5). The lowest number of shoots to develop were those from shoot tips (A), with the highest number of shoots developing from buds at the base (D). The buds from LD- and SD-treated trees also differed in their capacity to form adventitious buds. Shoots that developed from buds of SD-treated trees produced adventitious buds more frequently [Fig. 5 (SD, D_1)]. The adventitious shoots were formed at the base of the explant without any intervening callus. After four days, cell

Tab. 1. In vitro phenolic exudation and development of buds at different positions from LD- and SD-treated trees.

			Days in culture				
			7	7	14	21	21
Stock plant treatment	Bud position	Number of excised buds	Estimated phenolic exudate	Number of green buds	Number of open buds	Number of buds developed to a shoot	Number of shoots with adventitious buds
LD	A	40	0-X	5	5	2	1
	B	40	X	15	11	7	0
	C	35	0-X	13	10	6	3
	D	38	XX-XXX	35	34	32	5
SD	B	40	0-X	35	30	26	23
	C	40	0-X	33	26	20	17
	D	40	X-XX	39	39	39	28

Note: Exudate score 0 (no exudate) to XXX (heavy exudate). The values are based on two separate experiments. For explanation of A-D, see legend to Fig. 1. Data are average values from duplicate experiments.

LD SD

A B C D B C D

Fig. 3. Development of buds in vitro that were excised from different positions on LD- and SD-treated trees. The cultures are five weeks old. See legend to Fig. 1 for description of bud positions corresponding to A, B, C, and D. Scale bar = 10 mm.

division could be observed (Fig. 6A), and after 15 days developed meristems with leaf primordia were visible (Fig. 6B). Later during shoot development, adventitious buds were also formed from leaves, irrespective of their origin.

In Tab. 2, shoot multiplication during four subcultures is shown for the different types of buds on trees exposed to LD and SD. Irrespective of bud position and stock plant treatment, the number of axillary shoots produced from the original explant was very low (1.0 to 2.4) and declined steadily. After three subcultures the shoots became necrotic and ceased to grow except for C and D buds from SD-treated trees and D buds from LD-treated trees. However, buds of adventitious origin produced a larger number of shoots (2.5 to 5.7) continuously. The shoots of axillary and adventitious origin also differed in morphology. The shoots of axillary origin were vigorous and compact with short internodes and large leaf lamina whereas the shoots of adventitious origin were thinner with longer internodes and smaller leaf lamina (Fig. 7).

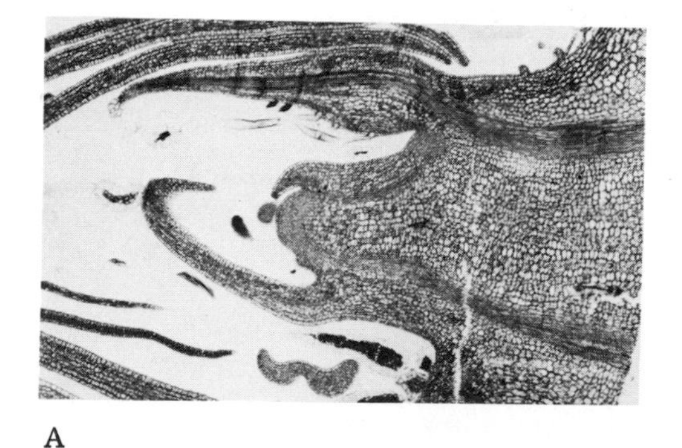

A

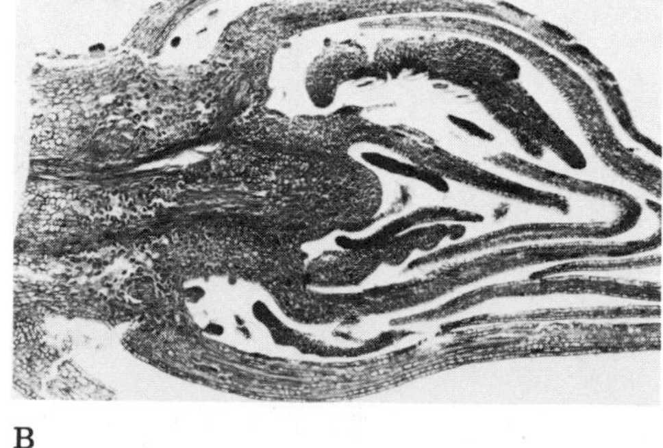

B

Fig. 4. Longitudinal sections of B buds from LD- and SD-treated trees showing differences in the number of leaf primordia surrounding the meristem. (A) Bud from LD plant (X50.4). (B) Bud from SD plant (X50.4).

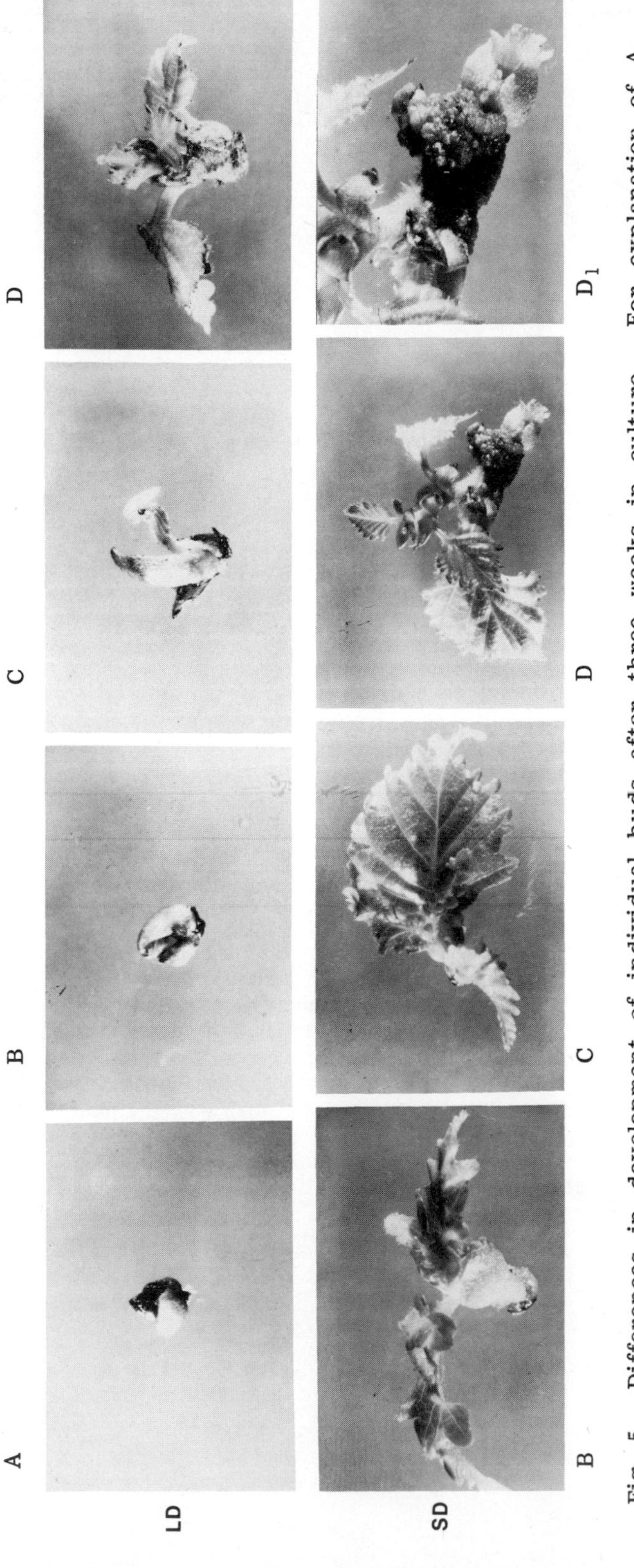

Fig. 5. Differences in development of individual buds after three weeks in culture. For explanation of A through D, see legend to Fig. 1. SD, C and D, X2.5; the others, X5.0. D_1 is a higher magnification (X5.0) of SD, D, showing adventitious buds at the base of the explant.

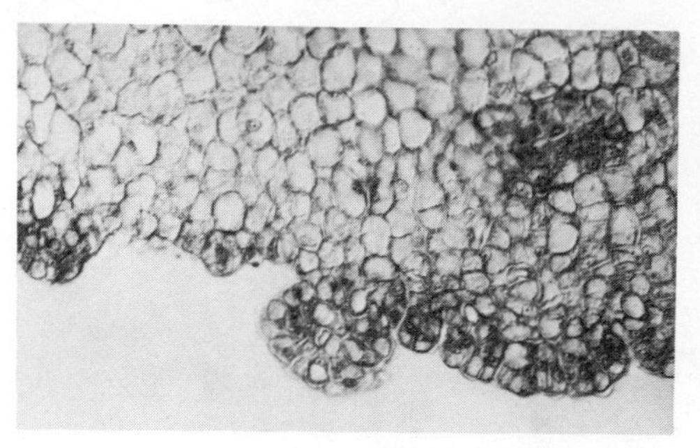

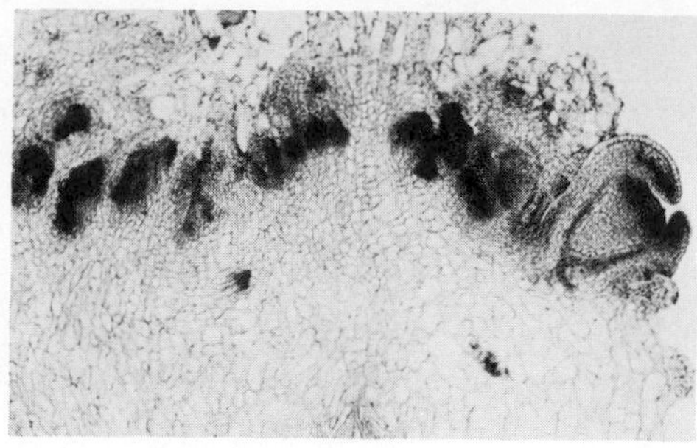

A B

Fig. 6. Development of adventitious buds after different culture periods. (A) Longitudinal section of a D bud from LD-treated tree showing initiation of adventitious buds at the base of the original explant after four days (X200). (B) Adventitious bud plus meristem showing leaf primordia after 15 days (X50.4).

Data from the rooting experiments are given in Tab. 3. The rooting experiments were performed at subcultures 5, 6, and 7. Data are given only for subcultures 5 and 6, since similar results were obtained for subcultures 6 and 7 except that 100% rooting was obtained at subculture 7 for D_{ad} buds from SD-treated trees after 14 days. There were no consistent differences in rooting ability between shoots of axillary or adventitious origin or due to bud position. However, shoots from SD required a longer time in the rooting medium and an additional subculture before reaching values of rooting percentage and number of roots similar to those of LD-treated shoots. At subculture 6, after 14 days most of the shoots from both LD- and SD-treated trees rooted to 100%.

Adult Material

Results of studies of in vitro phenolic exudation and development of buds excised from adult material are shown in Tab. 4. Buds from the dormant stage and flush I secreted large amounts of phenolics into the medium whereas shoot tips from flush II did not produce any exudate (Fig. 8). Due to the initial browning it was necessary to transfer these buds to new medium after two days. After this step, the development of buds in the initial stage was not affected. However, more shoots survived after prolonged culture time when buds originated from flush II. A severe problem with adult material is elongation of the shoot in order to be able to use nodal segments for shoot multiplication. In order to improve shoot elongation, several media were tested, including those with gibberellin (3 μM) or reduced BAP content (0.4 μM), or with no hormones at all, but without success.

Analysis of Carbohydrate and Nitrogen Content

Seedlings. Contents of glucose, sucrose, starch, and nitrogen in buds at different positions from LD- and SD-treated stock plants are given in Tab. 5. In all buds, the content of glucose is three to five times higher than that of sucrose. Both the glucose content and the sucrose content

Tab. 2. Shoot multiplication rate during four subcultures from buds at different positions from LD- and SD-treated trees.

	Subculture:	1		2		3		4	
		Number of shoots per shoot tip							
Stock plant treatment	Bud position	Ax	Ad	Ax	Ad	Ax	Ad	Ax	Ad
LD	A	1.0 ± 0.0	-	1.0 ± 0.0	Nodules	Dead	3.0 ± 0.0	Dead	3.0 ± 0.8
	B	1.0 ± 0.0	-	1.0 ± 0.0	Nodules	Dead	2.8 ± 1.0	Dead	2.5 ± 1.0
	C	1.4 ± 0.5	-	2.5 ± 0.5	Nodules	Dead	Dead	Dead	Dead
	D	2.0 ± 0.9	Nodules	1.8 ± 0.7	3.8 ± 2.9	1.6 ± 0.5	3.5 ± 3.0	1.6 ± 0.9	2.7 ± 1.6
SD	B	1.4 ± 0.9	Nodules	1.4 ± 0.9	5.7 ± 1.5	1.3 ± 0.9	4.2 ± 1.4	1.2 ± 0.5	3.3 ± 0.9
	C	1.0 ± 0.0	Nodules	1.0 ± 0.0	4.5 ± 2.0	Dead	5.1 ± 2.0	Dead	3.8 ± 1.3
	D	2.2 ± 0.2	Nodules	2.3 ± 0.3	5.6 ± 2.2	2.4 ± 1.1	3.5 ± 1.1	1.3 ± 0.5	2.9 ± 0.6

Note: Shoots from axillary (Ax) and adventitious (Ad) origin are kept separated. For explanation of A-D, see legend to Fig. 1. Data, given as mean ± S.D., are average values from duplicate experiments.

Fig. 7. Morphology of shoots with axillary (A_x) and adventitious (A_d) origin of D bud from an SD-treated tree. Scale bar = 10 mm.

are slightly higher in buds from SD- compared to LD-treated trees. When expressed per dry weight the differences increase, since the dry weight of buds from SD- and LD-treated trees amounts to 50% and 40%, respectively, of the fresh weight. No differences due to bud position were observed except for A buds, which contain the lowest amount of glucose. The most pronounced differences were observed in the content of starch. Buds exhibiting the best development in vitro also contained the highest amount of starch. Buds from SD-treated trees contained almost twice as much nitrogen as buds from LD-treated trees, but no differences due to bud position were observed.

Adult material. Contents of glucose, sucrose, starch, and nitrogen in buds excised on different occasions from adult material are given in Tab. 6. More glucose and sucrose was stored in dormant buds. After flushing, the content of both decreased, with the lowest amounts found in growing shoot tips. As in juvenile material, soluble carbohydrates were not correlated to growth in vitro. However, shoot tips with a higher survival rate after prolonged culture time also contained the highest amount of nitrogen.

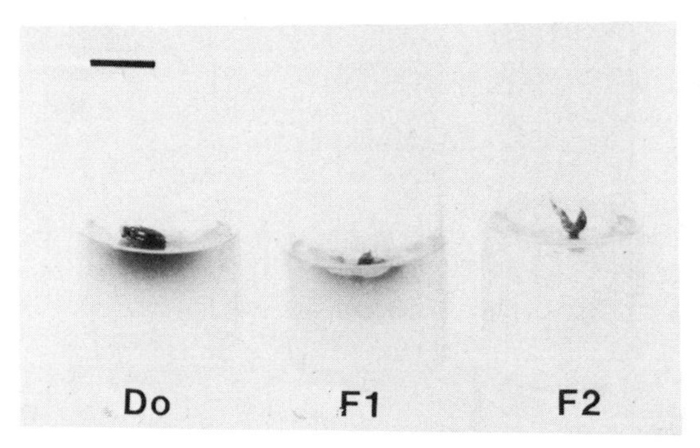

Fig. 8. Differences in phenolic exudation from adult material taken at dormant stage (Do) and at flush I (F1) and II (F2). For further explanation of the different developmental stages, see Fig. 2. Scale bar = 10 mm.

Tab. 3. Rooting of shoots of different origin derived from buds at different positions on LD- and SD-treated trees.

	Subculture:	5			6		
		Percent rooted shoots		Number of roots per rooted shoot	Percent rooted shoots		Number of roots per rooted shoot
Stock plant treatment	Bud position and origin of shoots	7d	14d	14d	7d	14d	14d
LD	A_{ad}	77	77	3.1 ± 2.0	80	90	3.3 ± 1.0
	B_{ad}	50	70	4.4 ± 1.9	33	100	2.3 ± 0.6
	D_{ax}	83	94	3.9 ± 0.7	73	93	4.3 ± 2.2
	D_{ad}	86	86	3.5 ± 1.7	36	100	4.3 ± 1.0
SD	B_{ax}	0	9	1.0 ± 0.0	0	100	2.6 ± 1.5
	B_{ad}	10	24	2.6 ± 1.5	21	100	2.4 ± 1.2
	C_{ad}	9	35	1.5 ± 0.5	13	100	3.3 ± 1.6
	D_{ax}	0	7	2.0 ± 0.0	50	87	3.9 ± 2.3
	D_{ad}	0	14	2.3 ± 1.5	0	45	2.8 ± 1.6

Note: Rooting percentage was recorded after seven and 14 days and the number of roots per rooted shoot, after 14 days in the rooting medium. Between 10 to 20 shoots were included in each assay. For explanation of A–D, see legend to Fig. 1; ax, axillary origin; ad, adventitious origin. Subcultures 5 and 6 are continuation of subcultures 1, 2, 3, and 4 which are analyzed in Tab. 2.

Tab. 4. In vitro phenolic exudation and development of buds excised on different occasions from adult material.

		Days in culture			
		7	14	21	90
Type of bud development	Number of excised buds	Estimated phenolic exudate	Number of open buds	Number of buds developed to a shoot	Number of surviving shoots
Buds from dormant stage	100	XXX	75	52	6
Buds from flush I	85	XXX	73	41	0
Shoot tips from flush II	77	0	62	62	15

Note: For explanation of type of bud development, see legend to Fig. 2. Data are average values of duplicate experiments.

Analysis of Cytokinins

In Tab. 7, the levels of cytokinins are given for buds taken from position B in LD- and SD-treated trees and from position D in LD-treated trees. The cytokinins found in the different buds all appeared as zeatin riboside or zeatin glucoside. No activity of zeatin was identified. There was no difference in the levels of zeatin riboside but there was an enhanced level of zeatin glucoside in B buds from SD-treated trees.

Analysis of Phenolic Compounds

Seedlings. High-pressure liquid chromatography separations of water and ethanol extracts from buds at position B from LD- and SD-treated

Tab. 5. Contents of glucose, sucrose, starch, and nitrogen in buds at different positions from LD- and SD-treated trees.

Stock plant treatment	Bud position	Glucose (mg/g FW)	Sucrose (mg/g FW)	Starch (mg/g FW)	Nitrogen (mg/g DW)
LD	A	4.9 ± 0.6	1.5 ± 1.7	NA*	NA
	B	9.5 ± 1.7	1.5 ± 1.1	0.0 ± 0.0	15.9 ± 2.1
	C	8.1 ± 0.4	1.9 ± 1.1	0.0 ± 0.3	14.7 ± 3.3
	D	8.7 ± 2.2	3.1 ± 1.8	13.2 ± 2.4	14.8 ± 0.6
SD	B	13.7 ± 4.9	4.4 ± 1.9	11.7 ± 3.1	26.6 ± 1.6
	C	10.2 ± 2.8	3.1 ± 2.0	14.5 ± 2.7	27.4 ± 1.8
	D	10.3 ± 1.1	3.3 ± 1.1	11.8 ± 3.0	21.8 ± 1.9

*Not analyzed.

Note: For explanation of A-D, see legend to Fig. 1. Data, given as mean ± S.D., are average values from duplicate experiments. FW, Fresh weight; DW, dry weight.

Tab. 6. Contents of glucose, sucrose, starch, and nitrogen in buds excised on different occasions from adult material.

Type of bud development	Glucose (mg/g FW)	Sucrose (mg/g FW)	Starch (mg/g FW)	Nitrogen (mg/g DW)
Buds from dormant stage	10.7 ± 0.8	2.94 ± 1.6	5.80 ± 0.7	12.9 ± 2.2
Buds from flush I	6.2 ± 0.3	0.03 ± 0.04	0.00 ± 0.0	17.5 ± 1.4
Shoot tips from flush II	5.4 ± 0.2	0.00 ± 0.00	0.00 ± 0.0	32.3 ± 2.2

Note: For explanation of type of bud development, see legend to Fig. 2. Data, given as mean ± S.D., are average values from duplicate experiments.

trees are shown in Fig. 9. Both number and concentration of phenolic compounds were higher in buds from LD-treated trees. Data from TLC separations (Tab. 8) give further indications of differences in phenolic metabolism in buds at different positions and from different stock plants. Interestingly, buds showing very good growth in vitro (B, C, D in SD and D in LD) have a similar pattern of phenolic compounds.

The results from an assay of different extracts in regard to their effect on growth and development of buds in vitro are shown in Fig. 10. All concentrations of the chloroform extract stimulated growth in vitro. The highest concentration especially stimulated elongation of the shoots. Addition of ethyl acetate or water extracts caused browning of the medium in the same way as when dormant buds alone were placed on the medium. The lowest concentration of ethyl acetate promoted in vitro growth whereas amounts corresponding to five and ten buds were inhibitory. The water extract did not produce a specific inhibition.

Tab. 7. Cytokinin levels in buds at position B from LD- and SD-treated trees and at position D from LD-treated trees.

		Cytokinin levels ($ng \cdot g^{-1}$ fresh weight)		
Sample	Position	[9G]Z	[9R]Z	Z
SD	B	55	11.5	0
LD	B	8.0	10.3	0
LD	D	6.1	9.4	0

Note: [9G]Z, zeatin glucoside; [9R]Z, zeatin riboside; Z, zeatin. Measurements of [9R]Z-like immunoreactive material were performed with antibodies against [9R]Z in an enzyme-linked immunosorbent assay. For explanation of positions B and D, see legend to Fig. 1.

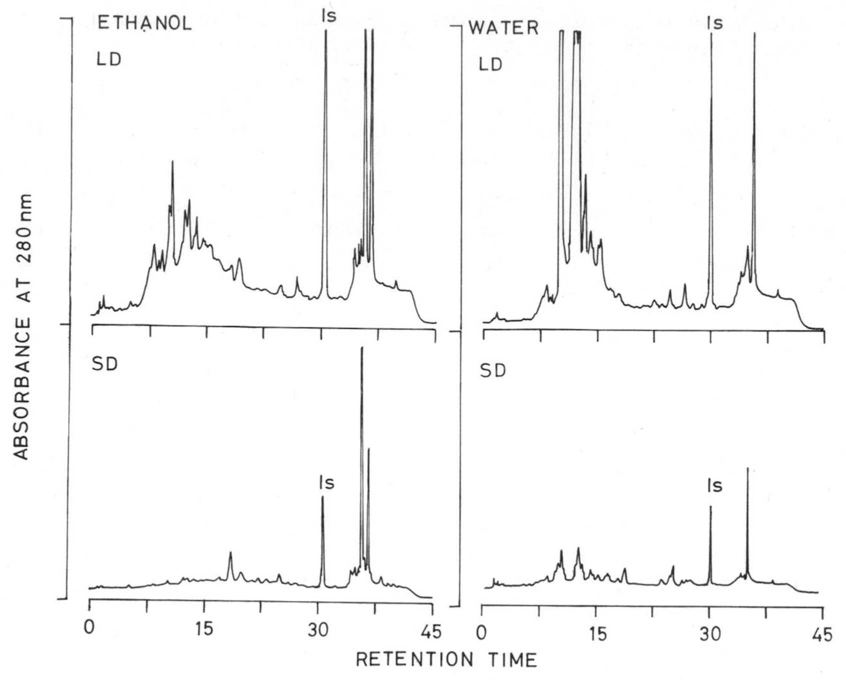

Fig. 9. Separation by HPLC (see text for details) of phenolic compounds in ethanol and water extracts of B buds from LD- and SD-treated trees. Is, Internal standard (2,4-dichlorophenol).

Tab. 8. Thin-layer chromatography separation of ethanol extracts from buds at different positions from LD- and SD-treated trees.

Stock plant treatment	Bud position	Phenolic compounds			
		a	b	c	d
LD	B	XXX	t	X	XX
	C	XXX	t	X	XX
	D	XX	t	XXX	t
SD	B	X	XX	XXX	t
	C	X	X	XXX	t
	D	X	t	XXX	t

Note: Spots were detected by spraying with diazotized sulpanilic acid plus sodium carbamate, and after spraying with sulfuric acid. Thin-layer chromatography was performed on Merck's silica gel plates with fluorescent indicator and solvent system (chloroform:methanol:water). For explanation of B-D, see legend to Fig. 1. Score: t (trace) to XXX (highly stained).

Fig. 10. The effect of chloroform (Ch), ethyl acetate (Et), and water (W) extracts on growth and development of buds in vitro. The extracts were obtained from buds of adult material in dormant stage. The buds tested originated from adult material flush II established one week on the initiation medium before being subjected to the different extracts. The amounts of extract added correspond to one, five, and ten buds. C, Control.

DISCUSSION

The results obtained in this investigation show that the initial growth and further development of explants from seedlings as well as from adult material depend on several factors. In birch seedlings, controlled photoperiod as well as bud position can be used to optimize in vitro conditions. This has also been shown for oak (13). Interestingly, SD treatment, which induced dormancy in birch (36), resulted in the most suitable explants for in vitro growth. Buds from SD-treated trees also developed more rapidly in vitro. This means that resting cells do not lack useful physiological properties. In SD, bud scales are formed and vertical growth ceases. The anatomical studies revealed that buds from branches of SD-treated trees contain a substantially larger number of leaves compared to buds at the same position on LD-treated trees. For conifers, it has been shown that the apical dome itself remains active and is actually followed by a period of accelerated production of leaf primordia (6). This might be one cause of the more rapid development in vitro after SD treatment.

The capacity to form adventitious buds was more influenced by stock plant treatment than by bud position. The adventitious shoots differed from the axillary ones in a number of characteristics. The main difference was the multiplication rate. Axillary shoot production was very low and after four subcultures most of the initial explants died except for B and D buds from SD-treated trees. In birch, which so easily produces adventitious buds, it can be difficult to distinguish between shoots of different origin. Although birch plants derived from callus seem to be genetically stable (17,30,32), it is important to ascertain whether the plants remain true to type. For example, in sandalwood a high frequency of somaclonal variation was observed in callus-derived plants (28).

The biochemical analyses showed that the most pronounced differences were found in starch and nitrogen content. Buds from SD-treated trees contained the highest amount of starch and nitrogen. Interestingly, D buds from LD-treated trees also contained large amounts of starch but the same quantity of nitrogen as other buds from LD-treated trees. It has always been claimed that the good growth of buds from the lower part of the tree is due to juvenility. The question is if these buds instead exhibit a useful metabolic state. The present results are supported by the numerous studies showing that in deciduous trees, both the nitrogen content and reserve food in the form of starch accumulate during the onset of dormancy and decrease when growth begins (18).

In adult material, the content of glucose, sucrose, and starch was highest in buds from the dormant stage. After the first flush, when the female catkins had developed and vegetative buds were still closed, the amount of both glucose and sucrose decreased. After the second flush, when the vegetative buds had burst, sucrose and starch were not detectable in the buds. Sauter and Ambrosius (31) showed very similar changes in the soluble carbohydrates in the wood of branches of Betula pendula during bud burst. However, they found a marked increase in the content of starch in the wood during the same period which could not be detected in the buds in the present investigation. During the same period, the content of nitrogen increased three-fold in the buds. When comparing the results from seedlings and adult material, it seems as if good growth in vitro can be obtained if the buds contain large amounts of starch and nitrogen or at least one of the two.

Allocation of carbon and nitrogen is probably dependent upon hormonal stimuli (12). There is a vast literature showing changes in levels of growth inhibitors and growth promoters in association with bud dormancy (25). There are some indications that cytokinins are involved in regulation of dormancy (1,15). Few cytokinin analyses have been performed in the present investigation and the results are difficult to interpret. In a soybean callus test, it was shown that the most active cytokinin is zeatin riboside whereas the glucosides of zeatin are very stable and inactive and are regarded as storage forms of cytokinin (26). Whether the accumulation of zeatin glucoside in B buds during SD treatment is related to the capability of in vitro growth is impossible to decide at this stage.

A serious problem encountered in tissue culture of woody species is the brown exudate secreted into the medium (11). It is well known that phenolic compounds can be produced as a result of wounding after excision (29). The initial exudation depends on age and kind of tissue and might be prevented by different pretreatments (9). The present results showed that age, bud position, and stock plant treatment influenced phenol exudation. More phenolics were produced in buds from LD-treated trees than in those from SD-treated trees. This result is supported by the findings that most enzymes involved in phenolic biosynthesis are stimulated by light (22). Although the analyses of the phenolic compounds in the present investigation are preliminary, there are strong indications that buds showing good growth in vitro have the same pattern of phenolic compounds. Due to TLC comparisons with commercial substances and with the flavonoid apigenin-4',7-dimethylether isolated in earlier experiments (33), it is very likely that p-coumaric acid and ferulic acid are present as glucosides in the extracts as well as the flavonoid. However, it is too early to relate these compounds with the spots on TLC. The results with extracts from dormant buds show that most of the inhibitory effect is due to substances that are present in the ethyl acetate extract, and that some promoting substances are present in the chloroform phase.

ACKNOWLEDGEMENTS

This work was supported by grants from the Swedish Council for Forestry and Agricultural Research. The following persons are acknowledged for their contributions to the work: Elisabeth Andersson, Eva Jansson, Hans Lindquist, Mats Hallberg, and Kärstin Sunnerheim-Sjöberg.

REFERENCES

1. Altman, A., and R. Goren (1974) Growth and dormancy cycles in citrus bud cultures and their hormonal control. Physiol. Plant. 30:240-245.
2. Anonymous (1984) Methods of enzymatic food analysis. In Boehringer Mannheim Biochemica, Boehringer Mannheim GmbH, Mannheim, Federal Republic of Germany.
3. Ball, E. (1978) Cloning in vitro of Sequoia sempervirens. In The Propagation of Higher Plants through Tissue Culture, K.W. Hughes, R. Henke, and M. Constantin, eds. Technical Information Center, U.S. Department of Energy, Oak Ridge, Tennessee, p. 259.
4. Bonga, J.M. (1982) Vegetative propagation in relation to juvenility, maturity and rejuvenation. In Tissue Culture in Forestry, J.M. Bonga and D.J. Durzan, eds. Martinus Nijhoff/Dr. W. Junk Publishers, Amsterdam, pp. 387-412.
5. Boulay, M. (1979) Propagation in vitro du Douglas par micropropagation de germination aseptique et culture de bourgeons dormants [Pseudotsuga menziesii (Mirb.) Franco]. In Micropropagation d'Arbres Forestiers, Etudes et Récherches No. 12, AFOCEL, Nangis, France, pp. 67-75.
6. Cannell, M.G.R., and C.M. Cahalan (1979) Shoot apical meristems of Picea sitchensis seedlings accelerate in growth following bud-set. Ann. Bot. 44:209-214.
7. Chalupa, V. (1981) Clonal propagation of broad-leaved forest trees in vitro. Comm. Inst. Forestalis Cechosloveniae 12:255-271.
8. Chu, C.C., C.C. Wang, C.S. Sun, C. Hsü, K.C. Yin, C.Y. Chu, and F.Y. Bi (1975) Establishment of an efficient medium for anther culture of rice through comparative experiments on the nitrogen sources. Sci. Sin. 18:659-668.
9. Compton, M.E., and J.E. Preece (1986) Exudation and explant establishment. IAPTC Newsletter No. 50, pp. 9-18.
10. Croes, A.F., T. Creemers-Molenaar, G. Van den Ende, A. Kemp, and G.W.M. Barendse (1985) Tissue age as an endogenous factor controlling in vitro bud formation on explants from the inflorescence of Nicotiana tabacum L. J. Exp. Bot. 36:1771-1779.
11. Durand-Cresswell, M. Boulay, and A. Franclet (1982) Vegetative propagation of Eucalyptus. In Tissue Culture in Forestry, J.M. Bonga and D.J. Durzan, eds. Martinus Nijhoff/Dr. W. Junk Publishers, Amsterdam, pp. 150-181.

12. Essiamah, S., and W. Eschrich (1985) Changes of starch content in the storage tissues of deciduous trees during winter and spring. IAWA Bull. (N.S.) 6:97-106.
13. Favre, J.M., and B. Juncker (1987) In vitro growth of buds taken from seedlings and adult plant material in Quercus robur L. Plant Cell Tissue Organ Culture 8:49-60.
14. Franclet, A. (1979) Rajeunissement des arbres adultes en vue de leur propagation végétative. In Micropropagation d'Arbres Forestiers, Etudes et Réchérches No. 12, AFOCEL, Nangis, France, pp. 3-18.
15. Hewett, E.W., and P.F. Wareing (1973) Cytokinins in Populus x robusta: Changes during chilling and bud burst. Physiol. Plant. 28:393-399.
16. Hocking, T.J., and J.R. Hillman (1975) Studies on the role of abscisic acid in the initiation of bud dormancy in Alnus glutinosa and Betula pubescens. Planta 125:235-242.
17. Huhtinen, V.O., and Z. Yahyaoglu (1973) Das frühe blühen von aus Kalluskulturen herangezogenen Pflanzen bei der Birke (Betula pendula Roth). Silvae Genet. 23:32-34.
18. Kramer, P.J., and T.T. Kozlowski (1979) Physiology of Woody Plants, Academic Press, Inc., New York.
19. Lenton, J.R., V.M. Perry, and P.F. Saunders (1972) Endogenous abscisic acid in relation to photoperiodically induced bud dormancy. Planta 106:13-22.
20. Litz, R.E., and R.A. Conover (1981) Effect of sex type, season, and other factors on in vitro establishment and culture of Carica papaya L. explants. J. Am. Soc. Hort. Sci. 106:792-794.
21. Lloyd, G., and B. McCown (1980) Commercially feasible micropropagation of Mountain laurel, Kalmia latifolia, by use of shoot tip culture. Int. Plant Prop. Combined Proc. 30:421-426.
22. McClure, J.W. (1975) Physiology and function of flavonoids. In The Flavonoids, J.B. Harborne, T.J. Mabry, and H. Mabry, eds. Chapman and Hall, London.
23. Murashige, T., and F. Skoog (1962) A revised medium for rapid growth and bioassays with tobacco tissue cultures. Physiol. Plant. 15:473-497.
24. Nelson, D.W., and L.E. Sommers (1973) Determination of total nitrogen in plant material. Agron. J. 65:109-112.
25. Noodén, L.D., and J.A. Weber (1978) Environmental and hormonal control of dormancy in terminal buds of plants. In Dormancy and Developmental Arrest: Experimental Analysis in Plants and Animals, M.E. Cutter, ed. Academic Press, Inc., New York, pp. 221-268.
26. Palni, L.M.S., M.V. Palmer, and D.S. Letham (1984) The stability and biological activity of cytokinin metabolites in soybean callus tissue. Planta 160:242-249.
27. Quoirin, M., P.H. Lepoivre, and P.H. Boxus (1977) Un Premier Bilan de Dix Annees de Réchérches sur le Cultures de Meristemes et la Multiplication "In Vitro" de Fruitiers Ligneux--Compte Rendu des Rechérches 1976-1977 et Rapports de Synthèse, Station des Cultures Fruitières et Maraichères, Centre de Réchérches Agrinomiques de l'Etat, Gembloux, Belgium.
28. Rao, P.S., V.A. Bapat, and M. MHatre (1984) Regulatory factors for in vitro multiplication of sandalwood tree (Santalum album Linn.). II. Plant regeneration in nodal and internodal stem explants and occurrence of somaclonal variations in tissue culture raised plants. Proc. Indian Natl. Sci. Acad. B-50:196-202.
29. Ripley, K.P., and J.E. Preece (1986) Micropropagation of Euphorbia lathyris L. Plant Cell Tissue Organ Culture 5:213-218.

30. Ryynanen, L., and M. Ryynanen (1986) Propagation of adult curly-birch succeeds with tissue culture. Silvae Fennica 20:139-147.
31. Sauter, J.J., and T. Ambrosius (1986) Changes in the partitioning of carbohydrates in the wood during bud break in Betula pendula Roth. J. Plant Physiol. 124:31-43.
32. Simola, L.K. (1985) Nitrogen metabolism of leaf and microspore callus of Betula pendula. In Primary and Secondary Metabolism of Plant Cell Cultures, K.-H. Neumann, W. Barz, and E. Reinhard, eds. Springer-Verlag, Berlin, pp. 74-84.
33. Summerheim, K., T. Palo, O. Theander, and P.G. Knutsson (1987) Chemical defense in birch. Platyphylloside: A phenol from Betula pendula inhibiting digestibility. J. Chem. Ecol. (in press).
34. Sutter, E.G., and P.B. Barker (1985) In vitro propagation of mature Liquidambar styraciflua. Plant Cell Tissue Organ Culture 5:13-21.
35. Tran Thanh Van, K. (1980) Control of morphogenesis or what shapes a group of cells? In Advances in Biochemical Engineering. Plant Cell Cultures II, A. Fiechter, ed. pp. 152-171.
36. Wareing, P.F., and M. Black (1958) Photoperiodism in seeds and seedlings of woody species. In The Physiology of Forest Trees, K.V. Thimann, W.B. Critchfield, and M.H. Zimmerman, eds. Ronald Press, New York, pp. 539-556.
37. Welander, M. (1985) In vitro shoot and root formation in the apple cultivar. Åkerö. Ann. Bot. 55:249-261.

PHYSIOLOGICAL GENETICS OF ORGANOGENESIS IN VITRO

M.L. Christianson and D.A. Warnick

Zoecon Research Institute
Sandoz Crop Protection Corporation
Palo Alto, California 94304

ABSTRACT

The recovery of plants from cell culture proceeds by one of two pathways: somatic embryogenesis or shoot organogenesis. The now classic experiments of Skoog and Miller demonstrated that organogenesis was controlled by the phytohormones in the medium. Shoot-inducing medium is relatively low in auxin and high in cytokinin, root-inducing medium is high in auxin and low in cytokinin, and callus-inducing medium has intermediate levels of auxin and cytokinin. A series of experimental manipulations demonstrates that the process of shoot organogenesis can be divided into three physiological phases: the acquisition of competence for induction (phase 1), induction per se (phase 2), and morphological differentiation and growth (phase 3). These phases can be further subdivided. For example, induction includes five transient sensitivities to inhibitors. Such stage-specific inhibitions reflect phenocritical times in development rather than general metabolic toxicities. The phenocopying agents are TIBA, sorbitol, ribose, ammonium ion, and ASA. A number of species or cultivars will not produce shoots in response to any of a large number of phytohormone combinations; in some cases, this can be shown to be the result of a block in the acquisition of competence (phase 1) rather than a block in the induction of shoots. Close attention to the physiological genetics of the regeneration process can lead to more efficient regeneration from responsive cultivars and regeneration from otherwise nonresponsive cultivars.

INTRODUCTION

Whole plants have a remarkable propensity for asexual or vegetative propagation, and it is not surprising that this ability extends to plant cells or tissues cultured in vitro (50). This "regenerative response" in vitro includes somatic embryogenesis as well as organogenesis (the formation of shoots or roots from cultured tissues). This chapter will only consider organogenesis. The phenomenon of organogenesis in vitro is not simple. Explants placed in vitro may have the ability to give rise to shoots, roots, or floral structures when cultured on a medium that supplies mineral salts, vitamins, and a carbon source. In most cases, exogenously supplied plant hormones not only facilitate these processes, but are an essential requirement for organogenesis.

Skoog and Miller (39) first showed that organogenesis was governed by the balance of auxin and cytokinin in the tissue culture medium. Media with a relatively large auxin/cytokinin ratio induce roots, those with a low auxin/cytokinin ratio induce shoots, and those with an intermediate auxin/-cytokinin ratio induce unorganized growth as callus tissue. Once induction occurs, it is followed by morphological differentiation and development. Our interest focuses on the process of induction rather than on the process of morphological differentiation and development. Specifically, we are interested in two questions. Does the phytohormone balance really _induce_ the formation of organs, in the sense that zoologists use the word _induce?_ How much of the substantial amount of time between placing an explant into culture and seeing the newly formed organs is devoted to the induction process and how much is devoted to morphological differentiation and growth?

Viewing morphogenesis in vitro as a developmental process is the complement to more usual approaches which treat it as a single physiological event--a response to the right combinations of plant hormones and nutrients. This complementary viewpoint allows certain insights and solutions to problems that have not yet fallen before physiological approaches; it means adopting the terminology of animal developmentalists, and the precise meanings they attach to such concepts as induction, competence, and determination.

THE EXPERIMENTAL SYSTEM

Shortly after Skoog and Miller's (39) demonstration of the controls on organogenesis from cultures of tobacco, Earle and Torrey (15) demonstrated the formation of shoots in vitro from plated suspensions of friable callus of field bindweed, _Convolvulus arvensis_ L. We find that small pieces of the lamina of leaves from greenhouse-grown plants will form organs readily when cultured on quite simple media: shoots on shoot-inducing medium (SIM); roots on root-inducing medium (RIM); and callus on callus-inducing medium (CIM). (For details of the culture system and exact formulations of media, see Ref. 7.)

LOOKING AT "INDUCTION"

The word "induce" is often used to mean "results in" or simply "produces"; exposure to auxins or to heat shock "induces" sets of proteins or mRNAs, for example. Animal developmentalists have a restricted sense in which they use the words "induce," "inducer," and "induction." In this restricted and precise sense, induction is a change in the fate or destiny of a cell or group of cells. This change occurs because of exposure to an inducer or to inductive conditions. Most importantly, once induction has occurred, the inducer or the inductive conditions are no longer needed. This is in sharp contrast to heat shock-"induced" proteins, auxin-"induced" RNAs, or even the "induction" of growth in _Avena_ coleoptile segments by exogenous auxin; in all these cases, removal of the inducer leads to cessation of the response.

In the case of in vitro shoot organogenesis from leaf explants, is there actually an "induction" by the exogenous phytohormones that results in cells of leaves giving rise to cells or groups of cells fated or determined for shoot formation? Such fating or determination has been demonstrated in several whole plant systems (see Ref. 7 for examples). The artificial nature of tissue culture, most especially the dependence on exogenous hormones and nutrients, makes cell and tissue culture phenomena remarkably amenable to such demonstration. In the case of in vitro shoot organogenesis, acquiring this specific developmental state (i.e., determination) has an experimentally recognizable endpoint; the explant will produce shoots even if the inductive medium is replaced by basal medium.

DETERMINATION FOR SHOOT PRODUCTION

Walker et al. (46) described a system for the study of induction of organogenesis in competent cultures of alfalfa. We developed a similar protocol to ascertain the time of determination in shoot regeneration from leaf explants of *Convolvulus*. Determination is defined operationally: determined tissue will go on to produce shoots on basal media (salts, vitamins, and carbon source) with no need for supplemental hormones or plant growth factors. In the experimental protocol, leaf explants are placed on SIM for various lengths of time. When the explants are removed and placed on basal medium, only those explants containing cells or groups of cells determined for shoot formation will go on to produce shoots. Table 1 shows a set of representative data. After ten days on inductive medium, explants of genotype 23 are determined and produce shoots even if transferred to basal medium. Replicate experiments give the same or closely similar values for the time of determination.

The phytohormone balance in SIM, then, does *induce* the formation of shoots in the restricted sense of animal developmentalists. Exposure to inductive medium leads to a change in fate; some cells in the leaf explant give rise to cells now fated for shoot formation. Once this change in fate has occurred, the inductive conditions are no longer necessary.

Tab. 1. Mean numbers of shoots produced from leaf explants exposed to SIM for various lengths of time: genotype 23. From Ref. 7, with permission of Academic Press, Inc.

	Day of transfer from inductive medium									
	0	3	5	7	10	12	14	17	19	21
Expt. 1	0.00*	0.44	0.00	0.22	1.78	4.56	6.63	5.38	9.67	8.89
Expt. 2	0.00	0.00	0.00	0.00	2.78	3.67	6.22	7.44	8.89	5.89

*Numbers are means of counts taken after three, four, and five weeks in culture; n = 3 x 3.

Note: Explants are transferred from SIM to basal medium.

COMPETENCE FOR THE INDUCTION OF SHOOTS

On a gross scale, shoot organogenesis from leaf explants is preceded by the formation of small amounts of callus at the cut margins of the explant. Indeed, histological investigation shows this callus to be the tissue from which shoots arise (10). The elaboration of callus tissue is commonly referred to as the "dedifferentiation of the explant" and is widely assumed to be a prerequisite for subsequent differentiation.

Animal developmentalists distinguish a very similar phenomenon called "competence." Competence is the ability of a cell or tissue to respond to an inducer. Now that we have shown that SIM is actually an inducer in the restricted sense of the word, we would like to know whether the induction of shoots in vitro includes a period of time during which the explant becomes competent to respond to SIM. In the case of shoot organogenesis from leaf explants, we have chosen an operationally defined meaning for competence. The dedifferentiation process can be initiated by RIM, SIM, or CIM. The time at which CIM or RIM can no longer substitute for SIM is the time at which the specific induction of shoots begins.

MEASURING COMPETENCE

Measurements of the time at which determination has occurred with our adaptation of the protocol of Walker et al. (46) include both the time required for the specific induction of shoots as well as the time required to produce tissue competent to be induced. A modification of our protocol can distinguish these two events. Callus-inducing medium induces and supports the growth of callus from a wide variety of *Convolvulus* genotypes. Preculture on CIM prior to transfer to SIM can shorten the length of time required for culture on SIM to make explants determined for shoot formation. Table 2 shows the results of such an experiment. Explants are precultured on CIM for various times, then cultured for various lengths of time on SIM, and finally transferred to basal medium. The first row in Tab. 2 is exactly our protocol for ascertaining the time of determination (Tab. 1). Genotype 30 requires 14 days for explants to become determined for shoot formation. Preculture on CIM media for three days shortens the time required on SIM by "three" days; increasing amounts of time on CIM do not further shorten the required length of culture on SIM. This minimum time requirement for SIM is the time actually required for the *induction* of roots. Subtraction of that time from the total time of culture necessary to result in explants determined for shoot formation (Tab. 1) estimates the time required for the explant to become competent for induction. Genotype 30 takes four days to become competent ("dedifferentiated") and ten days to be induced; other genotypes may take seven or more days to acquire competence (10).

Elegant experiments by Walker et al. (46) had shown that competence for embryogenic induction in alfalfa (*Medicago sativa* L.) is a function of the size of the cellular aggregates. In contrast, competence for organogenesis is not directly related to the extent of callus proliferation (nor is determination). Documenting callus proliferation by measuring fresh weights of explants from two widely divergent *Convolvulus* genotypes shows that growth of both is very similar, being exponential over the first 14 days on SIM, and that competence is not a simple function of "explant mass" (10). In a similar vein, histological examination of explants killed and fixed after various lengths of time in culture does not reveal the

Tab. 2. Mean numbers of shoots produced from leaf explants exposed to CIM, then SIM, before transfer to basal medium: genotype 30. From Ref. 7, with permission of Academic Press, Inc.

Days on CIM	Days on SIM 0	3	5	7	10	12	14	17	19	21	
0	0.00*	0.00	0.00	0.00	0.00	0.00	1.56	4.67	6.89	6.11	Expt. 1
	0.00	0.00	0.00	0.00	0.00	0.33	1.11	2.89	3.89	4.67	Expt. 2
3	0.00	0.00	0.00	0.00	1.00	1.44	1.22	5.00	10.00	8.00	
	0.00	0.00	0.00	0.00	1.11	1.11	1.89	6.89	8.89	7.89	
5	0.00	0.00	0.00	0.00	0.89	2.56	3.22	6.22	6.11	7.89	
	0.00	0.00	0.00	0.00	0.00	3.44	2.56	1.89	-	10.00	
7	0.00	0.00	0.00	0.00	2.22	4.00	3.78	6.00	6.22	7.78	
	0.00	0.00	0.00	0.33	0.89	0.33	7.11	9.89	4.33	12.00	
10	0.00	0.00	0.00	0.00	0.00	0.33	3.67	6.78	3.89	5.33	
14	0.00	0.00	0.00	0.00	2.44	4.67	5.11	3.56	6.56	6.78	

*Numbers are means of counts taken after three, four, and five weeks in culture; n = 3 x 3.

Note: Explants are precultured on CIM, cultured on SIM, and removed to basal medium.

presence of organized shoot apices until approximately two days after the time that transfer experiments first demonstrate the presence of cells or groups of cells determined for shoot formation.

RECAPITULATION

These simple transfer experiments have subdivided the process of shoot organogenesis into three distinct physiological phases (7). Explants placed into culture first become competent to respond to SIM. Competence for shoot induction can be acquired under the influence of SIM, RIM, or CIM. Competent tissue then responds to the specific phytohormone balance in SIM, and undergoes shoot induction, sensu stricto. The end product of this induction process is an explant that contains cells or groups of cells determined for shoot formation. The third phase is the process of morphological differentiation and growth. Determined cells will grow out to become shoots on basal medium or in the presence of SIM, RIM, or CIM.

The phytohormone balance in SIM exerts its control of the kind of organ produced from the explant during the second phase, from the time the explant becomes competent to the time it becomes determined for organ development. This is a period of substantial duration, five to ten days in most genotypes we have examined. We would like to know whether this time can be subdivided into an enchainment of steps and subprocesses.

A CLOSE LOOK AT INDUCTION

The developmental process leading the determined state almost certainly involves gene action and, as such, should be amenable to further analysis through this induction, recovery, and characterization of mutants. Enrichment selection techniques allowed the application of such an approach to the dissection of the process of somatic embryogenesis in carrot (4). We have been unable to recover similar ts mutants of shoot organogenesis in Convolvulus.

Just as mutant genes can result in the disruption of the integrated functions in wild-type physiology, certain external agents, both chemical and physical, can intervene in otherwise gene-determined developmental events and lead to a mutant-like phenotype (29). This mutant-like phenotype is a phenocopy (20), and the agents that induce them are phenocopying agents. In a few cases, the period of sensitivity to a phenocopying agent has been implicated as coincident with the effective period of the mutant gene it mimics (19,21,23). Whether or not the action of phenocopying agents is an exact epigenetic substitute for mutant gene action (28,29,44,45), the use of such agents identifies distinct, sensitive points in development. Classically, there is a single, limited time of sensitivity for the induction of each type of phenocopy.

In the case of shoot organogenesis in vitro, the phenocopy we want to induce is "no shoots." There are many reports of substances which inhibit organogenesis in vitro. Some of these substances are general toxicants, but some should represent specific interferences with the process of organogenesis. With ts mutants, shifts from permissive to nonpermissive temperatures can identify the time in development when the mutant gene acts (of fails to act) (40); similarly, shifts to and from permissive and nonpermissive media can identify stage-specific inhibitors of the process of shoot organogenesis (8). These transient sensitivities to certain inhibitors identify the phenocritical time in development--the epigenetic crises sensu Waddington (44). They allow us to demonstrate that although a single phytohormone balance controls shoot induction, the induction process itself is composed of discrete steps.

The experimental protocol is quite simple. Preliminary experiments reveal the concentration of a compound that just gives complete inhibition of shoot regeneration, but allows the explant to stay green, produce callus, and appear "healthy." Explants are placed on SIM with or without the inhibitor, and shifted to the other medium after various lengths of time in culture. After an appropriate amount of time (three weeks in our system), explants are observed and the number of shoots are counted and recorded. When the test compound is a general toxicant, only the "no-treatment control" explants produce shoots. When the test compound is a stage-specific inhibitor, a reciprocal pattern of shoots and no shoots results from the two series of treatments.

For example, tri-iodobenzoic acid (TIBA) is one such stage-specific inhibitor of shoot regeneration from Convolvulus explants (8). Explants of genotype 30 moved to SIM plus TIBA after the tenth day in culture will form shoots (Tab. 3). Explants moved from SIM plus TIBA by day 7 will also form shoots. But any transfer sequence that has the explant exposed to SIM plus TIBA on day 10 results in no shoots. This is evidence for an event in the regeneration process, sensitive to inhibition by TIBA, which occurs on the tenth day of culture. Independent estimates of the time that explants of genotype 30 become competent for induction and become determined for shoot formation reveal that this transient sensitivity to TIBA occurs during the induction process.

Tab. 3. Mean numbers of shoots produced from leaf explants exposed to TIBA at different times: genotype 30 at three weeks. From Ref. 8, with permission of Academic Press, Inc.

Media sequence	Day of transfer									
	0	3	5	7	10	12	14	17	19	21
To SIM plus TIBA	0.00	0.00	0.00	0.00	0.10	1.29	2.40	1.93	2.60	7.13
From SIM plus TIBA	4.78	1.40	1.71	1.33	0.00	0.00	0.00	0.00	0.00	0.00

Note: n = 15, 6 μM TIBA.

In addition to the sensitivity to TIBA, we have described stage-specific sensitivities to sorbitol, ribose, ammonium ion, and acetylsalicylic acid (ASA) in the process of in vitro shoot organogenesis from leaf explants of C. arvensis (8). All these sensitivities occur between the time the tissue becomes competent for induction and the time the tissue becomes determined for shoot production (7). As such, they identify steps or subprocesses in the process of organogenic induction. Although we have not located each of the five sensitivities in every genotype, all evidence to date suggests that shoot organogenesis in any genotype of Convolvulus includes all five phenocritical times (8). Experiments transferring explants from SIM plus Inhibitor A to SIM plus Inhibitor B demonstrate that these sensitivities are truly distinct events (6).

We examined a large number of compounds for their effect on shoot regeneration from Convolvulus leaf explants (Tab. 4). The five phenocopying agents we reported illustrate the range of the kinds of compounds that can affect regeneration: hormones, carbohydrates, nitrogen sources, osmotica, and others. Clearly, shoot organogenesis can be disrupted by perturbations in any area of metabolism; like whole plant development, this in vitro process involves the close coordination of cellular metabolism.

Not all compounds reported in the literature as inhibitors of shoot regeneration turn out to be (a) effective in Convolvulus, (b) stage-specific in their action, or (c) effective on both root and shoot organogenesis. This certainly does not limit the usefulness of this approach in dissecting organogenesis in any particular species; it also means that compounds examined and found ineffective in C. arvensis may be quite useful in other species.

We have grouped the compounds listed in Tab. 4 by suspected mode of action; certain compounds could fit in any of several categories. Data on the effects of these compounds on the regeneration of shoots from tobacco callus are drawn from the literature. We have given a reference for each compound; this reference is either a report of an inhibiting effect on tissue cultures or a report of a mode of action. Those few cases where we merely confirmed that a compound was an inhibitor of shoot regeneration from Convolvulus explants but did not examine that inhibition for stage-specificity are indicated by an appropriate superscript.

Tab. 4. Compounds tested for effects on organogenesis in vitro.

Compound	Concentration range*	Convolvulus Shoots	Convolvulus Roots	Nicotiana Shoots	Reference(s)
Hormones/Antihormones					
Trans-cinnamic acid	To 3.4 x 10^{-5}	NE			13
Coumarin	10^{-5} to 10^{-3}	Toxic			14
Tri-iodobenzoic acid	1 to 14 μM	SSI	SSI	I	34
Naphthylpthlamic acid	1 to 24 μM	SSI			7
Gibberellic acid	To 100 mM	NE		SSI	32, 33
LiCl	To 10^{-3}	NE			11
CCC	10^{-6} to 10^{-3}	NE		I	34
Amo 1680				I	34
Phosphon	10^{-6} to 10^{-4}	I		I	34
Abscisic acid	10^{-7} to 10^{-5}	I	I	SSI	43
Amino-cyclopropane-carboxylic acid	10 to 1,000 μM	I		SSI	25
Calcium metabolism					48
Chlorpromazine	10^{-7} to 10^{-3}	I	Toxic		27, 38
Inosine	10^{-5} to 10^{-3}	NE	NE		27, 38
Specific protein inducers					
Salicylic acid	10 to 100 μM	SSI	NE		12, 26
Acetylsalicylate	10 to 100 μM	SSI	I		24
Dinitrophenol	10 to 100 μM, 1 or 2 hr	Toxic	Toxic		30
Canaline	To 1 mM	Toxic			F.S. Wu, pers. comm.
Hydroxy-norvaline	To 1 mM	Toxic			F.S. Wu, pers. comm.
5-Azacytidine	10^{-5} to 10^{-3}	Toxic			F.S. Wu, pers. comm.
Phenylboronic acid	10^{-7} to 10^{-3}	I	Toxic		22
Heat shock	Acute/Chronic	Toxic			1
Miscellaneous					
Malate	To 10^{-2}	NE			35
Ammonium citrate	40 mM	SSI		I	5

*Effective range for compounds having effects; total range tested for compounds with lethal or no effects.

Key: NE = no effect (no substantial reduction in numbers of organs); Toxic = explants without shoots are brown and dead; SSI = reciprocal, stage-specific inhibitor; NRI = nonreciprocal inhibition; I = inhibitor (explants are green and alive; includes inhibitions not yet tested for stage specificity).

Almost all compounds known to affect shoot regeneration from tobacco callus cultures inhibited shoot formation from leaf explants of Convolvulus (Tab. 5). The ineffectiveness of gibberellic acid and gibberellin antagonists was surprising.

A much larger number of compounds were tested in the parallel (and alternate) organogenic pathways from Convolvulus leaf explants (Tab. 6). Only one compound, TIBA, was a stage-specific inhibitor in both processes; significantly, the TIBA-sensitive step in shoot organogenesis occurs during induction (8), while the TIBA-sensitive step in root organogenesis occurs after induction (47). We have previously suggested that root and shoot organogenesis, although governed by alternate phytohormone balances in otherwise identical culture systems, do in fact represent two

Tab. 4 (continued).

Compound	Concentration range*	Convolvulus Shoots	Convolvulus Roots	Nicotiana Shoots	Reference(s)
Miscellaneous (continued)					
Cortisone	10^{-8} to 10^{-4}	NE	I		
Cholesterol	10^{-8} to 10^{-4}	NE	I		
Testosterone	10^{-8} to 10^{-4}	NE	I		
Estradiol	10^{-8} to 10^{-4}	NE	I		
Carbohydrates/Metabolism					
Ribose	10 to 100 mM	SSI	I		42
Erythrose	33 mM	SSI			
Methylene blue	10^{-7} to 10^{-3}	SSI			2
Sorbitol	20 to 25 g/l	SSI	I		41
Galactose	5 to 10 g/l	I	I	Toxic	49
3-Methyl glucose	20 to 40 g/l				31
Mannitol	20 to 60 g/l	I	I		5
Sucrose	100 g/l	I	I		5
Dye molecules					51
Bromthymol blue	10^{-8} to 10^{-4}	NE	NE		
Janus green	To 10^{-4}	I	NE		
Methyl blue	10^{-5} to 10^{-4}	NE	I		
Methyl green	10^{-5} to 10^{-4}	I	I		
Methyl orange	To 10^{-4}	NE	I		
Methyl red	10^{-8} to 10^{-4}	NE	NE		
Methyl violet	10^{-6} to 10^{-5}	I	I		
Phenol red	To 10^{-4}	NE	I		
Herbicides					
Glyphosate	10^{-6} to 10^{-5}	I	SSI		
Maleic hydrazide	10^{-6} to 10^{-4}	NRI	I		
Chlorflurenol methyl ester	10^{-6}	NRI	NRI		18, 36, 37
Norflurazon	To 10^{-4}	NE†			41

*Effective range for compounds having effects; total range tested for compounds with lethal or no effects.
†Shoots had no chlorophyll at higher concentrations.

Key: NE = no effect (no substantial reduction in numbers of organs); Toxic = explants without shoots are brown and dead; SSI = reciprocal, stage-specific inhibitor; NRI = nonreciprocal inhibition; I = inhibitor (explants are green and alive; includes inhibitions not yet tested for stage specificity).

processes distinct from their inception (9). The distinct patterns of reaction to the "inhibitors" listed in Tab. 6 emphasize this basic and fundamental difference between root and shoot organogenesis.

SECOND RECAPITULATION

Transfer experiments between various hormone-containing media and basal medium subdivided the process of shoot organogenesis into three phases: (a) acquisition of competence for induction, (b) induction per se, and (c) morphological differentiation and growth. Transfer experiments using SIM with and without various inhibitors reveal that the induction phase itself is composed of several substeps. For shoot formation from

Tab. 5. Compounds tested for effects on shoot organogenesis in vitro using Convolvulus and Nicotiana explants.

Compound	Convolvulus	Nicotiana
Effective in Convolvulus and Nicotiana		
Amino-cyclopropane-carboxylic acid	NRI	SSI
Tri-iodobenzoic acid	SSI	I
Phosphon	I	I
Abscisic acid	I	SSI
Ammonium citrate	SSI	I
Galactose	I	Toxic
Sorbitol	SSI	I
Mannitol	I	I
Sucrose	I	I
Effective in Nicotiana alone		
Gibberellic acid	NE	SSI
CCC	NE	I
Effects on organogenesis in Convolvulus alone		
(No examples in this class due to experimental design.)		
Effective in neither species		
(No examples in this class due to experimental design.)		

*Effective range for compounds having effects.

Key: NE = no effect (no substantial reduction in numbers of organs); Toxic = explants without shoots are brown and dead; SSI = reciprocal, stage-specific inhibitor; NRI = nonreciprocal inhibition; I = inhibitor (explants are green and alive).

leaf explants of Convolvulus, induction includes a time which is sensitive to inhibition by salicylates, followed by a time sensitive to TIBA, which is followed in turn by a time sensitive to sorbitol, and culminates in a cell or groups of cells determined for shoot formation. This process also includes a time sensitive to inhibition by ribose, although its place in the order of events is not yet firmly assigned. There is also a sensitivity to ammonium ion (or lack of nitrate) at or near the time the explant becomes determined for shoot production.

Beyond the fact that the process of organogenesis is comprised of subprocesses and sub-subprocesses, we also see temporal restrictions. Most, if not all, of the phenocritical periods and the two developmental phenomena, competence and determination, occur at times that are not only genotype-specific but extremely limited in duration. Development, at least in this tissue culture system, has an associated momentum. In the presence of an inhibitor, the explant is unable to continue on the path to shoot organogenesis; development still proceeds, however, but along an alternate pathway--callus growth. Although we call the effect of TIBA, for example, an inhibition of shoot production, this is neither a reversible inhibition nor an accumulation or arrest of cells at a TIBA-sensitive step. It may be more descriptive to refer to divertors, rather than inhibitors, of shoot production.

Tab. 6. **Compounds tested for effects on root and shoot organogenesis in vitro from explants of *Convolvulus arvensis*.**

Compound	Shoots	Roots
Effects on root and shoot regeneration		
Tri-iodobenzoic acid	SSI	SSI
Acetylsalicylate	SSI	I
Ribose	SSI	I
Sorbitol	SSI	I
Maleic hydrazide	NRI	I
Chlorflurenol methyl ester	NRI	NRI
Chlorpromazine	I	Toxic
Phenylboronic acid	I	Toxic
Abscisic acid	I	I
Mannitol	I	I
Sucrose	I	I
Galactose	I	I
Methyl green	I	I
Methyl violet	I	I
Glyphosate	I	SSI
Effects on root regeneration alone		
Cortisone	NE	I
Cholesterol	NE	I
Testosterone	NE	I
Estradiol	NE	I
Methyl blue	NE	I
Methyl orange	NE	I
Phenol red	NE	I
Effects on shoot regeneration alone		
Salicylic acid	SSI	NE
Janus green	I	NE
Effects on neither		
Inosine	NE	NE
Bromthymol blue	NE	NE
Methyl red	NE	NE

*Effective range for compounds having effects; total range tested for compounds with lethal or no effects.

Key: NE = no effect (no substantial reduction in numbers of organs); Toxic = explants without shoots are brown and dead; SSI = reciprocal, stage-specific inhibitor; NRI = nonreciprocal inhibition; I = inhibitor (explants are green and alive; includes compounds yet to be examined for stage specificity).

Given the complexity of the regeneration process, it is even more amazing that single "consensus" media are possible for a large number of species and that Skoog and Miller's (39) observation is so widely applicable.

PROBLEM SOLVING

There are a number of plant species that do not reliably produce shoots in vitro when cultured on media of known phytohormone balances.

Such species or cultivars could represent cases where regeneration is not controlled by phytohormone balance, or they could simply represent cases where one ("consensus") medium does not satisfy the requirements of all three phases in the regeneration process.

There are seed-derived individuals of *C. arvensis* that will not produce shoots when leaf explants are cultured on SIM (and individuals that will not produce roots from leaf explants cultured on RIM). It is possible to show that some of these represent not the inability of explants to be *induced* by SIM or RIM, but the inability of explants to develop *competence* for induction under standard conditions (9). In such genotypes, reliable shoot formation occurs only if the explants are first cultured on RIM in order to develop competence for induction. Before culture on RIM has led to the commitment to produce roots, the explants can be moved to SIM for the induction of shoot formation within the now competent tissue.

We have previously noted that this concept of competence is similar to the idea of the dedifferentiation of the explant. It is not synonymous with the extensive proliferation of callus tissue, however; competence for shoot induction develops in *Convolvulus* explants within the very first days of culture. In Tab. 2, we illustrate a genotype where competence is quite stable; explants precultured on CIM for three days and explants precultured on CIM for 14 days are equally responsive to SIM. This is not generally true. Just as prolonged culture on RIM leads to a commitment to produce roots, prolonged culture on CIM can lead to a commitment for growth as unorganized callus. The time of transfer of an explant from one medium to another can be as important as the composition of those media. Two of our co-workers have used this observation to achieve rapid and efficient shoot regeneration from leaf explants of *Arabidopsis* (17).

Previous reports of shoot regeneration from explants of *Arabidopsis* have involved the use of a first medium to induce a proliferating callus and the transfer of pieces of this callus to a second, low-auxin, high-cytokinin medium for shoot formation. In contrast, culturing explants on the CIM PG1 for 0, 3, 5, 7, or 14 days followed by transfer of the explants to low-auxin, high-cytokinin medium reveals a sharp peak of responsiveness and shoot production. Both 0 and 14 days on CIM resulted in no shoots from the explants (n=32); three days on CIM resulted in approximately 0.6 shoots per explant, while five to seven days on CIM resulted in approximately 2.6 shoots per explant (17).

These real but abstract phenomena of competence, determination, and other fleeting developmental events are not unique to the nonwoody plants studied in our laboratory. Both Beaty and Schwartz (3) and Ellis and Bilderbach (16) have looked for and found "windows of opportunity" for the culturability of gymnosperms. The take-home message is that shoot organogenesis is a developmental process and that distinct events are occurring at distinct times. Viewing regeneration as a dynamic process can provide insights and ways to solutions; the static view of regeneration as an all-or-none response to a particular combination of phytohormones is inaccurate and often ineffective.

REFERENCES

1. Altschuler, M., and J.P. Mascarenhas (1982) Heat shock proteins and the effects of heat shock in plants. *Plant Mol. Biol.* 1:103-115.

2. Barron, E.S.G., and G.A. Harrop, Jr. (1928) Studies on blood metabolism. II. The effect of methylene blue and other dyes upon the glycolysis and lactic acid formation of mammalian and avian erythrocytes. J. Biol. Chem. 79:65-87.
3. Beaty, R.M. (1987) M.S. Thesis, Department of Botany, University of Tennessee, Knoxville, Tennessee.
4. Breton, A.M., and Z.R. Sung (1982) Temperature-sensitive carrot variants impaired in somatic embryogenesis. Dev. Biol. 90:58-66.
5. Brown, D.C.W., D.W.M. Leung, and T.A. Thorpe (1979) Osmotic requirement for shoot formation in tobacco callus. Physiol. Plant. 46:36-41.
6. Christianson, M.L. (1987) Causal events in morphogenesis. In Plant Tissue and Cell Culture, C.E. Green, D.A. Somers, W.P. Hackett, and D.D. Biesboer, eds. Alan R. Liss, Inc., New York, pp. 45-55.
7. Christianson, M.L., and D.A. Warnick (1983) Competence and determination in the process of in vitro shoot organogenesis. Dev. Biol. 95:288-293.
8. Christianson, M.L., and D.A. Warnick (1984) Phenocritical times in the process of in vitro shoot regeneration. Dev. Biol. 101:382-390.
9. Christianson, M.L., and D.A. Warnick (1985) Temporal requirement for phytohormone balance in the control of organogenesis in vitro. Dev. Biol. 112:494-497.
10. Christianson, M.L., and D.A. Warnick (1987) Organogenesis in vitro as a developmental process. HortScience (in press).
11. Claire, A. (1982) Augmentation de l'activité gibberellique chez les tiges volubiles d'Ipomoea purpurea. Effects d'un traitement au chlorure de lithium. Physiol. Veg. 20:11-22.
12. Cleland, C.E., and Y. Ben-Tal (1982) Influence of giving salicylic acid for different time periods on flowering and growth in the long day plant Lemna gibba G3. Plant Physiol. 70:287-290.
13. Cross, J.W., and W.R. Briggs (1979) Solubilized auxin-binding protein. Planta 146:263-270.
14. Dhawan, R.S., and K.K. Nanda (1982) Stimulation of root formation on Impatiens balsamina L. cuttings by coumarin and the associated biochemical changes. Biol. Plant. 24:177-182.
15. Earle, E.D., and J.G. Torrey (1965) Morphogenesis in cell colonies grown from Convolvulus cell suspensions plated on synthetic media. Am. J. Bot. 52:891-899.
16. Ellis, D. (1986) Ph.D. Dissertation, Department of Botany, University of Montana, Missoula, Montana.
17. Feldmann, K.A., and M.D. Marks (1986) Rapid and efficient regeneration of plants from explants of Arabidopsis thaliana. Plant Sci. 47:63-69.
18. Gabara, B. (1982) Effect of morphactin (chlorflurenol IT 3456) on the mitotic activity and cell growth in roots of Pisum sativa L. Acta Soc. Bot. Pol. 51:39-50.
19. Gloor, H. (1947) Phanokopie-Versuche mit Ather an Drosophila. Rev. Suisse Zool. 54:637-713.
20. Goldschmidt, R.B. (1935) Gen und Ausseneigenschaft. Z. Indukt. Abstamm. Vererbungsl. 69:38-69.
21. Goldschmidt, R.B. (1957) Problematics of the phenomenon of phenocopy. J. Madras Univ. B 27:12-24.
22. Haccius, B., and D. Wilhelm (1966) Mutationen kopierende Bluten-Anomalien bei Pisum sativum nach Phenylborsaure-behandlung. Planta 69:288-291.
23. Hadorn, E. (1961) Developmental Physiology and Lethal Factors, John Wiley and Sons, New York.

24. Han, P.F., G.Y. Han, H.C. McBay, and J. Johnson, Jr. (1978) Alteration of the regulatory properties of chicken liver fructose-1,6-bisphosphatase by treatment with aspirin. Biochem. Biophys. Res. Comm. 85:747-755.
25. Huxter, T.J., T.A. Thorpe, and D.M. Reid (1981) Shoot initiation in light- and dark-green tobacco callus: The role of ethylene. Physiol. Plant. 53:319-326.
26. Kumar, S., and K.K. Nanda (1981) Gibberellic acid- and salicylic acid-caused formation of new proteins associated with extension growth and flowering of Impatiens balsamina. Biol. Plant. 23:321-327.
27. Lado, P., R. Cerana, A. Bonetti, M.T. Marre, and E. Marre (1981) Effects of calmodulin inhibitors in plants. I. Synergism with fusicoccin in the stimulation of growth and H^+ secretion and in the hyperpolarization of the transmembrane potential. Plant Sci. Lett. 23:253-262.
28. Landauer, W. (1957) Phenocopies and genotype, with special reference to sporadically-occurring developmental variants. Am. Naturalist 91:79-90.
29. Landauer, W. (1958) On phenocopies, their developmental physiology and genetic meaning. Am. Naturalist 92:201-213.
30. Mitsuhasi-Kato, M., and H. Shibaoka (1981) Effects of actinomycin-D and 2,4-dinitrophenol on the development of root primordia in azuki bean stem cuttings. Plant Cell Physiol. 22:1431-1436.
31. Moore, D. (1981) Effects of hexose analogues on fungi: Mechanisms of inhibition and of resistance. New Phytol. 87:487-515.
32. Murashige, T. (1961) Suppression of shoot formation in cultured tobacco cells by gibberellic acid. Science 134:280.
33. Murashige, T. (1964) Analysis of the inhibition of organ formation in tobacco tissue culture by gibberellin. Physiol. Plant. 17:636-643.
34. Murashige, T. (1965) Effects of stem-elongation retardants and gibberellin on callus growth and organ formation in tobacco tissue culture. Physiol. Plant. 18:665-673.
35. Plumb-Dhindsa, P.L., R.S. Dhindsa, and T.A. Thorpe (1979) Non-autotropic CO_2 fixation during shoot formation in tobacco callus. J. Exp. Bot. 30:759-767.
36. Ram, H.Y.M., and G. Mehta (1982) Regeneration of plantlets from cultured morphactin-induced barren capitula of African marigold (Tagetes erecta L.). Plant Sci. Lett. 26:227-232.
37. Rucker, W. (1982) Morphactin-induced changes in the cytokinin effect on tissue and organ cultures of Nicotiana tabacum. Protoplasma 113:103-109.
38. Saunders, M.J., and P.K. Hepler (1982) Calcium ionophore A23187 stimulates cytokinin-like mitosis in Funaria. Science 217:943-945.
39. Skoog, F., and C.O. Miller (1957) Chemical regulation of growth and organ formation in plant tissues cultured in vitro. Symp. Soc. Exp. Biol. 11:118-140.
40. Suzuki, D.T. (1970) Temperature-sensitive mutations in Drosophila melanogaster. Science 170:695-706.
41. Thorpe, T.A. (1974) Carbohydrate availability and shoot formation in tobacco callus cultures. Physiol. Plant. 30:77-81.
42. Thorpe, T.A., and D.D. Meier (1972) Starch metabolism, respiration, and shoot formation in tobacco callus cultures. Physiol. Plant. 27:365-369.
43. Thorpe, T.A., and D.D. Meier (1973) Effects of gibberellic acid and abscisic acid on shoot formation in tobacco callus cultures. Physiol. Plant. 29:121-124.

44. Waddington, C.H. (1956) Principles of Embryology, Allen and Unwin, London.
45. Waddington, C.H. (1961) Genetic assimilation. Adv. Genet. 10:257-292.
46. Walker, K.A., M.L. Wendeln, and E.G. Jaworski (1979) Organogenesis in callus tissue of Medicago sativa. The temporal separation of induction processes from differentiation processes. Plant Sci. Lett. 16:23-30.
47. Warnick, D.A. (1985) Developmental biology of rhizogenesis in vitro in Convolvulus arvensis. M.A. Thesis, San Jose State University, San Jose, California.
48. Wu, F.S., Y.C. Park, D. Roufa, and A. Martonosi (1981) Selective stimulation of the synthesis of an 80,000 dalton protein by calcium ionophores. J. Biol. Chem. 256:5309-5312.
49. Yamamoto, R., N. Sakurai, and Y. Masuda (1981) Inhibition of auxin-induced cell elongation by galactose. Physiol. Plant. 53:543-547.
50. Yusufov, A.G. (1982) Origin and evolution of the phenomenon of regeneration in plant (problem of evolution ontogenesis). Usp. Sovrem. Biol. 93:89-104 (translated from Russian by Leo Kanner Associates).
51. Zatyko, J., F. Kiss, and I. Simon (1980) Indikatorok es mikrotechnikai festekek hatasa szovet- es szervtenyeszetekre. Bot. Kozlem. 67:97-101.

TISSUE CULTURE OF CONIFERS USING LOBLOLLY PINE AS A MODEL

H.V. Amerson, L.J. Frampton, Jr.,
R.L. Mott, and P.C. Spaine*

Departments of Forestry and Botany
North Carolina State University
Raleigh, North Carolina 27695-7612

ABSTRACT

Loblolly pine (Pinus taeda) is a good model for conifer tissue culture, since studies include three methods of in vitro propagation, long-term field evaluations of tissue-culture plantlets, and the development of in vitro trait selection methods. Abbreviated protocols are outlined for the following: (a) organogenesis of adventitious shoots and roots; (b) shoot micropropagation via fascicular and axillary shoots obtained from juvenile, adolescent, and mature explants; and (c) embryogenesis via immature zygotic embryos. Regulatory features of these processes, especially adventitious organ induction and shoot and root elongation, are examined. Field data on four- to six-year height growth, morphological characteristics, and fusiform rust resistance of tissue-culture plantlets derived from cotyledon explants are summarized from multiple plantings. Histological and immunological studies on fusiform rust resistance, evaluated in vitro in loblolly pine embryos, are presented as examples of in vitro trait selection.

INTRODUCTION

The diversity of tissue culture procedures currently being applied for research on loblolly pine (Pinus taeda L.), as well as field evaluations of plantlets produced in culture, makes this species a good model system. Research topics presently being pursued with loblolly pine include: (1) cotyledon system propagation and regulation of adventitious shoot and root development; (2) field performance of tissue culture plantlets; (3) axillary or fascicular shoot micropropagation; (4) embryogenesis; (5) in vitro trait selection; (6) callus growth and development; (7) protoplast systems; and (8) in vitro genetic manipulations. The first five topics will be

---

*Present address: U.S. Department of Agriculture Forest Service, Athens, Georgia.

considered in the present chapter. The consideration of research topics 1, 3, and 4 will summarize current tissue culture methods for loblolly pine and, where appropriate, will discuss regulatory controls within these methods. The coverage of field performance (topic 2) will examine morphological traits of plantlets, plantlet field resistance to fusiform rust, and four to six years of height growth data. The consideration of in vitro trait selection will examine studies that histologically and immunologically evaluated fusiform rust resistance of seedlings in vitro.

COTYLEDON SYSTEM PROPAGATION AND THE REGULATION OF ADVENTITIOUS SHOOT AND ROOT DEVELOPMENT

The induction of adventitious shoots in loblolly pine tissue cultures has been reported from a number of explants, including cotyledons (26), hypocotyls (20), and needle fascicles (20,28). Of these explants, cotyledons are by far the most commonly used, and they will be the starting point for this consideration.

The sequence for tissue culture propagation of loblolly pine covered in this section was originally reported in 1981 (2,21) and is organized around the concept of pulse timing for growth regulator applications. Central to this concept is the idea that growth regulators which are necessary to initiate organogenic events may subsequently be inhibitory to further development. Therefore, both the timely application and removal of growth regulators are necessary to progress from organ induction to organ growth and ultimately to whole plant propagation. This sequence was updated in 1985 (3) with amendments in media, environments, and pulse durations, and further amendments will be included here. The approach to this section will be first to present an update of the propagation sequence and then to examine regulatory controls such as growth regulator pulses, media regulation, and environmental parameters.

Propagation Sequence

To begin the culture process, loblolly pine seeds are scarified at the micropylar end and germinated in H_2O_2 at 30°C until the radicle and hypocotyl extend from the seed coat. A typical H_2O_2 exposure would entail daily changes, with three days at 1% H_2O_2 and two days at 0.03% H_2O_2. Next the female gametophyte is removed from the seed coat and surface-sterilized prior to aseptic excision of the embryonic axis. The cotyledons are then surgically removed from the embryo and planted horizontally on a shoot initiation medium (GD_1, BLG, or LMG*) which is rich in cytokinin, typically 44 µM benzylaminopurine (BAP). In the presence of cytokinin and light the peripheral regions of the cotyledon will undergo cell divisions to produce a warty meristematic surface. Cotyledons are removed from the cytokinin after the appearance of meristematic tissues but prior to the actual visualization of shoots. The time of removal is dependent upon researcher judgment but typically occurs after 14 to 28 days of cytokinin

---

*GD_1 = Gresshoff and Doy_1 medium (11) modified according to Mott and Amerson (21). BLG = Brown and Lawrence (5) medium with 10 mM glutamine, 1 mM KNO_3 and 10 mM KCl substituted for the original nitrogen components and potassium balance. LMG = Litvay medium (17) substituted in the same manner as BLG with glutamine, KNO_3, and KCl.

exposure, depending upon specific environmental variables and medium. Cotyledons are next transferred to a hormone-free medium containing activated charcoal to further aid cytokinin removal. On this differentiation and growth medium shoot apices become recognizable and shoot elongation begins. Shoot growth continues with the shoots attached to the cotyledons during additional subcultures on hormone-free medium minus charcoal. Shoots are excised from the cotyledons when they become crowded or reach a length of about 5 mm. These are then cultured individually on hormone-free medium to promote further growth. Shoots with well-formed stems may be placed on hormone-free medium containing activated charcoal to further accelerate growth.

Following the shoot growth phase, shoots that have elongated to a length of approximately 1 to 2 cm are placed on a root induction, auxin-rich medium (GD 1/2) which is a one-half dilution of GD_1, typically containing either α-naphthaleneacetic acid (NAA) or indole-3-butyric acid (IBA) at 2.5 μM or 50-250 μM, respectively. For root induction, shoots are cut at the base of the stem to expose healthy tissues, implanted upright, and pulsed on the auxin medium until basal swelling produces rips in the epidermis (typically six to nine days). The basal swelling results primarily from preroot divisions in the cambial area. The preroot divisions may begin to organize into primordia on the auxin medium; but to facilitate organization and rapid root growth, shoots at the end of the pulse are transferred either directly to soil in a misted greenhouse area or to hormone-free medium. Root growth on hormone-free medium is rapid in the presence of light (mixed fluorescent and incandescent), and plantlets with elongating roots should be quickly transferred to soil in a misted greenhouse area. Plantlets are typically maintained in the mist area for three to six weeks until new vigorous growth appears, and then subsequently grown without mist in the greenhouse for approximately six months. Before field planting, plantlets are gradually adapted to outside conditions (3).

Regulation of Shoot Initiation and Growth

In loblolly pine, the induction of preshoot cell divisions to yield cotyledon meristematic tissues, which ultimately produce shoots, requires the presence of a cytokinin in the induction medium. This is true of many other conifers (7). Attempts to form shoots on cytokinin-free medium typically result in cotyledon elongation. In the absence of cytokinin, meristematic tissues do not form along the length of the cotyledons, and shoots are only occasionally observed at the apex of a few cotyledons. Several cytokinins, including zeatin, kinetin, 2-isopentenyl adenine, and BAP, have been used to induce shoots in loblolly pine cotyledons (20). The cytokinin most commonly used is BAP, and concentrations tested between 2.2 and 111 μM have all produced shoots (3,20).

Although a cytokinin (BAP in our protocol) is needed to stimulate shoot initiation, the BAP pulse (exposure time) is critically important in obtaining growth-competent shoots. Excess exposure to BAP can suppress shoot growth; this point is illustrated in Tab. 1 which shows the results of an examination of shoots still attached to the cotyledons. The growth of shoots initiated with a 28-day exposure to 44 μM BAP was significantly less than that of shoots given a 14-day exposure. Similar observations related to cytokinin exposure and shoot growth have been made in the case of lodgepole pine (23) and in a review of conifer tissue culture (7).

Tab. 1. Comparison of shoot initiation and growth as a function of initiation time. From Ref. 3, with permission.

Initiation time (days)	Mean number of shoots per embryo	Percent shoots <2 mm	Percent shoots ≥2 but <5 mm	Percent shoots ≥5 mm
14	46	18	34	48
	NS	*	NS	*
28	41	48	45	7

* = Significant difference within column at $P \leq 0.05$ level as determined by T-test.
NS = No significant difference within column.

Note: Shoots were initiated on BLG medium containing 44 μM BAP. BAP initiation and subsequent differentiation on hormone-free medium plus charcoal occurred at 23°C with 24 hr illumination (∿2,200 lux cool white fluorescent and 800 lux incandescent). Elongation occurred at 22 ± 2°C with 24 hr illumination (∿2,700 lux cool white fluorescent and 200 lux incandescent). Counts were made at four months total.

Further evidence that BAP suppresses shoot growth is presented in Tab. 2, which compares the growth of individually excised, actively growing shoots placed back onto BAP-supplemented medium. Inhibition occurred in all three BAP treatments and appeared proportional to overall BAP exposure. The highest concentration of BAP, 1.3 μM, supplied continuously gave the most suppression. The nine-day pulse treatment with 1.3 μM BAP or the continuous application of 0.44 μM BAP permitted intermediate growth, with suppression between these two treatments being equivalent.

Tab. 2. The influence of exogenous BAP supplied to adventitious shoots growing on GD 1/2 medium. From Ref. 3, with permission.

Treatment	Number shoots tested	Mean growth (in mm) at eight weeks*
1. GD 1/2 control	50	6.4 a
2. GD 1/2 + 0.44 μM BAP	49	4.8 b
3. GD 1/2 + 1.3 μM BAP (pulse)	50	4.7 b
4. GD 1/2 + 1.3 μM BAP	50	2.0 c

*Growth values followed by the same letter are not significantly different as determined by Waller-Duncan Test at $P \leq 0.05$.

Note: BAP was supplied continuously in the medium for eight weeks except in pulse treatment (no. 3). This treatment received a nine-day pulse on GD 1/2 + 1.3 μM BAP followed by transfer to GD 1/2. Shoots were grown at 22 ± 2°C with 16 hr mixed incandescent light (∿200 lux) and cool white fluorescent light (∿3,200 lux) and 8 hr dark.

The influence of auxin on shoot initiation and growth is less defined than the influence of cytokinin, but exogenously supplied auxins are certainly not required for shoot induction. Mehra-Palta et al. (20) showed that the percentage of embryos forming shoots from their excised cotyledons was ⩾80%, with cotyledons exposed to 9 μM zeatin and NAA ranging from 0 to 2.5 μM. In our laboratories we have routinely initiated shoots on media containing 44 μM BAP with either no NAA or 0.05 μM NAA and see no apparent differences. Although low levels of auxin do not appear harmful to shoot production, Tab. 3 shows that auxin can, at high concentrations, interfere with shoot induction. The inclusion of NAA at ⩾5 μM in shoot induction medium reduces the production of meristematic tissues along the length of the cotyledon; and at concentrations of ⩾2.5 μM NAA, friable callus, which is detrimental to shoot development, is fostered. Indeed, at 5 μM NAA it was observed that originally produced meristematic tissues were converted to friable callus. The negative effects of high concentrations of auxin on meristematic tissue development should ultimately depress shoot production.

A variety of media, such as BLG, LMG, GD_1, and GD 1/2, without growth regulators have been routinely used to grow adventitious shoots of loblolly pine. A comparison of BLG, LMG, and GD_1 (Tab. 4), with their wholesale differences in composition, shows that growth rates may vary according to medium; the 50 to 60% improvement in growth obtained with BLG and LMG over an eight-week period reflects the importance of media selection.

Of all the media factors that influence growth on a permissive medium, however, the most pronounced thus far recognized for loblolly pine is the presence of activated charcoal. Activated charcoal has been used to supplement both LMG- and GD_1-based media, with impressive growth being produced in all cases. Table 5 shows the extent of growth enhancement obtained with activated charcoal added to GD_1 or GD 1/2 medium.

Tab. 3. The influence of NAA on meristematic tissue and callus production in cotyledons.

NAA (μM)	Percent of clones with meristematic tissue	Length of cotyledons with meristematic tissue	Percent of clones with callus	Length of cotyledons with callus
0.05	73*	50%*	0	0
0.5	93	50%	0	0
2.5	100	50%	100	30%
5.0	71	10%**	100	50%
50.0	93	10%	93	90%

*Normal values (control treatment).

**NAA treatment of 5.0 μM had good meristematic tissue to 50% of cotyledon length at 22 days; this meristematic tissue was lost to callus with the additional six days of exposure.

Note: Cotyledons were cultured for 28 days on BLG medium supplemented with 44 μM BAP and variable levels of NAA (0.05-50 μM). N=14 clones, with data at 28 days.

Tab. 4. A comparison of shoot elongation on BLG, LMG, and GD_1 media.

Medium	Mean shoot growth* (mm)
LMG	11.6 a
BLG	10.6 a
GD_1	7.1 b

*Means followed by different letters are significantly different at P≤0.05 as determined by Waller-Duncan Test.

Note: Shoots were cultured at 23°C with 24 hr mixed cool white fluorescent and incandescent illumination at 2,500 lux. N=60 shoots/treatment with five shoots/treatment derived from each of 12 clones. Shoots were subcultured once during the experiment and data were taken at eight weeks.

Although growth of the no-charcoal GD_1 treatment (Tab. 5) was unusually good, the charcoal treatments grew significantly better than their corresponding no-charcoal treatment, with mean growth improvements of 150 to 270% over a one-month period.

The use of charcoal to stimulate shoot elongation is a common practice with conifer tissue cultures, but explanations for its benefit are few (7). An experiment conducted to examine the influence of media changes and maintenance of fresh basal cuts on stem growth (Tab. 6) can be interpreted to suggest that charcoal growth benefits may be related to healthy callus which develops on the base of stems in charcoal medium. Results from that experiment, as well as experimental format, are presented in Tab. 6.

Tab. 5. The influence of charcoal on the elongation of individually excised shoots on GD-based media.

Treatment	Mean growth (in mm) at one month*
GD_1, 2% sucrose, 1% charcoal	20.3 a
GD 1/2, 1% sucrose, 1% charcoal	16.0 b
GD_1, 2% sucrose, 0% charcoal	7.9 c
GD 1/2, 1% sucrose, 0% charcoal	4.3 d

*Means followed by different letters are statistically different at P≤0.05 level as determined by Waller-Duncan Test.

Note: Shoots were maintained at 23 ± 1°C with 24 hr mixed warm white fluorescent and incandescent illumination at 2,700 to 3,800 lux. Shoots were grown in 11 cm tall containers, and growth was measured after one month. N=30 shoots/treatment with five shoots/treatment derived from each of six clones.

Tab. 6. The influence of basal cuts and media changes on the growth of adventitious shoots.

	Treatment		BLG		GD_1 ch	
	Fresh cuts	New media	Number shoots tested	$\bar{x}$ Stem growth* (mm)	Number shoots tested	$\bar{x}$ Stem growth* (mm)
A	No	No	10	8.2 a	7	24.9 a
B	No	Yes	10	6.8 a	10	20.2 b
C	Yes	No	10	6.7 a	10	13.5 c
D	Yes	Yes	10	8.5 a	12	11.3 c

*Within a column, values followed by different letters are significantly different at $P \leq 0.05$ as determined by Waller-Duncan test.

Note: Shoots were cultured on BLG no charcoal (ch) medium with 0.5% sucrose or on GD_1, 1% ch medium with 2% sucrose for eight weeks. All shoots were initially cut to a stem length of 1 cm and given a two-week acclimation period on each base medium. They then received six weeks in the various treatments. Measurements represent growth over the six weeks of treatment and exclude the acclimation period. Measurements are stem length only, excluding needles. Cultures were maintained at 24 ± 2°C in a 16/8 photoperiod. Light period was 2,500 to 3,000 lux mixed cool white fluorescent and incandescent illumination. All shoots received a fresh basal cut at the start of the experiment. Treatment A shoots subsequently received no fresh cuts and no fresh medium. Treatment B received no additional cuts and new medium biweekly. Treatment C received fresh cuts biweekly but no fresh medium. Treatment D received fresh cuts and fresh medium biweekly.

From Tab. 6 it is clear that neither basal cuts nor media changes significantly affected stem growth on BLG medium. However, on the GD_1-charcoal medium, stem growth in treatments A and B in which stems did not receive basal cuts significantly exceeded that in treatments C and D in which the stems were cut biweekly. A plausible hypothesis to explain why growth differences related to basal cutting occurred on GD_1-charcoal medium and not on BLG medium requires inspection of the shoot bases. On the charcoal medium the shoot bases produced small, whitish, healthy calli within two weeks of cutting, and large, healthy calli within four weeks. In treatments C and D, these calli were removed on a biweekly basis just as they were developing; but in treatments A and B, basal calli were allowed to enlarge. We now propose that the healthy basal calli continuously available in treatments A and B provided an excellent surface for nutrient uptake whereas the biweekly callus removal in treatments C and D somewhat negated potential uptake benefits and reduced growth. The same pattern of growth was not seen on the BLG medium without charcoal because neither cut (C,D) nor uncut (A,B) treatments produce callus. The lack of basal callus is a typical response on elongation media without charcoal, whether BLG-, LMG-, or GD_1-based. On elongation media without charcoal, shoot bases often become rusty-brown in color within a few days of routine subculture, which involves a fresh basal cut. In contrast, the production of whitish basal callus is a common feature on charcoal medium. Further investigations on the relationships of charcoal and basal callus to shoot growth are in progress.

The light environment plays an important role in the regulation of adventitious shoots developed from cotyledons. Both the induction of preshoot divisions and subsequent shoot development are light-dependent (Fig. 1). Examination of Fig. 1 reveals that preshoot cell divisions occurred on cotyledons when light was present (treatments 2 and 3) but not when cotyledons were cultured in the dark (treatment 1). When cotyledons with preshoot divisions were transferred to a lighted environment (treatment 3), shoots developed; however, cotyledons with preshoot divisions failed to initiate shoots when placed in darkness (treatment 2). Thus, in the shoot initiation process neither cell divisions nor shoot organizations will occur in constant darkness.

Regulation of Root Initiation and Growth

The regulatory features of adventitious root initiation and growth have been investigated with both adventitious shoots derived from cotyledons and with seedling explants, referred to as hypocotyl explants. Adventitious shoots suitable for rooting are those that have reached a size ⩾1 cm. Hypocotyl explants are one- to two-week-old in vitro-grown seedlings which were surgically derooted, making them suitable for rooting experiments. In several rooting trials, hypocotyl explants and adventitious shoots have performed similarly, although hypocotyl explants generally root approximately 15 to 20% better and display faster root growth. Since their overall responses are similar, both explant types will be used in this consideration of adventitious rooting.

Spontaneous rooting of adventitious shoots of loblolly pine on auxin-free medium is rare, occurring at a rate of <1%. In contrast to the cytokinin dependency noted for shoot initiation, the rooting process is auxin-dependent. Similar to shoot production, however, the rooting process involves a distinct preroot division phase followed by root grow-out (21), and occurs best in response to pulses (Tab. 7). Continuous exposure of adventitious shoots to a low level of auxin (0.5 µM NAA) yielded only 17%

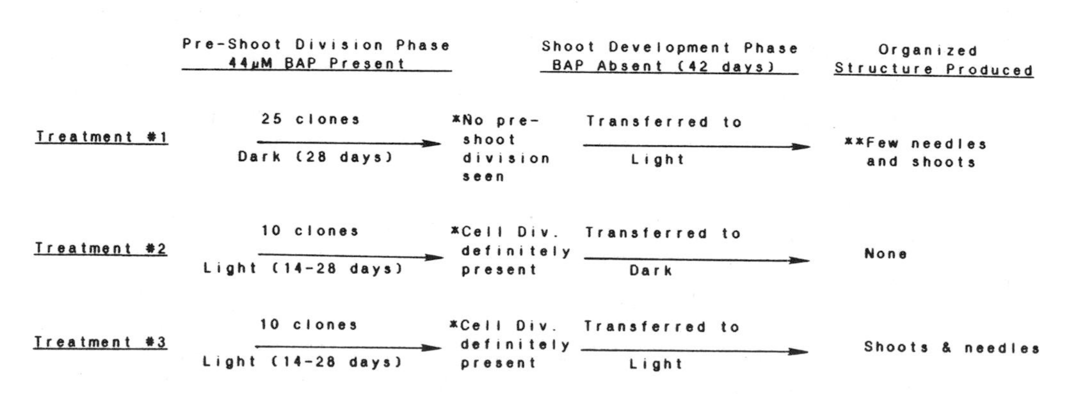

Fig. 1. Effect of light and dark on adventitious shoot development from cotyledons. Light = 24 hr illumination at approximately 1,250 lux cool-white fluorescent and 750 lux incandescent. Temperature = 21°C. From Ref. 21a, with permission.

Tab. 7. Comparison of various rooting treatments (auxin applications) using loblolly pine adventitious shoots. From Ref. 3, with permission.

NAA (μM)	Number shoots tested	Auxin time (days)	Growth medium	Percent rooting
0.5	90	42	-	17
0.5	30	12	No auxin (5 weeks)	50
2.5	128	6-9	No auxin	71

rooting, whereas the same 0.5 μM NAA gave 50% rooting when supplied as a 12-day pulse (Tab. 7). In both cases the auxin application was sufficient to stimulate preroot divisions, but the continued presence of auxin throughout the 42-day period apparently hindered actual root grow-out in a manner analogous to cytokinin depression of shoot grow-out. The rooting percentage was further improved to 71% by a shorter-term (six- to nine-day), higher-concentration (2.5 μM NAA) auxin pulse (Tab. 7). Hypocotyl explants exposed to continuous auxin applications for 31 days or to ten-day auxin pulses followed by hormone-free medium have rooted in both cases at high percentages (40 to 100%), but roots formed with continuous auxin were severely stunted relative to those in pulse treatments (H.V. Amerson, unpubl. data). Thus, the functional importance of pulses is evident with both adventitious shoots and hypocotyl explants.

Short-term (six- to nine-day) pulses using NAA (2.5 μM) or IBA (50 to 250 μM) are now routinely used as a general guide for root induction, but one must remember that evaluation of basal swelling is a better guide than absolute times. Environmental variations and differences in shoot quality or morphology may influence the time required for a proper pulse. We consider a proper auxin stimulation to be one that induces a basal swelling and splitting of the epidermis similar to that shown in Fig. 2.

Attempts to root loblolly pine shoots in continuous darkness have failed, showing that at least one portion of the overall rooting process requires light; but this lack of rooting does not indicate whether both root initiation and root growth are light-dependent. The dependency of root growth on light is shown in Fig. 3. Irrespective of the presence or absence of light during the root initiation period, root growth occurred in the light (treatments 1, 3, and 4). In contrast, those shoots undergoing

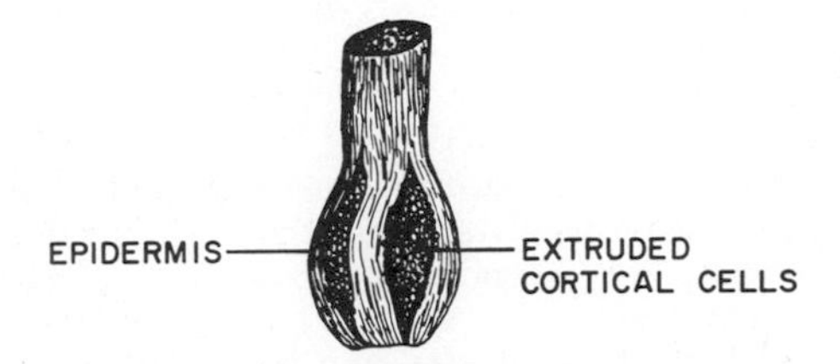

Fig. 2. A schematic diagram showing a shoot base with basal swelling deemed ideal for a root-stimulating pulse. Note the split epidermis and extrusion of underlying cortical cells.

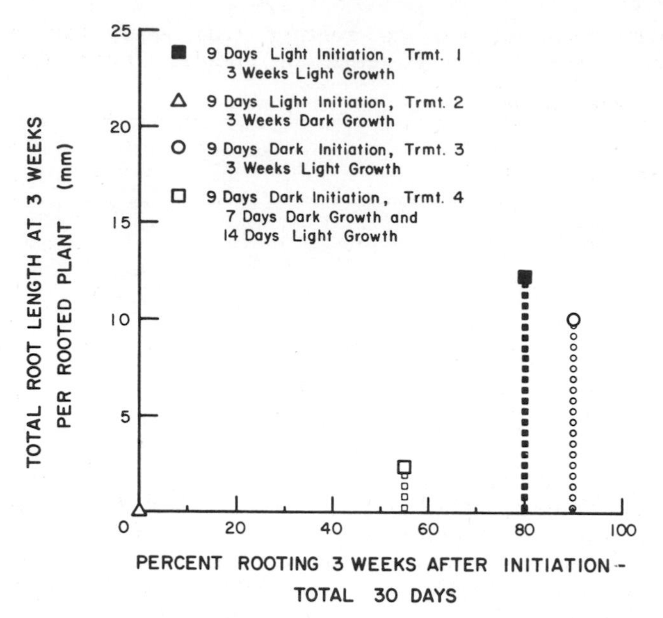

Fig. 3. Comparison of root initiation and growth in light vs dark environments. Data collected from hypocotyl explants that were derooted and then cultured on GD 1/2 medium plus 0.44 μM BAP and 2.5 μM NAA for nine days, followed by three weeks growth on GD 1/2 without growth regulators. Light = 24 hr illumination at approximately 1,830 lux warm-white fluorescent and 320 lux incandescent. Temperature = 25 ± 2°C. N = 9 to 11. From Ref. 21a, with permission.

treatment 2 failed to root after receiving a light initiation period when a dark period followed for root growth, thus demonstrating the need for light in root growth. It also appears (Fig. 3) that root initiation does not require light, since shoots undergoing treatments 3 and 4, which involved dark initiation periods, produced roots when transferred to light to allow root grow-out. Microscopic examination of hypocotyl bases taken from treatment 4 after the seven-day dark growth period revealed that preroot cell divisions had formed in darkness, but organized root apices were not observed. In summary, preroot cell divisions that start the rooting process are not light-dependent, but root organization and continued growth require light.

Not only is root growth dependent on the presence of light, but light quality also regulates the rate of root growth (Fig. 4). Figure 4 is a comparison of root growth on hypocotyl explants in three different light environments. Growth in the fluorescent environment is significantly less than growth in the two environments containing incandescent light. Improved root growth in environments containing incandescent light has been observed in many rooting experiments using both hypocotyl explants and adventitious shoots. Indeed, with adventitious shoots, roots can be initiated readily in fluorescent-only environments, but little if any root growth occurs, making the importance of incandescent light even more pronounced than that seen in Fig. 4 with hypocotyl explants.

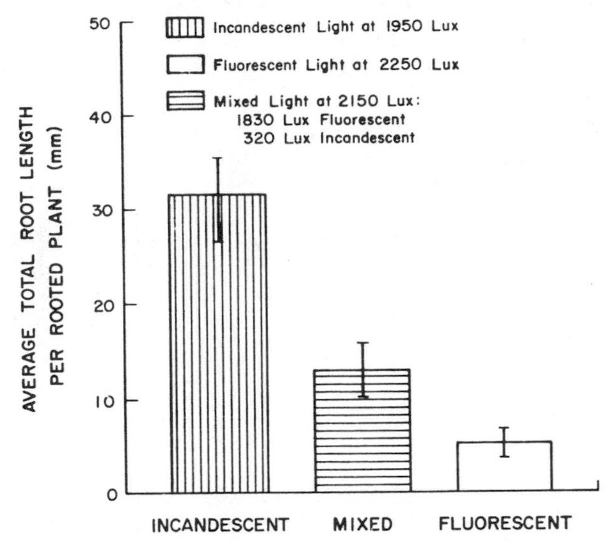

Fig. 4. Root growth in three different 24-hr light regimes (warm-white fluorescent light, warm-white fluorescent/incandescent light, and incandescent light). Data were collected from hypocotyl explants that were derooted and rerooted in vitro. Plants were grown on GD 1/2 medium at 25 ± 2°C, and data were collected after three weeks growth. Error bars represent 95% confidence intervals for root growth means. N = 52 plants per treatment. From Ref. 3, with permission.

FIELD PERFORMANCE

Plantlets produced from cotyledon explants via adventitious shoots and roots can be routinely established in a greenhouse and, after sufficient growth, transferred to the field (3). The first field planting of tissue culture-produced loblolly pine derived from cotyledon explants was established in 1978 in Raleigh, North Carolina. Between 1981 and 1987, more than 3,000 loblolly pine plantlets produced from cotyledon explants of approximately 25 different half-sib families were outplanted by the North Carolina State University Tissue Culture Project, at 17 locations in the southeastern United States. In these research trials plantlets from several half-sib families were planted in paired row plots along with an equal number of seedlings, representing the same half-sib families from which plantlets were obtained. Data from four growing seasons have been collected and analyzed for eight trials established in 1981 (8). These 1981 trials will be the primary focus of this section.

Data for height growth of the tissue culture plantings established in 1981 are shown in Fig. 5. Figure 5A plots the mean overall heights for plantlets and seedlings over four years of growth. At the time of establishment and in each subsequent year, the seedling height significantly exceeded that of the plantlets. Examination of Fig. 5B, which expresses growth as a ratio of plantlet to seedling heights, is helpful in understanding how the 0.66-m differential evident at year 4 (Fig. 5A) occurred. Despite efforts to outplant plantlets and seedlings of equal size, the plantlets were only 83% the height of the seedlings when planted. After one year of growth the plantlets fell further behind, to only 75% the size of seedlings.

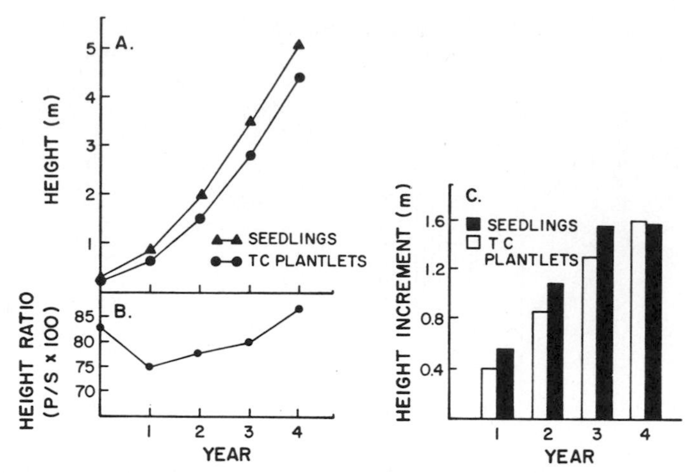

Fig. 5. Four-year height growth data for loblolly pine tissue culture plantlets and seedlings averaged over eight field plantings established in 1981. (A) Mean height growth curves. (B) Plantlet-seedling height ratios. (C) Annual mean height increments. Mean heights of plantlets and seedlings at the time of planting and at all four years were significantly different ($P \leq 0.05$). From Ref. 8, with permission.

In the three subsequent years, the ratio consistently increased over the previous year's value, and after four years of growth plantlets were 87% of the seedling height. If the plantlet-to-seedling height ratio had remained constant from one year to the next, then the percent height increase for plantlets and seedlings would have been equal. However, the increasing ratio seen after year 1 shows that the percent height increase for plantlets actually exceeded that of seedlings in years 2 through 4. Thus, the overall height differential seen at year 4 is largely attributable to differences in initial planting size and to a distinct first-year lag in plantlet growth.

Despite an increasing plantlet-to-seedling height ratio in years 2 through 4, Fig. 5C shows that actual growth increments were smaller for plantlets than for seedlings in years 1 through 3. This resulted in yearly increases in absolute height differentials between seedlings and plantlets in years 1 through 3. However, fourth-year height increments (Fig. 5C) were essentially equal even though the plantlets began the growing season shorter than the seedlings. Equal growth increments have also been observed in sixth-year data obtained from the planting originally established in 1978 (L.J. Frampton, Jr., unpubl. data). Should comparable height increments continue, the absolute height difference would not change, and the percent height difference at rotation age would be insignificant.

The incidence of fusiform rust, an economically devastating disease caused by the fungus Cronartium quercuum (Berk.) Miyabe ex. Shirai f. sp. fusiforme (Cumm.) Burds. et Snow, has also been monitored in the eight plantings established in 1981. When present, rust infection has been consistently lower in plantlets than in seedlings at every location. Six of the plantings are located in high-hazard rust zones where significant amounts of infection routinely occur. Fourth-year infection percentages (incidences) were obtained for five of the six locations and reported (8). Those data are presented in Tab. 8.

Tab. 8. Mean fourth-year incidence of fusiform rust in five loblolly pine field plantings established in 1981. From Ref. 8, with permission.

Location	Fusiform rust infection (year 4) Plantlets		Seedlings
Bogalusa, Louisiana	41		72
Summerville, South Carolina	51	*	82
Yulee, Florida	26	*	66
Jesup, Georgia	57		64
Monroeville, Alabama	48		69
Overall	46	*	71

*Significantly different at $P \leq 0.05$ level as determined by paired T-test.

At two of the five locations the infection percentages were significantly different; overall, the infection rates of 46% for plantlets and 71% for seedlings were also significantly different. Since the plantlets and seedlings at each location are derived from the same half-sib families, it is unlikely that infection differences would be related to genetic differences. Several hypotheses related to the physiology and development of the plantlets have been postulated to account for the infection differences (19), but none of these have actually been shown to be causal.

Although overall height differences, as already noted, can be easily seen between plantlets and seedlings, the plantlets in our field studies do not display grossly abnormal developmental features. However, subtle differences in plantlet and seedling shoot morphology are clearly recognizable. One planting established at Jesup, Georgia in 1981 has been extensively measured to assess differences in plantlet and seedling shoot morphology. Characteristics such as bud length, bud diameter, needle dry weight, number of branches per year, number of growth cycles per year, and other morphological features have been monitored. Based on two-year data, McKeand (18) noted that although plantlets were of embryonic origin, many of their morphological characteristics were more mature-like than comparably aged seedlings at the same location. Fourth-year observations (L.J. Frampton, Jr., unpubl. data) showed that only slight morphological differences existed in three of five traits measured. One can only speculate about the continuation of this mature-like appearance in these plantlets; however, it is noteworthy that when similar morphological traits were measured at years 4 and 6 in the 1978 planting, only one difference was detectable at each time (S.E. McKeand and L.J. Frampton, Jr., unpubl. data).

PERSPECTIVES ON COTYLEDON SYSTEM PROPAGATION

Studies on cotyledon system propagation and the plantlets produced from it in loblolly pine have served good experimental purposes. Basic understanding of organ induction and growth has been obtained, and methodology for establishing plantlets in soil has been elaborated (3,19). Multiyear field data have shown both potential advantages and disadvantages associated with plantlet growth. The experimental value of this system remains high, and field plantings will undoubtedly continue to yield new findings on plantlet growth. However, in loblolly pine the practical application of this system is limited since clones typically contain a small number of plantlets (approximately 20 or less) and by necessity are of juvenile origin.

Alternative methods of propagation are now being developed for loblolly pine. These methods include micropropagation a.f. (micropropagation from axillary or fascicular shoots) and embryogenesis. These methods potentially offer large clone sizes, and in the case of micropropagation a.f. the utilization of mature explants is now possible, although still only experimental. These emerging methods are covered in subsequent sections.

MICROPROPAGATION a.f.

Micropropagation from meristematic cells in leaf axils and/or the apex of needle fascicles has been reported in a number of conifers (see Ref. 7). In loblolly pine, axillary and/or fascicular shoot formation (micropropagation) in vitro has been obtained with explants ranging in age from a few months to 6.5 years (1,20,28).

In our laboratories we began developing micropropagation a.f. procedures in 1986 to propagate from fully mature, 11-year-old explants maintained as grafts in a greenhouse. The procedures developed in this work have allowed us to successfully multiply shoots obtained from mature explants, and these procedures have also been successfully applied to younger specimens ranging in age from one to seven years.

Micropropagation Procedures

Figure 6 presents a schematic outline of the protocol used for mature tree micropropagation. The procedure begins with an extremely severe hedging of the donor (parent) tree. This is done to stimulate the production of shoots exhibiting juvenile morphology with elongated primary needles. The juvenile-appearing shoots are harvested, surface-sterilized, and dipped intact for 1 to 2 min in 440 μM BAP dissolved in 70% ethanol. The ethanol is allowed to evaporate from the shoots and, after receiving a fresh basal cut, they are placed into culture on GD_1 medium containing 0.5 to 1% charcoal and 2% sucrose. Throughout the process, shoots are maintained on this medium at 24 ± 2°C with 16/8 hr photoperiod having a mixture of cool-white fluorescent and incandescent light at 2,000 lux during the 16-hr light phase. Within two to four weeks of the BAP dip, responsive explants will produce swollen buds in many of their leaf axils. Nonresponsive or poorly responsive shoots redipped in BAP may subsequently produce swollen buds.

After six to eight weeks of culture new shoots begin to emerge from swollen buds or small fascicles, and at that time the parent shoots are segmented. These segments with emerging shoots attached are placed on fresh GD_1-charcoal medium to permit further shoot growth (Fig. 7). Once shoots reach a size of approximately 1 to 2 cm, they are individually excised and further elongated on GD_1-charcoal medium. Shoots that elongate to a length of several centimeters may be recycled through the same BAP dip and in vitro procedures to multiply the clone, or presumably they may be rooted, but this step is as yet untested with mature tree shoots.

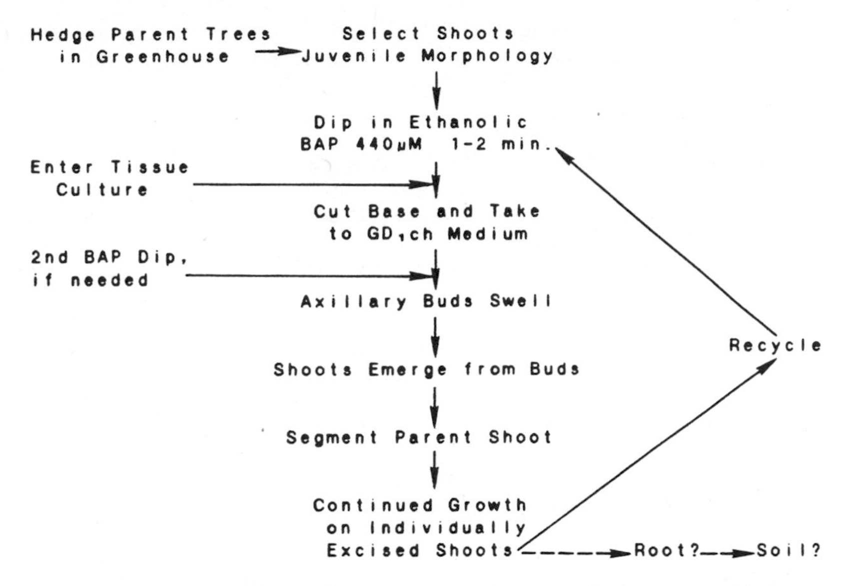

Fig. 6. Schematic outline for mature tree micropropagation protocol. Solid arrows represent steps already achieved. Dashed arrows represent steps not yet attempted with mature tree shoots.

When this system was used in 1986 with four 11-year-old clones, seven of 22 parent shoots, with at least one representative from each clone, successfully produced micropropagated shoots. The physical appearance of the micropropagated shoots was highly variable. All new shoots initially produced primary needles, and some of the shoots elongated to a length of several centimeters, with continued primary needle production. However, many of the shoots that initially formed primary needles elongated only 1 to 2 cm, then produced apical fascicles and little or no subsequent growth. Second- and/or third-generation micropropagated shoots are now being produced from two of the four original clones. Attempts to root shoots derived from 11-year-old plants have not yet been undertaken; however,

Fig. 7. Micropropagated shoots growing attached to a stem segment of an 11-year-old tree. OS = Original stem; NS = new shoots.

second-generation micropropagated shoots derived from five-year-old explants have been rooted (250-μM IBA pulse) and are now in soil.

Regulation of Micropropagation

Our emphasis to date on micropropagation has focused primarily on the development of mature tree propagation procedures, and studies to examine regulation of the process are just now beginning. It appears that the use of charcoal in the medium is very important. In an experiment conducted with one-year-old explants dipped in 440 μM BAP and placed into culture on GD_1 medium plus or minus charcoal, we found that 20 of 30 explants produced new shoots when charcoal was present in the medium. In contrast, only one of 30 explants produced any new shoots when charcoal was absent. Interestingly, all 30 of the shoots on charcoal developed healthy basal callus masses. Shoot bases on the no-charcoal medium were typically rusty-brown in color and none established healthy basal callus, although five of 30 specimens did have a meager amount of brown callus. Given the poor status of explant bases on the no-charcoal medium, it is understandable that a nutritional relationship capable of supporting significant micropropagation would be difficult to establish. This poor basal condition, contrasted with the healthy callused bases formed on medium containing charcoal, could perhaps explain the propagative differences seen here, but this is only speculation.

In 1986, we attempted in vitro micropropagation with four clones from 11-year-old trees and eight clones from five-year-old trees. Many shoots from these clones were nonresponsive, but all 12 clones were successfully micropropagated (produced new shoots) from at least one parent shoot. This suggests that the ability to micropropagate is not permissive vs prohibitive genetics among clones. Rather, it is more likely that the ability to propagate depends on selection of a shoot in an appropriate developmental state, since some shoots within each clone were successfully cultured while others from the same clone or even from the same plant failed. As yet, we cannot define what constitutes an appropriate developmental state for micropropagation, but in general shoots with very juvenile morphology (i.e., very long primary needles) appear to be the most responsive. Considerably more work is needed in this area.

EMBRYOGENESIS

Embryogenesis in tissue-cultured conifers was first obtained in 1985 for _Picea abies_ (15). Since that time, several species, including _Picea glauca_ and _Picea mariana_ (14), _Pinus lambertiana_ (12), and _Pinus strobus_ (S. Wann and M. Becwar, pers. comm.), have been cultured from zygotic embryo explants (inclusive of attached embryonal suspensors) or, in the case of _Larix decidua_ (22), from megagametophyte explants. Most recently, embryogenesis was obtained from immature seeds of _Pinus taeda_ (13), with embryogenic cultures arising from the embryonal suspensor mass (D. Durzan, pers. comm.). That study has carried embryogenesis for loblolly pine all the way through to establishment of plantlets in soil. Also, researchers at the Institute of Paper Chemistry (IPC), Appleton, Wisconsin, have obtained embryogenic loblolly pine cultures from zygotic suspensors attached to excised immature embryos (S. Wann and M. Becwar, pers. comm.; also see Wann et al., poster abstract, this Volume). The procedures used by Gupta and Durzan (13) and at IPC will be briefly summarized and apparently important controls highlighted.

Cultural Protocols

The tissue culture process reported by Gupta and Durzan (13) began with the culture of female gametophytes inclusive of suspensors and young embryos. These explants were obtained from seeds four to five weeks past fertilization, and were cultured on a highly modified, agar-solidified one-half Murashige and Skoog (MS) medium (see Ref. 13 for additional details) containing 50 μM 2,4-dichlorophenoxyacetic acid (2,4-D), 20 μM BAP, and 20 μM kinetin. Within three to four weeks, proliferating embryonal suspensor masses containing proembryos formed in darkness on approximately 10% of the explants. Transfer of the masses to the same medium but with a ten-fold reduction in growth regulators fostered completion of early embryogeny in darkness. Next, globular-stage embryos, transferred to filter paper supports in liquid medium without growth regulators, elongated and developed cotyledons under continuous light. Complete plantlets were formed in a lighted environment when elongated embryos were transferred to one-half-strength basal medium (presumably agar-solidified) with 0.25% activated charcoal.

The culture process currently being developed at IPC has not yet progressed past the original induction of embryogenic cultures, so developmental details are not available. Embryogenic cultures were initiated with a frequency of approximately 5% from zygotic suspensors attached to precotyledonary zygotic embryos excised from the gametophyte. Cultures were obtained in darkness on either BLG medium supplemented with 9 μM 2,4-D and 4.4 μM BAP, or DCR medium with glutamine and casein hydrolysate (12) supplemented with 13.5 μM 2,4-D and 2.2 μM BAP.

Regulation of Embryogenesis

Given the newness of the procedures noted above, regulatory details of loblolly embryogenesis need continued investigation; however, some controlling features have been noted. Gupta and Durzan (13) report that 2,4-D is necessary for the induction and continued proliferation of embryonal suspensor masses.

The presence or absence of light is recognizably important in loblolly embryogenesis. White light inhibits initial culture establishment (Ref. 13; S. Wann and M. Becwar, pers. comm.) but is stimulatory to late-stage embryo development (13). Further evidence that dark initiation conditions should be used is provided by more than 3,000 unsuccessful attempts to establish loblolly embryogenic cultures in a lighted environment using procedures that were otherwise similar to those used at IPC (H.V. Amerson, unpubl. data).

The developmental stage of the explant cultured currently appears to be important. Gupta and Durzan (13) reported successful culture establishment only from seeds that were four to five weeks postfertilization. These should contain immature embryos and healthy suspensors, which would make them very similar to the excised precotyledonary explants used at IPC. Researchers at IPC have been unable to obtain embryogenic cultures from explants once cotyledon development is recognizable (S. Wann and M. Becwar, pers. comm.). However, as methodology improves we may find that older explants of loblolly pine can be successfully used, as is the case for sugar pine (12).

IN VITRO TRAIT SELECTION: FUSIFORM RUST

Fusiform rust is the most economically damaging disease in southern pines (24), commonly infecting loblolly pine and slash pine (Pinus elliottii var. elliottii), the two most widely planted species. At present, the best strategy to combat rust in commercial plantations is planting of resistant stock (25). However, field testing to evaluate resistance in a given family or seedlot is a slow process, requiring at least four to five years, and the slowness thus hinders this strategy. Greenhouse tests are available to evaluate resistance in seedlings of slash pine (29) and loblolly pine (4) in approximately six to seven months. However, these procedures only moderately (r^2=0.6) account for variation in family field resistance expressed in long-term field trials (6,29). Procedures that could potentially yield both rapid and predictive assays for fusiform rust disease resistance in loblolly pine embryos or seedlings cultured in vitro have been reported (9,10,27).

Gray and Amerson (10) evaluated the development of incompatible necrosis in embryos of three control-pollinated families of loblolly pine inoculated in vitro with basidiospores of C. quercuum f. sp. fusiforme, in order to assess resistance. These families had known field performance rust resistance ratings of susceptible, intermediate, and highly resistant. In this study, three parameters of necrosis were histologically examined at 1.5, 7, and 18 days after inoculation. At day 7, the family with high field resistance displayed significantly more necrosis for all three parameters measured than did the susceptible and intermediately resistant families, which did not separate from each other. Thus on the basis of necrosis alone, a single highly resistant family was recognizable in vitro.

In a subsequent study (9), embryos from 24 full-sib families, obtained from a 3 x 8 factorial mating design, were inoculated in vitro with basidiospores of C. quercuum f. sp. fusiforme and evaluated for three measures of necrosis (10) and two additional staining characteristics. Known rust field performance levels (16) for eight half-sib families were regressed on the same eight paternal half-sib family means for each of the five in vitro traits. When the best two or three trait models were considered in the regression, coefficients of determination (r^2) of 0.86 and 0.94, respectively, were obtained. This indicates that evaluation and analysis of just two or three traits could explain 86 to 94% of the variation in field performance for rust resistance among the eight half-sib families. Thus in the limited population studied, in vitro assessment was highly predictive of field performance. This particular study clearly shows that in vitro assessment has the potential to quickly and accurately screen for rust resistance.

Although present histological assessments certainly appear promising, they are labor-intensive, applicable to only a few specimens at one time, and subject to variable interpretations among individual observers. Development of less labor-intensive and more objective measures based on a biochemical evaluation would be highly desirable. One effort in this direction has been made through development of an enzyme-linked immunosorbent assay (ELISA) for the detection of fusiform rust infection in specimens inoculated in vitro (27). This method was used to examine rust infection in embryos of three control pollinated families of loblolly pine having known field resistance ratings of susceptible, intermediate, and resistant. The ELISA evaluations of infection severity showed that the known resistant family was significantly less infected than the susceptible and intermediate families, which were evaluated as equivalent. Thus, as was true of early necrosis evaluations (10), immunological assessment was capable of recognizing a single resistant family in three widely divergent seedlots. This initial study suggests that ELISA methods have potential for development into a rapid and objective screening procedure, but much additional work is needed to make reliable screening a reality.

CONCLUSIONS

Tissue culture-related studies currently in progress with loblolly pine are very diverse and include propagation research, field evaluations of tissue culture-produced plantlets, and in vitro selection. Three methods of loblolly pine tissue culture propagation are now achievable. These include cotyledon system propagation, axillary or fascicular shoot micropropagation, and embryogenesis. The cotyledon system relies upon adventitious shoot production followed by adventitious root formation. Both adventitious shoot and root production occur best in response to growth regulator pulses. Illumination is required for both shoot and root organization and continued growth. However, preshoot cell divisions leading to shoots require light whereas preroot cell division can occur in darkness. The addition of charcoal to shoot elongation media greatly enhances shoot growth, and root growth is significantly improved when the environment contains an incandescent light component.

More than 3,000 loblolly pine plantlets obtained from cotyledon explants are now in the field. Four years of field data showed that plantlets generally were not as tall as seedling comparisons. However, analyses of the growth patterns revealed that height differences were primarily due to differences in starting size and a first-year plantlet growth lag. After year 1, plantlets grew slightly faster than seedlings on a relative (percent height increase) basis, and fourth-year growth increments for plantlets and seedlings were equal in absolute measure (meters) even though plantlets started the growing season smaller than seedlings. Four-year data also revealed that plantlets were significantly more fusiform rust-resistant than seedlings.

Axillary and fascicular shoot in vitro micropropagation methods for loblolly pine are developing with both juvenile and mature tree explants. The procedures begin with juvenile-appearing shoots obtained via hedging, and depend on BAP dips for shoot multiplication. Micropropagated shoots obtained from 11-year-old trees are being recycled to increase their numbers, and micropropagated shoots obtained from five-year-old trees are now rooted and in soil.

Loblolly pine embryogenesis has been obtained from suspensor cells at the base of immature zygotic embryos. The developmental status of the original explant and the initial culture in darkness are recognizably important for successful culture establishment.

Methods for in vitro detection of fusiform rust resistance in loblolly pine seedlings are being developed. An immunological evaluation has identified a highly resistant family in a comparison of three seedlots with widely divergent resistance ratings. A histological study conducted with multiple families showed a very strong correlation between in vitro resistance assessment and known field resistance rankings for the families tested.

REFERENCES

1. Abo-El-Nil, M.M. (1982) Method for asexual reproduction of coniferous trees. U.S. Patent No. 4,353,184.
2. Amerson, H.V., S.E. McKeand, and R.L. Mott (1981) Tissue culture and greenhouse practices for the production of loblolly pine plantlets. In Proceedings of the 16th Southern Forest Tree Improvement Conference, Blacksburg, Virginia, pp. 168-173.
3. Amerson, H.V., L.J. Frampton, Jr., S.E. McKeand, R.L. Mott, and R.J. Weir (1985) Loblolly pine tissue culture: Laboratory, greenhouse, and field studies. In Tissue Culture in Forestry and Agriculture, R.R. Henke, K.W. Hughes, M.J. Constantin, and A. Hollaender, eds. Plenum Press, New York, pp. 271-287.
4. Anderson, R.L., C.H. Young, T.D. Triplett, and T.L. Knight (1982) Resistance Screening Center Procedures Manual: A Step by Step Guide to Materials and Methods Used in Operational Screening of Southern Pines for Resistance to Fusiform Rust, Forest Pest Management Report No. 82-1-18, U.S. Department of Agriculture, Asheville, North Carolina, 55 pp.
5. Brown, C.L., and R.H. Lawrence (1968) Culture of pine callus on a defined medium. For. Sci. 14:62-64.
6. Carson, M.J. (1983) Breeding for resistance to fusiform rust in loblolly pine. Ph.D. Thesis, North Carolina State University, Raleigh, North Carolina, 122 pp.
7. David, A. (1982) In vitro propagation of gymnosperms. In Tissue Culture in Forestry, J.M. Bonga and D.J. Durzan, eds. Martinus Nijhoff/Dr. W. Junk Publishers, The Hague, pp. 72-104.
8. Frampton, Jr., L.J. (1986) Field performance of loblolly pine tissue culture plantlets. In Proceedings of the IUFRO Conference, Williamsburg, Virginia, pp. 547-553.
9. Frampton, Jr., L.J., H.V. Amerson, and R.J. Weir (1983) Potential of in vitro screening of loblolly pine for fusiform rust resistance. In Proceedings of the 17th Southern Forest Tree Improvement Conference, Athens, Georgia, pp. 325-333.
10. Gray, D.J., and H.V. Amerson (1983) In vitro resistance of embryos of Pinus taeda to Cronartium quercuum f. sp. fusiforme. Ultrastructure and histology. Phytopathology 73:1492-1499.
11. Gresshoff, P.M., and C. Doy (1972) Development and differentiation of haploid Lycopersicon esculentum (Tomato). Planta 107:161-170.
12. Gupta, P.K., and D.J. Durzan (1986) Somatic polyembryogenesis from callus of mature sugar pine embryos. Bio/Technology 4:643-645.
13. Gupta, P.K., and D.J. Durzan (1987) Biotechnology of somatic polyembryogenesis and plantlet regeneration in loblolly pine. Bio/Technology 5:147-151.
14. Hakman, I., and L.C. Fowke (1987) Somatic embryogenesis in Picea glauca (white spruce) and Picea mariana (black spruce). Can. J. Bot. 65:656-659.
15. Hakman, I., L.C. Fowke, S. von Arnold, and T. Eriksson (1985) The development of somatic embryos in tissue cultures initiated from immature embryos of Picea abies (Norway spruce). Plant Sci. 38:53-59.
16. Hatcher, A.V., F.E. Bridgewater, and R.J. Weir (1981) Performance level--A standardized score for progeny test performance. Silvae Genet. 30:184-187.
17. Litvay, J.D., M.A. Johnson, D. Verma, D. Einspahr, and K. Weyrauch (1981) Conifer suspension culture medium development using analytical data from developing seeds. Institute of Paper Chemistry Technical Paper Series No. 115, Appleton, Wisconsin, 17 pp.

18. McKeand, S.E. (1985) Expression of mature characteristics by tissue culture plantlets derived from embryos of loblolly pine. J. Am. Soc. Hort. Sci. 110:619-623.
19. McKeand, S.E., and L.J. Frampton, Jr. (1984) Performance of tissue culture plantlets of loblolly pine in vivo. In Proceedings of the International Symposium on Recent Advances in Forest Biotechnology, J. Hanover, D. Karnowsky, and D. Keathley, eds. Traverse City, Michigan, pp. 82-91.
20. Mehra-Palta, A., R.H. Smeltzer, and R.L. Mott (1978) Hormonal control of induced organogenesis: Experiments with excised plant parts of loblolly pine. TAPPI J. 61:37-40.
21. Mott, R.L., and H.V. Amerson (1981) A tissue culture process for the clonal production of loblolly pine plantlets. N.C. Agric. Res. Serv. Tech. Bull. No. 271, 14 pp.
21a. Mott, R.L., and H.V. Amerson (1984) Role of tissue culture in loblolly pine improvement. In Proceedings of the International Symposium on Recent Advances in Forest Biotechnology, J. Hanover, D. Karnowsky, and D. Keathley, eds. Traverse City, Michigan.
22. Nagmani, R., and J.M. Bonga (1985) Embryogenesis in subcultured callus of Larix decidua. Can. J. For. Res. 15:1088-1091.
23. Patel, K.R., and T.A. Thorpe (1984) In vitro differentiation of plantlets from embryonic explants of lodgepole pine (Pinus contorta Dougl. ex. Loud.). Plant Cell Tissue Organ Culture 3:131-142.
24. Powers, Jr., H.R., J.F. Kraus, and H.J. Duncan (1979) A Seed Orchard for Rust Resistant Pine--Progress and Promise, Georgia Forest Research Report No. 1, Georgia Forestry Commission, 8 pp.
25. Schmidt, R.A., H.R. Powers, Jr., and G.A. Snow (1981) Application of genetic disease resistance for the control of fusiform rust in intensively managed southern pine. Phytopathology 71:993-997.
26. Sommer, H.E., C.L. Brown, and P.P. Kormanik (1975) Differentiation of plantlets in longleaf pine [Pinus palustris (Mill.)] tissue cultured in vitro. Bot. Gaz. 136:196-200.
27. Spaine, P.C. (1986) The development and application of an ELISA for fusiform rust disease screening in vitro in loblolly pine seedlings. Ph.D. Thesis, North Carolina State University, Raleigh, North Carolina, 110 pp.
28. Stomp, A.M. (1985) Approaches to regeneration in mature pine tissue: Shoot production from needle fascicles and effect of light on callus growth and organization. Ph.D. Thesis, North Carolina State University, Raleigh, North Carolina, 147 pp.
29. Walkinshaw, C.H., T.R. Dell, and S.D. Hubbard (1980) Predictive Field Performance of Slash Pine Families from Inoculation of Greenhouse Seedlings, U.S. Department of Agriculture Forest Service Research Paper SO-160, 7 pp.

USE OF LEAF PETIOLES OF *HEDERA HELIX* TO STUDY REGULATION OF ADVENTITIOUS ROOT INITIATION

Wesley P. Hackett, Robert L. Geneve,*
and M. Mokhtari**

Department of Horticultural Science
and Landscape Architecture
University of Minnesota
St. Paul, Minnesota 55108

ABSTRACT

Two experimental systems for studying the regulation of root initiation in easy- and difficult-to-root tissues are described. One system utilizes in vitro cultured, debladed petioles of juvenile and mature leaves, and the other involves approach grafting of detached leaves to form cuttings composed of the four reciprocal combinations of juvenile and mature petioles with juvenile and mature lamina. The results of experiments with these systems lead to the following conclusions: (a) the morphogenetic process of root initiation is very different in easy- and difficult-to-root tissues; (b) auxin is required for root initiation in debladed juvenile petioles but stimulates only unorganized cell division in mature debladed petioles; (c) exogenously applied auxin has similar distribution patterns in juvenile and mature debladed petioles; (d) differences in ethylene metabolism do not appear to be causally related to differences in rooting potential in juvenile and mature debladed petioles; (e) root formation is a function of the rooting potential of cells localized in the petiole; (f) there is no evidence of a rooting inhibitor being transported from mature lamina; and (g) a translocatable substance(s) formed in juvenile lamina can either induce root initiation in mature petioles or increase rooting potential of new cells formed as a result of auxin treatment.

INTRODUCTION

There is a loss of adventitious root initiation potential during development of many perennial species from seed. This loss in rooting potential

---

*Present address: Department of Horticulture and Landscape Architecture, University of Kentucky, Lexington, Kentucky.
**Present address: Institut Agronomique et Veterinaire Hassan II, Agadir, Morocco.

is one of the characteristics of a syndrome known as maturation or ontogenetic aging (5). The loss is particularly severe in many long-lived tree species and limits the success or efficiency in clonally propagating desirable mature individuals after the expenditure of much time and effort in the evaluation and selection process. In vitro microcutting propagation as well as conventional cuttage propagation is limited by this decreased potential for rooting (11).

Numerous methods have been used to try to enhance the rooting of cuttings from mature individuals. In most cases, treatments with plant growth substances such as auxins have not been successful in enhancing rooting (1,4,7). The most successful methods are those that apparently retard or reverse the aging or maturation process (5). Even these methods are only successful for some species and not others.

The ultimate solution to the problem of loss of rooting potential with maturation may be found in an understanding of the physiological or morphogenetic basis of this loss. A considerable amount of research has been done in this area. However, the conclusions that can be drawn from much of the research are limited to a large degree by the inadequacies of the experimental systems used (1,4,7). Many approaches have been taken but much of the research focuses on the search for evidence of a positive or negative correlation between endogenous levels of rooting inhibitors or rooting promoters, respectively, with loss in rooting potential; the assumption being that an extractable factor is involved. In almost all cases tissue with high rooting potential has been used to assay for rooting activity of extracts and compounds (8,9,12). The use of tissue with high rooting potential may be a sensitive and specific assay for attempting to relate endogenous levels of rooting inhibitors to loss of rooting potential (12), but this method is much less sensitive and specific for studies of levels of endogenous rooting promoters in relation to rooting potential (4). This is because it is not known whether the same factors control rooting in easy- and difficult-to-root tissues (9). For greatest specificity the assay tissue should be from the species from which the extracts were made (1). The exclusive use of easy-to-root tissue as an assay system does not take into account the possibility that loss in rooting potential may be related to some nonextractable factor (4,9).

The nearly universal use of easy-to-root tissue as a bioassay for rooting activity is based mainly on the ease and speed with which experiments may be accomplished. Based on the above discussion, two important and minimal criteria for an experimental system to study loss of rooting potential are: (i) use of both juvenile easy-to-root and mature difficult-to-root tissues, and (ii) ease and convenience of use and speed of rooting.

Mullins (11) and his co-workers have developed such a system using in vitro subcultured microcuttings derived from mature shoot apices of apple clones. These microcuttings have varying degrees of rooting potential depending on the number of times they have been subcultured and thereby provide direct comparison of root initiation in easy- and difficult-to-root tissues of the same genotype. In my laboratory, we have been developing experimental systems involving the dimorphic species *Hedera helix* (3,10). In this species, the juvenile seedling, nonflowering stage has a phenotype that is very different from the mature, flowering stage phenotype (6). Characteristics that differ are morphological ones such as leaf shape and phyllotaxis, and physiological-biochemical ones such as flowering potential, stem anthocyanin accumulation, and adventitious root initiation

(ARI) potential. These distinct phenotypes are very stable and can coexist on the same plant or through an asexual method of propagation (stem cuttage) as separate individual plants (5,6). Thus, through use of asexual propagation, the same genotype can be maintained as separate individuals with very different phenotypes for long periods of time (years).

The ARI potential of the juvenile and mature phenotypes is so different that stems of the juvenile form have aerial roots at each node while those of the mature form have no aerial roots. This difference in ARI potential is reflected in the ease with which stem cuttings of the two phenotypes can be rooted. Hess (8) found that auxin-treated juvenile stem cuttings root at a 100% frequency while mature stem cuttings root at a 16% frequency. Recently, we have discovered that detached, intact leaves and delaminated petioles of leaves of the two-phase phenotypes reflect as great or greater differences in rooting potential as the juvenile and mature stems from which they were detached. In the remainder of this chapter I will describe how we have used these two systems to characterize root initiation in easy- and difficult-to-root tissues.

DELAMINATED LEAF PETIOLES IN VITRO

Petioles from mature and juvenile stock plants were selected for assay use when the new lamina were determined to be fully expanded, The lamina were immediately detached from the petioles and the debladed petioles were maintained in distilled water until prepared for assay use. After rinsing in distilled water, petioles were surface-sterilized by agitation in a solution of 0.5% sodium hypochlorite from household bleach and 0.1% Alconox detergent for 10 min. They were then rinsed three times with autoclaved deionized water. Each petiole was aseptically trimmed to a standard length of 2.3 cm prior to aseptic culture on filter paper saturated with the nutrient medium (4) in a 25-ml Erlenmeyer flask. Each flask was fitted with a translucent plastic retainer to hold the petioles upright. The plastic retainer was made by cutting a 4 cm x 2 cm segment from polypropylene Nalgene bottles. Five 1/8-inch diameter holes were drilled into the central portion of the plastic segment. The plastic segment was folded at a right angle 7 mm from each end of its long dimension to form a bridge shape (Fig. 1). The plastic retainer was forced into the 25-ml Erlenmeyer flask. Whatman #1 filter paper was cut into 2.5-cm^2 pieces and three pieces were inserted under the retainer in each Erlenmeyer flask. A 1-ml aliquot of modified Romberger medium (4) including the test component was added to each flask. Each Erlenmeyer flask was closed with an aluminum foil cap and autoclaved at 121°C. Delaminated petioles were aseptically inserted into the retainer with the morphological base of the petiole making contact with the medium-saturated filter paper (Fig. 1).

The Erlenmeyer flasks were placed in a growth chamber maintained at a constant day/night temperature of 21°C and approximately 80% relative humidity. A day length of 16 hr of light and 8 hr of darkness was maintained with fluorescent lights giving a quantum flux density of 200 $\mu moles \cdot sec^{-1} \cdot m^{-2}$ at the level of the flasks. The experiments were terminated after 18 days when adventitious roots were visible through the epidermis in juvenile petioles (Fig. 1).

Dose Response to Auxin

Juvenile. Both naphthaleneacetic acid (NAA) and indoleacetic acid (IAA) stimulated adventitious rooting in juvenile petioles. The average

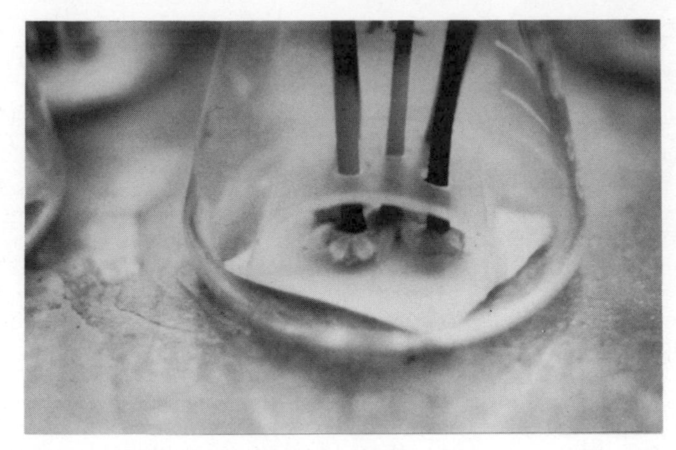

Fig. 1. Debladed petiole rooting system. Shown is the plastic retainer bridge holding rooted juvenile petioles treated with 100 μM NAA.

number of roots increased with increasing concentration of auxin from 10 to 100 μM. There was no stimulation of rooting at 0, 0.1, or 1 μM concentrations of auxin. Concentrations of both IAA and NAA that were greater than 100 μM depressed rooting, and at a concentration of 1,000 μM no observable rooting occurred. Naphthaleneacetic acid was more stimulatory to rooting than IAA at any given concentration.

Mature. Mature petioles did not form roots in response to auxin. During the course of repeated experiments with mature petioles treated with NAA at 100 μM, only 15 of 815 petioles (1.8%) showed any capacity to form adventitious roots, and in these cases only one or two roots formed per petiole.

Histological Observations

Juvenile. The first cell divisions in response to auxin occurred at day 6 and were first visible in the epithelial cells associated with ducts that were adjacent to the vascular bundles. Observations at day 9 showed localized cell division in the inner cortical parenchyma cells associated with the vascular bundle. By day 12, a root primordium with a well-defined meristem and vascular system was observable. Cell divisions were also initiated in the outer cortical parenchyma cells basipetal to the plane of cell divisions involved in root initiation. These divisions were not directly involved in root primordia formation. During the assay period, no cell division was apparent in juvenile petioles that were not treated with NAA.

Mature. Mature petioles initiated root primordia at a very low frequency, but they did show a cell division response to auxin. The first cell division was observable at day 6 in the epithelial cells associated with vascular bundles. By day 9, unorganized cell division had occurred throughout the cortical parenchyma cells. The cell divisions appeared to lack orientation and did not result in an organized meristem or subsequent root initial. No cell divisions were evident in mature petioles not treated with NAA. Histological investigation of 18-day-old mature petioles that did initiate root primordia revealed that the root initial lacked vascular connections to the vascular bundle and appeared to originate from cell divisions in cortex-derived callus.

Naphthaleneacetic Acid Uptake and Distribution in Petioles

There was no difference in the uptake and localization of NAA within juvenile and mature petioles expressed on a dry weight or per segment basis. The amount of ^{14}C from NAA within the petiole increased with increasing exposure time. More than 90% of the NAA was localized in the basal one-third of the juvenile and mature petioles.

Ethylene Metabolism and Adventitious Root Initiation

Geneve (3) has shown that there are significant differences in the time course of ethylene evolution from petioles of the two forms treated with the optimal concentration of NAA for rooting of the juvenile petioles. Ethylene evolution from juvenile petioles subsided from a maximum level at 24 hr after treatment to near control (nontreated) petiole levels after six days. Ethylene evolution in mature petioles was similar to that in juvenile petioles after 24 hr. In contrast to juvenile petioles, evolution in mature petioles continued to increase slowly over the 14-day experimental period. The correlation of the timing of subsidence of ethylene evolution in juvenile petioles with the period when root initials were being formed suggested that root initiation potential might be related to ethylene synthesis or action.

Numerous methods, including treatments with ethylene or ethylene precursors, inhibitors of ethylene synthesis or action, and ethylene scrubbers, were used to test the hypothesis that ethylene metabolism or action was causally involved in the difference in rooting potential of petioles of the two phases. The results of these experiments are summarized in Tab. 1 and 2. Note that elevated ethylene levels from exogenous treatments inhibited rooting of juvenile petioles when pulsed at days 6 to 9 but had no effect when pulsed at days 1 to 3 or 3 to 6, indicating that ethylene inhibits root initial outgrowth but not root initiation. Also note that

Tab. 1. Influence of ethylene (C_2H_2), ethephon, and 1-aminocyclopropane-1-carboxylic acid (ACC) on the rooting response of juvenile and mature *H. helix* petioles with and without auxin treatment.

		ACC	Ethephon	C_2H_2 Pulse
Juvenile*	Auxin	-	-	- Days 6-9 0 Days 1-3 or 3-6
	No auxin	0	0	0
Mature**	Auxin	0	0	0
	No auxin	0	0	0

*Auxin- and nonauxin-treated juvenile control petioles formed 8-10 and 0 roots, respectively.
**Auxin- and nonauxin-treated mature control petioles formed 0 roots.
Key: -, Inhibition of rooting; 0, no effect on rooting.

Tab. 2. Influence of inhibitors of ethylene synthesis (AVG) and action (NDE and silver) and an ethylene scrubber ($KMNO_4$) on the rooting responses of juvenile and mature H. helix petioles with and without auxin treatment.

		AVG	AVG + C_2H_2	AVG + ACC	NDE	Silver	$KMNO_4$ Scrub
Juvenile*	Auxin	–	–	–	0	0	0
	No auxin	0	NA	NA	0	0	0
Mature**	Auxin	0	NA	NA	0	0	0
	No auxin	0	NA	NA	0	0	0

*Auxin- and nonauxin-treated juvenile control petioles formed 8-10 and 0 roots, respectively.
**Auxin- and nonauxin-treated mature control petioles formed 0 roots.

Key: ACC, 1-aminocyclopropane-1-carboxylic acid; AVG, aminoethoxyvinylglycine; C_2H_2, ethylene; NDE, 2,5-norbornadiene; silver = silver thiosulfate; -, inhibition of rooting; 0, no effect on rooting; NA, not applied.

aminoethoxyvinylglycine (AVG), an inhibitor of 1-aminocyclopropane-1-carboxylic acid (ACC) and ethylene biosynthesis, inhibited rooting of juvenile petioles, and that the inhibition could not be counteracted by treatment with ACC or ethylene. This indicates that the inhibitory effect of AVG on rooting is a nonspecific effect and not the result of the reduced ethylene and ACC levels that were observed. We found no evidence that reduction of ethylene levels or interference with ethylene action would promote rooting in mature petioles. We also found that ethylene treatment of either juvenile or mature petioles not treated with auxin had no effect on rooting or cell division. We therefore rejected the hypothesis that ethylene metabolism or action was causally involved in the difference in rooting potential of petioles of the two phases. We also concluded that auxin promotion of rooting in juvenile petioles is not mediated through its induction of ethylene synthesis.

The work discussed above on the relation of ethylene metabolism to root initiation illustrates the importance of using both easy- and difficult-to-root tissues in order to confirm or reject the causal significance of correlative data on a given parameter (ethylene in this case) in relation to control of root initiation. Table 3 summarizes the responses of delaminated juvenile and mature petioles to auxin.

RECIPROCALLY GRAFTED DETACHED LEAVES AS CUTTINGS

Leaves from mature and juvenile stock plants were selected for use when the lamina were determined to be fully expanded. These detached leaves could be used intact as cuttings or as components of composite cuttings formed by approach grafts (Fig. 2). When used for approach grafting, appropriate lamina or petioles were excised after unions were anatomically complete (15 days) to give all four combinations of juvenile and mature petioles with juvenile and mature lamina. During graft healing

Tab. 3. Summary of the response of delaminated juvenile and mature *H. helix* petioles to auxin.

Morphogenetic stimulus	Juvenile Auxin			Mature Auxin		
Physiological and morphogenetic responses	Polar auxin distribution	Transient C_2H_4 evolution	Cell division ↓ Root initiation	Polar auxin distribution	Continued C_2H_4 evolution	Cell division ↓ Callus ↓? Root initiation

and the experimental period, the leaf cuttings were maintained in damp vermiculite, in a high humidity environment in a greenhouse at a minimum temperature of 21°C. Experiments were initiated by removing the basal 0.5 cm of petiole and dipping the cut end in an auxin (IBA, 2,000 mg/l) or control solution for 5 sec. Treated cuttings were returned to the same environmental conditions.

Mokhtari (10) used such composite cuttings consisting of all four combinations of juvenile and mature petioles as receivers and juvenile and mature lamina as donors to obtain evidence regarding the possible translocation of substances involved in the control of root initiation and/or the acquisition of rooting potential. Auxin-treated juvenile petioles with either juvenile or mature lamina were maximally rooted in 24 days with a mean of 12 roots/petiole. At 24 days, auxin-treated mature petioles with either juvenile or mature lamina had not rooted. However, 21 days later, 100% of the mature petioles with juvenile lamina had rooted, with a mean of eight roots per petiole, while mature petioles with mature lamina exhibited only callus formation. Microscopic observation showed that root initials in graft combinations involving juvenile petioles form directly from cortex parenchyma, while those in graft combinations involving mature petioles with juvenile lamina form only in callus.

These data strongly suggest that root initiation is a function of rooting potential of cells localized in the petiole and that, at least initially, root initiation is affected little by the type of lamina. That is, there is no evidence of a rooting inhibitor being translocated from mature lamina. However, the results also indicate that with time newly formed (and apparently dividing) cells in callus at the base of mature petioles with juvenile lamina have an increased rooting potential as compared with cells in callus at the base of mature petioles with mature lamina. This increase in rooting potential of apparently dividing cells in response to a juvenile lamina may be similar to rejuvenation in terms of morphological and physiological characteristics of mature *H. helix* shoot scions when grafted onto leafy juvenile rootstocks (2).

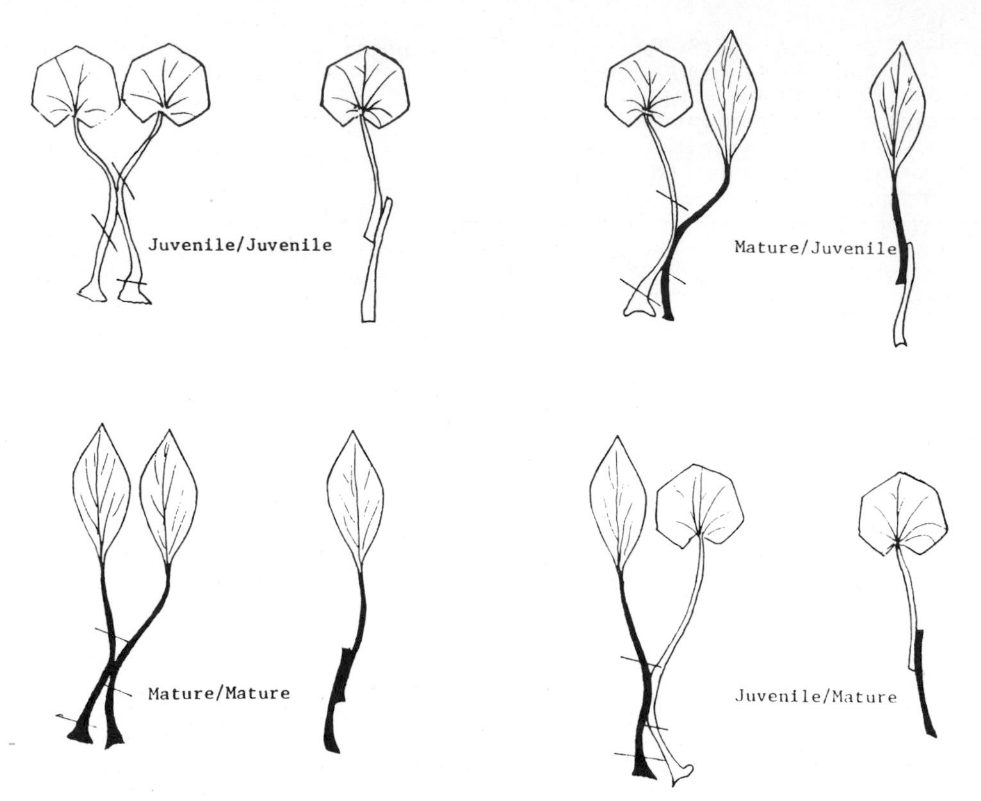

Fig. 2. Schematic representation of the detached leaf approach grafting procedure used to obtain cuttings with reciprocal combinations of juvenile and mature petioles and juvenile and mature lamina.

CONCLUSIONS

The results of experiments with these systems lead to the following conclusions: (a) the morphogenetic process of root initiation is very different in easy- and difficult-to-root tissues; (b) auxin is required for root initiation in debladed juvenile petioles but stimulates only unorganized cell division in mature debladed petioles; (c) exogenously applied auxin has similar distribution patterns in juvenile and mature debladed petioles; (d) differences in ethylene metabolism do not appear to be causally related to differences in rooting potential in juvenile and mature debladed petioles; (e) root formation is a function of the rooting potential of cells localized in the petiole; (f) there is no evidence of a rooting inhibitor being transported from mature lamina; and (g) a translocatable substance(s) formed in juvenile lamina can either induce root initiation in mature petioles or increase rooting potential of new cells formed as a result of auxin treatment.

The in vitro cultured, delaminated petioles of the juvenile and mature phases of H. helix have several characteristics important for the experimental analysis of the morphogenetic, physiological-biochemical, and genetic basis of root initiation and root initiation potential. These characteristics include: (a) high and low rooting potential in tissues having very similar physiological ages and identical anatomical organization and genetic make-up; (b) relatively simple, uniform, fully differentiated tissue systems in which endogenous sources of plant growth substances are minimized; (c) easy in vitro manipulation of environment and precise provision of nutrients and plant growth substances as pulses or continuous feed from proximal or distal ends; (d) relatively rapid, specific, and morphogenetically distinct responses to auxin; and (e) ready availability of tissues in large amounts on a year-round basis.

Delaminated mature petioles also provide an easy-to-use bioassay for discovery of natural or synthetic substances that (a) promote root initiation in difficult-to-root tissue, or (b) cause a change in rooting potential of newly formed cells. Delaminated juvenile petioles likewise provide a bioassay for discovery of substances that inhibit root initiation. This system is particularly suitable as a bioassay since it is done aseptically, requires only 1 ml of solution per assay, and involves only minimal endogenous sources of active substances.

The reciprocally grafted, detached leaves of _H. helix_ have provided evidence that a translocatable substance(s) formed in juvenile lamina can either induce root initiation in difficult-to root mature petioles or increase the rooting potential of new cells formed as a result of auxin treatment. This provides an impetus for trying to isolate and characterize such a substance(s) from juvenile lamina. The in vitro debladed mature petiole bioassay system provides a means to detect biological activity as the isolation and purification procedure is carried out.

REFERENCES

1. Batten, D.J., and P.B. Goodwin (1978) Phytohormones and related compounds--A comprehensive treatise. In _Phytohormones and the Development of Plants_, Vol. II, D.S. Letham, P.B. Goodwin, and T.J.V. Higgins, eds. Elsevier/North-Holland Biomedical Press, Amsterdam, pp. 137-173.
2. Doorenbos, J. (1954) Rejuvenation of _Hedera helix_ in graft combinations. _Proc. Koninkl. Ned. Akad. Wetenschap_ (Ser. C) 57:99-102.
3. Geneve, R.L. (1985) The role of ethylene in adventitious root initiation in debladed petioles of the juvenile and mature phase of _Hedera helix_ L. Ph.D. Thesis, University of Minnesota, St. Paul, Minnesota.
4. Hackett, W.P. (1970) The influence of auxin, catechol and methanolic extracts on root initiation in aseptically cultured shoot apices of the juvenile and mature forms of _Hedera helix_. _J. Am. Soc. Hort. Sci._ 95:398-402.
5. Hackett, W.P. (1985) Juvenility, maturation and rejuvenation in woody plants. _Hort. Rev._ 7:109-155.
6. Hackett, W.P., R.E. Cordero, and C. Srinivasan (1987) Apical meristem characteristics and activity in relation to juvenility in _Hedera_. In _Manipulation of Flowering_, J.G. Atherton, ed. Butterworths, London, pp. 93-99.
7. Haissig, B.E. (1986) Metabolic processes in adventitious rootings of cuttings. In _New Root Formation in Plants and Cuttings_, M.B. Jackson, ed. Martinus Nijhoff Publishers. Dordrecht, The Netherlands, pp. 141-189.
8. Hess, C.E. (1964) Naturally occurring substances which stimulate root initiation. In _Regulateurs Naturels de la Croissance Végétale: Fifth International Conference on Plant Growth Substances_, J.P. Nitsch, ed. C.N.R.S., Gif-sur-Yvette, France, pp. 517-527.

9. Heuser, C.W. (1976) Juvenility and rooting cofactors. Acta Hort. 56:251-259.
10. Mokhtari, M. (1985) Use of petiole grafting to test translocatable factors influencing root initiation in Hedera helix (Araliaceae). M.S. Thesis, University of Minnesota, St. Paul, Minnesota.
11. Mullins, M.G. (1985) Regulation of adventitious root formation in microcuttings. Acta Hort. 166:53-61.
12. Paton, D.M., R.R. Willing, W. Nicholls, and L.D. Pryor (1970) Rooting of stem cuttings of Eucalyptus: A rooting inhibitor in adult tissues. Austral. J. Bot. 18:175-183.

NODULE CULTURE: A DEVELOPMENTAL PATHWAY WITH HIGH POTENTIAL FOR REGENERATION, AUTOMATED MICROPROPAGATION, AND PLANT METABOLITE PRODUCTION FROM WOODY PLANTS

Brent H. McCown,[1] Eric L. Zeldin,[1]
Hamilton A. Pinkalla,[2] and Richard R. Dedolph[3]

[1]Department of Horticulture
University of Wisconsin
Madison, Wisconsin 53706

[2]Bell Flavors and Fragrances
Northbrook, Illinois 60062

[3]Gravimechanics Corporation
Naperville, Illinois 60540

ABSTRACT

Nodules are independent, spherical, dense cell clusters which form a cohesive unit and display a consistent internal cell/tissue differentiation. At a minimum, three cell types (meristematic cells, plastid-dense parenchyma, and vascular elements) and two cell layers (epidermal and internal cortex/vascular) can be distinguished in nodules of poplar. Although nodules have been randomly observed by many researchers working with a myriad of plant species, they most commonly are seen in cultures of woody plant species being differentiated from dedifferentiated cells. We have developed liquid culture systems in which nodules are the predominant structures. Such cultures grow via nodule enlargement (a three-stage process) and nodule multiplication (via two general pathways). In general, nodules display a high capacity for plant/organ regeneration via organogenesis. The nodular developmental pathway parallels that of the embryogenic developmental pathway; a theoretical comparison of the two pathways as bridges between totipotency and competence is discussed. Although the research is still very preliminary, nodule cultures have apparent applications in regeneration strategies, automated micropropagation, and in vitro phytochemical production.

INTRODUCTION

Over the past several decades, our laboratory has been developing microculture systems for horticultural and forestry crops, particularly woody species (15). When attempting to differentiate cells (calli and suspension cultures) or organs (leaves, roots, and internodes), occasionally we would observe the formation of dense cell clusters that were highly meristematic (actively growing by cell division), while appearing to be well-differentiated (plastid-dense, vascularized, and multilayered). We were certainly not the only nor even the first research group to observe such structures. For example, herbaceous plants such as carrots ("vascularized nodules") (6,7,8,21) and daylily ("nubbins") (11), as well as woody plants such as citrus ("nodular callus") (4), spruce ("nodules") (1), and pine ("meristematic tissue") (J. Aitken-Christie et al., this Volume), have all been recorded to produce such dense cellular masses. In reality, most any group actively engaged in cell differentiation work has seen similar phenomena. However, it was not until Ms. Julie Russell in our laboratory perfected a procedure for differentiating shoots from protocalli of *Populus* (16) that we began to give much more serious attention to these structures. The reason for this new awareness was that in the vast majority of cases, shoot differentiation from protocalli proceeded through an intermediate stage that was typified by these structures. In addition, Eric Zeldin, working on a project involving mass cell cultures of woody species grown in bioreactors, also questioned the potential importance of these formations after having induced them in several treatments.

Many researchers have termed these structures "nodules." Realizing this precedence, the appropriateness of the definition of the term [a small mass of rounded or irregular shape (24)], and the superficial analogy to both mineral nodules in geology and root nodules of legumes, we propose that this term be formally adopted to describe these structures. This chapter will detail some of the characteristics of nodules; however, for clarity and to distinguish nodules from other cell aggregates, we propose that nodules be anatomically characterized by the following:

(a) Dense cell masses which are independent and which form spherical, cohesive units. That is, the cells in an individual nodule will not fall apart even with rather strong physical disruption (except that cells may readily slough off the surface layers or the nodule may break up into smaller nodules).

(b) Prominent tissue differentiation is present, with at least two cell layers and three cell types being distinguished.

(c) Vascularization is present except in the first stages of nodule development.

We further propose that cultures that consist predominately of nodules be termed "nodule cultures" (14).

As will hopefully be made clear in this chapter, nodule cultures are as distinct a cultural system as are suspension, callus, and shoot cultures. As importantly, nodules themselves may represent a morphogenetic pathway of importance equal to that of embryogenesis.

IN VITRO CULTURE ASPECTS

Establishment and Initial Culture of Nodules

Figure 1 gives a generalized scheme that has proven successful for the establishment of nodule cultures with Populus and a number of other woody plant species. For reproducibility and speed of response, tissues taken from stabilized shoot cultures (16) are the most appropriate (13). In addition, reliance on this source allows the repeatable reestablishment of nodule cultures as needed.

We have tended to minimize the total time that tissues are kept in the callus or suspension stages. With poplar, the total time for callusing and suspension culture establishment can be less than two months. Not only does this give a more reproducible response, but the development of genetic and/or epigenetic aberrations should be minimized.

From the suspensions through the induction of shoot differentiation, a roller-bottle-type culture system is employed. Because the cells and cell clusters stick to the liquid film on the surface of the bottles (Fig. 2), as the bottles turn the tissues are alternatively bathed in medium and air. This may promote growth in ways analogous to Steward's nipple flasks which were used so successfully for carrot embryogenesis (22). As importantly for nodule cultures, the rolling action tends to promote clumping of cells in spherical aggregations and thus may actually promote nodule formation (although we have little direct evidence for this assumption).

We have not done extensive analysis of the environmental variables necessary for nodule formation. With poplar, cultures are grown under 24-hr fluorescent lighting and at temperatures of 25 to 28°C.

One of the important variables is the hormone level in the medium. At least with poplar, the cultures are more responsive to the auxin [naphthaleneacetic acid (NAA)] levels than to the cytokinin [benzyladenine (BA)] levels. The specific levels vary widely with both the species and

STABILIZED SHOOT CULTURE
|
INTERNODE, LEAF, ROOT CALLUS ON SEMI-SOLID MEDIUM (HIGH AUXIN/CYTOKININ)
|
SHORT-TERM SUSPENSION CULTURE
|
SCREEN SUSPENSION AND SUBCULTURE NODULAR CLUMPS (IN LIQUID)
|
NODULE PRODUCTION (LOWER AUXIN/CYTOKININ RATIO) (IN LIQUID)
|
SUBCULTURE AS NEEDED USING UNIFORMLY SIZED NODULES
|
INDUCTION OF SHOOT DIFFERENTIATION (POSSIBLY WITH PULSE OF THIDIAZURON)
|
SHOOT DEVELOPMENT ON SOLID-SUPPORT MEDIUM (NO HORMONE)
|
ADVENTITIOUS ROOTING (POSSIBLY WITH AUXIN PULSE)
|
PLANTLET DEVELOPMENT AND ACCLIMATION

Fig. 1. A generalized schematic for production of nodule cultures and the subsequent shoot organogenesis and utilization for mass micropropagation.

Fig. 2. Roller bottles used for suspension and nodule cultures (suspension cultures shown here). Bottles are rolled at about 1 rpm so that a film of liquid is maintained on the surfaces; plant cells then are entrapped in this film and are pulled around as the bottle turns, thus alternately immersing them in medium and air.

the specific genotype. The general strategy that we adopt is to reduce the auxin concentrations to levels below that necessary to maintain loose suspension cultures, while at the same time maintaining levels that are adequate for the general stimulation of cell division and growth. With the Populus hybrid NC5339, this is around 10 µM NAA (Tab. 1).

Developmental Sequence and Anatomy of Nodules

Poplar nodules appear to have at least three developmental stages. The first stage is characterized by small (less than 2 mm diameter) cell

Tab. 1. The response of nodule cultures to the naphthaleneacetic acid (NAA) concentration in the culture medium.

NAA Level (µM)	Refractive index	Wet weight (g)	Number of nodules	Number of loose cells
1	1.00 (0.01)	0.6 (0.02)	20.7 (1.2)	Few
4	0.9 (0.06)	1.2 (0.11)	20.0 (1.1)	Few
10	0.7 (0.02)	3.2 (0.06)	139.7 (77)	Few
40	0.01 (0.01)	9.6 (0.46)	189.3 (134)	Many

Note: Cultures were started with a 20-nodule initial charge and grown in a liquid WPM-based medium with 0.4 µM BA. Data are means with S.E. with three replicates per treatment. The tissue was Populus hybrid NC 5339.

clusters. In cross-section, such clusters do not show prominent internal tissue differentiation but have a diffuse surface which is dominated by linear arrays of rapidly dividing cells (Fig. 3). This stage will be present any time nodules are forming from undifferentiated cells, as from suspensions or from cells sloughing off of other nodules.

In appropriate media, such nodules will continue to increase in size and will rapidly differentiate a more defined internal structure. Prominent tissues will be a central area of vascularization ("unicenter nodules") (Fig. 4 and 5), an epidermal layer that consists of either dense, highly phenolic cells (Fig. 4) or linear arrays of rapidly dividing cells (Fig. 3), and an intervening cortical-like area of highly plastid-dense cells (Fig. 5). Often, an area of parenchymatous cells having a much lower density of plastids will develop immediately around the vascularized center and internal to the plastid-dense layer (Fig. 6).

The third stage of development is typified by nodules that have multiple internal centers of vascularization ("polycenter" nodules) (Fig. 7). Such nodules will have a diversity of developmental fates depending on the medium/environment in which they are grown. Nodules can remain whole and continue to increase in size to what appears to be an almost indefinite limit (Fig. 8). In such "meganodules," cell division occurs around the vascularized centers and on surfaces of voids created by differential expansion of the nodules as a result of such cell division (Fig. 9).

Polycenter nodules can also be induced to form other independent nodules. This may be either by a break-up of the nodule itself by splitting off from the vascularized centers at the interface of dividing cells produced around each center (Fig. 7), or by the development and budding off of small, unicenter nodules at the surface of the larger "mother" nodule (Fig. 10).

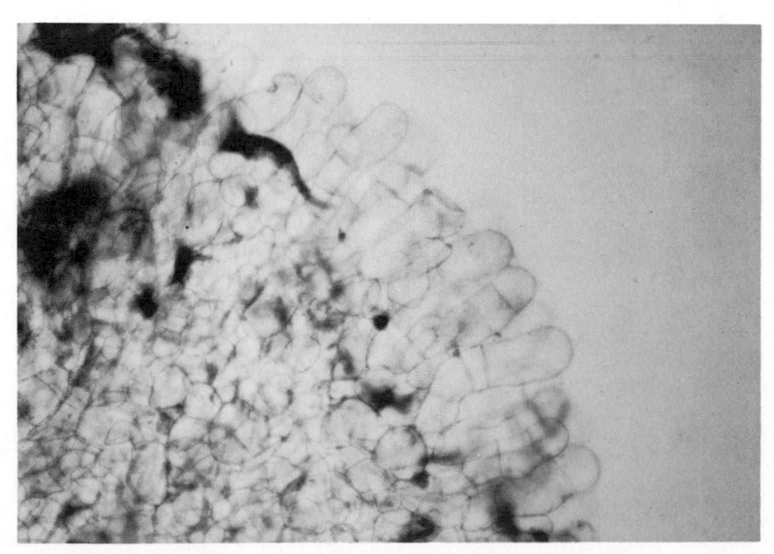

Fig. 3. A fresh, unstained cross-section of the surface of a young nodule showing linear arrays of rapidly dividing cells. Such cells can slough off and generate new, independent nodules.

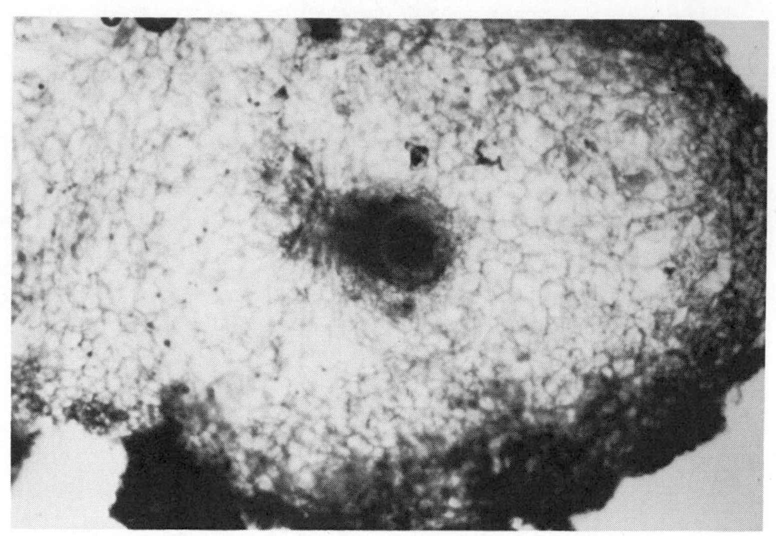

Fig. 4. A fresh, unstained cross-section of a newly formed nodule. Note the vascular elements just beginning to form in the center of the nodule. This nodule has an epidermal layer composed of dense, phenolic-rich cells.

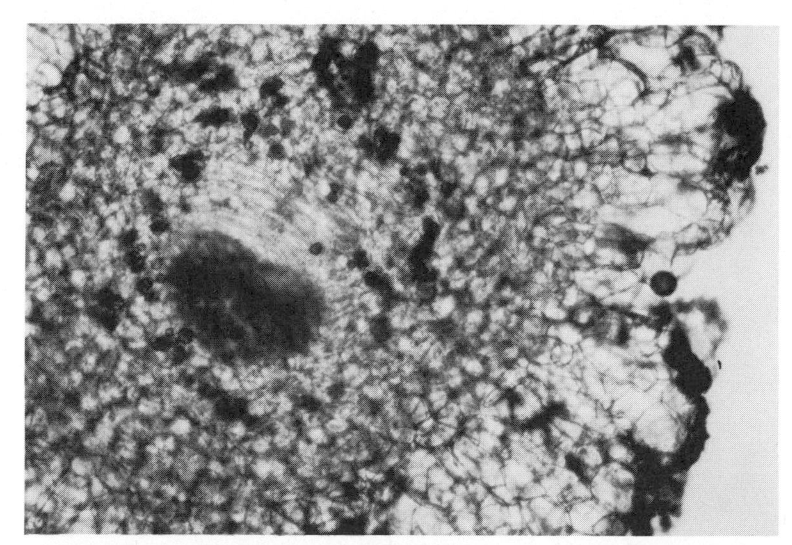

Fig. 5. A fresh, unstained cross-section of a nodule with a clearly developed vascularized center ("unicenter nodule"). This nodule has an epidermal layer that is in rapid division and expansion. Note the intervening cortical-like layer which is composed of plastid-rich cells.

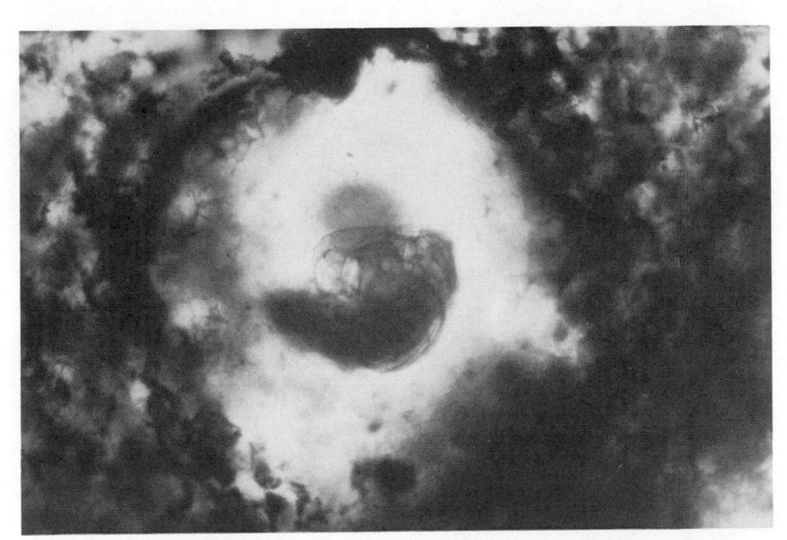

Fig. 6. A close-up of the center of a unicenter nodule showing the layer of visually undifferentiated parenchyma cells around the vascularized center. This fresh cross-section was stained with phloroglucinol to emphasize the tannin/lignified cells.

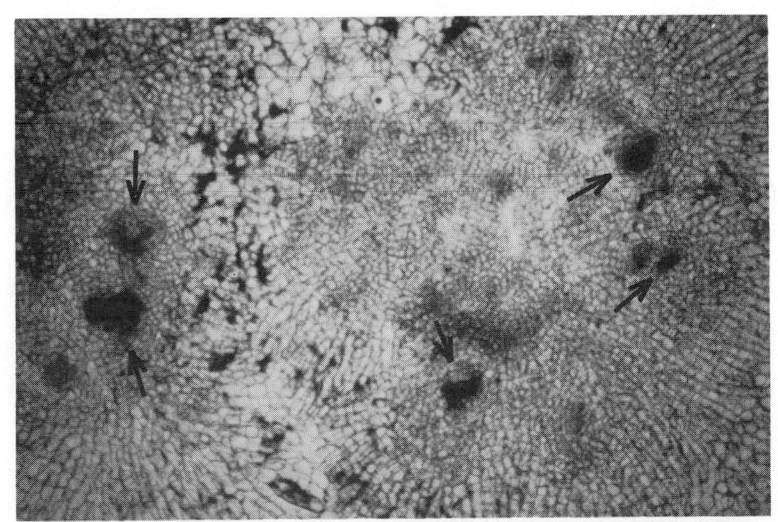

Fig. 7. A fresh, unstained cross-section of a mature nodule showing multiple vascularized centers ("polycenter nodule"). Note the layers of cells radiating around each vascularized center (arrows); these cells are in active division and are thus causing the nodule to expand in diameter. Under some culture conditions, the nodule will break up along the surfaces where the dividing cell layers from different centers meet, thus multiplying the numbers of independent nodules.

Fig. 8. A set of freshly harvested nodules that have been allowed to increase in size but not multiply. Such large nodules are termed "meganodules" and will have a complex internal structure (as in Fig. 7).

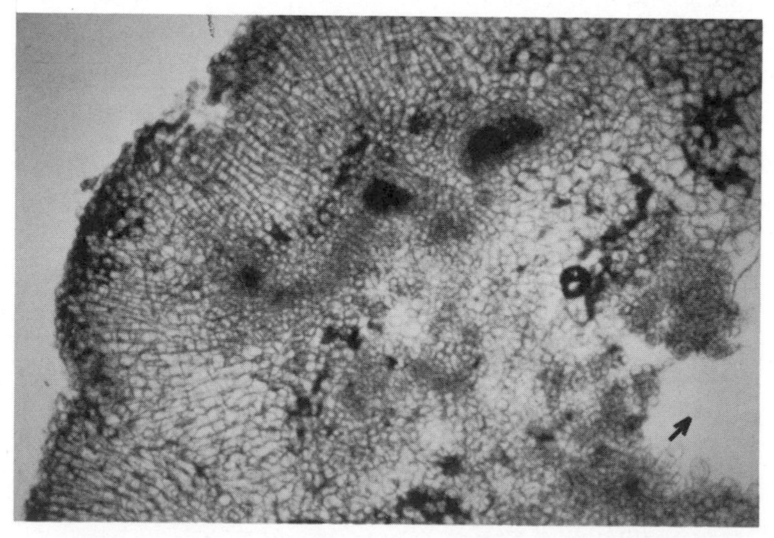

Fig. 9. A fresh, unstained partial cross-section of a meganodule showing the open center (arrow) created by the expansion of the surrounding layers of cells and vascularized centers.

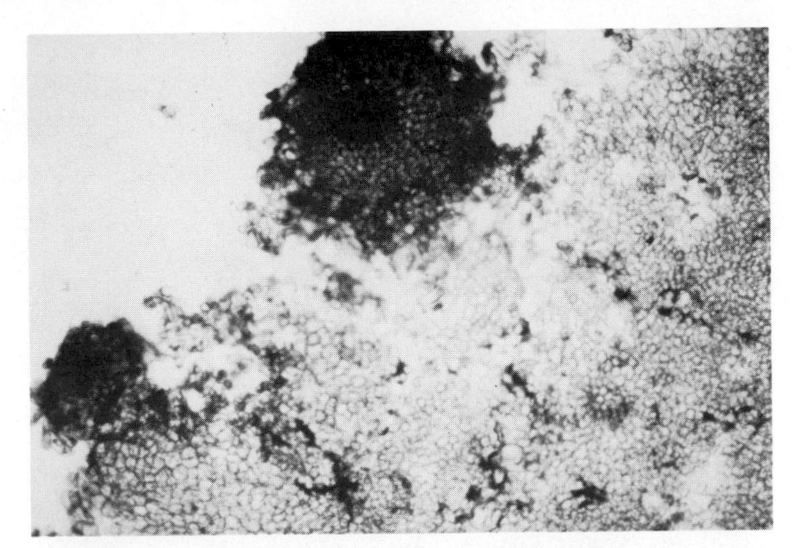

Fig. 10. A fresh, unstained cross-section of the surface of a large nodule showing the "budding-off" of several unicenter nodules originally formed near the surface of the mother nodule. This is another form of multiplication of nodules in addition to that described in Fig. 3 and 7.

Thus, there are two ways in which nodule cultures can be perpetuated and multiplied indefinitely: (a) stock nodules slough cells from their surfaces, which then proceed to develop into independent nodules; and (b) budding or breaking-up of the stock nodules into independent nodules. Uniformity of development can best be controlled by sizing and grading of the nodules at subculture (Fig. 11).

Organogenesis of Nodules

Nodules of poplar appear to have substantial totipotency. Nodules can be differentiated directly in the liquid culture medium (Fig. 12) or on solid supports such as floating webs (19) or agar (Fig. 13). Shoot and/or root development can be induced (Fig. 12, 13, and 14). In poplar nodule cultures, organ development appears to be by organogenesis (Fig. 15) associated with the vascularized centers, possibly in the parenchymatous, plastid-poor cell layer surrounding such centers (Fig. 6). However, organogenesis may also occur from nonvascularized nodules (J. Aitken-Christie et al., this Volume), such as those in the first stage of development.

We have yet to conduct extensive research on the factors controlling organogenesis. With poplar, nodule shoot differentiation is stimulated by thidiazuron (Tab. 2; Ref. 20). Since this compound appears to be inhibitory to subsequent shoot development, thidiazuron may best be administered as a pulse treatment (20).

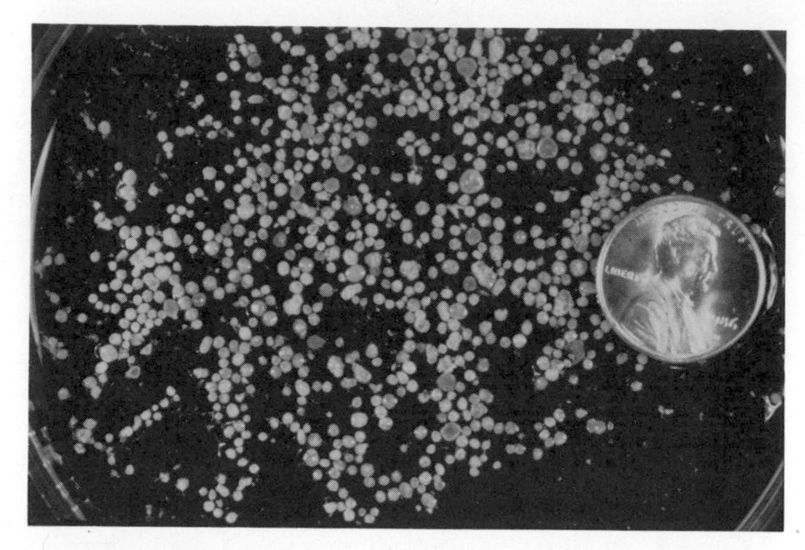

Fig. 11. A newly harvested set of nodules that has been screened for size uniformity.

Fig. 12. A culture of nodules showing all the phases of shoot differentiation. These nodules were differentiated in liquid roller-bottle culture. The most developed shoot has also formed an adventitious root.

Fig. 13. A nodule differentiating a shoot after transfer to agar medium.

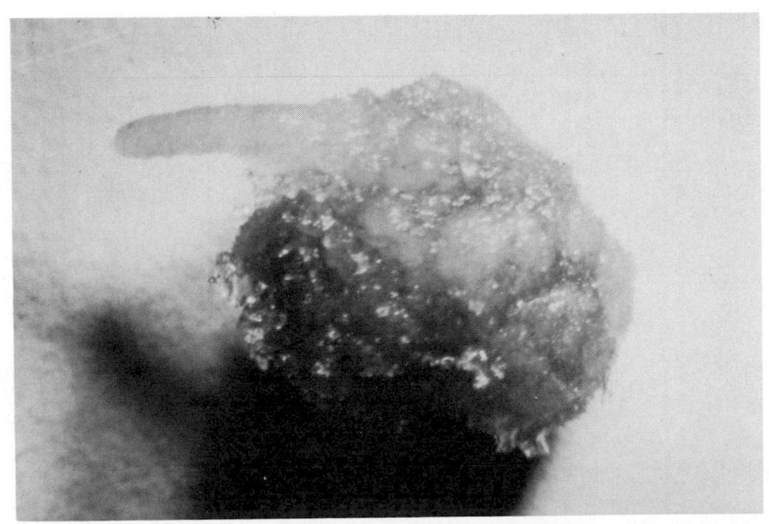

Fig. 14. A nodule differentiating a root. No shoot development is evident.

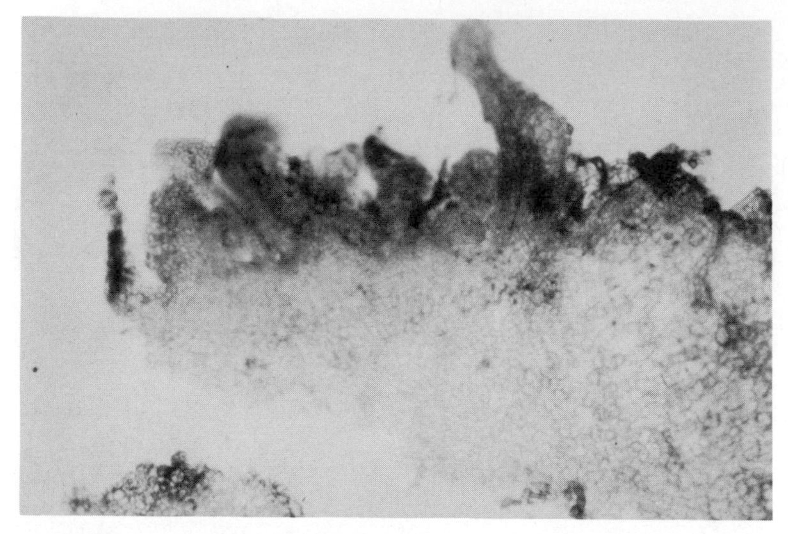

Fig. 15. A fresh, unstained cross-section of the area of a nodule differentiating a shoot meristem.

THEORETICAL CONSIDERATIONS FOR THE NODULAR DEVELOPMENTAL PATHWAY

The position in morphogenetic theory of the nodular developmental pathway is not clear. However, the pathway does fit nicely into a number of theoretical concepts.

The idea of the meristemoid was originally developed by Bunning in 1952 (3), and Torrey (23) extended its use. A meristemoid can be considered a group of cells that act together as a meristematic center. Thus, nodules in their simplest form can be considered meristemoids. Such meristemoids should theoretically be totipotent. However, as discussed so well in previous papers (2,6,9), there is a major bridge between totipotency and competence (the reliable formation of differentiated structures). There appear to be at least three steps by which totipotent cells

Tab. 2. Nodule production and shoot differentiation using a poplar roller-bottle culture system (see Fig. 1).

Replicate culture	Percent increase in nodule number (six-week period)	Percent of nodules differentiating shoots six weeks after shift to differentiation medium
1	504	31
2	379	33
3	483	27

Note: Nodules were differentiated with a two-week pulse of 0.1 μM thidiazuron. Cultures were established using a 300-nodule charge in 150 ml modified WPM medium.

become competent (2): (a) cell dedifferentiation; (b) cell interaction; and (c) reactivity to specific signals. Nodules would appear to possess all three capabilities: visually dedifferentiated cells are apparent in the meristematic regions of nodules; vascular/meristematic centers at least physically interact (Fig. 7), and, as evidenced by the uniform developmental sequence of the cell layers in nodules, there are probably considerable other tissue interactions; and nodules are responsive to auxin levels (Tab. 1) and thidiazuron (Tab. 2). Thus, the nodular developmental pathway may form an important bridge between cell totipotency and cell competence.

Another bridge between cell totipotency and cell competence is embryogenesis. From the limited literature on the topic, nodulation as defined and described above and embryogenesis are distinctly different but highly parallel. Although both pathways lead from undifferentiated states to full competence, there appear to be distinct differences between embryogenic cell clusters and nonembryogenic clusters. Embryogenic cell clusters have never been reported to be vascularized (6), while a prominent feature of nodules is vascularization (Fig. 5; Ref. 6), or at least the potential for vascularization. Some nodule systems do not show vascularization (e.g., J. Aitken-Christie et al., this Volume), but this may be a result of the nodules being maintained in the first stage of development and on solidified media. In our observations with nodules, especially with protocalli where close observation of small nodules is facilitated, we have not identified embryogenesis as such. All shoot development appeared to be by organogenesis (Fig. 15). However, it is feasible and has been observed by both us and others (6,9) that embryogenesis and nodule development can occur in the same culture.

The nodular developmental pattern may not be restricted to in vitro culture. Nodules closely resemble spheroblasts (1) in their vascularization and competency. In addition, nodules loosely resemble protocorms of orchids (18) and the more primitive Lycopodiaceae (5). In fact, nodules may relate to the "protocorm theory" in which tuberous structures (protocorms) are developed to carry young sporophytes over unfavorable periods by forming a support for subsequent plant differentiation.

At present, it is our hypothesis that the nodule developmental pathway and the embryogenic pathway are parallel developmental pathways. This hypothesis is detailed in Fig. 16.

PRACTICAL APPLICATIONS OF NODULE CULTURE

We can see three important applications of nodule cultures:

(a) Differentiation of woody plants. Considering the competency of nodular systems and the prominence of nodule formation in woody genera, the definition of a theoretical and manipulative basis for the nodular developmental pathway should offer successful strategies for plant differentiation for many woody species now considered recalcitrant. The success with pine reported by Aitken-Christie (see J. Aitken-Christie et al., this Volume) and recent successes in differentiating spruce in our laboratory (Fig. 17) using the nodular approach support the idea of the importance of this system in forest trees.

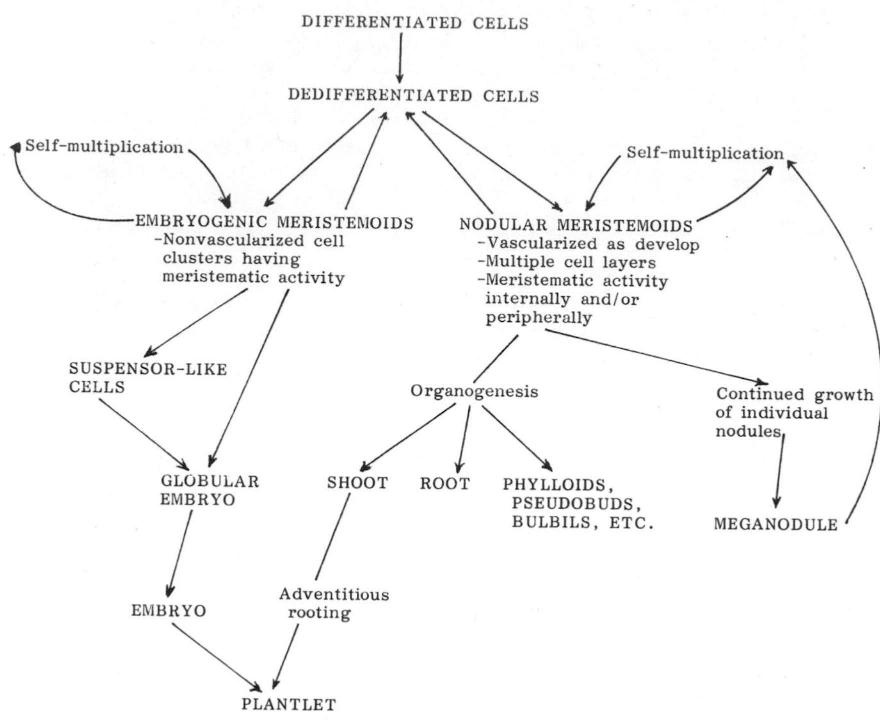

Fig. 16. A flowchart showing the parallelism between two in vitro plant developmental pathways--somatic embryogenesis and nodular development.

(b) Automated micropropagation. Nodules are uniform in shape and can be sized for uniformity. In addition, they can be cultured in mass in self-replicating systems. Nodules may act like protocorms and support the delicate developing plantlet during the early stages of differentiation. These features make nodule cultures a very attractive alternative for automated mass culture and transfer to ex vitro growth systems such as transplant plugs (17). Work on this aspect is currently being pursued in our laboratories. One encouraging observation is that the micropropagule derived from nodule organogenesis is much more substantial and developed than embryos derived from somatic embryogenesis (Fig. 18), and thus may be more readily moved into a greenhouse or field environment. An important question here is the genetic/physiological uniformity and stability of plantlets derived from nodules. Work with daylily by Krikorian and colleagues (10) using "nubbins" supports the proposition that nodular systems may be relatively stable.

(c) In vitro phytochemical production. A major problem with producing many chemical products in in vitro plant systems is a conflict between the desirable cellular dynamics of cultures and the biochemical dynamics of complex molecules such as secondary metabolites. The most economical cell

Fig. 17. The surface of a callus of white spruce that had been induced to nodulate, showing the differentiation of a shoot meristem. Such meristems develop into normal shoots (see Fig. 18). Original callus compliments of the Institute of Paper Chemistry (IPC), Appleton, Wisconsin.

cultures consist of cells kept at or near the log phase of growth; however, many if not most secondary metabolites are produced in quantity in highly differentiated cells. Nodule cultures may be an alternative solution to this paradox, since they possess the ability to replicate and increase in biomass in continuous culture while being highly differentiated. Meganodules (Fig. 8) may be especially attractive for this application. This approach is also being pursued in our laboratories. Luckner and Diettrich (12) have reported that nonembryoid nodules have the capacity to produce the secondary metabolite cardenolide, while cell suspensions of the same genotype produce little or no cardenolide.

CONCLUSIONS

Nodules should be considered a culture system in themselves. They are yet another alternative to other culture systems such as suspension cultures and shoot cultures. The work on nodule cultures is still very preliminary but exciting potentials are already apparent in both morphogenetic theory and practical applications.

There are a number of topics that need immediate addressing:

(a) What is the theoretical basis of the nodular developmental pathway? Is this pathway a basic developmental feature common to most plants?

(b) What are the controls on uniformity (growth and differentiation)?

Fig. 18. A comparison of a somatic embryo (top) and a nodule-derived organogenic shoot of white spruce. Both were differentiated from the same callus originally obtained from the IPC.

(c) How high a rate of differentiation can be achieved? Rates of 90% or more will need to be achieved on a uniform basis before this technology can be applied in micropropagation on a commercial scale.

(d) What methods are needed for ex vitro establishment of nodule-derived plantlets? Again, rates of 90% or more will be necessary for commercial utilization.

(e) How widely applicable are nodule cultures in regard to crop species? Might this be an avenue for the economical micropropagation of some agronomic species? vegetables? forest species? fruits? plantation crops?

(f) What is the genetic and physiological uniformity of nodule-derived plants?

These are not easy questions to address but the results will be intellectually exciting, if not revolutionary.

ACKNOWLEDGEMENTS

This work has emerged from a large number of seemingly independent projects. Major support was received from the Agricultural Experiment Station, University of Wisconsin-Madison (HATCH and McIntire-Stennis funding), the University of Wisconsin University-Industry Research Program, the U.S. Department of Agriculture-Forest Service through the North Central Forest Experiment Station, and the two industrial cooperators co-authoring this chapter. The progress would not have been possible without the stimulation and hard work of the students, technicians, and cooperators in our research group.

REFERENCES

1. Bornman, C.H. (1987) Picea abies. In Cell and Tissue Culture in Forestry. Vol. 3. Case Histories: Gymnosperm, Angiosperm and Palms, J.M. Bonga and D.J. Durzan, eds. Martinus Nijhoff Publishers, Boston, pp. 2-29.
2. Brown, D.C.W., and T.A. Thorpe (1986) Plant regeneration by organogenesis. In Cell Culture and Somatic Cell Genetics of Plants. Vol. 3. Plant Regeneration and Genetic Stability, I.K. Vasil, ed. Academic Press, Inc., Orlando, Florida, pp. 49-65.
3. Bunning, E. (1952) Morphogenesis in plants. In Survey of Biological Program, C.E. Avery, Jr., ed. Academic Press, Inc., New York, pp. 105-138.
4. Chatwvedi, H.C., and G.C. Mitra (1975) A shift in morphogenetic pattern in Citrus callus tissue during prolonged culture. Ann. Bot. (London) (N.S.) 39:683-687.
5. Eames, A.J. (1977) Morphology of Vascular Plants, Robert E. Krieger Publishing Company, Huntington, New York, pp. 18-23 and 397-398.
6. Halperin, H.W. (1986) Attainment and retention of morphogenetic capacity in vitro. In Cell Culture and Somatic Cell Genetics of Plants. Vol. 3. Plant Regeneration and Genetic Variability, I.K. Vasil, ed. Academic Press, Inc., Orlando, Florida, pp. 3-47.
7. Halperin, W. (1966) Alternative morphogenetic events in cell suspensions. Am. J. Bot. 53:443-453.
8. Halperin, W. (1970) Embryos from somatic plant cells. In Control Mechanisms in the Expression of Cellular Phenotypes, H. Padykula et al., eds. Academic Press, Inc., New York, pp. 169-191.
9. Jones, L.H. (1974) Factors influencing embryogenesis in carrot cultures (Daucus carota L.). Ann. Bot. (London) (N.S.) 38:1077-1088.
10. Krikorian, A.D., and R.P. Kann (1986) Regeneration in lidiaceae, iridaceae, and amaryllidaceae. In Cell Culture and Somatic Cell Genetics of Plants. Vol. 3. Plant Regeneration and Genetic Stability, I.K. Vasil, ed. Academic Press, Inc., Orlando, Florida, pp. 187-205.
11. Krikorian, A.D., S.A. Staicu, and R.P. Kann (1981) Karyotype analysis of a daylily clone reared from aseptically cultured tissues. Ann. Bot. (London) (N.S.) 47:121-131.
12. Luckner, M., and B. Diettrich (1985) Formation of cardenolides in cell and organ cultures of Digitalis lanata. In Primary and Secondary Metabolism of Plant Cell Cultures, D. Neumann et al., eds. Springer-Verlag, Berlin, Heidelberg, pp. 154-163.
13. McCown, B.H. (1985) From gene manipulation to forest establishment: Shoot cultures of woody plants can be a central tool. TAPPI J. 68:116-119.
14. McCown, B.H. (1986) Application of nodule culture to forest crop improvement and production. In Proceedings of the 1986 Research and Development Conference, TAPPI Press, Atlanta, Georgia, pp. 87-88.
15. McCown, B.H. (1986) Woody ornamentals, shade trees, and conifers. In Tissue Culture as a Plant Production System for Horticultural Crops, R.H. Zimmerman et al., eds. Martinus Nijhoff Publishers, Dordrecht, The Netherlands, pp. 333-342.
16. McCown, B.H., and D.D. McCown (1987) North American hardwoods. In Cell and Tissue Culture in Forestry. Vol. 3. Case Histories: Gymnosperm, Angiosperm and Palms, J.M. Bonga and D.J. Durzan, eds. Martinus Nijhoff Publishers, Boston, pp. 247-260.

17. McCown, D.D. (1986) Plug system for micropropagation. In Tissue Culture as a Plant Production System for Horticultural Crops, R.H. Zimmerman et al., eds. Martinus Nijhoff Publishers, Dordrecht, The Netherlands, pp. 53-60.
18. Morel, G.M. (1974) Clonal multiplication of Orchids. In The Orchids, Scientific Studies, C.C. Withner, ed. John Wiley and Sons, New York, pp. 169-222.
19. Russell, J.A., and B.H. McCown (1986) Culture and regeneration of Populus leaf protoplasts isolated from non-seedling tissue. Plant Sci. 46:133-142.
20. Russell, J.A., and B.H. McCown (1986) Thidiazuron-stimulated shoot differentiation from protoplast-derived cells of Populus. In Abstracts, VI International Congress of Plant Tissue and Cell Culture, D.A. Somers et al., eds. University of Minnesota, Minneapolis, Minnesota, p. 49.
21. Sharp, W.R., M.R. Sondahl, L.S. Caldas, and S.B. Maraffa (1980) The physiology of in vitro asexual embryogenesis. Hort. Rev. 2:268-310.
22. Stewart, F.C. (1963) The control of growth in plant cells. Sci. American 209:104-110.
23. Torrey, J.G. (1966) The initiation of organized development in plants. Adv. Morphol. 5:39-91.
24. Webster's Third New International Dictionary (1961) G.C. Merriam Co. Publishers, Springfield, Massachusetts.

PHYSIOLOGY OF BUD INDUCTION IN CONIFERS IN VITRO

Trevor A. Thorpe

Plant Physiology Research Group
Department of Biological Sciences
University of Calgary
Calgary, Alberta, Canada T2N 1N4

ABSTRACT

Using excised cotyledons of *Pinus radiata* as an experimental system, it was shown that while cytokinin was required for bud induction, both cytokinin and light were required for subsequent primordium formation. Both ethylene and carbon dioxide stimulated bud formation, and autotrophic and nonautotrophic CO_2 fixation took place during shoot initiation. Using histochemical, autoradiographic, and precursor incorporation approaches, it was found that DNA, RNA, and protein synthesis was preferentially in the epidermal and subepidermal cell layers in contact with the medium. In these cell layers organized development occurred. Feeding the tissue with [^{14}C]glucose and [^{14}C]acetate led to the release of $^{14}CO_2$ and the production of both ethanol-soluble and ethanol-insoluble fractions. Most of the label went into the ethanol-soluble fraction, which was further fractionated into lipids, amino acids, organic acids, and sugars. Further studies with [^{14}C]glucose showed that only some amino acids and organic acids were labeled and that some turnover, particularly in glutamate and malate, occurred during bud initiation. The importance of these data in relation to bud induction is discussed.

INTRODUCTION

The process of de novo organogenesis in cultured tissues is a complex one, in which extrinsic and intrinsic factors play a role. Organized development can be regulated through manipulation of the culture medium and the culture environment, and by judicious selection of the inoculum. Manipulation of these factors allows cells that are quiescent or committed only to cell division to undergo a transition, which at the molecular level involves selective gene activity. This activity is reflected by biochemical, biophysical, and physiological changes, which lead to structural organization within the cultured tissues. Much is known about the manipulation of the factors regulating de novo organogenesis, particularly in herbaceous plant species, but virtually nothing is known about regulation at the molecular level (8,28,30). In this respect, even less is known about conifers than angiosperms.

Understanding the process of de novo organogenesis is not only of interest to developmental morphologists and plant physiologists, but also to those involved in the exploitation of tissue culture technology for plant improvement because regeneration of plantlets is central to this activity. To date, the application of tissue culture technology to the vegetative propagation of plants is the widest use of this technology. While asexual multiplication can be achieved by various routes, the most generally applicable method for conifers is via adventitious budding (31). As a result, the process is a multistaged one, consisting of at least four distinct phases, the first of which is the induction of shoot buds on the inoculum. In conifers this is usually an organized explant such as a mature embryo, seedling part, or vegetative bud. It has been found that one key to success is the selection of the explant.

The most prolific conifer regeneration system presently known is that in Pinus radiata, a system which demonstrates the importance of selection of the proper inoculum (1). When excised mature embryos are used as explants, ten to 20 rootable shoots are formed per embryo. When the cotyledons are separated from the embryos after one week in culture, the number of shoots per embryo increases to 20-50, but when approximately one-week-old cotyledons excised from aseptically germinated seeds are used as explants, the number of shoots per seed further increases to 180-250. Both younger and older cotyledons produce fewer shoots. Some clones produce in excess of 1,000 shoots using the latter explants (1). The cotyledons must be chosen at a particular developmental state, as this influences the shoot-forming capacity of the excised cotyledons in culture (2). Ideal cotyledons were found to have epidermal and subepidermal layers with less-developed stomatal complexes, thinner cell walls, no epicuticular wax, partially depleted protein reserves, and unhydrolyzed lipid reserves; whereas older cotyledons had more-developed stomatal complexes, thicker cell walls, epicuticular wax, depleted protein reserves, and nearly depleted lipid reserves. We have used this cotyledon explant system to study the physiology of bud induction.

THE COTYLEDON EXPLANT SYSTEM

The cotyledon explant system of P. radiata D. Don (radiata or Monterey pine) consists of cotyledons excised from aseptic, dark-germinated seed approximately one day after radicle emergence. This corresponds to approximately eight days after seed imbibition for most seeds, of which two days are needed for stratification. Radicle emergence usually occurs five to ten days after imbibition. Radicle length at excision varies between 0.5 and 2.0 cm and the cotyledons, between 3 and 5 mm (Fig. 1). Each radiata pine seed has seven to ten cotyledons that respond uniformly in culture. About 90% of the seeds from different seedlots behave identically in culture, although they were collected from open-pollinated cones. (Seeds were obtained from the Forest Research Institute, Rotorua, New Zealand.)

In all of our studies these excised cotyledons are cultured in sterile plastic petri dishes on modified Schenk and Hildebrandt medium (20) containing 3% sucrose and 25 μM N^6-benzyladenine (BA) in the light (16 hr, ∿80 $\mu mol \cdot m^{-2} \cdot s^{-1}$ from Sylvania Gro-lux fluorescent tubes) at 27 ± 1°C (1).

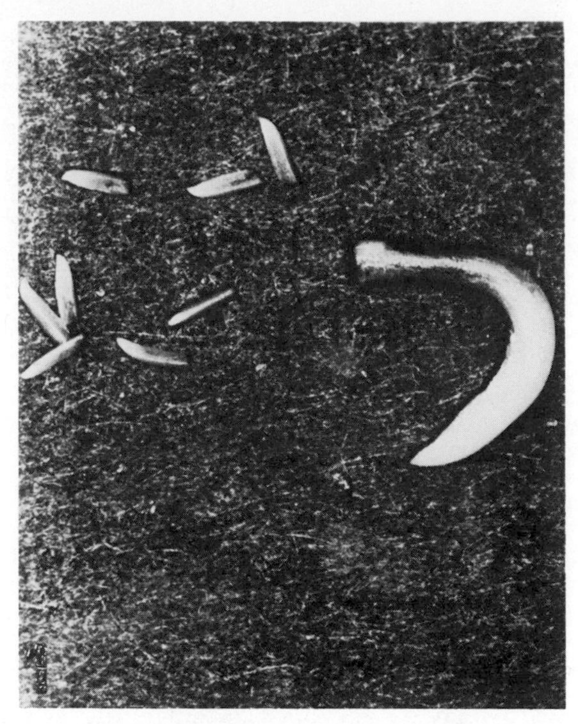

Fig. 1. Cotyledons and radicle excised from radiata pine seed one day postgermination (5X). (From Ref. 2, with permission of the University of Chicago Press.)

Concentrations of BA above 50 μM inhibited subsequent shoot elongation, while concentrations below 5 μM reduced shoot formation but enhanced elongation (6). The optimum time of exposure was 21 days. A few shoots were ultimately formed by exposure to BA for three and seven days, but only about 50% of the cotyledons formed shoots with 14 days exposure to BA (see below). For any shoot formation to take place, BA must be present during the first three days of culture (6,34); however exposure to light could be delayed at least until day 10, but after 21 days in darkness no shoots were formed upon transfer to light (33). These data suggest that cytokinin is directly involved in the induction of shoot initials and that both light and cytokinin are required for the development of meristematic tissue and subsequent shoot formation. At the end of 21 days in culture, meristematic tissue is formed along the entire length of the cotyledon in contact with the medium (Fig. 2 and 3). For shoot development, these cotyledons must be transferred to BA-free medium with 2% sucrose. During the initial 21 days in culture, cotyledons cultured in the absence of BA have served as control. These cotyledons do not form meristematic tissue, but elongate rapidly until about day 10 (Fig. 4A). Changes in fresh weight during this 21-day culture period are shown in Fig. 4B.

HISTOLOGY OF BUD FORMATION

At the time of excision, numerous mitotic figures could be seen throughout the cotyledon, and all cell divisions were anticlinal (38).

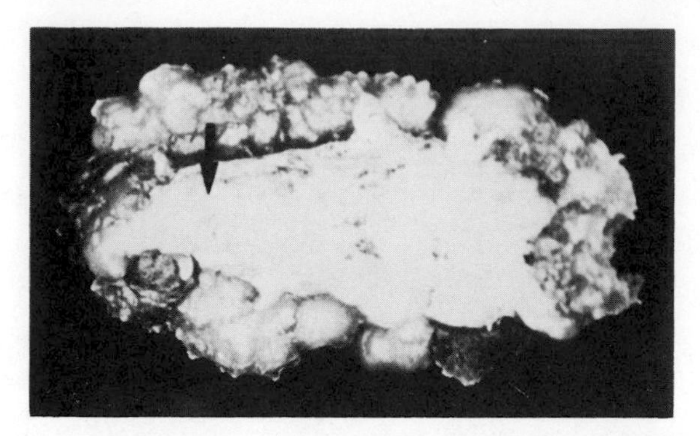

Fig. 2. Responses of excised cotyledons after three weeks in culture on the shoot-initiation medium (15X). Upper surface of cotyledon not in contact with medium does not form meristematic tissue (arrow). (From Ref. 2, with permission of the University of Chicago Press.)

Large and prominent nuclei were present, the cytoplasm stained densely, and reserve substances were abundant, with numerous starch grains and protein bodies being observed. In the absence of cytokinin, mitotic activity stopped by day 2 in culture. Small vacuoles, probably derived from protein bodies, gradually fused, resulting in large vacuoles. Protein bodies disappeared by day 2, the plastids enlarged, and there was a concomitant reduction of starch within them. Stomatal complexes began to differentiate at day 1 and the stomata were fully developed by day 5. Large intercellular air spaces gradually appeared in the mesophyll, the cells of which became more vacuolated, so that the chloroplasts and cytoplasm were confined to the periphery of the cells (10). The chloroplasts were fully developed by day 3 (12,34,35).

In contrast to the above, the pattern of cell divisions leading to organized structures in the subepidermal regions of the cotyledonary face

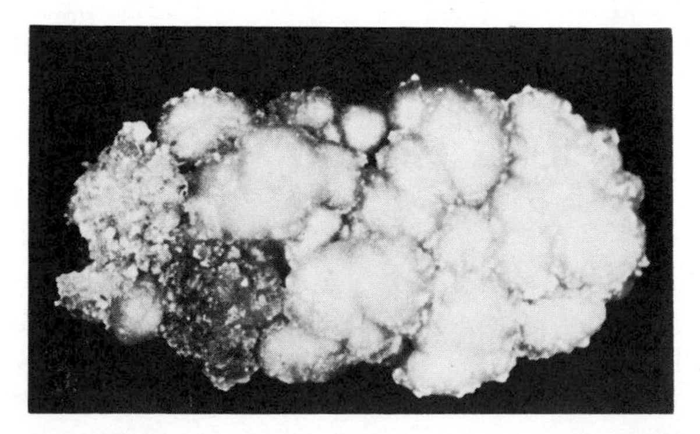

Fig. 3. Responses of excised cotyledons after three weeks in culture on the shoot-initiation medium (15X). Lower surface of cotyledon in contact with the medium forms meristematic tissue along the entire length of the cotyledon. (From Ref. 2, with permission of the University of Chicago Press.)

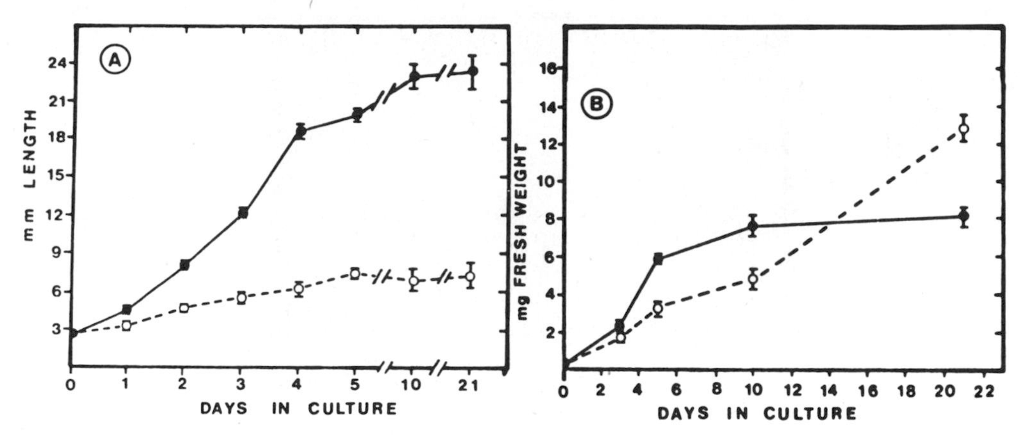

Fig. 4. Growth of excised cotyledons of radiata pine after three weeks in culture in the presence (o) or absence (•) of BA. (A) Length. (B) Fresh weight.

in contact with the medium became apparent very early in culture (34,35). Initially, cell division in cotyledons cultured in the presence of BA was random. However, by day 3, mitotic activity was concentrated in the epidermis and subepidermal parenchyma cells in contact with the medium. At the same time, organized structures could be detected. Each of these organized structures had its origin from a single subepidermal cell that began to divide in a periclinal direction after the third day in culture. Subsequent anticlinal and periclinal divisions led to the formation of a six- to eight-celled organized structure by day 5, which we have termed a promeristemoid (Ref. 34; Fig. 5A). Cells within each organized structure were tightly packed together, with little or no intercellular spaces between them.

Ultrastructurally, after two days in culture in the presence of BA, the subepidermal cells had a dense cytoplasm with prominent nuclei and lipoidal storage products were still present. Vacuoles gradually started to appear in the cytoplasm at this time. At day 3, just prior to cell division, some of the vacuoles had coalesced to form larger ones. Chloroplasts were prominent and randomly distributed throughout the cytoplasm. Concomitant with the breakdown of storage lipids, starch granules appeared in the stroma of the plastids. As the subepidermal cells underwent further organized cell divisions, the most notable ultrastructural features were that plasmodesmata were easily discerned in walls between daughter cells and that the cell walls were relatively thin as compared with the parent cell wall. In contrast, plasmodesmata were absent between different organized clusters. The starch content gradually decreased as these developed.

In the presence of BA, these organized structures continued to develop. After the tenth day of culture, the cotyledons had a nodular appearance as a result of the increase in size of the organized structures underneath the epidermis. By day 21, leaf primordia were evident. From the scanning electron microscope study, it was clear that epidermal cells of the cotyledon explants did not rupture; instead they became the protoderm of the shoot primordium (Fig. 5B). Judging from the cell pattern of the epidermal cells immediately adjacent to the organized structures, evidence of cell division in the explant could be observed This indicated that the

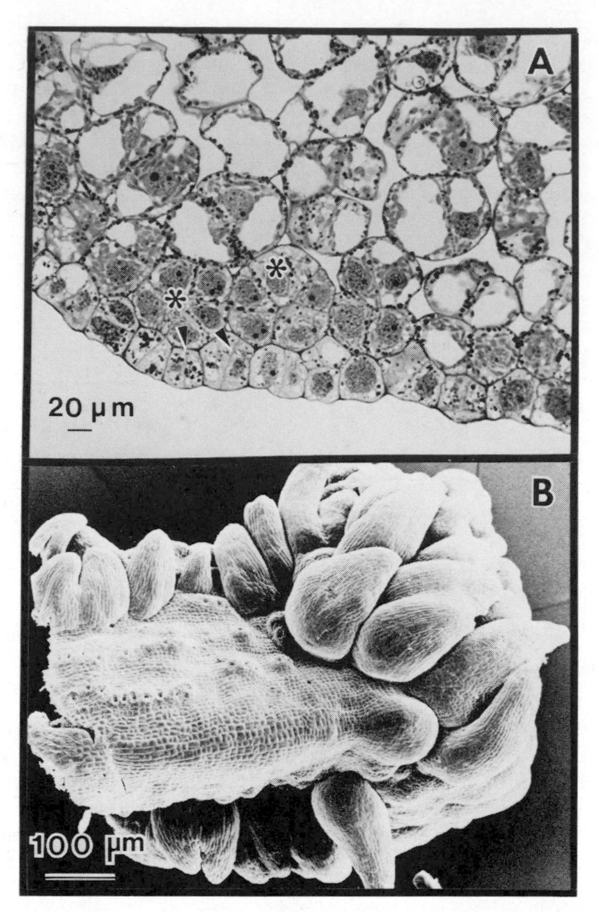

Fig. 5. Excised cotyledon of radiata pine on shoot-forming medium. (A) Light micrograph showing promeristemoids (*) in a subepidermal cell layer after five days in culture. The cell patterns of the epidermal cells immediately adjacent to the organized structures suggest anticlinal cell division activity (arrowheads). (B) Scanning electron micrograph showing shoot primordium formation after 21 days in culture. Note that the newly formed organized structures maintained the original epidermis (arrowhead). (From Ref. 34, with permission of the National Research Council of Canada.)

epidermal cells of radiata pine cotyledons could adjust to the size increases and changes in shape of the explant. This observation differs significantly from other systems, e.g., bud-forming cotyledons of Douglas fir, in which developing shoot primordia rupture the explant epidermis.

The position at which promeristemoids develop is probably influenced in part by physiological gradients of nutrients (including cytokinin) from the medium into the tissue. It has been suggested that such gradients are operative during shoot initiation in tobacco callus (21). The promeristemoids, which were spherical or polarized, did not arise from a typical meristematic cell. The cells contained vacuoles and had abundant chloroplasts. Densely plasmatic cells, as found in tobacco (22) and carrot (26) during organized development, were conspicuously absent in radiata pine cotyledons. The cells became densely plasmatic later in culture.

The developmental sequence leading to multiple shoot formation is quite similar in various conifer explants, such as mature embryos, cotyledons, and epicotyls (32). The sequence includes the formation of (a) meristemoids (also referred to as meristematic bud centers or meristematic tissue), (b) bud primordia, and (c) adventitious shoots with well-organized apical domes and needle primordia. In radiata pine an important structural feature is the promeristemoid, which gives rise to the meristemoids. Such structural entities can also be observed in other tissues, e.g., in Torenia (9) and Picea abies (7). The observations made on the origin of the promeristemoid support the idea that organized development in vitro begins with changes in a single cell which then becomes activated (28,30). These activated cells become "mitosis-determined" [in the sense of Stebbins (25)] and give rise to meristemoids (via promeristemoids) and, later, shoot primordia. The common features of the developmental sequence for shoot primordium formation in conifers (as well as angiosperms) lead to the conclusion that the radiata pine system, which is apparently the most prolific in vitro bud-forming system, can be used in a general way to study the physiology of in vitro organogenesis.

PHYSIOLOGY OF BUD FORMATION

Concentration, physiological, or diffusion gradients of materials from the medium into the tissue have been implicated in determining the loci at which primordium initiation begins (21,34). This mechanism, however, cannot explain how individual cells become activated and induced to undergo an altered pathway of differentiation. This process has at least three requirements: (a) cell dedifferentiation, (b) cell interaction, and (c) reaction to specific signals (28,30). The net result is specific changes in the tissue, which precede and are, presumably, causative to organized development. Limited study has been undertaken to document these changes during de novo organogenesis in conifers. In my laboratory we have begun to do so, and our findings to date are outlined below.

Role of Phytohormones and other Factors

Exogenous cytokinin is required for both bud induction and primordium development as indicated earlier. The addition of other phytohormones or growth regulators to the medium during the first 21 days in culture tended to reduce cytokinin-induced meristematic tissue formation and to promote callus production (6). The effect was more pronounced with increasing concentration (in the range of 10^{-8} to 10^{-4} M) of the test compound in the medium. The auxins naphthaleneacetic acid and 2,4-dichlorophenoxyacetic acid (but not indolebutyric acid), abscisic acid, and cAMP all caused callus formation. Transfer of cotyledons to BA-free medium after day 21 led to additional callus proliferation, reducing the number of shoots ultimately formed. The substances tested that were least inhibitory to shoot formation were the growth retardants CCC and AMO-1618, the aromatic amino acids phenylalanine and tyrosine, and gibberellic acid (6). Gibberellic acid (GA_3) caused some reduction in meristematic tissue, when the tissues were exposed to it between days 0-4 and 0-7, but not if they were exposed for longer periods or after day 4. The above results do not mean that the endogenously produced phytohormones play no role in bud induction. As a matter of fact, we have preliminary data indicating that the contents of endogenous indoleacetic acid and abscisic acid change during bud induction in excised radiata pine cotyledons (D.E. Macey, D.M. Reid, and T.A. Thorpe, unpubl. results).

The role of ethylene and its interaction with CO_2 in bud induction has been examined (16). The excised cotyledons were cultured in Erlenmeyer flasks and three experimental approaches were used to study the effects of these gases. The flasks were sealed with serum caps on various days in culture, and the caps were also removed at different stages of the morphogenetic process. Second, the gases within the flasks were absorbed by traps, singly or in combination, during the various stages of differentiation. Third, the cultures were incubated under continuous flow of constant gas mixtures.

From these experiments it was found that shoot-forming cotyledons produced considerable amounts of C_2H_4 and CO_2, and the frequency of organogenesis could be correlated with the concentrations of these two gases. The highest number of buds per explant was obtained when the flasks had accumulated about 5 to 8 $\mu l\ l^{-1}$ of C_2H_4 and about 10% CO_2 in the headspace during the first 15 days in culture. A large proportion of the CO_2 and C_2H_4 in bud-forming cultures was produced between days 15 and 21, after the key events leading to bud primordia had occurred. The removal of these gases at that time had no effect on morphogenesis.

From the experiments dealing with serum capping and gas trapping it was found that these two gases exerted their influence on morphogenesis even after they were released from the tissue into the culture vessel. When one of the gases was absorbed from the atmosphere, differentiation was adversely affected. Also, when both the gases were eliminated from the flasks, differentiation was completely inhibited. When C_2H_4 and CO_2 were allowed to accumulate in the flasks during the first ten to 15 days, the differentiation proceeded normally, even if, subsequently, they were absorbed by the traps or allowed to diffuse out by replacing the serum caps with foam bungs. Therefore, the effects of C_2H_4 and CO_2 are possibly synergistic and necessary in order for the cytokinin (BA) in the medium to bring about the switch in morphogenesis from the normal maturation of the cotyledons to shoot bud differentiation.

The first ten days in culture (when C_2H_4 and CO_2 exert their influence on differentiation) coincided with a period of intense, localized cell division leading to the formation of meristematic domes under shoot-forming conditions (34,38). An earlier study had indicated that C_2H_4 can bring about partial synchrony in cell suspension cultures (10). Thus, the stimulatory action of these gases on morphogenesis could be mediated through modification of the cell division process. However, if excessive amounts of these two gases were allowed to accumulate within the flasks after the first 15 days in culture, the process of organogenesis was partially reversed. This was clear from the observed dedifferentiation of the shoot buds, browning of the cotyledons at the cut ends, and also from the inability of these buds to elongate and form normal shoots. Browning of the cotyledons at the cut ends in the serum capped flasks kept under a continuous flow of $C_2H_4 + O_2 + N_2$ indicates that this may be a result of the high concentration of C_2H_4 to which the cotyledons were exposed. Oxygen (O_2) appeared to have no direct role in bringing about morphogenesis, but was probably required to maintain normal oxidative metabolism.

From the known interactions of CO_2 and C_2H_4, three possible roles of CO_2 can be proposed. First, in the early stages it might be acting primarily to enhance the biosynthesis of C_2H_4; second, at later stages (beyond day 10 or 15) it might be important in antagonizing the action of C_2H_4; and third, CO_2 might have a totally independent role in metabolism, which was essential for changing the morphogenic response of the tissues to phytohormones. For example, it would increase the nonphotosynthetic carbon fixation. We have evidence indicating that the activity of phosphoenolpyruvate carboxylase is high during the process of differentiation (P.P. Kumar, L. Bender, and T.A. Thorpe, unpubl. results), as it also is in shoot-forming tobacco callus (19) and in exponentially growing carrot cell cultures (3).

Metabolic Studies

We have approached this topic through the use of autoradiographic, histochemical, and biochemical (including precursor incorporation) techniques. In most cases, bud-forming (+BA) cotyledons were compared and contrasted to nonbud-forming (-BA) tissues, in an attempt to determine what aspects of metabolism could be directly correlated with and could presumably be causative to shoot bud formation.

Autoradiograms of precursor incorporation of [^{3}H]uridine into RNA, [^{3}H]leucine into protein, and [^{3}H]thymidine into DNA in the epidermal and subepidermal cells from the cotyledonary face that was in contact with the medium during the first five days in culture are shown in Fig. 6 (36). At the time of excision (day 0), incorporation of the precursors, as judged by the pattern of silver grain deposition, was randomly distributed among the various cells in the explant (Fig. 6A, B, and C). The patterns of incorporation at days 1 and 2 in culture were similar. By days 3 to 5 in culture, there was not only a difference in the appearance of the cells due to the presence or absence of BA, but the incorporation patterns of the precursors were also different. In the presence of cytokinin, labeling became concentrated in the epidermal and subepidermal cells (Fig. 6D, E, and F), while those cells cultured in the absence of cytokinin showed very little incorporation (Fig. 6G, H, and I). Cytoplasmic (plastidic) incorporation of [^{3}H]thymidine was also evident (Fig. 6F and I), and only in BA-treated cotyledons were labeled nuclei observed after day 2.

The BA-free cotyledons showed a dramatic increase in the rate of RNA and, to a lesser extent, protein synthesis during the first 24 hr in culture (36). During that period the rate of synthesis in the BA-treated cotyledons decreased. After 24 hr the rate also decreased in the BA-free cotyledons, so that by day 3 both BA-free and BA-treated cotyledons exhibited the same rate of RNA and protein synthesis. In contrast, the DNA synthetic rates decreased, the decrease being greater in the BA-treated tissues. In both cultures a secondary rise in synthetic rate was observed at day 3. Thus the change in the rates of macromolecular synthesis preceded the histological localization of labeling patterns and the subsequent elongation/maturation of the BA-free cotyledons and the differentiation of the BA-treated tissues.

The methods used above do not allow for a determination of the type of RNA that was being synthesized, or of how much of the RNA was being processed and subsequently translated. Although more RNA and protein were being synthesized in the elongating cotyledons, qualitative differences in the type of RNA and protein being produced could be taking place in

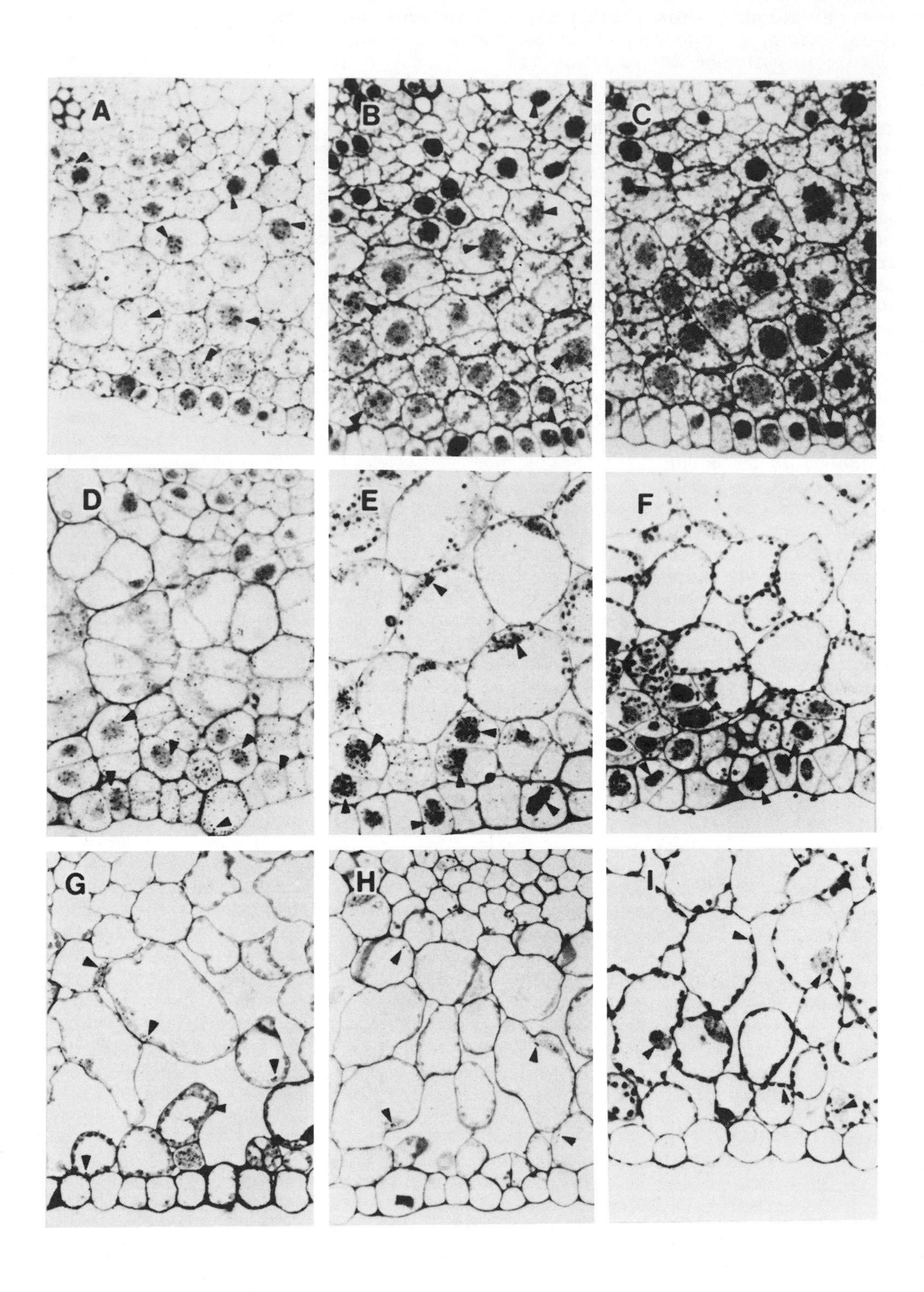
A
B
C
D
E
F
G
H
I

the bud-forming cells. Such studies are now underway in my laboratory. Some evidence is available to indicate that specific proteins are produced during de novo shoot formation (13,27,37).

A histochemical analysis of the process of de novo bud formation in the excised cotyledons confirmed that the initial explant was mitotically active (18). It also showed intense staining for nucleic acids and for cytoplasmic as well as nuclear basic proteins. The cells of initial explants were rich in reserves such as lipids, proteins, and starch, and also showed uniform localization of activity of various enzymes. The cotyledons cultured without BA showed a rapid decline in their reserves. The localization of nucleic acids, proteins, and enzyme activity was random, and the intensity of staining decreased in all cells. In contrast, in the cotyledons cultured in the presence of BA, the shoot-forming layers in contact with the medium showed increased staining for DNA, RNA, and cytoplasmic and nuclear proteins. Starch reserves declined in most cells of the explant, but the epidermal and subepidermal layers maintained abundant starch grains, which disappeared only as the meristematic tissue was formed. Lipase activity was confined to the shoot-forming layers during the initial stages of shoot formation but during the later stages was detected in cells underlying the meristematic tissue. Increased staining intensities for acid phosphatases, adenosine triphosphatase, succinate dehydrogenase, and peroxidase were found in shoot-forming regions of the cotyledons. The above findings suggest that the newly synthesized proteins in the shoot-forming radiata pine cotyledons (36) could be enzymatic, related at least in part to energy metabolism.

The excised cotyledons of radiata pine are packed with reserve material, and as indicated above these disappear during de novo organogenesis. Lipid and free sugars were depleted approximately ten- and six-fold, respectively, and there was also a steady decline in free amino-N levels (6). However, the protein-N pool remained relatively large during the 21-day culture period. Since lipids were the major polymeric storage substance, and since they declined most rapidly during organogenesis, their fate was followed in greater detail (12). Fatty acid and sterol analyses indicated that there were both quantitative and qualitative changes in the different classes of lipids. The most dramatic change in lipid content was the rapid and nearly linear degradation of triglycerides. Steryl/wax esters, the other major neutral lipid component in the cotyledons, showed a similar, but less pronounced decrease during bud induction. While the neutral lipids were decreasing, there was an increase in the content of

Fig. 6. Autoradiogram of cotyledonary tissues of radiata pine from surfaces in contact with the media at various times in culture. Cotyledons labeled with [^{3}H]uridine at day 0 (A), after three days on BA-supplemented medium (D), and after three days on BA-free medium (G). (B), (E), and (H) are tissues labeled with [^{3}H]leucine at the same time intervals. (C), (F), and (I) correspond to tissue labeled with [^{3}H]thymidine at day 0 (C), after five days on BA-supplemented medium (F), and after five days on BA-free medium (I). Arrowheads indicate sites of silver grain deposition. All photographs are at the same magnification. (From Ref. 36, with permission.)

polar lipids, indicating that at least one fate of the fatty acid, glycerol, and sterol components was to produce new membranes. The polar lipid fraction was composed predominately of C16:0, C18:1, and C18:2 fatty acids, which are the major components of phospholipids in many plant species. However, only the C18:3 fatty acid, which is a major component of glycolipids and is particularly abundant in chloroplast lamellae, increased markedly. This increase probably reflected plastid biogenesis, which was also confirmed ultrastructurally (12,35).

The actual amounts of triglyceride fatty acids utilized during cotyledon culture were far in excess of those needed for the observed increases in polar lipids. This suggests that the excised cotyledons, like germinating seeds, rely heavily upon the stored lipid reserves for energy production, particularly during the first three days in culture when respiration rates are highest (6). Sterols are structural and functional components of plant cell membranes, and the specific changes observed in the spectrum of free 4-desmethylsterols suggested that the excised cotyledons behaved similar to those of germinating seeds (12). There was a lower content of chlorophylls and carotenoids in the cultured cotyledons in comparison to developing seedlings. This lower rate of pigment synthesis was probably a consequence of the high level of sucrose (3% w/v) in the medium (11). Nevertheless, it appears that changes in lipids (quantity and type) and pigments are similar to those that occur in developing seedlings. This would suggest that the observed changes are not directly involved in the initiation of organized development in vitro, but play only an indirect role through energy production, membrane proliferation, etc. (12).

All the above findings are consistent with the hypothesis that the initiation of organized development involves a shift in metabolism, which precedes and is coincident with the process (28,30). To gain greater insight into the nature of this metabolic shift, we have begun an in-depth analysis of primary metabolism in the cultured cotyledons of radiata pine, as well as in tobacco callus, during shoot initiation. Initial studies with radiata pine involved feeding the tissues with [^{14}C]glucose, [^{14}C]acetate, or [^{14}C]bicarbonate on different days in culture (17).

When the cotyledons were fed [^{14}C]glucose and [^{14}C]acetate, $^{14}CO_2$ was produced (Tab. 1). (No CO_2 measurement was made for [^{14}C]bicarbonate feeding.) Label from these precursors was incorporated into ethanol-soluble and ethanol-insoluble fractions. The largest percentage of radioactivity was associated with the ethanol-soluble portion, which was further fractionated into lipids, amino acids, organic acids, and sugars (Fig. 7). The amount of label and the pattern of labeling associated with each of the above classes of metabolites varied with time in culture and morphogenetic behavior of cotyledons. In general, there was a tendency towards a high rate of incorporation of label in elongating cotyledons during the period of rapid elongation (day 3). Also, a high rate of incorporation of label in shoot-forming cotyledons coincided with the period of meristematic tissue and shoot primordium formation (days 10 and 21).

Further studies have been carried out in which [^{14}C]glucose was supplied to the cotyledons at different days in culture for 3 hr, followed by a chase of 3 hr with [^{12}C]glucose (4,5). The incorporation of ^{14}C into individual soluble metabolites, as well as into protein, was followed. The major individually labeled metabolites were malate, citrate, glutamate, glutamine, and alanine (Fig. 8). No labeled citrate could be detected at day 0, however, a flow of ^{14}C from glucose to glutamate/glutamine

Tab. 1. Radioactivity (± one-half difference) present in various fractions of excised cotyledons of radiata pine, cultured in the absence or presence of benzyladenine (BA) and incubated in [^{14}C]glucose (A), [^{14}C]acetate (B), and [^{14}C]bicarbonate (C) for 3 hr on day indicated. (From Ref. 17, with permission.)

Fraction	Treatment	Days in culture, Bq* (cotyledon)$^{-1}$			
		0	3	10	21
A. CO_2	-BA	44.1 ± 5.2	76.7 ± 10.9	23.0 ± 1.6	21.9 ± 5.5
	+BA	49.7 ± 4.3	37.6 ± 4.6	57.6 ± 7.9	277.2 ± 29.2
Ethanol-soluble	-BA	324.3 ± 30.5	375.2 ± 11.4	215.5 ± 31.6	202.7 ± 29.2
	+BA	408.9 ± 62.9	154.1 ± 3.7	257.4 ± 45.4	800.1 ± 76.2
Ethanol-insoluble	-BA	82.2 ± 7.4	118.6 ± 0.6	65.8 ± 0.3	59.9 ± 13.1
	+BA	98.6 ± 13.1	44.0 ± 1.2	79.2 ± 0.8	301.4 ± 40.7
B. CO_2	-BA	46.5 ± 2.5	49.1 ± 3.2	32.9 ± 1.5	54.8 ± 3.0
	+BA	39.1 ± 0.8	26.5 ± 2.3	42.7 ± 1.6	283.4 ± 3.0
Ethanol-soluble	-BA	483.0 ± 105.0	762.0 ± 112.7	489.2 ± 15.0	755.6 ± 23.6
	+BA	362.4 ± 64.2	361.7 ± 1.4	516.2 ± 30.9	1,916.9 ± 126.5
Ethanol-insoluble	-BA	15.3 ± 0.6	34.0 ± 5.0	14.8 ± 3.5	26.9 ± 1.3
	+BA	12.9 ± 3.5	12.1 ± 0.3	19.3 ± 3.1	61.2 ± 3.2
C. Ethanol-soluble	-BA	104.4 ± 30.7	2,202.7 ± 435.3	7,379.3 ± 506.7	7,243.9 ± 2,509.4
	+BA	110.8 ± 21.1	738.1 ± 18.6	3,142.3 ± 252.7	3,545.6 ± 230.1
Ethanol-insoluble	-BA	71.1 ± 22.9	723.6 ± 91.6	2,756.5 ± 336.2	2,175.8 ± 1,086.8
	+BA	63.6 ± 13.3	295.0 ± 34.2	1,515.2 ± 256.8	1,544.6 ± 20.3

*Becquerel.

occurred. During this stage the synthesis of glutamine strongly increased in the BA-treated cotyledons. This suggests a positive influence of this cytokinin on nitrogen incorporation prior to differentiation. After three days of culture, large amounts of labeled citrate were detected (5). Recently completed work has indicated that there is a high rate of turnover in citrate at day 1 and that the pool size is very small (L. Bender, R.W. Joy, and T.A. Thorpe, unpubl. results).

At days 10 and 21, it was found that general metabolic patterns were qualitatively the same in shoot-forming and nonshoot-forming cotyledons (4). However, metabolism leading to respiration and amino acid synthesis was strongly enhanced in the shoot-forming cultures. An increased incorporation of label into protein due to the cytokinin treatment was detected during the early culture period (days 0 and 3). Labeled amino acids were incorporated into protein to different degrees, but this was not influenced by the hormonal treatment. Later in culture, protein synthesis was strongly enhanced in the shoot-forming cultures. The radioactivity present in the lipid fraction at days 0 and 3 indicated the synthesis of metabolically stable lipids, since the radioactivity of this fraction increased during the chase period. In contrast, later in culture there was considerable turnover in the lipid fraction, indicating that lipid synthesis for both structural and nonstructural components was taking place.

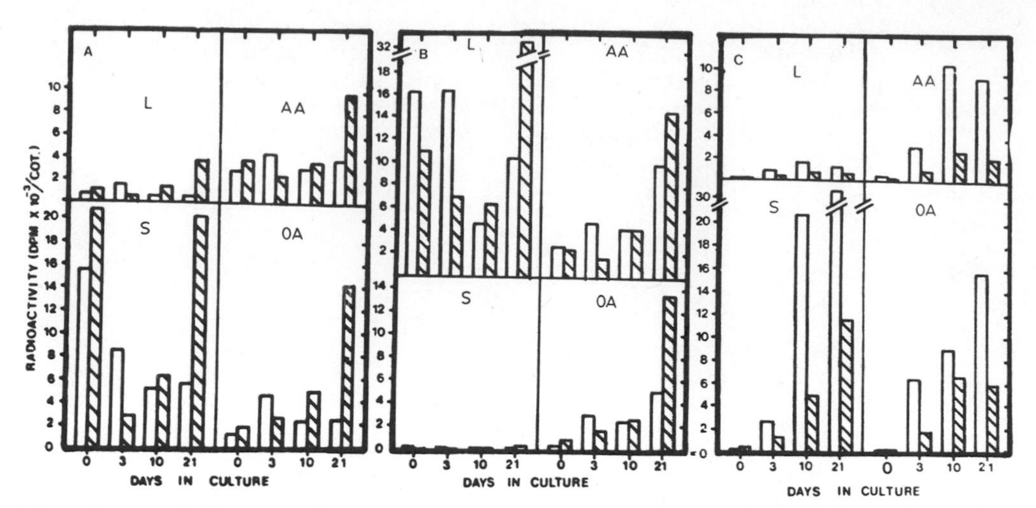

Fig. 7. Radioactivity present in various ethanol-soluble compounds of excised cotyledons of radiata pine, cultured in the absence ☐ or presence ▧ of BA and incubated on [^{14}C]glucose (A), [^{14}C]acetate (B), or [^{14}C]bicarbonate (C) for 3 hr on day indicated. L, lipids; AA, amino acids; S, sugars; OA, organic acids.

Although the findings above are not dramatic, and in many respects predictable, it is clear that at day 0, during the first 3 hr of excision, BA had an effect on the cotyledonary metabolism (glutamine) and, as indicated earlier, major differences in the rate of RNA and protein synthesis were evident by 24 hr (36). Although we have shown that the BA must be present during the first 24 hr of culture for subsequent bud formation (33), it would seem that the cotyledons must receive the cytokinin signal from the start of culture.

CONCLUSIONS

Work carried out on in vitro organogenesis in tobacco has indicated the complex involvement of phytohormones in the shoot-forming process, both exogenously (24) and endogenously (28,29,30). The use of modified Ti-plasmids has added support to this view (23). The data suggest that different gene products independently suppress shoot and root formation and that the T-DNA gene products act in an analogous way to auxin- and cytokinin-like growth regulators (14,23). Furthermore, it appears that plants have separate genetic programs for shoot and root development and that both of these programs must be internally coordinated, since they can be controlled by the products of single genes.

Carbohydrate has been shown to have a dual role in both energy (ATP) production and osmotic adjustment in tobacco (28,29,30). Various biochemical measures have supported the idea that primordium formation is a high energy-requiring process, in which accumulated starch and free sugars from the medium are utilized. The process is metabolically very demanding, as there is also a greater requirement for reducing power (NADPH) compared to growing tissue. Studies carried out on nitrogen assimilation and amino acid (particularly aromatic amino acid) metabolism also

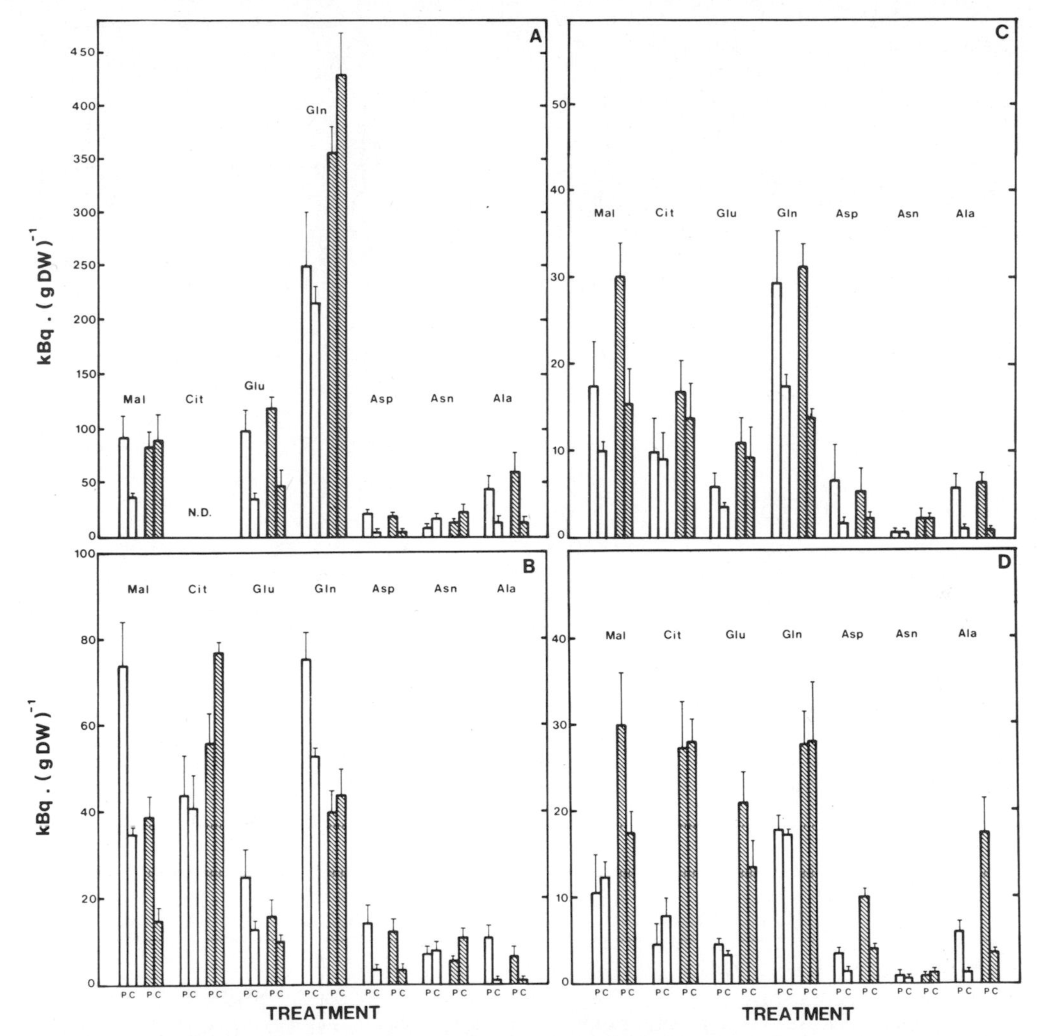

Fig. 8. Radioactivity of some labeled metabolites in cultured radiata pine cotyledons after 3 hr of pulse (P)-labeling with [^{14}C]glucose and a subsequent chase (C) with unlabeled glucose at days 0 (A), 3 (B), 10 (C), and 21 (D) in culture in the absence ☐ or presence ▨ of BA. Mal, malate; Cit, citrate; Glu, glutamate; Gln, glutamine; Asp, aspartate; Asn, asparagine; Ala, alanine; N.D., not detectable. Data are the mean ±S.D. of two independent experiments with two replicates per experiment. [(A) and (B) reproduced from Ref. 5, with permission.]

indicate, not unexpectedly, that N-metabolism is also important in the differentiation process. The findings we have obtained using excised cotyledons of radiata pine support those obtained with tobacco callus, with obvious differences, e.g., the utilization of stored lipid in energy production in cotyledons, as opposed to the accumulation of starch and its utilization in tobacco. Nevertheless, the requirement for the production of energy is clear in both tissues. However, much more remains to be done.

Much less is known about de novo organogenesis in conifers than in herbaceous or even woody angiosperms. This is not surprising since less is also known about the physiology of trees in general, and conifers in particular. However, things are beginning to change and more research is being carried out on the physiology of woody species. As evidence of this belief is the appearance of two new journals (Tree Physiology and Trees: Structure and Function) in the last two years. With expanded interest in the application of tissue culture and recombinant DNA technologies to woody plants, the need to understand the physiology of trees increases. Furthermore, the exploitation of these technologies for tree modification and improvement demands the ability to regenerate plantlets from the modified cells. While manipulation of the cultures through empirical studies will continue to be a main approach, understanding of the process is likely to make such experimentation more rational. This is likely to bring success earlier with recalcitrant tissues, such as subcultured conifer callus, which to date has shown little capacity to undergo de novo organogenesis.

ACKNOWLEDGEMENTS

The author expresses thanks to present graduate students Michael R. Thompson and Richard W. Joy IV for their assistance with the manuscript. He also acknowledges with gratitude the contributions of colleagues, graduate students, postdoctoral fellows, visiting scientists, and research assistants to the research reported here. The author also gratefully acknowledges research funding from the Natural Sciences and Engineering Research Council of Canada (Operating and Strategic Grants), and the Weyerhaeuser Company, Tacoma, Washington. Finally, he acknowledges with thanks the gift of the radiata pine seed from the Forest Research Institute, Rotorua, New Zealand.

REFERENCES

1. Aitken, J., K.J. Horgan, and T.A. Thorpe (1981) Influence of explant selection on the shoot-forming capacity of juvenile tissue of Pinus radiata. Can. J. For. Res. 11:112-117.
2. Aitken-Christie, J., A.P. Singh, K.J. Horgan, and T.A. Thorpe (1985) Explant developmental state and shoot formation in Pinus radiata cotyledons. Bot. Gaz. 146:196-203.
3. Bender, L., A. Kumar, and K.H. Neumann (1985) On the photosynthetic system and assimilate metabolism of Daucus and Arachis cell cultures. In Primary and Secondary Metabolism of Plant Cell Cultures, K.H. Neumann, W. Barz, and E. Reinhard, eds. Springer-Verlag, Berlin, pp. 24-42.
4. Bender, L., R.W. Joy IV, and T.A. Thorpe (1987) [^{14}C]-Glucose metabolism during shoot bud development in cultured cotyledon explants of Pinus radiata. Plant Cell Physiol. (in press).
5. Bender, L., R.W. Joy IV, and T.A. Thorpe (1987) Studies on [^{14}C]-glucose metabolism during shoot induction in cultured cotyledon explants of Pinus radiata. Physiol. Plant. 69:428-434.
6. Biondi, S., and T.A. Thorpe (1982) Growth regulator effects, metabolite changes and respiration during shoot initiation in cultured cotyledon explants of Pinus radiata. Bot. Gaz. 143:20-25.
7. Bornman, C.H. (1983) Possibilities and constraints in the regenerations of trees from cotyledonary needles of Picea abies in vitro. Physiol. Plant. 57:5-16.

8. Brown, D.C.W., and T.A. Thorpe (1986) Plant regeneration by organogenesis. In Cell Culture and Somatic Cell Genetics of Plants, Vol. 3, I.K. Vasil, ed. Academic Press, Inc., New York, pp. 49-65.
9. Chlyah, H. (1974) Formation and propagation of cell division centres in the epidermal layer of internodal segments of Torenia fournieri grown in vitro. Simultaneous surface observations of all the epidermal cells. Can. J. Bot. 52:867-872.
10. Constabel, F., W.G.W. Kurz, K.B. Chatson, and J.W. Kirkpatrick (1977) Partial synchrony in soybean cell suspension cultures induced by ethylene. J. Cell Res. 105:263-268.
11. Dalton, C.C., and H.E. Street (1976) The role of the gas phase in the greening and growth of illuminated cell suspension cultures of spinach (Spinacea oleracea L.). In Vitro 12:485-494.
12. Douglas, T.J., V.M. Villalobos, M.R. Thompson, and T.A. Thorpe (1982) Lipid and pigment changes during shoot initiation in cultured explants of Pinus radiata. Physiol. Plant. 55:470-477.
13. Hasegawa, P.M., T. Yasuda, and T.Y. Cheng (1979) Effect of auxin and cytokinin on newly synthesized proteins of cultured Douglas fir cotyledons. Physiol. Plant. 46:211-217.
14. Inze, D., A. Follin, M. Van Lijsebelens, C. Simoens, C. Genetello, M. Van Montague, and J. Schell (1984) Genetic analysis of the individual T-DNA genes of Agrobacterium tumefaciens; further evidence that two genes are involved in indole-3-acetic acid synthesis. Mol. Gen. Genet. 194:265-274.
15. Kirby, E.G., and M.E. Schalk (1982) Surface structural analysis of cultured cotyledons of Douglas-fir. Can. J. Bot. 60:2729-2733.
16. Kumar, P.P., D.M. Reid, and T.A. Thorpe (1987) The role of ethylene and carbon dioxide in differentiation of shoot buds in excised cotyledons of Pinus radiata in vitro. Physiol. Plant. 69:244-252.
17. Obata-Sasamoto, H., V.M. Villalobos, and T.A. Thorpe (1984) ^{14}C-Metabolism in cultured cotyledon explants of radiata pine. Physiol. Plant. 61:490-496.
18. Patel, K.R., and T.A. Thorpe (1984) Histochemical examination of shoot initiation in cultured cotyledon explants of radiata pine. Bot. Gaz. 145:312-322.
19. Plumb-Dhindsa, P.L., R.S. Dhindsa, and T.A. Thorpe (1979) Non-autotrophic CO_2 fixation during shoot formation in tobacco callus. J. Exp. Bot. 30:759-767.
20. Reilly, K.J., and J. Washer (1977) Vegetative propagation of radiata pine by tissue culture; plantlet formation from embryonic tissue. New Zealand J. For. Sci. 7:199-206.
21. Ross, M.K., and T.A. Thorpe (1973) Physiological gradients and shoot initiation in tobacco callus cultures. Plant Cell Physiol. 14:473-480.
22. Ross, M.K., T.A. Thorpe, and J.W. Costerton (1973) Ultrastructural aspects of shoot initiation in tobacco callus cultures. Am. J. Bot. 60:788-795.
23. Schell, J., M. Van Montague, M. Holsters, J.P. Hernalsteens, P. Dhaese, H. De Greve, J. Leemans, H. Joos, D. Inzel, L. Willmitzer, L. Otten, A. Wostemeyer, and J. Schroeder (1982) Plant cells transformed by modified Ti plasmids: A model system to study plant development. In Biochemistry of Differentiation and Morphogenesis, L. Jaenicke, ed. Springer-Verlag, Berlin and New York, pp. 65-73.
24. Skoog, F., and C.O. Miller (1957) Chemical regulation of growth and organ formation in plant tissues cultured in vitro. Symp. Soc. Exp. Biol. 11:118-131.

25. Stebbins, G.L. (1965) Some relationships between mitotic rhythm, nucleic acid synthesis, and morphogenesis in higher plants. Brookhaven Symp. Biol. 18:204-221.
26. Street, H.E., and L.A. Withers (1974) The anatomy of embryogenesis in culture. In Tissue Culture and Plant Science, H.E. Street, ed. Academic Press, Inc., London, pp. 71-100.
27. Syono, K. (1965) Changes in organ forming capacity of carrot root callus during subcultures. Plant Cell Physiol. 6:403-419.
28. Thorpe, T.A. (1980) Organogenesis in vitro: Structural, physiological and biochemical aspects. Int. Rev. Cytol. (Suppl.) 11A:71-111.
29. Thorpe, T.A. (1983) Morphogenesis and regeneration in tissue cultures. In Genetic Engineering: Applications to Agriculture, L.D. Owens, ed. Rowman and Allanheld, Totowa, New Jersey, pp. 285-303.
30. Thorpe, T.A., and S. Biondi (1981) Regulation of plant organogenesis. Adv. Cell Cult. 1:213-239.
31. Thorpe, T.A., and S. Biondi (1984) Conifers. In Handbook of Plant Cell Culture, Vol. 2, W.R. Sharp, D.A. Evans, P.V. Ammirato, and Y. Yamada, eds. Macmillan Company, Inc., New York, London, pp. 435-470.
32. Thorpe, T.A., and K.R. Patel (1987) Comparative morpho-histological studies on the rates of shoot initiation in various conifer explants. New Zealand J. For. Sci. (in press).
33. Villalobos, V.M., D.W.M. Leung, and T.A. Thorpe (1984) Light cytokinin interactions in shoot formation in cultured cotyledon explants of radiata pine. Physiol. Plant. 61:497-504.
34. Villalobos, V.M., E.C. Yeung, and T.A. Thorpe (1985) Origin of adventitious shoots in excised radiata pine cotyledons cultured in vitro. Can. J. Bot. 63:2172-2176.
35. Villalobos, V.M., E.C. Yeung, S. Biondi, and T.A. Thorpe (1982) Autoradiographic and ultrastructural examination of shoot initiation in radiata pine cotyledon explants. In Plant Tissue Culture (Proceedings of the Fifth International Congress on Plant Tissue and Cell Culture), A. Fujiwara, ed. Japanese Association for Plant Tissue Culture, Tokyo, pp. 41-42.
36. Villalobos, V.M., M.J. Oliver, E.C. Yeung, and T.A. Thorpe (1984) Cytokinin-induced switch in development in excised cotyledons of radiata pine cultured in vitro. Physiol. Plant. 61:483-489.
37. Yasuda, T., P.M. Hasegawa, and T.Y. Cheng (1980) Analysis of newly synthesized proteins during differentiation of cultured Douglas fir cotyledons. Physiol. Plant. 48:83-87.
38. Yeung, E.C., J. Aitken, S. Biondi, and T.A. Thorpe (1981) Shoot histogenesis in cotyledon explants of radiata pine. Bot. Gaz. 142:494-501.

USE OF PROTOPLASTS AND CELL CULTURES FOR PHYSIOLOGICAL AND GENETIC STUDIES OF CONIFERS

Edward G. Kirby[1] and Alain David[2]

[1]Department of Biological Sciences
Rutgers University
Newark, New Jersey 07102

[2]Laboratoire de Biologie et Physiologie Végétales
Université de Bordeaux I
Talence-Cedex, France

ABSTRACT

Potential applications of protoplast technology to programs for the genetic improvement of forest trees include somaclonal and gametoclonal variation, gene transfer, cell fusion, and cloning of selected phenotypes from single cells. The status of research involving conifer protoplasts is reviewed, including analysis of the starting materials, preconditioning procedures, isolation techniques, culture media, and culture systems. Protoplasts and cell suspension cultures have also been used for physiological and genetic studies regarding nitrogen nutrition and assimilation and the effects of simulated water stress on osmotic adjustment. Recent results are discussed.

INTRODUCTION

Evolving protocols in plant biotechnology hold particular promise for the improvement of forest tree species (2,22,24,28,32,37). Biotechnological procedures with direct application to tree improvement include mass clonal propagation, in vitro selection utilizing both natural variation and variation induced in culture, somatic hybridization, somaclonal and gametoclonal variation, and genetic transformation using suitable vectors. Fundamental to these research fronts are successful procedures for the isolation and culture of protoplasts of important forest species and the regeneration of trees from protoplast-derived cell and callus cultures.

POTENTIAL USES OF PROTOPLASTS IN TREE IMPROVEMENT AND BIOTECHNOLOGY

Somaclonal and Gametoclonal Variation

Simply defined, somaclonal variation is phenotypic (genetic) variation detected in plants derived from cell culture. Gametoclonal variation is variation detected in plants derived from cultures of gametophytic (haploid) tissues (11). Techniques for isolation, culture, and plant regeneration from protoplasts are an essential part of the application of somaclonal variation to genetic improvement. Techniques enabling regeneration of whole plants from single protoplasts can insure that regenerated individuals will be genetically uniform, so that the incidence of chimeras can be greatly reduced.

Applications of somaclonal variation have been reported for a number of economically important species, including carrot, geranium, tobacco, tomato, alfalfa, sugarcane, wheat, rapeseed, pepper, rice, oats, corn, and potato (11,26). Most exciting is the frequency of observed induced variation. The frequency of single gene mutations in tomato can be as high as one mutant in every 20 to 25 regenerated plants (11). Currently, investigators are making use of variation that is induced in culture to produce improved varieties of horticultural and agronomic crops. The application of somaclonal variation to tree biotechnology will be forthcoming as new procedures become established which allow regeneration of plants from protoplasts and cell cultures.

Somatic Hybridization

Somatic hybridization is potentially an extremely powerful technique. Much recent research has centered on the transfer of cytoplasmically encoded characteristics or on the transfer of a limited number of nuclear genes for which good selection systems have been established (10,31).

Procedures for protoplast fusion first involve the preparation of protoplasts followed by chemically or physically induced fusion. The polyethylene glycol (PEG) method for fusion (21) has produced consistently good results with a variety of species. An alternative to the use of PEG is the use of electric field-induced fusion (electrofusion) (15). Cell cultures derived from fusion experiments consist of heterogeneous populations of parental cells, homokaryotic fusion products, and heterokaryotic fusion products. It is necessary, therefore, to establish definitive procedures for selection of heterokaryotic fusion products. This can be accomplished by the use of genetic markers or by microselection, in which fusion products are visually identified and cultured independent from other cells.

Protoplast Applications in Transformation Technology

Clearly, the technology for production of transgenic forest trees has arrived. Introduction of an herbicide-resistance gene (*aroA*) into *Populus* NC5339 using a binary *Agrobacterium* vector has been achieved (13). Elsewhere in this Volume, applications of techniques in genetic transformation to woody plants will be addressed, including electroporation and *Agrobacterium*-mediated transformation. Development of techniques for the isolation, culture, and regeneration of forest trees from protoplasts is in many cases a prerequisite to development of efficient transformation protocols.

Current Directions and Applications

In order to facilitate the use of protoplasts in forest biotechnology, efficient systems for regeneration of trees from protoplasts must be developed. At present, relatively few plants, including herbaceous crops, can be regenerated from protoplasts. There are no reports of reliable regeneration of conifer forest trees from protoplasts. Potentially the most useful contribution of protoplast and cell culture technology to forest genetics perhaps lies in the area of somaclonal variation, where specific traits can be efficiently recognized in variant plants regenerated from somatic cells and protoplasts. By placing particular emphasis on selection of monogenic and allelogenic traits, such as disease resistance and herbicide tolerance, extremely desirable phenotypes of forest trees can be produced.

THE STATUS OF PROTOPLAST RESEARCH WITH CONIFER SPECIES

Protoplasts have been isolated from a number of woody plant species, and recent advances in protoplast culture techniques have enabled the establishment of a complete protoplast system for *Populus* (34). Although the in vitro responses of *Populus* appear to be genotype-dependent, applications of protoplast techniques to genetic improvement and utilization of the species are apparent. Conifers comprise the most important group of forest trees, but unfortunately there are few reports of successful culture of conifer protoplasts (Tab. 1).

Starting Material for Conifer Protoplast Isolation

Haploid cells. In spite of the interest in haploid cells, there are only two reports dealing with the use of haploid material as a source of protoplasts. Duhoux (9) incubated pollen of *Cupressus arizonica* for more than 20 hr in a solution containing a high concentration [12.5% (w/v)] of protoplasting enzymes. Complete digestion of the intine layer of the pollen wall was particularly difficult, since it is actively resynthesized during enzymatic digestion.

Female gametophytes of *Picea abies* easily released haploid protoplasts provided that the tissue was used when cells of the gametophyte were devoid of storage materials, i.e., between the stages of archegonium initiation and the beginning of embryo development (19).

Tab. 1. Isolation and culture of conifer protoplasts.

Species	Reference(s)
Biota orientalis	5
Pinus contorta	18
Pinus coulteri	30
Pinus lambertiana	17
Pinus pinaster	4, 6, 8, 12
Pinus taeda	36
Pseudotsuga menziesii	23, 25

Diploid cells. Cell suspension cultures have been used as starting material for conifer protoplast isolation. Cell suspensions are initially derived from callus cultures. Callus is induced by culture of organ explants (cotyledons or needles) on agar media. Actively growing cells are then suspended in liquid medium and routinely subcultured for several weeks.

Hakman and von Arnold (18) isolated protoplasts from cell suspensions initiated from three-month-old callus derived from mature embryos of *Pinus contorta*. Teasdale and Rugini (36) obtained protoplasts from cell suspensions initiated from hypocotyl callus that was derived from two-week-old seedlings of *Pinus taeda*. Protoplasts have also been isolated from callus initiated from both cotyledons and needles of adult trees of *Pinus lambertiana* (17).

Pine roots have served as starting material for protoplast isolation. Root segments 20 mm long, excluding 7 mm of tip, were collected from 12- to 15-day-old seedlings of *Pinus pinaster* (12). These segments represent the differentiating region of the root. In spite of seed germination under sterile conditions, an additional sterilization of the root segments was necessary to obtain axenic protoplast cultures. This sterilization was performed as follows. Root segments were immersed for 1 to 2 min in 0.1% (w/v) mercuric chloride in 50% (v/v) ethanol, then washed three times with sterile water. Enzyme penetration was facilitated by splitting the root segments longitudinally and by cutting transversely.

Cotyledons obtained from two- to four-week-old seedlings have also been used as a source of protoplasts for Douglas fir (*Pseudotsuga menziesii*) (25), *Pinus pinaster* (4), *Pinus coulteri* (30), and *Biota orientalis* (5). By careful examination of the physiological age of cotyledon source materials, three- to five-fold increases in protoplast yields have been obtained when young cotyledons were used. Protoplasts isolated from young cotyledons were also capable of regenerating cells with increased mitotic activity (6).

Elongating primary leaves from seedlings of *Pinus pinaster* germinated under sterile conditions and leaves from axillary buds induced in vitro have provided a good source of protoplasts (8). Leaf (needle) protoplasts were smaller than cotyledon protoplasts isolated at the same osmolality. Protoplasts have also been isolated from needles excised from sprouting apical and axillary buds of mature *Pinus lambertiana* (17).

Preconditioning Treatments

Depending on the protoplast source, various preconditioning treatments have been performed in order to increase the yield of viable protoplasts and to stimulate subsequent mitotic activity. For isolation of protoplasts from mature pollen of *Cupressus arizonica*, the pollen must be hydrated in sterile tapwater prior to enzyme digestion. As a result of hydration, the exine layer swells, bursts, and is shed from the pollen grain. Upon transfer of the pollen to the enzyme mixture, the cellulosic intine is digested (9).

Cell suspension cultures in the exponential phase of growth provide a superior source of starting material for protoplasts. Moreover, the formulation of the medium used for cell suspension culture can affect the cell wall composition and, as a consequence, the efficiency of protoplast isolation. Teasdale and Rugini (36) reported that a medium characterized by high Mg^{2+}, low Ca^{2+}, and high levels of most mineral micronutrients resulted in suspension cultures consisting of small cell clusters capable of producing high yields of viable protoplasts. Hakman and von Arnold (18) reported that high rates of cell division are found in protoplasts isolated from fast-growing, rapidly dividing cell suspension cultures.

Preplasmolysis in 0.7 M sorbitol in the presence of 30 mM cysteine for 1 hr prior to protoplast isolation increased the yield of root protoplasts of *Pinus pinaster* (12). It has been suggested that the sulfhydryl groups of cysteine prevent phenolic inhibition of commercially available cell wall-degrading enzymes (40).

In order to minimize the physiological heterogeneity of the plant material, Kirby and Cheng (25) have used a preconditioning treatment. Cotyledons of Douglas fir cultured for eight days on a solid nutrient medium supplemented with 15 μM N^6-benzylaminopurine (BAP) and 500 nM α-naphthaleneacetic acid (NAA) produced high yields of protoplasts (Fig. 1). Such a preculture period in the presence of phytohormones may stimulate cell division in cotyledons prior to protoplast isolation or may alter the chemical nature of the cell wall so that it becomes more susceptible to enzymatic degradation.

In isolating protoplasts from *Pinus pinaster*, preconditioning of starting materials has also proven valuable. Roots were removed from young seedlings and the cut ends of the hypocotyls were placed in a solution containing half-strength macronutrients and full-strength micronutrients of Murashige and Skoog medium (29) supplemented with 15 μM BAP and 0.5 μM NAA for seven days in the dark. Protoplasts (Fig. 2) from cotyledons thus treated produced cell cultures with higher rates of mitosis than those derived from nonpretreated plant materials (6).

In isolating protoplasts from young pine needles, improved protoplast yields were obtained if cut pieces of the starting material were pretreated with 10% (w/v) mannitol in a nutrient solution for 1 to 2 hr prior to enzymatic digestion (17).

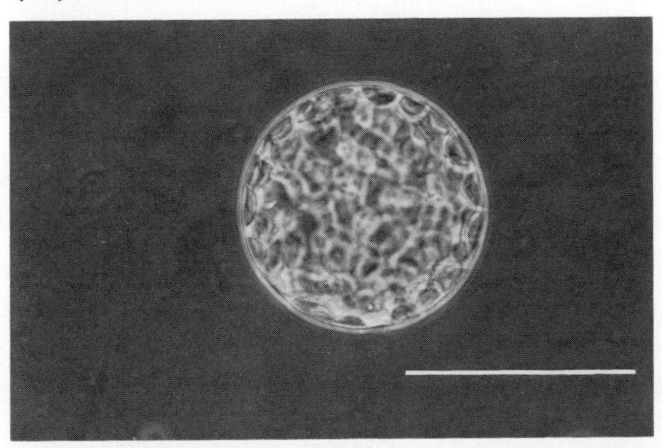

Fig. 1. Protoplast freshly isolated from cotyledons of Douglas fir (*Pseudotsuga menziesii*) cultured for seven days on medium containing 15 μM NAA and 500 nM BAP. Bar represents 0.05 mm.

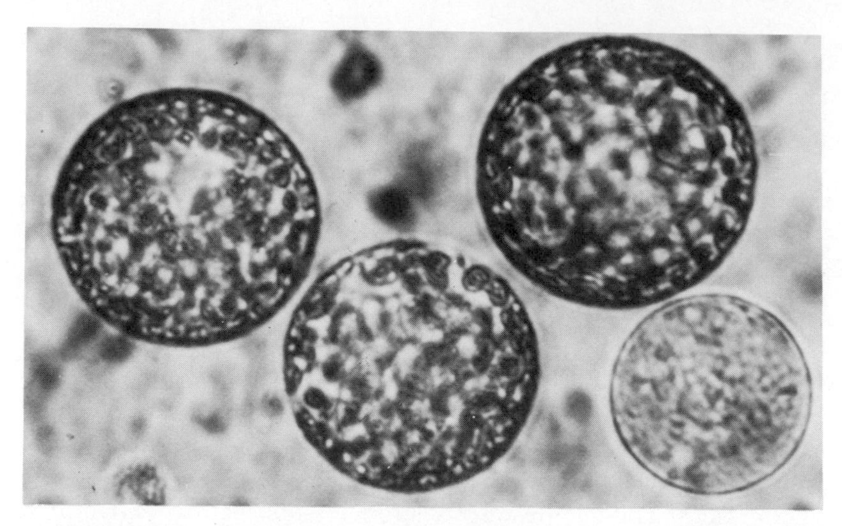

Fig. 2. Protoplasts isolated from cotyledons of pine (Pinus pinaster). Roots were removed from young seedlings, and cut ends of hypocotyls were placed in a solution containing half-strength macronutrients and full-strength micronutrients of Murashige and Skoog medium supplemented with 15 μM BAP and 0.5 μM NAA for seven days in the dark. Protoplasts isolated from cotyledons range in diameter from 35 to 50 μm.

Protoplast Isolation Techniques

Enzymes. Commercially available cell wall-degrading enzymes commonly used for protoplast isolation from herbaceous species (e.g., cellulases, hemicellulases, and pectinases) have been utilized to isolate conifer protoplasts. In isolating protoplasts from cotyledons of P. pinaster, an increase in yield and viability was obtained by using desalted pectinase and hemicellulase in combination with nondesalted cellulase (7). Desalting commercially available enzymes has not been necessary in isolating Douglas-fir protoplasts that are capable of cell division and callus formation (23,25).

The time required for enzymatic release of protoplasts differs with the species utilized, the nature of the starting material, and the enzymes employed. Protoplasts were produced in 3.5 hr from Douglas-fir cotyledons by using high concentrations of enzymes [up to 2.5% (w/v)] (25). Lower enzyme concentrations [0.2% (w/v)] required longer incubation periods (up to 24 hr) to release cotyledon protoplasts of P. pinaster (6). Within the same species, optimum enzyme concentration can vary for different donor tissues (8).

Plasmolyzing agents and isolation mixtures. Cells are plasmolyzed in order to facilitate isolation of protoplasts and to protect protoplast integrity. Plasmolyzing agents include compounds that can readily be metabolized (glucose and sucrose) and those that are metabolized to a far lesser extent (mannitol and sorbitol). Normal concentrations employed range between 0.5 and 0.6 M. The optimum concentration of sorbitol can, however, be dependent on the concentration of the enzymes employed. Higher enzyme levels [5% (w/v)] require lower concentrations of sorbitol (0.4 M) and vice versa (25).

Enzymes and osmoticum are dissolved in culture medium or in a solution containing various mineral components. Buffering agents, including 2-[N-morphilino]ethanesulfonic acid (MES), have been reported to be beneficial for the isolation of pine protoplasts (7,17). Viability of protoplasts of sugar pine (Pinus lambertiana) increased from 20 to 75% when 50 mM MES was included in the protoplast isolation mixture (17). Calcium chloride, usually in the range of 2 to 10 mM, has been shown to help maintain protoplast integrity. An increase in viability of protoplasts isolated from cell suspensions of loblolly pine was shown when 500 mg per ml bovine serum albumin (BSA) was added to the enzyme solution (36). The authors suggest that BSA acts as a competitive substrate for contaminating proteases and, thereby, protects membrane proteins.

Protoplast purification. Undigested materials and cell debris are routinely removed from protoplast isolation mixtures following Constabel's technique (3). In short, the mixture is filtered through a 40 to 50 μm sieve and centrifuged for 5 to 7 min at 100 to 150 xg. Pelleted protoplasts are washed a minimum of three times by resuspension in a washing solution of the same osmotic strength as the isolation mixture, followed by centrifugation. The washed pellet is finally resuspended in culture medium. In order to obtain purified protoplast suspensions, various modifications of this general procedure have been made. A brief enzymatic pretreatment (30 min) of sliced cotyledons of Douglas fir has been reported to remove much of the cell debris caused by the slicing procedure. Purity of conifer protoplast preparations can be improved by layering the protoplast pellet on a sucrose cushion [16% (w/v)] (19) or by suspending the pellet in 21% (w/v) sucrose (30). In both cases, tubes are centrifuged and the buoyant, purified protoplast layer is removed; then protoplasts are suspended in medium and cultured. Ficoll density gradient centrifugation has also been reported to be useful in obtaining clean conifer protoplast pellets (17).

Centrifugation procedures frequently disrupt fragile conifer protoplasts. If protoplasts have particularly high densities due to increased starch content, such as those isolated from cell suspensions of Pinus taeda, it is advisable to allow protoplasts to settle in tubes without centrifugation (33).

Determination of viability. Viability of freshly isolated protoplasts can be assessed by various techniques, including exclusion of the vital stain Evans blue (20) and staining with fluorescein diacetate (33).

Culture Techniques

From the few reports of regeneration of dividing cells from protoplasts of conifers (Tab. 1), it appears that the technique used for protoplast culture is secondary in importance to selection of proper starting materials. Plating efficiencies of conifer protoplasts are generally reported to be extremely low (below 10%). Plating efficiencies must be improved if protoplasts of conifer species are to be utilized efficiently in biotechnological applications.

Plating density. Many factors limit successful culture of conifer protoplasts. Of primary concern is plating density. Review of the literature reveals that plating densities for conifer protoplasts vary considerably, from 5 x 10^4 protoplasts per ml for loblolly pine (36) to 2 x 10^6 protoplasts per ml for maritime pine (4).

Culture media. Media used for culture of conifer protoplasts are based on those formulated for tissue culture of both herbaceous and woody species (14,16,29,39). The calcium concentration is usually increased to 3 to 5 mM, since this divalent cation improves protoplast stability and is important in cell wall formation. Dimethyl sulfoxide and MES encourage division in protoplast-derived cells of *Pinus lambertiana* (17). Culture media are supplemented with levels of auxin and cytokinin appropriate for callus induction (18,25).

Attention to nitrogen source and medium osmolality is fundamental in obtaining sustained cell divisions in cultures of conifer protoplasts. A supplemental source of reduced organic nitrogen, such as casein hydrolysate, arginine, asparagine, and glutamine, stimulates mitotic activity. Glutamine is a limiting factor in obtaining dividing cells from cotyledon protoplasts of Douglas fir (23,25). The optimum level of glutamine was shown to be 50 mM. Other reports of protoplast cultures of conifers have shown that glutamine supplements were very beneficial (7,8,36).

In culture of protoplasts of maritime pine, osmolality of the medium at the start of culture is adjusted to approximately 600 to 700 mOsm kg H_2O^{-1} with sorbitol and/or mannitol. Glucose and/or sucrose can be used as both carbon source and osmotic stabilizer. Cell division can be stimulated if the osmolality is gradually reduced by replacing part of the original medium with fresh medium of lower osmotic potential (7,8,17,25,36).

Culture systems. The most frequent culture technique reported for tree protoplasts has been liquid culture. Droplets (50 to 100 μl) of protoplasts suspended in culture medium are placed in plastic petri dishes which are then sealed with Parafilm strips (6,30). Alternatively, protoplast suspensions have been cultured in multiwell plates or in petri dishes of various diameters (17,36). A fabric support method has been reported beneficial for culture of cotyledon protoplasts of Douglas fir (25). One milliliter of protoplast suspension was pipetted into 35 mm x 10 mm plastic petri dishes layered with 100% polyester fleece (0.5 mm thickness). Under these conditions regenerated cells proliferated, leading to colony formation. The fabric culture system facilitates sequential lowering of the osmoticum concentration. Since cells are intermingled in the fabric, medium is easily removed. This culture method also allows free diffusion of inhibitory substances, including phenolics, away from living cells and provides a large surface area for diffusion of oxygen into the medium. In addition, the fleece forms a matrix within which protoplasts may grow independent of each other. A similar technique, the floating screen discs used for culture of poplar protoplasts (34), offers many of the benefits of the polyester fleece and has the advantage of facilitating protoplast observation.

Plating of protoplasts in low temperature agarose and use of alginate beads for protoplast culture have been reported to increase plating efficiencies significantly in some species (1,35,38). The use of alginate beads is particularly intriguing, since it allows protoplasts to be cultured with feeder suspensions of actively growing cells. Applications of these techniques to culture of conifer protoplasts may boost plating efficiencies.

Culture conditions. Reports of protoplast culture of conifer species indicate that protoplasts are best cultured at 22 to 25°C under high relative humidity either in darkness or in diffuse light.

Behavior of Protoplast-Derived Cells

Very low initial rates of cell division in protoplast-derived cells of conifers are the norm. In most cases, no divisions are maintained beyond the colony stage (ten to 20 cells) due to phenolic accumulation (17,30). Browning of cells and formation of tracheary elements were observed in cell cultures derived from protoplasts isolated from needles of Pinus pinaster. High-pressure liquid chromatography (HPLC) of soluble phenolic compounds has revealed that p-coumaric acid, p-hydroxybenzoic acid, (+)-catechin, and procyanidin B-3 are the major components. These products are precursors of condensed tannins. Vanillic acid, a precursor of lignin, has also been detected in small amounts (8).

When cotyledon protoplasts of Douglas fir were cultured in the presence of 50 mM glutamine on fabric supports combined with sequential lowering of the sorbitol level, regenerated cells proliferated, leading to colony formation (23,25) (Fig. 3). A combination of low osmolality (225 mOsm kg H_2O^{-1}) and elevated glutamine (40 mM) was essential in obtaining actively growing cell suspensions from cotyledon protoplasts of Pinus pinaster (7) (Fig. 4). Callus produced from protoplasts isolated from needles of this species (Fig. 5) maintained growth when subcultured onto an agar medium; however, reduction of the levels of calcium and carbon was shown to be essential (8). Total DNA contents of protoplasts freshly isolated from young seedling needles and DNA levels of three- and seven-month-old protoplast-derived callus lines have been determined. In three- and seven-month callus, 65% and 74% of the nuclei, respectively, have DNA levels within the range observed for nuclei isolated from needle protoplasts, indicating some genetic stability in protoplast-derived callus lines of conifers (8).

LESSONS FROM CONIFER CELL SUSPENSION CULTURES

In recent work, cell suspension cultures of Douglas fir have been used to characterize cellular responses to water stress as induced by PEG (27). Results of these studies are of key importance to protoplast work,

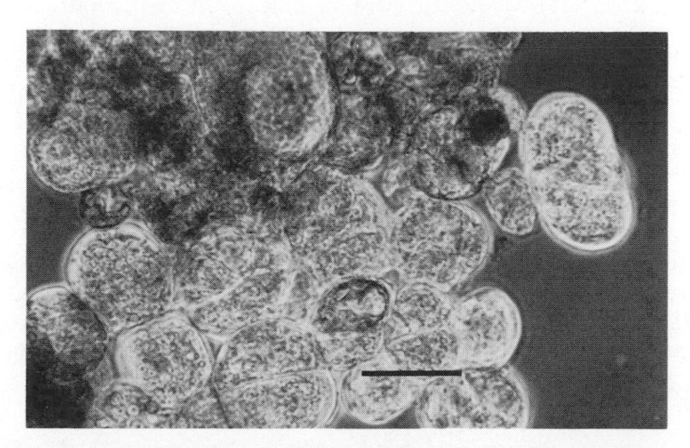

Fig. 3. Cotyledon protoplasts of Douglas fir cultured for 21 days in the presence of 50 mM glutamine, 15 μM NAA, and 5 μM BAP on a fabric support combined with sequential lowering of the sorbitol level. Bar represents 0.05 mm.

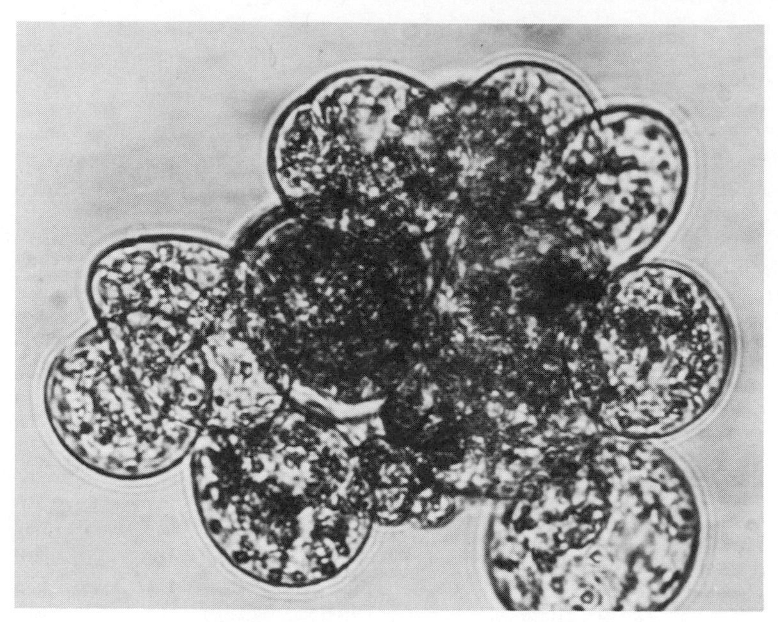

Fig. 4. Cell colony derived from cultured protoplasts of pine (Pinus pinaster) grown under conditions of low osmolality (225 mOsm kg H_2O^{-1}) and elevated glutamine (40 mM). Diameter of the colony is 250 to 300 µm.

since, as plasmolyzed cells, protoplasts are initially placed under water-stress conditions and are grown under water stress. The ability of cells to maintain viability and to produce actively growing cell cultures after subculture onto water-stress medium is directly related to the time in the growth cycle when the cells are first exposed to the induced osmotic stress.

Long-term growth of Douglas-fir cells on glutamine prior to PEG-induced water stress plays a significant role in increasing resistance to water stress. Levels of soluble amino acids in stress-adapted and control cell lines have been measured using HPLC. Proline levels in PEG-adapted cell lines were four times higher than in glutamine-grown controls, whereas glutamine levels were up to 12 times higher in PEG-adapted cells than in glutamine-grown controls.

The messages for conifer protoplast work are clear. The age of cells from which protoplasts are prepared (i.e., the point in the culture cycle when the stress is imposed) is critical in regenerating viable, growing cells and callus from protoplasts. In addition, conditions favoring accumulation of glutamine and proline in donor cells may facilitate high plating efficiencies of protoplasts derived from them.

CONCLUSIONS

Establishment of protoplast techniques for conifers is of crucial importance in developing forest biotechnology. The direction of current

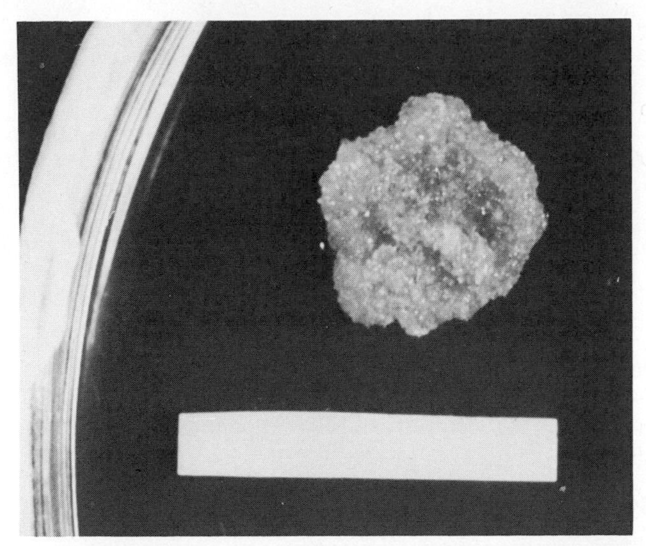

Fig. 5. Callus derived from protoplasts of pine (Pinus pinaster) grown on agar medium, as described in text. Bar represents 2 cm.

work is to improve plating efficiencies for conifer protoplasts and to use embryogenic cell cultures as a source of material for protoplasts capable of regeneration. It is clear, however, that applications of protoplast techniques for genetic improvement await development of a reliable system for the routine production of trees from protoplast-derived cell cultures.

REFERENCES

1. Adams, T.L., and J.A. Townsend (1983) A new procedure for increasing efficiency of protoplast plating and clone selection. Plant Cell Reports 2:165-168.
2. Ahuja, M.R. (1984) Protoplast research in woody plants. Silvae Genet. 33:32-37.
3. Constabel, F. (1975) Isolation and culture of plant protoplasts. In Tissue Culture Methods, O.L. Gamborg and L. Wetter, eds. National Research Council, Saskatoon, Saskatchewan, Canada, pp. 11-21.
4. David, A., and H. David (1979) Isolation and callus formation from cotyledon protoplasts of pine (Pinus pinaster). Z. Pflanzenphysiol. 94:173-177.
5. David, H., A. David, and T. Mateille (1981) Isolation and culture of protoplasts of two gymnosperms: Pinus pinaster and Biota orientalis. In Proceedings of the International Colloquium on In Vitro Culture of Tissues of Forest Species, M. Boulay, ed. AFOCEL, Nangis, France, pp. 339-347.
6. David, H., A. David, and T. Mateille (1982) Evaluation of parameters affecting the yield, viability, and cell division of Pinus pinaster protoplasts. Physiol. Plant. 56:108-113.
7. David, H., E. Jarlet, and A. David (1984) Effects of nitrogen source, calcium concentration and osmotic stress on protoplasts and protoplast-derived cell cultures of Pinus pinaster cotyledons. Physiol. Plant. 61:477-481.

8. David, H., M.T. de Boucaud, J.M. Gaultier, and A. David (1986) Sustained division of protoplast-derived cells from primary leaves of Pinus pinaster, factors affecting growth and change in nuclear DNA content. Tree Physiol. 1:21-30.
9. Duhoux, E. (1980) Protoplast isolation of gymnosperm pollen. Z. Pflanzenphysiol. 99:207-214.
10. Evans, D.A. (1983) Protoplast fusion. In Handbook of Plant Cell Culture, Vol. 1, D.A. Evans, W.R. Sharp, P.V. Ammirato, and Y. Yamada, eds. Macmillan Publishing Company, New York, pp. 291-321.
11. Evans, D.A., W.R. Sharp, and H.-P. Medine-Filho (1984) Somaclonal and gametoclonal variation. Am. J. Bot. 71:759-774.
12. Faye, M., and A. David (1983) Isolation and culture of gymnosperm root protoplasts (Pinus pinaster). Physiol. Plant. 59:359-362.
13. Fillatti, J.J., J.C. Sellmer, and B.A. McCown (1986) Regeneration and transformation of Populus. In Proceedings VI Congress of Plant Tissue and Cell Culture, D.A. Sommers et al., eds. IAPTC, Minneapolis, Minnesota, p. 127.
14. Gamborg, O. (1977) Culture media for plant protoplasts. In CRC Handbook Series in Nutrition and Foods, Vol. IV, M. Recheigl, Jr., ed. CRC Press, Inc., Cleveland, Ohio, pp. 415-422.
15. Gaynor, J.J. (1986) Electrofusion of plant protoplasts. In Handbook of Plant Cell Culture, Vol. 4, D.A. Evans, W.R. Sharp, P.V. Ammirato, and Y. Yamada, eds. Macmillan Publishing Company, New York, pp. 149-171.
16. Gupta, P.K., and D.J. Durzan (1985) Shoot multiplication from mature trees of Douglas-fir (Pseudotsuga menziesii) and sugar pine (Pinus lambertiana). Plant Cell Reports 4:177-179.
17. Gupta, P.K., and D.J. Durzan (1986) Isolation and cell regeneration of protoplasts from sugar pine (Pinus lambertiana). Plant Cell Reports 5:346-348.
18. Hakman, I.C., and S. von Arnold (1983) Isolation and culture of protoplasts from cell suspensions of Pinus contorta Dougl. ex. Loud. Plant Cell Reports 2:92-94.
19. Hakman, I., S. von Arnold, and H. Fellner-Feldegg (1986) Isolation and DNA analysis of protoplasts from developing female gametophytes of Picea abies (Norway spruce). Can. J. Bot. 64:108-112.
20. Kanai, R., and G.E. Edwards (1973) Purification of enzymatically isolated mesophyll protoplasts from C3, C4 and crassulacean acid metabolism plants using an aqueous dextran-polyethylene glycol two-phase system. Plant Physiol. 52:484-490.
21. Kao, K.N., and M.R. Micahyluk (1974) A method for high frequency intergeneric fusion of plant protoplasts. Planta 115:355-367.
22. Karnosky, D.F. (1981) Potential for forest tree improvement via tissue culture. BioScience 31:114-120.
23. Kirby, E.G. (1980) Factors affecting proliferation of protoplasts and cell cultures of Douglas-fir. In Plant Cell Cultures: Results and Perspectives, F. Sala et al., eds. Elsevier/North-Holland, Amsterdam, pp. 289-293.
24. Kirby, E.G. (1982) The use of in vitro techniques for genetic modification of forest trees. In Tissue Culture in Forestry, J.M. Bonga and D.J. Durzan, eds. Martinus Nijhoff/Dr. W. Junk Publishers, Amsterdam, pp. 369-386.
25. Kirby, E.G., and T.-Y. Cheng (1979) Colony formation from protoplasts derived from Douglas-fir cotyledons. Plant Sci. Lett. 14:145-154.

26. Larkin, P.J., R. Bretell, S. Ryan, and W. Scowcroft (1983) Protoplasts and variation from culture. In Proceedings of the International Protoplast Symposium, Basel, Switzerland, pp. 51-56.
27. Leustek, T., and E.G. Kirby (1987) The influence of glutamine on growth and viability of Douglas-fir cell suspension cultures after exposure to polyethylene glycol. Tree Physiol. (submitted for publication).
28. Libby, W.J., and R.M. Rauter (1984) Advantages of clonal forestry. For. Chron. 60:145-149.
29. Murashige, T., and F. Skoog (1962) A revised medium for rapid growth and bioassays with tobacco tissue cultures. Physiol. Plant. 15:473-497.
30. Patel, K.R., N.S. Shekhawat, G.P. Berlyn, and T.A. Thorpe (1984) Isolation and culture of protoplasts from cotyledons of Pinus coulteri. Plant Cell Tissue Organ Culture 3:85-90.
31. Pelleteir, G., and Y. Chapeau (1984) Plant protoplast fusion and somatic plant cell genetics. Physiol. Veg. 22:377-399.
32. Powledge, T.M. (1984) Biotechnology touches the forest. Bio/Technology 2:763-772.
33. Rotman, B.R., and R.D. Papermaster (1966) Membrane properties of living cells as studied by enzymatic hydrolysis of fluorogenic esters. Proc. Natl. Acad. Sci., USA 55:134-141.
34. Russell, J.A., and B.A. McCown (1986) Culture and regeneration of Populus leaf protoplasts isolated from non-seedling tissues. Plant Sci. 46:133-142.
35. Shillito, R.D., J. Paszkowski, and I. Potrykus (1983) Agarose plating and a bead type culture technique enable and stimulate development of protoplast-derived colonies in a number of plant species. Plant Cell Reports 2:244-247.
36. Teasdale, R.D., and R. Rugini (1983) Preparation of viable protoplasts from suspension-cultured loblolly pine (Pinus taeda) cells and subsequent regeneration to callus. Plant Cell Tissue Organ Culture 2:253-261.
37. Torrey, J.G. (1985) The development of plant biotechnology. Am. Scientist 73:354-363.
38. Tricoli, D.M., M.B. Hein, and M.G. Carnes (1986) Culture of soybean mesophyll protoplasts in alginate beads. Plant Cell Reports 5:334-337.
39. Verma, D.C., J.D. Litvay, M.A. Johnson, and D.W. Einspahr (1982) Media development for cell suspensions of conifers. In Proceedings of the Fifth International Congress of Plant Tissue and Cell Culture, Tokyo, pp. 59-60.
40. Wallin, A., K. Glimelius, and T. Eriksson (1977) Pretreatment of cell suspensions as a method to increase the protoplast yield of Haplopapaus gracilis. Physiol. Plant. 40:307-311.

PLANTLET REGENERATION IN VITRO VIA ADVENTITIOUS BUDS AND SOMATIC EMBRYOS IN NORWAY SPRUCE (*PICEA ABIES*)

Sara von Arnold and Inger Hakman

Department of Plant Physiology
University of Uppsala
S-751 21 Uppsala, Sweden

ABSTRACT

Methods are described for plant regeneration via adventitious bud formation and somatic embryogenesis from mature zygotic embryos of *Picea abies*. Adventitious buds are formed on embryos pulse-treated with cytokinin. A few epidermal and subepidermal cells adjacent to the stomata are stimulated to give rise to meristemoids which subsequently develop further into adventitious buds. Embryogenic callus is formed when the embryos are cultured on medium containing both auxin and cytokinin. The somatic embryos arise from the region between the hypocotyl and the cotyledons. The different developmental stages during initiation and development of adventitious buds and somatic embryos are described.

INTRODUCTION

During the last decade much success with plant regeneration of conifers under in vitro conditions has been achieved, as recently reviewed (16). Several coniferous species can now be regenerated in vitro via either adventitious buds or somatic embryos. However, before these tissue culture techniques can have practical importance, much more must be learned about regulation of differentiation processes under in vitro conditions. Conifer tissue culture research is only in its infancy, but in the future these techniques might become useful for vegetative propagation of juvenile and mature plant material, production and culture of haploids, freeze preservation of valuable genotypes, production of mutants and somatic hybrids, and in gene transfer studies. Furthermore, tissue culture techniques are useful for studying different aspects of plant development.

ADVENTITIOUS BUD FORMATION

The induction and growth of adventitious buds seems to be the most successful method today for multiplication of conifers in vitro. Adventitious buds can be formed either directly on the explant or indirectly through an intermediary callus. It is rather easy to obtain adventitious buds on isolated organs of conifers; the problem, however, is the development of buds into plantlets. In order to achieve a closer understanding of what difficulties exist, the various stages during bud initiation and development must be identified and carefully studied. This overview will summarize our present knowledge concerning the initiation and development of adventitious buds on embryos and vegetative buds in Picea abies.

Embryos

Cytokinin treatment. Adventitious buds are induced when newly isolated embryos are either cultured on medium containing cytokinin or pulse-treated with cytokinin and then cultured on medium lacking cytokinin. The final yield of adventitious buds is similar in both cases (Fig. 1). The advantages of the pulse treatment technique are that bud differentiation is more synchronized, adventitious buds develop faster, and the variation between different experiments is smaller. Therefore, we have chosen the pulse treatment technique for more detailed studies of the initiation and development of adventitious buds.

Adventitious buds are formed on embryos pulse-treated with N^6-benzyladenine (BA) for 1 min. However, the number of embryos forming adventitious buds increases significantly when the embryos are treated with

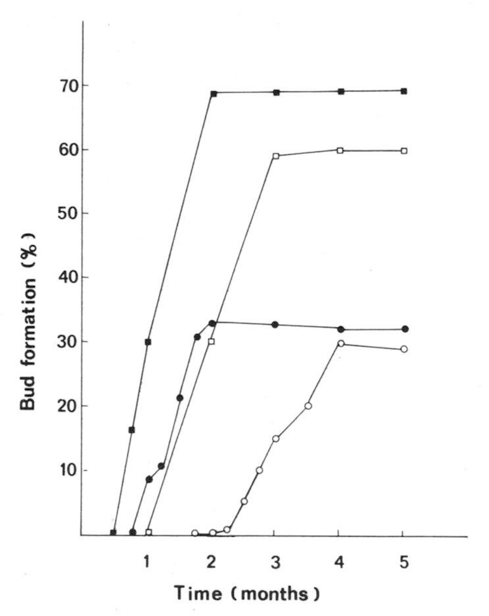

Fig. 1. Time-course for the development of adventitious buds on embryos and vegetative buds after two different types of cytokinin treatment. Embryos were either cultured on medium containing 5 μM BA for four weeks and then transferred to medium lacking BA (□), or pulse-treated with 250 μM BA for 2 hr and then cultured on a medium lacking BA (■). Vegetative buds were either cultured on medium supplied with 5 μM each of BA and kinetin for eight weeks and then transferred to medium lacking cytokinin (o), or pulse-treated with 250 μM BA for 3 hr and then cultured on a medium lacking BA (●).

BA for 30 min. Prolonged treatment of up to 4 hr has no significant effect on bud formation. Adventitious buds are induced to the same extent on embryos treated for 2 hr with 50 to 500 μM BA. However, after prolonged culture, the highest yield of adventitious shoots is obtained from embryos treated with 250 μM BA. Adventitious buds are formed to the same extent on embryos pulse-treated with BA at 4° and 20°C. In order to obtain a high yield of adventitious buds, it is necessary to adjust the pH of the BA solution to 5.5 (8).

Similar pulse treatment techniques can also be used for induction of adventitious buds on cotyledons of P. abies (14,15,44). Time-course studies with primary explants of P. abies incubated for various periods in solutions containing different concentrations of [^{14}C]-BA showed that uptake was linear for the initial 60 min, after which time linear uptake continued but at a much reduced rate (44). The amount of BA taken up by the explants saturated at a concentration about one-third of that of the medium. Concentration dependence experiments showed that BA uptake was directly proportional to the external concentration.

Embryos pulsed with BA release cytokinin to the culture medium for about 24 hr after the BA-pulse (11). After 24 hr the embryos contain less than 10% of the BA initially taken up. Roughly 50% of this BA is residual BA and 50% is nucleotide BA (40). We do not yet know if this nucleotide formation is important or necessary for bud formation.

Formation of meristemoids. Pulse treatment with BA induces changes in the developmental pattern of several cells in the embryos. Most of the responding cells are only stimulated to go through a limited number of divisions. However, a few epidermal and subepidermal cells are stimulated to develop further into meristemoids. These specific cells are located adjacent to the stomata (Fig. 2A) (13). Similar origin of meristemoids has been observed on cotyledons of P. abies (29).

Meristemoids appear on BA-pulsed embryos after one week in culture (Fig. 3B). Meristemoids can be initiated on embryos cultured on water agar in continuous light or in darkness at 10° to 30°C, but for further development of the meristemoids the embryos must be transferred to medium containing nutrients (Tab. 1). The nutrient reserves of the embryos are probably depleted after ten days, at which time additional nutrition is required for further development of the meristemoids. When the meristemoids are well developed (i.e., after two weeks), the embryos have a nodulated appearance (Fig. 3C). The developmental pattern of meristemoids is fairly synchronized between different embryos as well as within each embryo.

The cytokinin that is released to the culture medium after the pulse treatment influences embryo development (9). When the embryos are cultured on media containing released cytokinin for more than six days, elongation growth of cotyledons and roots is inhibited. In contrast, when the embryos are transferred, after the initial release of cytokinin (i.e., after about 24 hr), to fresh culture medium lacking cytokinin, the cotyledons and roots elongate to various extents (cf. Fig. 2C and D). Similar results are obtained irrespective of whether or not the embryos are rinsed in water after the BA treatment. The reason for the inhibited elongation growth of embryos cultured on medium containing released cytokinin is that a meristematic layer is formed from epidermal and subepidermal cells. Meristemoids develop within this meristematic layer (Fig. 2E). In contrast, when the embryos are not cultured on medium containing released cytokinin, only a few epidermal and subepidermal cells continue to divide and give rise to meristemoids (Fig. 2F). Despite these general differences between embryos transferred to fresh culture medium and those which are not, the formation of meristemoids appears to occur in a similar way. In both cases, the meristemoids arise from cells surrounding the stomata. The time-course for the development of adventitious buds is

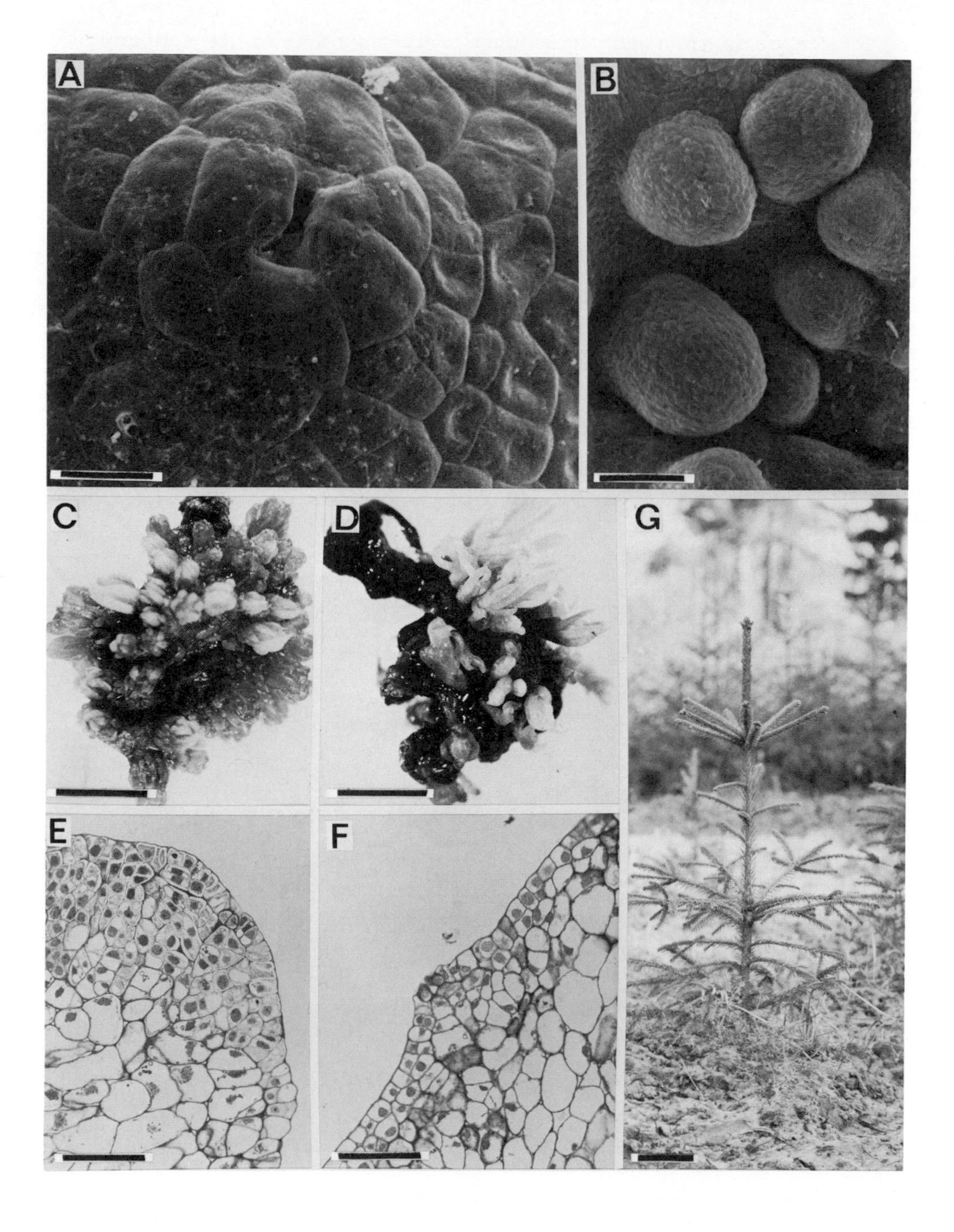

similar whether or not the embryos are cultured on medium containing released cytokinin. This shows that it is possible to follow initiation of adventitious buds without including many general cytokinin effects.

Formation of adventitious bud primordia. When the meristemoids have become more cone-shaped and the stomata are overgrown by meristematic cells, it can be considered that bud primordia have developed (Fig. 2B). In the dissecting microscope, bud primordia can only be observed as nodules on the surface of the embryos (Fig. 3D). The transition from meristemoids to adventitious bud primordia is gradual, but usually it occurs during the third week after the BA-pulse (13).

Adventitious bud primordia are formed provided that the embryos are cultured on medium containing mineral nutrients and sucrose (Tab. 1). Addition of auxin and/or cytokinin usually inhibits the transition. Neither light nor temperature seems to be critical for the formation of adventitious bud primordia.

Formation of adventitious buds. At earlier developmental stages, the adventitious buds can be observed in a dissecting microscope as scale-like organs (Fig. 3E). In sectioned material, a bud apex can be seen at the base of the scale-like organ. Later, needle primordia arise around the apex and normal-looking buds develop (Fig. 2C and D). The time for development of adventitious buds varies significantly among different embryos as well as within each embryo. Adventitious buds start to arise after three weeks and continue to form until the eighth week (Fig. 1).

Adventitious buds develop on embryos cultured in light on media containing mineral nutrients and sucrose (Tab. 1). The irradiance does not seem to be critical since we have not observed any significant differences between 15 and 260 $\mu E \cdot m^{-2} \cdot s^{-1}$. Although adventitious buds can develop at both 15° and 30°C, it is preferable to culture the embryos from this stage on at 20°C.

The DNA contents of interphase nuclei from adventitious buds of *P. abies* regenerated in vitro are in the same range and have the same distribution as those found in resting buds collected from field-grown trees (26).

Fig. 2. Various stages during the process of plant formation via adventitious buds. Embryos were pulse-treated with 250 μM BA for 2 hr, blotted on filter paper, and then cultured on medium lacking cytokinin. (A and B) Scanning electron micrographs of a meristemoid (A) and of bud primordia (B). (C and D) Embryos covered with adventitious buds after two months in culture. The embryo in (C) was transferred to fresh culture medium after one month; the embryo in (D) was transferred to fresh culture medium after 24 hr and then again after one month. (E and F) Anatomical appearance of the hypocotyl part of embryos after 12 days in culture. The embryo in (E) was treated as that in (C) and the embryo in (F), as that in (D). (G) Plant regenerated via an adventitious bud after two and a half years in field. Scale bars are as follows: (A), 0.02 mm; (B), (E), and (F), 0.1 mm; (C) and (D), 2 mm; and (G), 100 mm.

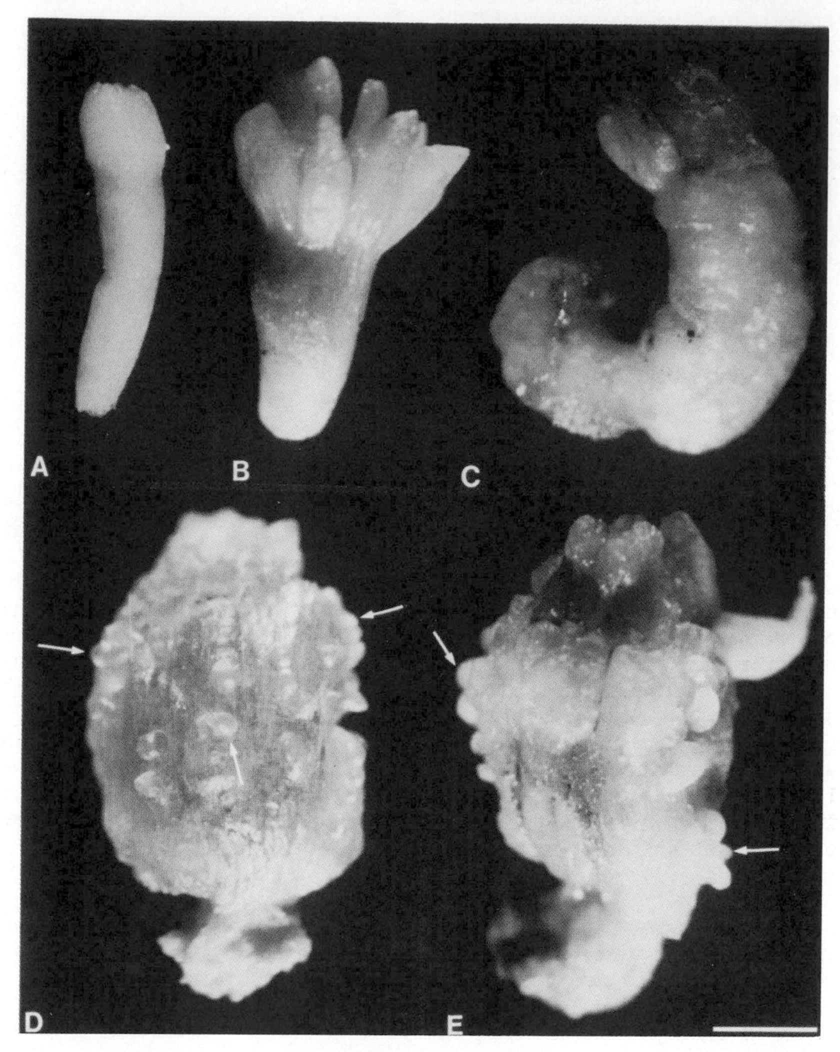

Fig. 3. Various stages of pulse-treated embryos during the induction of adventitious buds. Embryos were pulse-treated with 250 μM BA for 2 hr, blotted on filter paper, and then cultured on medium lacking cytokinin. (A) Embryo following BA treatment. (B) Embryo after one week in culture. (C) Embryo after two weeks in culture with nodulated surface. (D) Embryo after three weeks in culture with adventitious bud primordia (arrows). (E) Embryo after four weeks in culture with adventitious buds (arrows). Scale bar represents 1 mm [applies to (A) through (E)]. From Ref. 8, with permission.

Development of adventitious shoots. The method used for induction of adventitious buds strongly affects the ability of the induced buds to develop further into vigorous shoots (2). The most critical factors to consider are cytokinin treatment, basal medium concentration, and photoperiod. Best results are obtained when the embryos are pulse-treated with 250 μM BA for 2 hr and then cultured on half-strength LP medium

Tab. 1. **Stages during differentiation of adventitious buds on BA-pulsed mature zygotic embryos of Picea abies.**

Day	Characteristics	Requirements
0-3	None	No specific
4-6	Proliferation of cells surrounding the stomata	No specific
7-10	Appearance of meristemoids	No specific
11-16	Enlargements of meristemoids	LP x 1/2; 30 mM sucrose
17-49	Development of adventitious bud primordia	LP x 1/2; 30 mM sucrose
21-56	Development of adventitious buds and shoots	LP x 1; 60 mM sucrose; light; 20°C

(a modified MS medium) containing 30 mM sucrose. The embryos should be transferred to fresh culture medium of the same composition after one week. When the buds start to develop (i.e., after three weeks) the embryos are transferred to full-strength LP medium containing 60 mM sucrose. The embryos can be incubated at 20°C and under a 16-hr photoperiod during the entire procedure. Many of the adventitious buds formed on embryos treated in this way develop into shoots. When the stem of the shoot has reached a length of about 5 mm, the shoots can be isolated and cultured individually on full-strength LP medium containing 60 mM sucrose. The addition of plant growth regulators does not stimulate further growth of adventitious shoots.

The development of isolated adventitious shoots is affected by the agar concentration in the culture medium (7). Increasing the agar concentration from 0.5 to 2.0% decreases vitrification, but at the same time reduces shoot growth. Furthermore, the endogenous abscisic acid (ABA) concentration is much higher in shoots cultured on media containing 0.5 and 2% agar than it is in shoots cultured on medium containing 1% agar. In accordance with this, best growth has thus far been obtained with shoots cultured on medium containing 1% agar. The mesophyll of needles developed in vitro is interspersed with large air spaces; the lower the agar concentration, the larger the air spaces. After transfer to the greenhouse, the new needles from the acclimatized plantlets have an anatomy similar to that of plants growing in the field.

Rooting of adventitious shoots. Rooting of conifers is often difficult (especially under in vitro conditions), mainly because it takes a long time for the roots to develop. In order to solve this problem, we are now studying the various stages before and during initiation and development of adventitious roots. For practical reasons we have chosen hypocotyl cuttings for these studies. We have found that the cuttings often form a basal wound tissue complex from which roots develop. It takes from one month to one year for the cuttings to produce roots in this way (18). However, under specific culture conditions, it is possible to obtain roots directly on the cuttings without formation of a wound tissue complex. This type of rooting occurs within three weeks. Auxins are necessary for direct root formation (19). Preliminary results have shown that adventitious shoots of _P. abies_ can also develop roots directly on the stem within two weeks after an auxin treatment.

The rooting ability of auxin-treated adventitious shoots [24 hr with 10^{-4} M each of 3-indoleacetic acid (IAA) and 3-indolebutyric acid (IBA)] varies depending on the agar concentration in the culture medium. About 50% of shoots raised on 0.5% agar form roots while only 10% do so on 1% agar (7). Roots do not form on shoots cultured on a medium containing higher agar concentrations. The yield of rooted shoots increases if the photoperiod is shortened from 16 hr to 8 hr one month before the auxin treatment.

Plantlets transferred to field conditions continue to grow normally (Fig. 2G). Root tips from adventitious plantlets contain the diploid chromosome number (2n=24).

Biochemical and histochemical marker for bud formation. Cytokinin regulates many different processes in plants, such as cell division, enlargement, and differentiation. Cytokinin also stimulates peroxidase activity (17). However, the mechanisms of cytokinin action in bud differentiation are not yet understood. Variations in peroxidase activity and isozyme compositions have been proposed to indicate bud initiation in several plant systems (30,32,33,35,42). Both water- and BA-treated embryos of _P. abies_ have a peak of peroxidase activity after three days in culture (9). It is probable that this increased peroxidase activity results mainly from a stress reaction after embryo isolation. The peroxidase activity continued to be rather high in BA-treated embryos that developed a meristematic layer, and a second increase in peroxidase activity was obtained after 18 days, i.e., at the time when adventitious bud primordia developed. The importance of this second increase in peroxidase activity is unclear, but similar patterns have been found in _Cichorium_ (32) and _Nicotiana_ (33) during bud initiation. In contrast, in embryos of _P. abies_ which develop adventitious buds directly without formation of a meristematic layer, the peroxidase activity decreased markedly until day 9, after which it remained low. This shows that neither a prolonged, high peroxidase activity nor a second increase in peroxidase activity is necessary for bud formation in embryos of _P. abies_ (9). Therefore, total peroxidase activity is not useful as a biochemical marker for bud initiation in _P. abies_.

In several species, starch accumulation has been reported to be involved in organogenesis. During initial stages of meristemoid formation in tobacco callus, accumulation of large quantities of plastid starch occurred in the cells. The starch disappeared at later stages of meristemoid development (39). Similarly, one of the first manifestations of the establishment of organogenic domains on the surface of embryos of _Pinus coulteri_ was the accumulation of starch. Subsequently, the starch content declined as shoots were formed (36). Cells in the shoot-forming layers of cotyledons of _Pinus radiata_ contained abundant starch grains which were degraded as organogenic centers were formed (37). Embryos of _P. abies_ that were cultured on media containing sucrose started to accumulate starch during the first day (14). Starch accumulation occurred particularly in the cortex cells where starch grains were frequently present in the chloroplasts. The starch accumulation increased with higher concentrations of sucrose in the culture medium. The initial stages of bud formation could take place on culture medium lacking sucrose, but sucrose was required for further development of meristemoids into adventitious bud primordia and buds.

Embryos cultured on medium lacking sucrose did not accumulate starch before or during the formation of meristemoids. Therefore, starch accumulation in embryos of *P. abies* was not an early manifestation of bud formation. Although there seem to be differences in the pattern of starch accumulation during the initial stages of bud formation in different species and in different types of explants, a common phenomenon is that accumulated starch is not present in meristematic cells during later stages of bud formation, even though explants are cultured on media containing sucrose.

Buds

The most obvious use of in vitro techniques for conifers will be to propagate selected superior trees. Therefore, it is important that the techniques developed for juvenile plant material be applicable to mature plant material.

Cytokinin treatment. Adventitious buds can be formed on isolated resting vegetative buds from mature trees of *P. abies*. The time for development of adventitious buds varies depending on the cytokinin treatment (12). Buds pulse-treated with BA obtain their maximum yield of adventitious buds after two months, while buds cultured on cytokinin-containing medium do so after four months (Tab. 1). Similar time-courses for development of adventitious buds are obtained with embryos of *P. abies* that are either pulse-treated with cytokinin or cultured on medium containing cytokinin. This shows that the formation of adventitious buds can take the same amount of time, whether they are formed on vegetative buds from mature trees or on embryos.

An effective method for obtaining a high yield of adventitious buds within eight weeks is to pulse-treat the buds in 250 µM BA for 3 hr and then culture them on medium containing 5 µM each of BA and kinetin for one week. The requirement for additional cytokinin in the culture medium during one week cannot be reduced by increasing the pulse BA concentration or by prolonging the treatment time.

Initiation and development of adventitious buds. The formation of adventitious buds occurs on mature vegetative buds in a way similar to that on embryos. In sectioned buds it has been observed that many cells at the periphery of the needle primordia become meristematic. Some of these meristematic cells undergo repeated divisions and give rise to meristemoids from which adventitious bud primordia arise (6).

Depending on the genotype used, 30 to 60% of the buds develop adventitious buds. The number of adventitious buds per bud varies from two to 50, but usually each bud forms ten to 20 well-developed adventitious buds. Two to five of these buds usually start to elongate during the third month. However, it has not yet been possible to regenerate vigorously growing adventitious shoots.

Plant material. As trees go through the maturation process, it becomes more difficult to induce adventitious buds and to regenerate plantlets. To some extent, the bud-producing capability of *P. abies* can be increased after pregrowing cuttings or grafts from mature trees in a phytotron, but marked differences in the ability to form adventitious buds are observed among various clones (3). Similarly, the rooting potential of the tested clones increased after such preculturing, although the cuttings still exhibited mature growth characteristics (J. Dormling, pers. comm.).

Resting vegetative buds from various clones contain different amounts of endogenous cytokinin (12). The clone that contains the most endogenous cytokinin in its resting buds has the highest potential for adventitious bud formation but the lowest growth rate under field conditions and the lowest rooting potential of cuttings. In contrast, the clone that contains less endogenous cytokinin in its resting buds has the lowest rating for adventitious bud formation capacity and the highest ratings for growth and rooting potential. A similar correlation between endogenous cytokinin content and growth rate has been observed in Vitis vinifera (1). It is important to be aware of such correlations so that we do not make a negative selection during in vitro culture.

SOMATIC EMBRYOGENESIS

In plants, besides development from a zygote, embryos can be initiated in cells that are not products of gametic fusion. These embryos are called asexual or somatic embryos and originate in somatic cells or unfertilized gametic cells (see, for example, Ref. 43 for further details). Since the discovery of somatic embryogenesis in carrot tissue cultures (38,41), several species have been regenerated in vitro by this method (see Ref. 43). Although much effort has been directed towards coniferous species, as many conference and review articles give evidence of, only lately has any real progress within this area of research been achieved.

Somatic embryogenesis in conifers followed by plant regeneration was first reported for P. abies (23,28), where immature zygotic embryos were used for establishing the culture. Later, mature zygotic embryos were also used successfully for obtaining embryogenic cultures (5,10,20,31). These results are of great importance since they provide us with plant material for experimentation year around, and we are thus not restricted to the short period when immature embryos are available. This is not less important when working with spruce species, which generally show very irregular flowering habits. Other spruce species that have been regenerated in vitro by this method are Picea glauca and Picea mariana (25). Among the pines, Pinus lambertiana (21) and Pinus taeda (22) have proven successful for plant regeneration by somatic embryogenesis. Embryogenic cultures with the capacity for plant regeneration have also been established from female gametophytes of Larix decidua (34). The number of species that can be regenerated via somatic/asexual embryos will certainly increase as more species are investigated.

Although plantlets have been regenerated via somatic embryos in several conifers, the yield of fast-growing plantlets has so far been rather poor. The main reason for this is probably that enough efforts have not yet been made to optimize culture conditions that are required for somatic embryos as they pass through different stages during their growth into plantlets. During our work of devising a protocol that gives a high regeneration rate of plantlets in P. abies (11), we have classified somatic embryos into four different developmental stages, depending on certain features and growth requirements met by the embryos, as summarized in Tab. 2 and presented below.

Formation of Embryogenic Callus and Somatic Embryos

From our work with P. abies and with other Picea species, we have found that embryogenic cultures have the same appearance irrespective of

Tab. 2. Stages during differentiation of somatic embryos in cultures derived from mature zygotic embryos of Picea abies.

Stage	Characteristics	Requirements
1	Small somatic embryos consisting of an embryonic region of small, densely cytoplasmic cells subtended by a suspensor comprised of long, highly vacuolated cells (Fig. 4C).	LP x 1/2; pH 6.2; 30 mM sucrose; 15 mM NH_4NO_3; 5-20 µM NAA; 5 µM BA; darkness
2	Somatic embryos with a more prominent and dense embryonic region. The embryos are still attached to the callus with long suspensor cells (Fig. 4D).	LP x 1/2; 90 mM sucrose; 7.6 µM ABA
3	Somatic embryos with small cotyledons (Fig. 4E).	LP x 1; 60 mM sucrose
4	Germinating green somatic embryos (Fig. 4F).	LP x 1; 60 mM sucrose; light

species or of whether they have been induced on immature or mature zygotic embryos. In all cases, embryogenic callus is first seen to arise from the region just beneath the cotyledons of the cultured zygotic embryos (Fig. 4A), and within one month three types of tissues can be distinguished (10,28):

(i) A compact tissue with a rather smooth and nodulated surface. This tissue is green when cultured in the light and is similar to that obtained when embryos are treated with cytokinin for adventitious bud development.

(ii) Another nonembryogenic tissue which also turns green when cultured in the light. This unorganized callus consists of small, rounded cells of fairly uniform size, and continues to grow in the same way for months.

(iii) An embryogenic tissue that is easily recognized by its color and texture (Fig. 4B). It appears translucent to white under all growth conditions and consists of somatic embryos of different sizes embedded in a loose network of translucent cells.

A detailed description of the different tissues arising in Picea glauca cultures has been presented elsewhere (25). A single cultured zygotic embryo can produce one, two, or all three of these tissues. However, embryogenic callus can easily be isolated from the rest and cultured separately. Attempts have been made to characterize embryogenic cultures of P. abies by using biochemical methods (45).

The culture conditions for obtaining embryogenic callus on immature embryos of P. abies are not very critical as long as the medium contains both auxin and cytokinin. In contrast, many factors have to be considered when mature zygotic embryos are used (5). It is difficult to stress

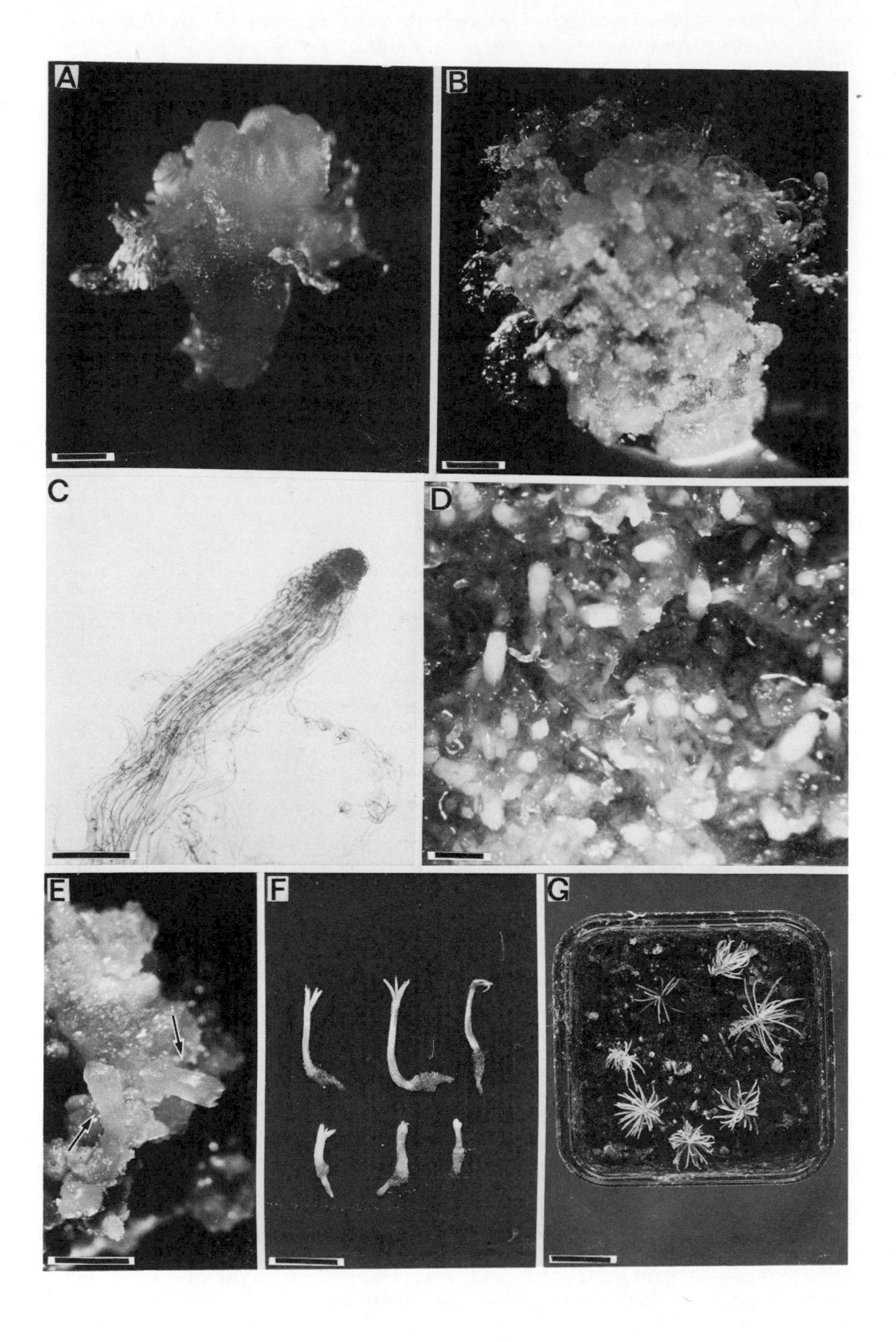
A
B
C
D
E
F
G

the importance of any single factor since they all depend upon each other. Important factors to consider are: (a) light regimes; (b) concentration of the basal medium, sucrose, and NH_4NO_3; and (c) pH of the medium, medium stiffness, and plant growth regulators. For example, when mature embryos of P. abies are cultured in darkness on half-concentrated LP medium containing 30 mM sucrose, 15 mM NH_4NO_3, 5-20 μM naphthaleneacetic acid (NAA) or 2,4-dichlorophenoxyacetic acid (2,4-D), and 5 μM BA with the pH adjusted to 6.2 (Tab. 2), about 50% of them produce embryogenic calli. If immature embryos are used instead and cultured under identical conditions, nearly 100% form embryogenic calli.

While embryogenic calli are proliferating on media containing both auxin and cytokinin (Tab. 2), small somatic embryos (stage 1, as is represented by the embryo depicted in Fig. 4C) are continuously produced. Callus cultures initiated from immature embryos seem capable of growing for a prolonged time period without apparent loss of embryogenic potential. However, after about two months in culture, callus cultures derived from mature embryos often start to turn brown. This is followed by a decline in both growth and production of new somatic embryos. When fast-growing embryogenic calli are transferred to liquid culture medium, suspension cultures that superficially resemble "normal" suspension-cultured cells can be established. A closer examination of such cultures reveals a more complex composition, where the most interesting feature is that these cultures contain numerous somatic embryos, essentially the same as those found in callus cultures (24,28). Such cultures can also be maintained for an extended time period with continuous production of somatic embryos.

Very little is yet known about the origin and growth of somatic embryos in cultures. There seem to be at least three different mechanisms which could account for their origin (see Ref. 27 for further details and discussion). First, embryos could arise from single cells or small cell aggregates by an initial asymmetric division which delimits the embryonic and suspensor region (24). Sustained division of the smaller cells could then produce the meristematic region of the embryo, which generates cells

Fig. 4. Various developmental stages during the process of plantlet formation via somatic embryogenesis from mature zygotic embryos. (A) Embryogenic callus protruding from the region between the cotyledons and the hypocotyl after about three weeks on half-strength LP medium containing 30 mM sucrose, 10 μM NAA, and 5 μM BA. (B) Embryogenic callus cultured as in (A) after about two months in culture. (C) Higher magnification of a somatic embryo at stage 1 in (B). (D) Stage 2 somatic embryos one month after transfer from callus medium to half-strength LP medium containing 90 mM sucrose and 7.6 μM ABA. (E) Stage 3 somatic embryos (arrows) two weeks after transfer from ABA-containing medium to full-strength LP medium containing 60 mM sucrose. (F) Germinating stage 4 somatic embryos after one month on full-strength LP medium containing 60 mM sucrose. (G) Plantlets regenerated via somatic embryogenesis two months after transfer to greenhouse conditions. Scale bars in (A), (B), (D), and (E) represent 1 mm; in (C), 0.3 mm; in (F), 5 mm; and in (G), 10 mm.

that elongate to form the suspensor. Somatic embryos may also develop from small meristematic cells within the suspensor. These initials could then arise by asymmetric division of suspensor cells or from meristematic cells of the embryonic region that have failed to elongate while being integrated into the suspensor. Interestingly, Gupta and Durzan (22) have recently reported about finding free-nuclear stages in suspension-cultured cells of Pinus taeda. These free-nuclear stages were then followed by early embryogeny. Finally, embryos could also arise by a mechanism similar to that found during early zygotic embryogenesis (cleavage polyembryony) in some conifers (e.g., Pinus), with the initial separation occurring in the embryonic region. Cleavage polyembryony has previously been reported to occur in embryogenic cultures of Larix decidua (34).

The general morphology of young somatic embryos and their development into plantlets seem to be quite similar in all coniferous species examined so far. These somatic embryos also show close resemblance to coniferous zygotic embryos. Unlike most angiosperm somatic embryos, those of conifers have an extensive suspensor consisting of very long, highly vacuolated cells extending from an embryonic region which is composed of small meristematic cells (Fig. 4C). Thus, two distinct areas can be discerned. The function of the long suspensor cells and their role in the developmental process of somatic embryos are not known. They exhibit active cytoplasmic streaming, with the major flow occurring longitudinally in the thin layer of peripheral cytoplasm lining the elongated cells. Examination of somatic embryos with the electron microscope reveals bundles of microfilaments in these cells. These bundles correspond to actin cables observed in light microscope preparations stained with rhodamine-labeled phalloidin (a phallotoxin with a high affinity for F-actin), and are oriented parallel to the direction of active streaming in these cells (27). These observations are in agreement with the general opinion regarding the function of actin microfilaments in the process of cytoplasmic streaming in plant cells. The small cells of the embryonic region are characterized by a high nucleus-to-cell volume ratio. They have thin cell walls with numerous plasmodesmata, relatively dense cytoplasm, and generally numerous small vacuoles. Divisions are frequent within this region, and the general appearance of the cytoplasm suggests a high metabolic activity (see Ref. 29).

Plant Regeneration

Although the requirements for initiation of embryogenic calli vary depending on the explant, those for the further development of calli into more advanced stages seem to be similar in all calli. Somatic embryos can develop into plantlets on medium lacking plant growth regulators. However, addition of 7.6 μM ABA to half-strength LP medium containing 90 mM sucrose (Tab. 2) greatly improves embryo development. The embryonic region of the embryos becomes firmer, assumes a glossy surface, and reaches stage 2 (Fig. 4D) after a month in culture. Lipids have been observed to accumulate in the embryos during ABA treatment in a way similar to what is found in mature seed embryos. Addition of auxin and/or cytokinin to the ABA-containing medium only inhibits the effect of ABA.

After transferring the cultures to medium lacking ABA (Tab. 2), the embryonic region of the embryos becomes more organized and cotyledons appear after about one month in culture, indicating that stage 3 has been reached (Fig. 4E). These somatic embryos resemble mature zygotic embryos in appearance, growth habits, and requirements under in vitro culture conditions. When calli containing stage 3 embryos are subcultured on full-strength LP medium containing 60 mM sucrose (Tab. 2), many embryos develop further into plantlets. However, the final yield of plantlets is much higher if embryos of this stage are first isolated from the callus and then cultured individually (Fig. 4F).

Plantlet regeneration can occur both in light and in darkness. We have found that the best results are obtained when isolated stage 3 embryos are first cultured in darkness (for a time period not exceeding three weeks) until roots have started to elongate, after which they are transferred to light conditions. Under these conditions, about 35% of the embryos reach the plantlet stage. Plantlets are then transferred to soil for further culture in the greenhouse. Under these conditions, they continue to grow into normal-looking plants (Fig. 4G).

By using this protocol (Tab. 2), we have also been able to regenerate plantlets at a high yield from embryogenic suspension cultures of *Picea glauca* (I. Hakman and S. von Arnold, ms. in prep.).

CONCLUSION

Plantlets can be regenerated from mature zygotic embryos of *P. abies* both via adventitious buds and somatic embryos. After culture in greenhouse conditions, such plants have a similar appearance irrespective of the method used for their regeneration. However, the developmental pattern as well as the culture requirements are different for the two processes (cf. Tab. 1 and 2). Adventitious buds arise from epidermal and subepidermal cells close to the stomata, all over the hypocotyl and cotyledons. Somatic embryos arise from cells in the region between the hypocotyl and cotyledons. Somatic embryos go through a proliferating callus stage, whereas no callus is formed during bud formation.

ACKNOWLEDGEMENTS

The authors are grateful to Prof. T. Eriksson for his encouragement during this work. Ms. Karin Bjelke and Ms. Ann-Charlotte Johansson are especially thanked for their experience and excellent handling of the cultures. Most of the studies with *Picea glauca* and *Picea mariana* presented in this chapter were carried out in collaboration with Prof. L.C. Fowke and Ms. Pat Rennie, who are gratefully acknowledged.

REFERENCES

1. Andonova, T.A., and D.T. Lilov (1977) Changes of the free cytokinins in the shoots and leaves in plants differing in their growth. *C.R. Acad. Bulg. Sci.* 30:905-908.
2. Arnold, S. von (1982) Factors influencing formation, development and rooting of adventitious shoots from embryos of *Picea abies* (L.) Karst. *Plant Sci. Lett.* 27:275-287.
3. Arnold, S. von (1984) Importance of genotype on the potential for in vitro adventitious bud production of *Picea abies*. *For. Sci.* 30:312-316.
4. Arnold, S. von (1987) Effect of sucrose on starch accumulation in and adventitious bud formation on embryos of *Picea abies*. *Ann. Bot.* 59:15-22.

5. Arnold, S. von (1987) Improved efficiency of somatic embryogenesis in mature embryos of Picea abies. J. Plant Physiol. (in press).
6. Arnold, S. von, and T. Eriksson (1979) Induction of adventitious buds on buds of Norway spruce (Picea abies) grown in vitro. Physiol. Plant. 45:29-34.
7. Arnold, S. von, and T. Eriksson (1984) Effect of agar concentration on growth and anatomy of adventitious buds of Picea abies (L.) Karst. Plant Cell Tissue Organ Culture 3:257-265.
8. Arnold, S. von, and T. Eriksson (1985) Initial stages in the course of adventitious bud formation on embryos of Picea abies. Physiol. Plant. 64:41-47.
9. Arnold, S. von, and R. Grönroos (1986) Meristematic zone formation and peroxidase activity during early stages of adventitious bud formation on embryos of Picea abies. Bot. Gaz. 147:425-431.
10. Arnold, S. von, and I. Hakman (1986) Effect of sucrose on initiation of embryogenic callus cultures from mature zygotic embryos of Picea abies (L.) Karst. (Norway spruce). J. Plant Physiol. 122:261-265.
11. Arnold, S. von, and I. Hakman (1987) Regulation of somatic embryo development in Picea abies, with emphasis on ABA effects. J. Plant Physiol. (submitted for publication).
12. Arnold, S. von, and E. Tillberg (1987) The influence of cytokinin pulse treatments on adventitious bud formation on vegetative buds of Picea abies. Plant Cell Tissue Organ Culture (in press).
13. Arnold, S. von, B. Walles, and E. Alsterborg (1987) Micromorphological studies of adventitious bud formation on Picea abies embryos treated with cytokinin. Ann. Bot. (submitted for publication).
14. Bornman, C.H. (1983) Possibilities and constraints in the regeneration of trees from cotyledonary needles of Picea abies in vitro. Physiol. Plant. 57:5-16.
15. Bornman, C.H., and T.C. Vogelman (1984) Effect of rigidity of gel medium on benzyladenine-induced adventitious bud formation and vitrification in vitro in Picea abies. Physiol. Plant. 61:505-512.
16. Dunstan, D.I., and T.A. Thorpe (1986) Regeneration in forest trees. In Cell Culture and Somatic Cell Genetics of Plants, Vol. 3, I.K. Vasil, ed. Academic Press, Inc., New York, pp. 223-241.
17. Gaspar, T., A.A. Khan, and D. Fries (1973) Hormonal control isoperoxidases in lentil embryonic axis. Plant Physiol. 51:146-149.
18. Grönroos, R., and S. von Arnold (1985) Initiation and development of wound tissue and roots on hypocotyl cuttings of Pinus sylvestris in vitro. Physiol. Plant. 64:393-401.
19. Grönroos, R., and S. von Arnold (1987) Initiation of roots on hypocotyl cuttings of Pinus contorta in vitro. Physiol. Plant. 69:227-276.
20. Gupta, P.K., and D.J. Durzan (1986) Plantlet regeneration via somatic embryogenesis from subcultured callus of mature embryos of Picea abies (Norway spruce). In vitro. Cell. Develop. Biol. 22:685-688.
21. Gupta, P.K., and D.J. Durzan (1986) Somatic polyembryogenesis from callus of mature sugar pine embryos. Bio/Technology 4:643-645.
22. Gupta, P.K., and D.J. Durzan (1987) Biotechnology of somatic polyembryogenesis and plantlet regeneration in loblolly pine. Bio/Technology 5:147-151.
23. Hakman, I., and S. von Arnold (1985) Plantlet regeneration through somatic embryogenesis in Picea abies (Norway spruce). J. Plant Physiol. 121:149-158.
24. Hakman, I., and L.C. Fowke (1987) An embryogenic cell suspension culture of Picea glauca (White spruce). Plant Cell Reports 6:20-22.
25. Hakman, I., and L.C. Fowke (1987) Somatic embryogenesis in Picea glauca (White spruce) and Picea mariana (Black spruce). Can. J. Bot. (in press).

26. Hakman, I., S. von Arnold, and A. Bengtsson (1984) Cytofluorometric measurements of nuclear DNA in adventitious buds and shoots of Picea abies regenerated in vitro. Physiol. Plant. 60:321-325.
27. Hakman, I., P.J. Rennie, and L.C. Fowke (1987) A light and electron microscope study of Picea glauca (White spruce) somatic embryos. Protoplasma (submitted for publication).
28. Hakman, I., L.C. Fowke, S. von Arnold, and T. Eriksson (1985) The development of somatic embryos in tissue cultures initiated from immature embryos of Picea abies (Norway spruce). Plant Sci. 38:53-59.
29. Jansson, E., and C. Bornman (1981) In vitro initiation of adventitious structures in relation to the abscission zone in needle explants of Picea abies: Anatomical considerations. Physiol. Plant. 53:191-197.
30. Kevers, S., M. Coumans, W. de Greffe, M. Jacobs, and T. Gaspar (1981) Organogenesis in habituated sugarbeet callus: Auxin content and protectors, peroxidase pattern and inhibitors. Z. Pflanzenphysiol. 101:79-87.
31. Krogstrup, P. (1986) Embryolike structures from cotyledons and ripe embryos of Norway spruce (Picea abies). Can. J. For. Res. 16:664-668.
32. Legrand, B., and J. Vasseur (1972) Evolution de l'acide ribonucleique, des protéines et de l'activité peroxydasique au cours de la culture in vitro de fragments de feuilles d'endive (Cichorium intybus L., var. Witloof). C.R. Acad. Sci. Paris (Series D) 275:357-360.
33. Mäder, M. (1975) Änderung der Isoperoxidase Isoenzymmuster in Kalluskulturen in Äbhängigkeit von der Differenzierung. Planta Med. (Suppl.) 155-162.
34. Nagmani, R., and J.M. Bonga (1985) Embryogenesis in subcultured callus of Larix decidua. Can. J. For. Res. 15:1088-1091.
35. Nakanishi, S. (1979) Peroxydases et bourgeonnement de fragments de racines d'Endive (Cichorium intybus L.). C.R. Acad. Sci. Paris (Series D) 289:695-698.
36. Patel, K.R., and G.P. Berlyn (1983) Cytochemical investigations on multiple bud formation in tissue cultures of Pinus coulteri. Can. J. Bot. 61:575-585.
37. Patel, K.R., and T.A. Thorpe (1984) Histochemical examination of shoot initiation in cultured cotyledon explants of radiata pine. Bot. Gaz. 145:312-322.
38. Reinert, J. (1958) Untersuchungen über die Morphogenese und Gewebekulturen. Ber. Deutsch. Bot. Ges. 71:15.
39. Ross, M.K., T.A. Thorpe, and J.M. Costerton (1973) Ultrastructural aspects of shoot initiation in tobacco callus cultures. Am. J. Bot. 60:788-795.
40. Staden, J. van, D. Forsyth, L. Bergman, and S. von Arnold (1986) Metabolism of benzyladenine by Picea abies embryos. Physiol. Plant. 66:427-434.
41. Stewart, F.C., M.O. Mapes, and J. Smith (1958) Growth and organized development of cultured cells. I. Growth and division of freely suspended cells. Am. J. Bot. 45:693-703.
42. Thorpe, T.A., and T. Gaspar (1978) Changes in isoperoxidases during shoot formation in tobacco callus. In Vitro 14:522-526.
43. Tisserat, B., E.B. Esan, and T. Murashige (1979) Somatic embryogenesis in Angiosperms. Hort. Rev. 1:1-78.
44. Vogelmann, T.C., C.H. Bornman, and P. Nissen (1984) Uptake of benzyladenine in explants of Picea abies and Pinus sylvestris. Physiol. Plant. 61:513-517.
45. Wann, S.R., M.A. Johnson, T.L. Noland, and J.A. Carlson (1987) Biochemical differences between embryogenic and non-embryogenic callus of Picea abies (L.) Karst. Plant Cell Reports 6:39-42.

DNA ANALYSIS AND MANIPULATION

ANALYSIS OF HOST RANGE IN TRANSFORMATION OF HIGHER PLANTS BY AGROBACTERIUM TUMEFACIENS

Eugene W. Nester

Department of Microbiology
School of Medicine
University of Washington
Seattle, Washington 98195

ABSTRACT

Agrobacterium tumefaciens transforms a wide variety of dicotyledonous plants by introducing a specific piece of DNA (T-DNA) of a large tumor-inducing plasmid (Ti-plasmid) into plant cells. A wide variety of dicotyledonous plants and some gymnosperms are susceptible to infection. However, the ease of infectability and the efficiency of transformation vary widely within these groups. As a general rule, most of the gymnosperms are infected only with difficulty if at all. The transfer of the T-DNA to plant cells depends upon the activity of a region of the Ti-plasmid termed the virulence (*vir*) region. We have shown that the efficiency of transfer and the host range properties of *Agrobacterium* are, in large part, due to the genes in the *vir* region. In particular, *virA*, a regulatory molecule which recognizes plant signals, seems to be especially important. In addition, we have identified other regions which seem to be associated with the efficiency of transformation.

INTRODUCTION

Agrobacterium tumefaciens is the vector of choice for introducing specific genes into a wide variety of higher plants. One of the main drawbacks in using this system is the fact that, although *Agrobacterium* can infect a very large number of different plants, many economically important plants appear to be generally resistant to *Agrobacterium* infection. These plants include most monocotyledonous plants such as wheat, corn, and oats, as well as many conifers such as Douglas fir. Therefore, to make this system for higher plant transformation more useful and versatile, it is important to understand the reasons why some plants are susceptible and others resistant to *Agrobacterium* infections. This chapter will attempt to provide some insight into this question.

OVERALL FEATURES OF CROWN GALL TUMOR FORMATION

Agrobacterium tumefaciens induces crown gall tumors in a wide variety of dicotyledonous plants by transferring a piece of DNA (T-DNA) from its tumor-inducing plasmid (Ti-plasmid) into plant cells where it becomes integrated into plant chromosomal DNA (for reviews see Ref. 17, 19, and 23). The expression of this newly integrated plasmid DNA confers new properties on the plant cells in culture. These include the ability to grow in the absence of exogenously added auxin and cytokinin and the ability to synthesize a class of nitrogenous compounds called opines which are not synthesized by either the bacteria or the untransformed plant cells. In order to transform plants, the *Agrobacterium* must contain a large Ti-plasmid, approximately 200 kilobases (kb) in size. This plasmid contains two regions that are important for plant cell transformation: the virulence (*vir*) region and the T-DNA (Fig. 1). The *vir* region functions in the transfer of the T-DNA into plant cells, although it is not integrated into the plant chromosome (10). This function includes the processing of the T-DNA and its transfer out of the bacterial cell into the plant cell.

The vir Region

The *vir* region can be divided into six operons (*virA*, *B*, *G*, *C*, *D*, and *E*) based on insertion mutagenesis by several transposons (22). This entire region has now been sequenced and analyzed and the relevant data

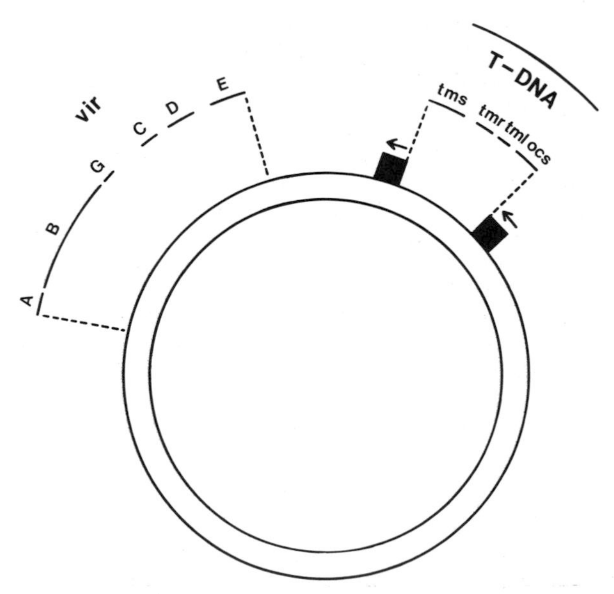

Fig. 1. Ti-plasmid. The two regions required for tumor formation are indicated. The loci are named according to the morphology of the tumor formed when the locus is mutated by a transposon insertion. The black squares on either side of the T-DNA are the imperfect direct repeats and define the T-DNA. The *vir* genes comprise six operons labeled A, B, G, C, D, and E. Key: *tms*, tumor morphology shooter; *tmr*, tumor morphology rooter; *tml*, tumor morphology large; *ocs*, octopine synthase.

are presented in Tab. 1. One interesting and important feature of this region is that, except for the *virA* gene, the *vir* genes are not expressed when cells of *Agrobacterium* grow in the absence of plant cells but are induced by a number of different phenolic compounds (2). The potent inducers include acetosyringone and sinnapinic acid (24), and chalcone synthase (P. Spencer and N. Towers, pers. comm.). These compounds are found in the vicinity of wound sites. The most active inducers are precursors of lignin (24).

Two of the six *vir* operons code for positive regulatory proteins which are concerned with the sensing of the plant signal and the subsequent activation of the other *vir* operons. These two loci, *virA* (sensor gene) and *virG* (regulatory gene), are a two-component system which very likely functions in the manner shown in Fig. 2 (15,28). The plant signal, the low-molecular-weight phenolic compound, associates with the *virA* gene product which is localized in the inner membrane such that the amino-terminal end is available to the exterior compartment of the cell in the paraplasmic space, and the carboxy-terminal end is in the cytoplasm. The association of the plant signal with the *virA* gene product activates the *virA* protein perhaps by an allosteric alteration of the protein's conformation. The activated carboxy-terminal portion of this protein then interacts with the regulatory protein, the *virG* gene product. This activated protein then binds to regulatory regions of the other *vir* genes where it activates their expression transcriptionally.

T-DNA

The T-DNA codes for a number of transcripts, the best understood and most important being those concerned with auxin, cytokinin, and opine biosynthesis. The relevant information is shown in Fig. 3. Transcripts 1

Tab. 1. Salient information on *vir* genes.

A → B → G → ← C D → E →

vir gene	Inducibility	Size (kb)	ORFs*	Size (kDa)	Mutant phenotype	Role	Function
B	+	9.5	11	-	Avirulent	Transfer	?
D	+	4.5	4	16 48 21 76	Avirulent	Transfer	Endonuclease
C	+	1.5	2	26 23	Attenuated	Accessory	?
E	+	2.2	2	7 60	Very attenuated	Accessory	?
A	-	2.8	1	92	Avirulent	Regulation	Phenolic sensor
G	+	1.0	1	30	Avirulent	Regulation	Transcriptional activator

*Open reading frames.

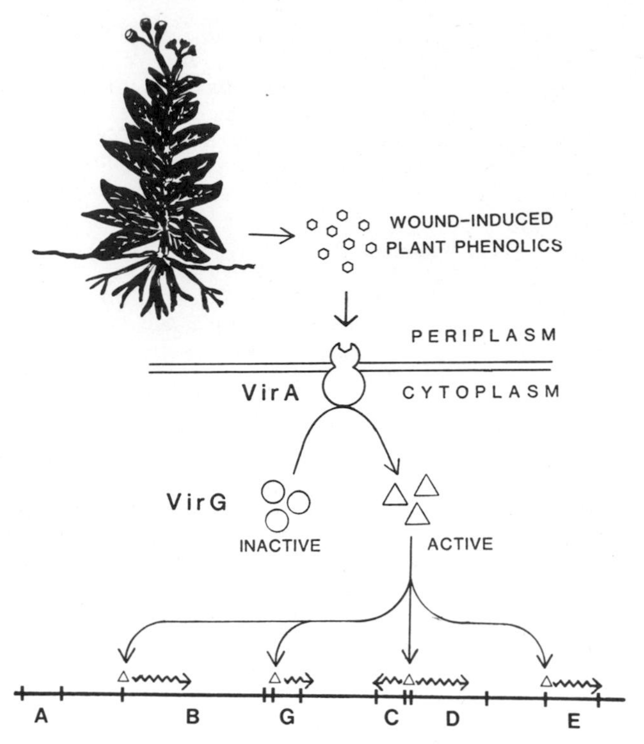

Fig. 2. Model for the transcriptional activation of the vir genes.

and 2 code for two enzymes of auxin synthesis (21,25,26). The first enzyme converts tryptophan to indoleacetamide; the second enzyme converts indoleacetamide to the auxin, indoleacetic acid. Transcript 4 converts adenosine monophosphate plus dimethylallyl pyrophosphate to isopentenyl adenosine monophosphate, a compound with cytokinin activity (1). Transcript 3 condenses pyruvate with arginine to form one of the opines called octopine (6). Other tumors synthesize other opines, depending on the strain of Agrobacterium that induces the tumor. All of the plant regulatory genes probably evolved from prokaryotic genes and were transferred to Agrobacterium from other bacteria, since they are highly homologous to plasmid-borne genes in Pseudomonas (29). All of these genes have eukaryotic regulatory regions.

GENERAL ASPECTS OF HOST RANGE

The host range of any pathogen reflects a complex interaction between the invading pathogen and the host plant. Agrobacterium is the most extensively characterized plant pathogen at the molecular level and is thus an attractive organism for studying the factors that contribute to host range. There are numerous reasons why Agrobacterium may not be able to form tumors on a plant. These include:

(a) The bacteria are unable to attach to plant cells.

(b) The bacteria cannot recognize the signal elicited by the plant and therefore the vir genes are not induced.

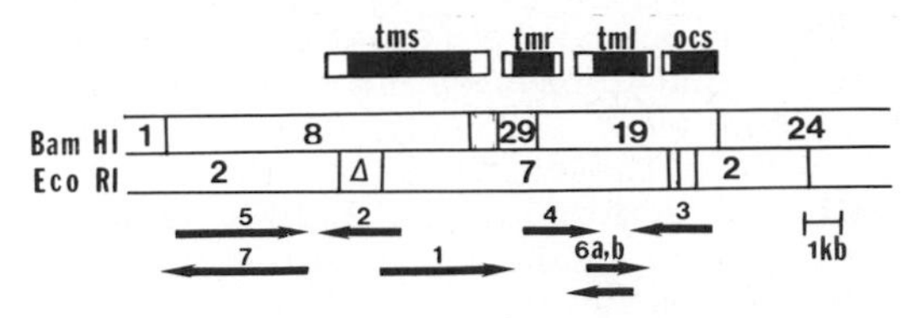

1 TRYPTOPHAN $\xrightarrow{\text{M.O.}}$ INDOLEACETAMIDE

2 INDOLEACETAMIDE $\xrightarrow{\text{HyD}}$ INDOLEACETIC ACID

3 PYRUVATE + ARGININE $\xrightarrow{\text{O.S.}}$ OCTOPINE

4 AMP + DIMETHYLALLYLPYROPHOSPHATES $\xrightarrow{\text{DMAPP TRANS.}}$ ISOPENTENYL ADENOSINE MONOPHOSPHATE

Fig. 3. The mRNAs and enzymes coded by the major T-DNA genes. The numbered arrows represent the transcripts coded by the T-DNA. The arrowheads indicate the direction of transcription. Abbreviations: M.O., monooxygenase; HyD, hydrolase; O.S., octopine synthase; DMAPP Trans, dimethyl allyl pyrophosphate transferase.

(c) The T-DNA of *Agrobacterium* is defective.

(d) Too much T-DNA may be introduced into the plant cell, resulting in a hypersensitive response.

(e) Too little T-DNA gets into the plant cell, presumably because of *vir* gene activity.

(f) The T-DNA is expressed inefficiently and not enough auxin and cytokinin are synthesized to result in tumor formation.

We will now consider each of these possibilities and discuss in what situations they might operate.

Inability to Attach

Although this is an attractive mechanism of resistance of plant cells, there are no unequivocal data which indicate that it, in fact, is an important mechanism of resistance. Data have been presented showing that a number of monocot species bind *Agrobacterium* very poorly (M. Hawes, pers. comm.; Ref. 16). However, the monocots bamboo and asparagus bind *Agrobacterium* as well as dicot species (4,5).

Unrecognized Plant Signals

Since *vir* gene functions are necessary for the processing and transfer of T-DNA, if the plant signal is not recognized by the *virA* protein, the remainder of the *vir* genes will not be induced and *Agrobacterium* will be unable to transfer its T-DNA. It seems quite reasonable that different plants might accumulate different compounds at their wound sites and this might contribute to the host range of *Agrobacterium*. At least two situations have been described which suggest that this, indeed, does operate in nature. Schafer et al. (20) recently reported that *Agrobacterium* could transform the monocotyledonous crop plant *Dioscorea bulbifera* (yam), if the bacterial cells were preincubated with wound substances from dicotyledonous plants. Without such preincubation, tumors did not develop. This strongly suggests that Dioscorea does not accumulate an inducing compound in amounts capable of activating the vir genes; consequently, it is necessary to activate these genes by the wound exudate from dicotyledonous plants. Once transferred, the T-DNA is integrated and expressed in Dioscorea just as it is in dicots, resulting in the production of an opine, nopaline, and the formation of crown gall tumors.

Another situation which is not quite as clear as the above example relates to strains of Agrobacterium that can induce tumors on only a limited number of dicotyledonous plants. These limited host range (LHR) strains, isolated from grapevine, can induce tumors on grapevine but on very few other plants (18). The virA gene in these strains is not homologous at the DNA level (30) to the virA gene on wide host range (WHR) strains (those able to induce tumors on most plants). This difference in the virA genes may result in their recognizing different plant signals. In support of this notion, it has been shown that when LHR strains are incubated in the presence of Nicotiana tabacum plant cells, the vir genes are induced very weakly. However, when a virA gene from the WHR strain is introduced into the LHR strain, there is very strong induction of the vir genes (Fig. 4) (15). These data suggest that the grapevine plants infected by the LHR strain synthesize a different plant signal which the LHR strains can recognize. However, this presumed plant signal has not yet been identified.

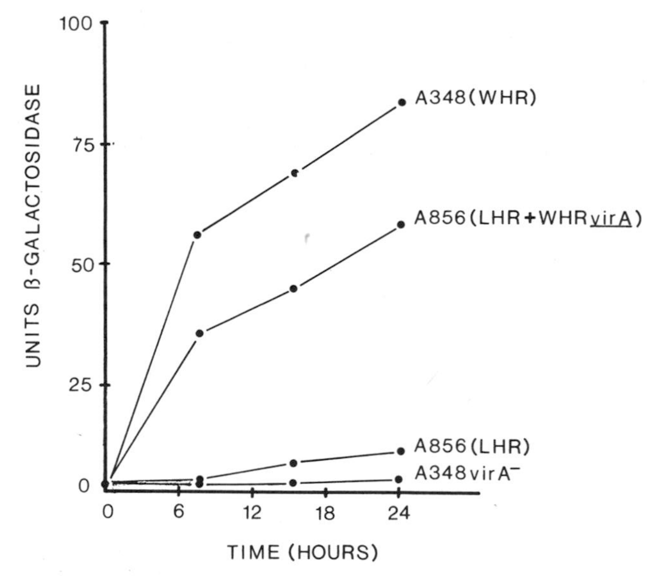

Fig. 4. Induction of vir gene expression by strains containing a wide host range (WHR) Ti-plasmid (A348), a limited host range (LHR) Ti-plasmid (A856), or an LHR Ti-plasmid together with a WHR virA gene. Expression of the vir genes was measured by the induction of a virB::lacZ transcriptional fusion in which the promoter of virB serves as the regulatory element for lacZ (β-galactosidase) activity.

Defective T-DNA

The LHR strains that are associated primarily with grapevine tumors do not have a functional tmr (cytokinin) gene (3,30). However, it appears that both the auxin and cytokinin genes (tms and tmr, respectively) are required for tumor formation in most dicotyledonous plants. It is possible to complement the defective cytokinin gene of the LHR strain by introducing a functional cytokinin gene from the WHR strain (3,30). The resulting strains now have an expanded host range and are able to infect many more plants (Tab. 2).

Integration of Too Much T-DNA: Hypersensitive Response

It appears that the introduction of too much T-DNA may also prevent tumor formation, perhaps by an excessive synthesis of T-DNA gene products, resulting in a hypersensitive response. Data in support of this possibility have also been obtained by studying the crown gall-grapevine system (30). It was observed that some grape varieties that are susceptible to tumor formation by the LHR strains are resistant to the WHR strain. Following inoculation with the WHR strain, the grapevine stem seems to undergo a hypersensitive response in which cells near the site of inoculation die. This hypersensitive response on the part of the plant could account for the inability of a tumor to develop. This reaction is a well-recognized defense mechanism to infection (7). If the amount of T-DNA transferred to the plant cell is reduced by a mutation in a vir gene which effectively reduces the efficiency of transfer of the T-DNA, then the WHR strain should induce a typical tumor. This, in fact, has been observed. One interpretation of these observations is that when less T-DNA

Tab. 2. The role of the tmr locus in host range.

Plant	Strains			
	WHR	LHR	LHR::tmr	LHR (T_L-DNA)
Vitis[a]	+	+	+	+
Nicotiana glauca	+	+	+	+
Nicotiana rustica	+	+[b]	+	+
Nicotiana tabacum	+	-	+	+
Tomato	+	-	+	+
Sunflower	+	-	+	+
Kalanchöe stems	+	-	-	-
Kalanchöe leaves	+	-	-	-

[a]Vitis sp. cv. Seyval.
[b]Roots from tumor.

Note: A comparison of the relative host ranges of the limited host range (LHR) and wide host range (WHR) strains is shown. Also shown is the result of introducing the WHR tmr gene into the LHR strain (LHR::tmr), and the result of introducing the WHR T-DNA into the LHR strain. Key: + = tumor formation; - = no tumor formation; T_L = left region of T-DNA, containing the tms, tmr, tml, and OCS genes.

is introduced, less of the T-DNA gene products will be synthesized and the hypersensitive response will not occur. The gene product(s) responsible for the hypersensitive response has not been identified.

Inefficient Expression of Growth Regulator Genes

In order for a tumor to develop, sufficient levels of auxin and cytokinin must be synthesized from the T-DNA. Further, the plant tissues must respond to these levels. Therefore, if either of these conditions is not met in any particular plant, tumors will not develop and there will be no evidence of transformation. This may very well account for the apparent inability of *Agrobacterium* to transform many different monocotyledonous plants as measured by tumor formation. Specifically, the levels of auxin required to effect a response in monocots are considerably higher than those necessary in dicots. Since this is the case, it seems likely that if additional copies of the T-DNA were introduced into apparently resistant plants, they might be infectable by *Agrobacterium*.

Supervirulent Strains

A strain has been isolated from nature that is more virulent than other strains of *Agrobacterium* and has a broader host range (9,14). It has now been shown that the reason for this increased virulence lies in the fact that all of the *vir* genes are expressed at higher levels in this strain (12). In turn, this leads to higher levels of T-DNA being processed and transferred into plant cells. The key region of the *vir* regulon which leads to the increased *vir* gene expression is the *virG* operon and the 3' end of the *virB* operon. Indeed, this region has been cloned on a WHR vector, and has been transferred into the common laboratory WHR strain. The new strain now becomes supervirulent (12). Not only are larger tumors formed in a shorter period of time, but this strain is able to induce tumors on plants that are not normally infectable. These include certain varieties of soybean (12) and alfalfa (L. Brown, pers. comm.).

DISCUSSION

It should be clear from the preceding descriptions that *Agrobacterium* may transfer T-DNA into plant cells without any apparent appearance of tumors. This raises the question of what the best procedure is to look for transfer and integration of T-DNA. One obvious approach is to look for transfer of a gene whose detection is extremely sensitive. One such gene currently being used is the bacterial gene specifying β-glucuronidase (11). There is no enzymatic background activity in uninfected plants, and so any biological activity detected strongly suggests that the gene has been transferred into the plant. However, final proof of integration of the gene into plant DNA requires that DNA of the putative transformed tissue hybridize with the bacterial gene under appropriate conditions (27).

Another important consideration in designing vectors is the choice of the plant promoter to engineer upstream of the gene to be assayed. The strongest plant promoter most commonly in use today is the 35S promoter isolated from cauliflower mosaic virus (CaMV). However, this promoter has the disadvantage that it is expressed to a measurable extent in bacteria. Thus, there is a background problem from the expression of DNA in the bacteria. The promoter for the synthesis of the opine mannopine, which is a component of the T_R-DNA of *Agrobacterium*, is also commonly used. It has the advantage of not being expressed in bacteria, and therefore not giving rise to background problems. Unfortunately, it is not nearly as strong a promoter as the 35S promoter of CaMV.

The use of the binary vector system is convenient (8), since a variety of vir gene combinations from different strains of Agrobacterium can easily be manipulated and used with different T-DNA genes. T-DNA processing seems to be the rate-limiting step in the virulence of many strains of Agrobacterium; so increasing the enzymatic activity of the site-specific endonuclease, coded by the virD operon, is one important way to increase processing and, consequently, virulence. This could be accomplished by increasing the survivability of Agrobacterium cells in close association with the plant surface, as well as by increasing the transcriptional activity of the genes concerned with T-DNA processing. One way is to increase the number of copies of the positive regulatory gene (virG) required for transcriptional activation (12). Further, it seems possible that the tzs locus which codes for cytokinin synthesis and excretion in some strains of Agrobacterium (13) might increase the infectivity of strains in which it is not already present. Similar considerations may hold for the PINF operon which may be concerned with the survivability of Agrobacterium in its natural environment (R. Kanemoto and A. Powell, unpubl. observ.). Thus, by altering the composition of the vir regulon and associated genes on the Ti-plasmid, it should be possible to alter the infectivity of Agrobacterium on a wide variety of plants.

REFERENCES

1. Akiyoshi, D.E., H. Klee, R.M. Amasino, E.W. Nester, and M.P. Gordon (1984) T-DNA of Agrobacterium tumefaciens encodes an enzyme of cytokinin biosynthesis. Proc. Natl. Acad. Sci., USA 81: 5994-5998.
2. Bolton, G.W., M.P. Gordon, and E.W. Nester (1986) Plant phenolic compounds induce expression of the Agrobacterium tumefaciens loci required for virulence. Science 232:983-985.
3. Buchholz, W.G., and M.F. Thomashow (1984) Host range encoded by the Agrobacterium tumefaciens tumor-inducing plasmid pTiAg63 can be expanded by modification of its T-DNA oncogene complement. J. Bacteriol. 160:327-332.
4. Douglas, C., W. Halperin, M. Gordon, and E. Nester (1985) Specific attachment of Agrobacterium tumefaciens to bamboo cells in suspension cultures. J. Bacteriol. 161:764-766.
5. Draper, J., A. MacKenzie, M. Davy, and J. Freeman (1983) Attachment of Agrobacterium tumefaciens to mechanically isolated asparagus cells. Plant Sci. Lett. 29:227-236.
6. Garfinkel, D.J., and E.W. Nester (1980) Agrobacterium tumefaciens mutants affected in crown gall tumorigenesis and octopine catabolism. J. Bacteriol. 144:732-743.
7. Goodman, R.N., Z. Kiraly, and K.R. Wood (1986) The Biochemistry and Physiology of Plant Disease, University of Missouri Press, Columbia, Missouri, 433 pp.
8. Hoekema, A., P.R. Hirsch, P. Hooykaas, and R.A. Schilperoort (1983) A binary plant vector strategy based on separation of vir and T-region of the Agrobacterium tumefaciens Ti-plasmid. Nature 303: 179-180.

9. Hood, E., G. Jen, L. Kayes, J. Kramer, R. Fraley, and M.-D. Chilton (1984) Restriction endonuclease map of pTiB0542, a potential Ti-plasmid vector for genetic engineering of plants. Bio/Technology 2:702-709.
10. Horsch, R.B., H.J. Klee, S. Stachel, S.C. Winans, E.W. Nester, S.G. Rogers, and R.T. Fraley (1986) Analysis of Agrobacterium tumefaciens virulence mutants in leaf discs. Proc. Natl. Acad. Sci., USA 83:2571-2575.
11. Jefferson, R.A., S. Burgess, and D. Hirsh (1986) Beta-glucuronidase from Escherichia coli as a gene-fusion marker. Proc. Natl. Acad. Sci., USA 83:8447-8451.
12. Jin, S., T. Komari, M.P. Gordon, and E.W. Nester (1987) Genes responsible for the supervirulent phenotype of Agrobacterium tumefaciens strain A281. J. Bacteriol. (in press).
13. Kaiss-Chapman, R.W., and R.O. Morris (1977) Trans-zeatin in culture filtrates of Agrobacterium tumefaciens. Biochem. Biophys. Res. Comm. 76:453-459.
14. Komari, T., W. Halperin, and E. Nester (1986) Physical and functional map of supervirulent Agrobacterium tumefaciens tumor-inducing plasmid pTiBo542. J. Bacteriol. 166:88-94.
15. Leroux, B., M.F. Yanofsky, S.C. Winans, J.E. Ward, S.F. Ziegler, and E.W. Nester (1987) Characterization of the virA locus of Agrobacterium tumefaciens: A transcriptional regulator and host range determinant. EMBO J. 6:849-856.
16. Matthysse, A., and R.H. Gurlitz (1982) Plant cell range for attachment of Agrobacterium tumefaciens to tissue culture cells. Physiol. Plant Pathol. 21:381-387.
17. Nester, E.W., M.P. Gordon, R.M. Amasino, and M.F. Yanofsky (1984) Crown gall: A molecular and physiological analysis. Ann. Rev. Plant Physiol. 35:387-413.
18. Panagopoulos, C., and P.G. Psallidas (1973) Characteristics of Greek isolates of Agrobacterium tumefaciens (E.F. Smith and Townsend) Conn. J. Appl. Bacteriol. 36:233-240.
19. Powell, A.L., and M.P. Gordon (1987) Plant tumor formation. In Biochemistry of Plants: A Comprehensive Treatise. Vol. II. Molecular Biology, S. Marcus, ed. Academic Press, Inc., Orlando, Florida (in press).
20. Schafer, W., A. Gorz, and G. Kahl (1987) T-DNA integration and expression in a monocot crop plant after induction of Agrobacterium. Nature 327:529-532.
21. Schroder, G., S. Waffenschmidt, E. Weiler, and J. Schroder (1984) The T-region of Ti-plasmids codes for an enzyme synthesizing indole-3-acetic acid. Eur. J. Biochem. 138:387-391.
22. Stachel, S.E., and E.W. Nester (1986) The genetic and transcriptional organization of the vir region of the A6 Ti-plasmid of Agrobacterium tumefaciens. EMBO J. 5:1445-1454.
23. Stachel, S.E., and P. Zambryski (1986) Agrobacterium tumefaciens and the susceptible plant cell: A novel adaptation of extracellular recognition and DNA conjugation. Cell 47:155-157.
24. Stachel, S.E., E. Messens, M. Van Montagu, and P. Zambryski (1985) Identification of the signal molecules produced by wounded plant cells that activate T-DNA transfer in Agrobacterium tumefaciens. Nature 318:624-629.
25. Thomashow, L., A. Reeves, and M. Thomashow (1984) Crown gall oncogenesis: Evidence that a T-DNA gene from the Agrobacterium Ti-plasmid pTiA6 encodes an enzyme that catalyzes synthesis of indoleacetic acid. Proc. Natl. Acad. Sci., USA 81:5071-5075.

26. Thomashow, M., W. Hugly, W.G. Buchholz, and L. Thomashow (1986) Molecular basis for the auxin-independent phenotype of crown gall tumor tissues. Science 231:616-618.
27. Thomashow, M.F., R. Nutter, K. Postle, M.-D. Chilton, F.R. Blattner, A. Powell, M.P. Gordon, and E.W. Nester (1980) Recombination between higher plant DNA and the Ti-plasmid of Agrobacterium tumefaciens. Proc. Natl. Acad. Sci., USA 77:6448-6452.
28. Winans, S.C., P.R. Ebert, S.E. Stachel, M.P. Gordon, and E.W. Nester (1986) A gene essential for Agrobacterium virulence is homologous to a family of positive regulatory loci. Proc. Natl. Acad. Sci., USA 83:8278-8282.
29. Yamada, T., C. Palm, B. Brooks, and T. Kosuge (1985) Nucleotide sequences of the Pseudomonas savastonoi indoleacetic acid genes show homology with Agrobacterium tumefaciens T-DNA. Proc. Natl. Acad. Sci., USA 82:6522-6526.
30. Yanofsky, M., B. Lowe, A. Montoya, R. Rubin, W. Krul, M. Gordon, and E. Nester (1985) Molecular and genetic analysis of factors controlling host range in Agrobacterium tumefaciens. Mol. Gen. Genet. 201:237-246.

DEVELOPMENT OF A DNA TRANSFER SYSTEM FOR PINES

Anne-Marie Stomp,[1] Carol Loopstra,[2] Ronald Sederoff,[1 2]
Scott Chilton,[3] JoAnne Fillatti,[4] Gayle Dupper,[2]
Patrick Tedeschi,[2] and Claire Kinlaw[2]

[1]Department of Forestry
North Carolina State University
Raleigh, North Carolina 27695-8002

[2]Pacific Southwest Forest and Range Experiment Station
U.S. Department of Agriculture Forest Service
Berkeley, California 94701

[3]Department of Botany
North Carolina State University
Raleigh, North Carolina 27695-7612

[4]Calgene, Inc.
Davis, California 95616

ABSTRACT

Methods are described for the transformation of sugar and loblolly pine cells by *Agrobacterium*. DNA transfer and expression is established by DNA-DNA and DNA-RNA hybridization data, opine identification, kanamycin resistance and neomycin phosphotransferase activity, growth characteristics of gall formation, and hormone autotrophy. Some of the difficulties inherent in conifer transformation systems are discussed, including strain selection, callus proliferation from galls, NPTII assays, and Southern hybridizations. A general strategy for the development of *Agrobacterium* transformation systems for woody plants is presented.

INTRODUCTION

Current methods in molecular biology provide a powerful approach for investigating the genomic structure of plants and the genetic regulation of phenotype. This approach is particularly important for forest tree species because their long breeding cycles hinder analysis of their genetic structure. The approach involves three steps: (i) the isolation and characterization of specific genes; (ii) the manipulation in vitro of DNA sequences by gene splicing or mutagenesis; and (iii) the transfer of genes into plants for the analysis of a modified phenotype. One key component of this approach, DNA transfer, involves the transfer and expression of genes from sources outside of the recipient tree.

This chapter reviews our work on DNA transfer in pines and discusses the strategy of the research and the technical problems we have encountered in developing our system. The DNA transfer system has four parts: (i) selection and use of an appropriate vector system for DNA transfer into plant cells; (ii) assay of integration and expression of transferred DNA (T-DNA) in cells; (iii) regeneration of transformed plants from the cells expressing the newly inserted DNA; and (iv) testing of transformed plants for the newly acquired phenotype.

Our eventual goal is the production of pine plants expressing newly integrated genes. Our first task was to select a delivery, or vector, system for DNA which would result in expression of transferred DNA. Ideally, this vector system should also be able to take advantage of existing tissue culture methods for plant regeneration.

SELECTION OF A VECTOR

Several DNA transfer systems have been used successfully in higher plants, specifically, DNA uptake into protoplasts by electroporation (7), direct DNA microinjection (10), and *Agrobacterium*-mediated gene transfer (3). Electroporation and microinjection require tissue culture methods for the production and regeneration of callus from protoplasts. Methodology must also exist to regenerate transformed plants from protoplast-derived callus, an approach which is currently unavailable in pines. The only method for the routine regeneration of pine plants is the production of shoots from cotyledons. If shoot regeneration from cotyledons could be combined with *Agrobacterium*-mediated DNA transfer, we could produce transformed plants. Such a system of "co-cultivation" exists for other plants (8), and we hypothesized that this method could be extended to pine.

DEVELOPMENT OF A DNA TRANSFER SYSTEM FOR PINES

Our strategy for producing transformed plants can be conceptualized as four steps, from the simplest system, i.e., gall formation on seedlings, to the most complex, i.e., the production of transformed plants. These four steps can be expressed in experimental terms as a set of four questions:

(a) Can wild-type *Agrobacterium* infect any pine species?

(b) Can wild-type *Agrobacterium* infect pine tissue in vitro?

(c) Can *Agrobacterium* infect pine cells in a dedifferentiated state?

(d) Can *Agrobacterium* infect pine cells in a meristematic state and allow subsequent shoot formation?

The answers to these questions will define methods for the production of both transformed pine callus cultures and plants.

INOCULATION OF PINE SPECIES

Agrobacterium strains that would induce gall formation in pines were identified by inoculating seedlings with a set of virulent Agrobacterium strains. Previous investigations of the host range of Agrobacterium had reported that pine could not be infected (5,14,15); however, a small number of strains were used in those screening experiments. Agrobacterium can show a high degree of host range specificity (1). Therefore, we reasoned that screening a larger number of wild-type strains might identify one which would infect pine. We selected loblolly pine (Pinus taeda L.) for the initial screening because of its economic importance.

Gall formation and opine synthesis were used to assay gene transfer in these early screening experiments. Two strains of Agrobacterium were found to infect loblolly pine (13). We have enlarged that initial survey using strains identified as infective in loblolly pine (M2/73 and U3) and three other broad host range strains (C58, A281, and 542) to inoculate nine pine species, Douglas fir (Pseudotsuga menziesii) and incense-cedar (Libocedrus decurrens) (A.-M. Stomp, ms. in prep.) (Tab. 1).

Age and the succulence of the plant tissue are important factors affecting infectivity. Inoculation of young, succulent seedlings, characterized by a low degree of woodiness, results in a high efficiency of gall formation. We inoculate seedlings as soon as the epicotyl is 2 to 4 cm high. Inoculation of stems that are not succulent and green results in a lower percentage of galls.

Galls begin to appear on stems about 8 to 10 weeks after inoculation, but many do not continue to grow. Scoring seedlings for gall formation at 12 weeks gives the highest levels of infectivity. However, subsequent scoring at six and 18 months after inoculation shows that most of these galls fail to increase in size and are usually sloughed off by the plant (Tab. 2). Continued gall growth results in galls of large size; the largest

Tab. 1. Maximum percent gall formation and opine synthesis in selected conifer species inoculated with Agrobacterium tumefaciens strains.

Conifer species	Strain	Percent gall formation	Opine present
Pinus eldarica	U3	11	Agropine
Pinus elliotti	A281	21	Agropine
Pinus jeffreyi	C58	2	Nopaline
Pinus lambertiana	U3	17	Agropine
Pinus ponderosa	M2/73	43	Nopaline
Pinus radiata	U3	38	Agropine
Pinus sylvestris	M2/73	24	Nopaline
Pinus taeda	542	39	Agropine
Pinus virginiana	U3	30	Agropine
Pseudotsuga menziesii	542	35	Agropine
Libocedrus decurrens	M2/73	61	Nopaline

Tab. 2. Continuation of gall growth in selected species of pine.

Species	Strain	Number of galls at six months	Number of galls at 18 months
P. eldarica	U3	2	0
P. elliotti	A281	10	1
P. jeffreyi	C58	1	0
P. lambertiana	U3	11	0
P. ponderosa	M2/73	21	0
P. sylvestris	M2/73	24	1
P. taeda	542	38	2
P. virginiana	U3	3	0

Note: Galls measured after 18 months are still growing and exceed 2 cm in diameter.

gall we have obtained is approximately 7 cm in diameter after two years. Presumably, many of the smaller galls are lost due to processes used by trees to wall off pathogenic attack. These observations suggest that both plant and bacterial genes regulate formation and growth of galls in pine species.

Gall tissue fails to proliferate callus cultures that grow well and, consequently, they cannot be subcultured for more than six months. Callus cultures have been produced from galls on loblolly pine (13), sugar pine (*Pinus lambertiana*), and Monterey pine (*Pinus radiata*), but these cultures grow slowly and show considerable browning. The cultures will not grow hormone autotrophically, which is in contrast to reports from other plants (11). Slow growth, considerable necrosis, and short lifespan make these cultures unsuitable for analysis of DNA integration into the pine cell genome. Such tissue can, however, be used to study expression of transferred genes, most notably opines.

Most know opines are modified amino acids whose biosyntheses are catalyzed by a family of enzymes, the opine synthetases. These enzymes are the products of a family of *Agrobacterium* genes transferred to and expressed in plant cells after transformation. Opines are strain-specific because each *Agrobacterium* strain carries genes for the synthesis of particular opines. Although small amounts of opines have been detected in untransformed plant cells under unusual culture conditions (4), opines are still considered good indicators of DNA transfer.

Gall tissue from several trees of different species has been assayed for opines (A.-M. Stomp et al., ms. in prep.) (Tab. 1). Opines were analyzed by paper electrophoresis of 80% ethanol extracts of gall tissue (13). The expected strain-specific opines were present at high levels in more than 90% of the gall tissue assayed. Therefore, *Agrobacterium* opine synthetase genes have been transferred and expressed in pine cells.

Inoculation of pine seedlings provides a means of identifying *Agrobacterium* strains that are capable of gall formation and opine synthesis. The advantage of this method is that highly infective strains can easily be identified. The disadvantages for pine are that the callus cultures derived from galls will not regenerate shoots and do not lend themselves to antibiotic marker selection (see below) or DNA analysis. Therefore, the next step is to use the highly infective strains to inoculate pine tissue in vitro, where plant regeneration is possible and DNA analysis might not be precluded.

INOCULATION OF PINE CELLS IN VITRO

The step with the highest probability for success would be inoculation of pine tissue that is similar to succulent seedling stems. Two such systems exist: (i) inoculation of aseptically grown seedlings and (ii) tissue-cultured shoots derived from cotyledons. In previous studies (13), inoculation of aseptically grown loblolly seedlings produced slow-growing galls after several months. Therefore, we inoculated shoots of both loblolly pine and sugar pine with *Agrobacterium* in hopes of quicker gall formation. These species were selected because of their economic importance and because they represent the two major pine groups--the white and yellow pines.

Compared to loblolly pine, inoculation of sugar pine shoots gives a higher frequency of gall formation, faster appearance of the galls, and more galls that continue to grow. In loblolly pine, gall formation is slow, taking several months, and it is difficult to proliferate callus from the gall tissue. In contrast, callus readily proliferates from galls of sugar pine, and this callus is hormone-autotrophic. The ease of growing this callus makes it amenable to antibiotic selection and DNA analysis.

Assay of Integration and Expression of Transferred DNA

Expression of transferred genes can be assayed through in vitro growth characteristics or by identifying RNA, protein, or enzyme products. Transfer of the genes can be assayed directly by DNA hybridization. The wild-type strains that we use have growth regulator genes and opine synthetase genes in their T-DNA. We also use binary strains with a wild-type plasmid and a disarmed plasmid containing one or two neomycin phosphotransferase chimeric genes in the T-DNA. These strains allow us to assay for gall formation or hormone-autotrophic callus growth in vitro (growth regulator genes), opines (opine synthetase), and resistance to the antibiotic kanamycin (neomycin phosphotransferase or NPTII). In addition, we have an *EcoRI* fragment (pCGn71) cloned from A6 (2), which allows us to probe RNA (northern) or DNA (Southern) blots for growth regulator genes.

Antibiotic Selection: Kanamycin

Antibiotic selection of transformed cells is a valuable component of the current DNA transfer technology. Integration and expression of chimeric genes for antibiotic resistance into plant cells allow for the selection of transformed cells on antibiotic-containing medium. Two antibiotics, kanamycin sulfate (and G418) and hygromycin, are commonly used for selection. To develop antibiotic selection as a useful tool in pine, the following questions were asked: (i) are pine cells sensitive to these antibiotics at any levels? and (ii) will the chimeric resistance genes be expressed in pine tissues at levels high enough to "rescue" cells from the antibiotic?

Sugar pine callus growth was measured on media with increasing concentrations of the antibiotic to determine the level of sensitivity to kanamycin sulfate (Fig. 1). Sugar pine cells are quite sensitive to kanamycin sulfate, with 50% inhibition of growth occurring at concentrations of 8 mg/l. Callus transferred to medium with lethal concentrations of kanamycin remain green but do not grow, in contrast to the "bleaching" or browning of callus seen with other species. Transfer of this callus to nonselective medium does not result in a resumption of growth; the callus starts to brown, indicating that, although green during the first subculture period, the callus cells were killed during kanamycin selection.

Inoculation of sugar pine shoots with a binary Agrobacterium strain, 542 kan^r, that harbors both a wild-type 542 plasmid (growth regulator and opine synthetase genes) and a miniplasmid (pEND4K) containing the chimeric gene for neomycin phosphotransferase under the control of the nopaline synthetase promoter (9), resulted in gall formation on several shoots. Hormone-autotrophic callus cultures were proliferated from these galls. After callus cultures were established, portions of the cultures were selected by transfer to medium containing 100 mg/l kanamycin. After several subcultures, callus lines resistant to kanamycin were established and tested for NPTII activity using a method modified from Reiss et al. (10).

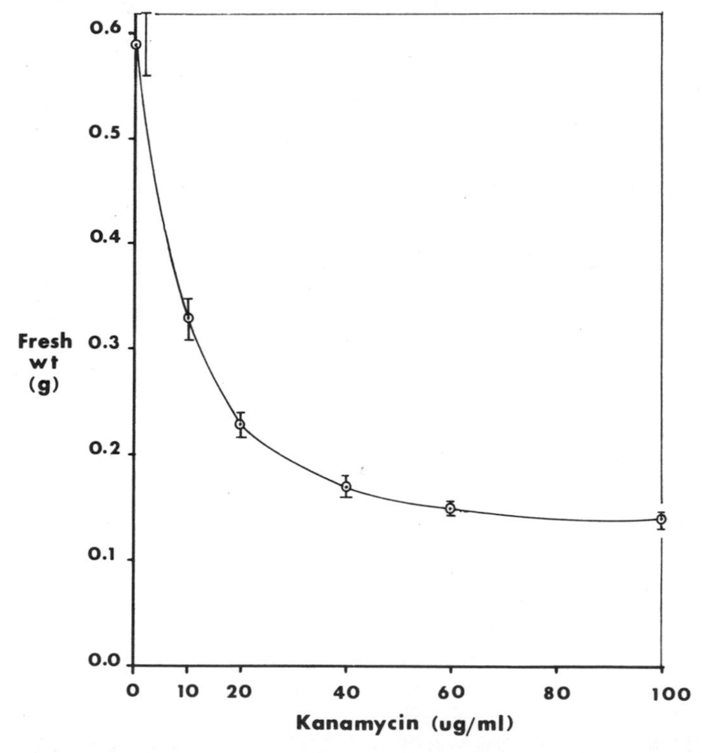

Fig. 1. Kanamycin killing curve for sugar pine callus. Pine callus was established on modified Litvay's medium (13) with increasing amounts of kanamycin sulfate added to the cooled medium before pouring. Fresh weights were determined after a two-week subculture from eight pieces of callus per kanamycin concentration. The experiment was replicated twice. Data are shown ± S.D.

Neomycin phosphotransferase activity was readily detectable in this callus (Fig. 2). However, this callus grew poorly and was transferred from selection to medium without antibiotic in order to grow enough for Southern blot analysis. After several months in culture, this callus was again tested for NPTII activity and none was found. Did the callus lose the gene or its activity or was something interfering with the transferase activity itself?

Tomato tissue known to be kanamycin-resistant (J. Fillatti, pers. comm.) was assayed for NPTII activity. When tomato tissue was homogenized with extracts from pine callus, tomato NPTII activity could not be detected. This result suggested that pine callus contained either an inhibitor of NPTII activity or protease activity which inactivated the enzyme. The solution to this problem came with the addition of ascorbic acid at 25 mg/ml to the homogenizing buffer. This restored activity to tomato-pine mixed extracts and allowed us to assay for NPTII activity in pine callus (Fig. 2).

Southern Blot Analysis: Growth Regulator Genes

The presence of T-DNA sequences in transformed pine cells has also been demonstrated by Southern blot analysis. The DNA of the transformed pine cells was isolated using cesium chloride gradients, and then cut into fragments with HindIII. These fragments were then sized by agarose gel electrophoresis. The DNA fragments were transferred and crosslinked by UV light to Hybond-N (a nylon membrane made by Amersham Corporation, Arlington Heights, Texas), and then probed with the large

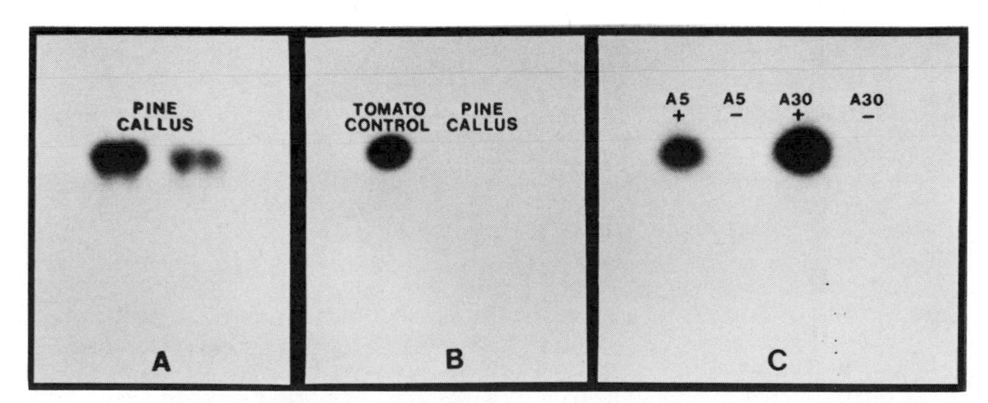

Fig. 2. Neophosphotransferase (NPTII) activity in pine callus cultures. (A) NPTII activity in a callus culture derived from a *Pinus lambertiana* shoot inoculated in vitro with *Agrobacterium tumefaciens* strain 542 kan^r and cultured on medium containing 100 mg/l kanamycin sulfate. The two lanes represent different amounts of callus extracts. (B) NPTII activity in an extract from a kanamycin-resistant tomato plant (left lane) and loss of NPTII activity in the sugar pine callus line depicted in (A). (C) NPTII activity in extracts of two sugar pine callus lines, A5 and A30, derived from shoots inoculated with *A. tumefaciens* strain 542 kan^r and growing on medium containing 10 mg/l kanamycin sulfate. The extracts were assayed with (+) and without (-) 25 mg/ml ascorbic acid.

EcoRI fragment containing growth regulator genes (the insert from pCGn71), radioactively labeled by the hexanucleotide random primer reaction.

The feasibility of using standard Southern blot analysis to find transferred T-DNA sequences in a genome as large as those found in Pinus [12 pg or more per haploid genome (6)] was determined by dilution experiments. Varying amounts of pCGn71 insert DNA were electrophoresed, blotted, and probed with radioactively labeled insert (Fig. 3). Amounts of DNA as low as 10 pg, equivalent to ten copies, could be detected on longer exposures.

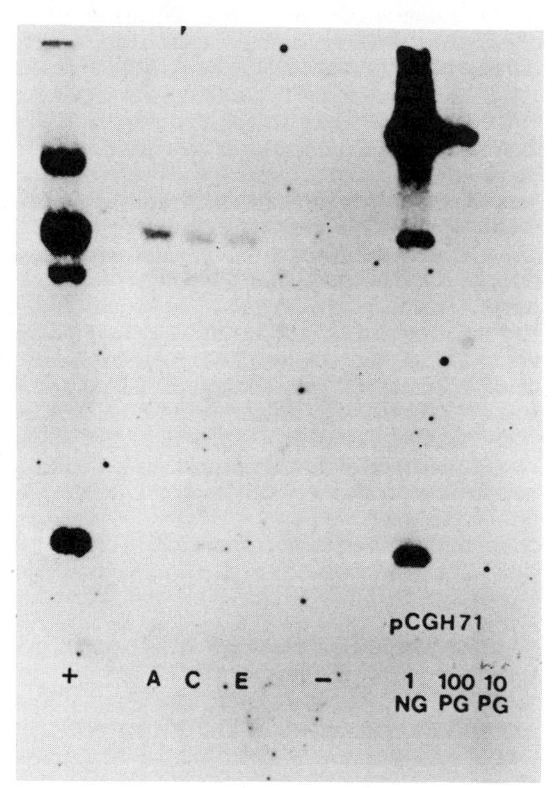

Fig. 3. Autoradiography of transformed pine callus DNAs probed with the EcoRI fragment from pCGn71. The DNAs were restricted with HindIII and sized on a 0.8% agarose gel. The gel was blotted to Hybond-N, cross-linked with UV light, and hybridized to radioactively labeled EcoRI fragment from pCGn71 (mislabeled as pCGH71) overnight at 65°C. Channels are as follows: "+": 0.5 μg of 542 kan^r DNA; A, B, and C: 10 μg, each, of DNA isolated from transformed callus cultures A, B, and C, respectively; "-": 10 μg of DNA isolated from untransformed callus; 1 ng, 100 pg, and 10 pg: DNA isolated from pCGn71, which was restricted and loaded on the gel in those respective amounts. The X-ray film was exposed at -70°C for four days using an intensifying screen.

Southern hybridization data from three independently transformed callus cultures indicates that T-DNA sequences can be clearly detected (Fig. 3). Comparison with the dilution experiment suggests that the T-DNA is present in multiple copies. We conclude that hormone-autotrophic growth in these callus lines results from expression of growth regulator genes from transferred T-DNA.

Transfer and expression of T-DNA genes for antibiotic resistance and growth regulator synthesis have been demonstrated by growth in culture, NPTII activity, and Southern analysis, in callus derived from sugar pine shoots inoculated with *Agrobacterium*. Knowing that *Agrobacterium* will work as a vector system in pine, we now can turn our attention to transforming callus cultures and meristematic cells which can regenerate plants.

INOCULATION OF UNDIFFERENTIATED PINE CELLS: COTYLEDONS

Cotyledons from newly germinating pine seeds can be induced either to make callus or to regenerate shoots by changing the medium concentration of growth regulators. Our strategy is to first optimize cotyledon transformation on medium formulated for callus proliferation to learn about the requirements for successful, high-efficiency transformation starting with cotyledons. We would then use these optimal methods to transform cotyledons on medium containing high concentrations of cytokinin and to regenerate transformed shoots.

Recently, several cotyledon co-cultivation experiments have apparently produced transformed callus which exhibits hormone-autotrophic growth and growth on medium containing kanamycin. Experiments are currently underway to assay molecular markers to confirm transformation. These results with co-cultivated cotyledons are encouraging and bring us closer to our goal of developing a method for the routine production of transformed pine plants.

DISCUSSION

The development of DNA transfer methods in conifers can be viewed from two diametrically opposed perspectives. One position assumes that transformation in conifers could be easily achieved, since *Agrobacterium*-mediated DNA transfer is believed to be quite similar in all plant species. The opposing view contends that the biological differences between gymnosperms and angiosperms are great. Therefore, recombinant DNA methods and DNA transfer techniques developed for angiosperms will not readily extrapolate to conifers, making molecular biological studies quite difficult. Our experience with DNA transfer methods fits neither of these views. The general principles derived from work with *Agrobacterium* on angiosperms are true for conifers, but the precise details differ.

First, host range specificity remains an important factor for any species of pine. We feel confident that most pine species can be infected by *Agrobacterium*, but several strains should be used when studying pathogenicity in an untested pine species. The frequency of gall formation, the rate of gall growth, and the subsequent defense response elicited in the plants also vary greatly depending on the strain of *Agrobacterium* and the species.

Growth of gall tissue in culture continues to be difficult if explants are derived from inoculated seedlings. In contrast, callus proliferation from explants derived from galls on tissue-cultured shoots of sugar pine occurs readily, but with large differences in callus appearance. Our studies are not yet extensive enough for us to know if ease of callus culture or callus morphology is a function of pine species and/or Agrobacterium strain. Alternatively, variation in callus growth characteristics could be a function of individual or family genetic differences within each pine species.

Our experience with kanamycin selection in conifer cells reinforces the need for careful interpretation when working with metabolic inhibitors. We find that under certain conditions callus tissue is killed by low concentrations of kanamycin but remains green and appears viable. Transfer for several subcultures must be done to establish transformed callus cultures. Therefore, correct interpretation of in vitro kanamycin selection requires prior studies on the toxicity of the antibiotic in the particular tissue (i.e., callus or cotyledon) to be transformed.

Interpretation of molecular data requires care as well. Our pine callus extracts prepared in the absence of ascorbate give false negative assays for NPTII activity. However, addition of ascorbate restores this activity. We believe that ascorbate acts as an antioxidant that inhibits reactions in the crude extracts, greatly reducing NPTII activity. Given the proper controls in both tissue culture and molecular studies, kanamycin selection can be a useful technique for DNA transfer studies in conifers. However, Southern hybridization still remains the best technique by which DNA transfer can be confirmed.

ACKNOWLEDGEMENTS

This work was supported by the U.S. Department of Agriculture (USDA) Forest Service, by North Carolina State University, and by a competitive grant from the Forest Biology Program of the USDA.

REFERENCES

1. Anderson, A.R., and L.W. Moore (1979) Host specificity in the genus Agrobacterium. Phytopathology 69:320-323.
2. Barker, R., K. Idler, D. Thompson, and J. Kemp (1983) Nucleotide sequence of the T-DNA region from the Agrobacterium tumefaciens octopine Ti plasmid pTi15955. Plant Mol. Biol. 2:335-350.
3. Barton, K.A., and M.-D. Chilton (1983) Agrobacterium Ti plasmid as vector for plant genetic engineering. Methods in Enzymology 101:527-539.
4. Christou, P., S.G. Platt, and M.C. Ackerman (1986) Opine synthesis in wild-type plant tissue. Plant Physiol. 82:218-221.
5. DeCleene, M., and J. DeLey (1976) The host range of crown gall. Bot. Rev. 42:389-466.
6. Dhillon, S.S. (1987) DNA in tree species. In Cell and Tissue Culture in Forestry, J.M. Bonga and D.J. Durzan, eds. Martinus Nijhoff Publishers, Dordrecht, The Netherlands, pp. 298-313.
7. Fromm, M.E., L.P. Taylor, and V. Walbot (1986) Stable transformation of maize after gene transfer by electroporation. Nature 319:791-793.

8. Horsch, R.B., J.E. Fry, N.L. Hoffman, D. Eichholtz, S.G. Rogers, and R.T. Fraley (1985) A simple and general method for transferring genes to plants. Science 227:1229-1231.
9. Klee, H.J., M.F. Yanofsky, and E.W. Nester (1985) Vectors for transformation of higher plants. Bio/Technology 3:637-642.
10. Morikawa, H., A. Iida, C. Matsui, M. Ikegami, and Y. Yamada (1986) Gene transfer into intact plant cells by electroinjection through cell walls and membranes. Gene 41:121-124.
11. Ooms, G., P.J. Hooykaas, G. Moleman, and R.A. Schilperoort (1981) Crown gall plant tumors of abnormal morphology induced by Agrobacterium tumefaciens carrying mutated octopine Ti plasmid; analysis of T-DNA functions. Gene 14:33-50.
12. Reiss, B., R. Sprengel, H. Will, and H. Schaller (1984) A new sensitive method for qualitative and quantitative assay of neomycin phosphotransferase in crude cell extracts. Gene 30:211-218.
13. Sederoff, R., A.-M. Stomp, W.S. Chilton, and L.W. Moore (1986) Gene transfer into loblolly pine by Agrobacterium tumefaciens. Bio/-Technology 4:647-649.
14. Smith, C.O. (1939) Susceptibility of species of Cupressaceae to crown gall as determined by artificial inoculation. J. Agric. Res. 59:919-925.
15. Smith, C.O. (1942) Crown gall on species of Taxaceae, Taxodiaceae, and Pinaceae, as determined by artificial inoculations. Phytopathology 32:1005-1009.

DEVELOPMENT OF GLYPHOSATE-TOLERANT *POPULUS* PLANTS THROUGH EXPRESSION OF A MUTANT *aroA* GENE FROM *SALMONELLA TYPHIMURIUM*

JoAnne J. Fillatti,[1] Bruce Haissig,[2] Brent McCown,[3]
Luca Comai,[1] and Don Riemenschneider[2]

[1]Calgene, Inc.
Davis, California 95616

[2]Forestry Sciences Laboratory
North Central Forest Experimental Station
U.S. Department of Agriculture Forest Service
Rhinelander, Wisconsin 54501

[3]Department of Horticulture
University of Wisconsin
Madison, Wisconsin 53706

ABSTRACT

The development of a plant regeneration and transformation system for the poplar hybrid NC5339 (*Populus alba* x *Populus grandidentata*) is described. A binary armed strain of *Agrobacterium tumefaciens* harboring three chimeric gene fusions was used as a vector. Genetic transformation was confirmed through western blot analyses. Employing this system, we have introduced into *Populus* NC5339 a bacterial *aroA* gene which confers tolerance to the herbicide glyphosate. Expression of this foreign gene into *Populus* is discussed.

INTRODUCTION

The genus *Populus* contains many fast-growing hardwood tree species which can be cultivated as short rotation crops for pulp production. During the establishment and management of short-rotation *Populus* plantations, losses due to weed competition are substantial. At present, weeds are controlled largely through mechanical means and by limited applications of selected herbicides during periods when trees are dormant (1). These methods, however, are expensive and only partially effective. The process of genetically engineering herbicide tolerance into *Populus* is therefore a goal of significant commercial value.

In the past few years, the technology for introducing foreign genes into plants has been developed for many plant species (7,12,18,19). This technology has facilitated the development of transgenic plants which have increasing tolerance to the herbicide glyphosate (8,22). In this chapter, we describe our strategy for developing herbicide-tolerant Populus trees.

ISOLATION AND CHARACTERIZATION OF THE MUTANT aroA GENE

The herbicide glyphosate inhibits 5-enolpyruvylshikimate (EPSP) synthase, an enzyme involved in the biosynthesis of aromatic amino acids in plants (2,24). Tolerance to glyphosate can be mediated by overproduction of the target enzyme (3) or by the presence of an altered enzyme (7,21). The objective of this work was to test whether expression of a mutant bacterial gene coding for an altered EPSP synthase enzyme confers resistance to the glyphosate when introduced into a hybrid clone of Populus, NC5339.

In Salmonella typhimurium, EPSP synthase is encoded by the aroA locus. Mutants resistant to the herbicide glyphosate were obtained by mutagenizing cultures of S. typhimurium with ethylmethanesulfonate and screening for the ability to grow on medium containing the herbicide. A mutant aroA gene was cloned from the resistant cultures. Plasmids containing this sequence were shown to complement strains of S. typhimurium with a mutation at the aroA locus. Sequence analysis demonstrated that the mutant allele of the aroA locus had a single cytosine-thymine transition which resulted in the substitution of serine for proline. This single base pair substitution caused a decrease in the affinity of the enzyme for glyphosate without affecting the kinetic efficiency of the enzyme (23).

In order to obtain expression of this gene in plants, the mutant aroA structural gene was fused with the 5' regulatory sequences of the mannopine synthetase gene from Agrobacterium tumefaciens (4). The chimeric aroA gene was then introduced into binary Agrobacterium vectors which will be described below.

PLANT TRANSFORMATION

Agrobacterium tumefaciens provides a natural gene transfer system for many dicotyledonous plants. During the natural infection cycle of A. tumefaciens, bacterial DNA (T-DNA) is integrated into the host plant chromosome, resulting in the production of tumors on plants (6). The tumor-inducing genes can be deleted and substituted for by heterologous genes without affecting the ability of A. tumefaciens to transfer T-DNA to plants (9). These modified strains of A. tumefaciens can be incubated with protoplasts, cell suspension cultures, or explants, and transformed plants that lack the oncogenic symptoms can be obtained.

Since transformation with A. tumefaciens involves the incubation of bacterial cells with plant tissue (co-cultivation) and subsequent regeneration of only selected transformed cells, an efficient regeneration system and an effective selectable marker are required. Our first objective was to develop a regeneration system for a hybrid clone, Populus alba x Populus grandidentata (NC5339). This clone was chosen because shoot cultures could be maintained in vitro and utilized for clonal propagation, thus providing a sterile source of explant material. After testing a series of media with varied hormonal regimes, salt composition, and carbon sources, regeneration from leaf explants of the hybrid poplar clone NC5339 was obtained on a medium containing Murashige and Skoog minimal salts (MS), sucrose (3%), benzylaminopurine (BAP) (1 mg/l), zeatin (1 mg/l), and phytoagar (0.6%) at pH 5.6 (MS1BZ medium). Shoots developed at the cut edges of leaf explants 20 to 30 days after planting. On the average, one to three shoots developed from 35% of the leaf explants plated. The preculture conditions and the subsequent handling of the leaf explants affected the regeneration rate. If shoot cultures were maintained on media containing BAP prior to explanting, shoot regeneration was inhibited. Populus was also very sensitive to H_2O stress; therefore, leaf explants had to be cut in the presence of H_2O or shoots did not develop. In addition, the size and age of the leaf explants had a small effect on the rate of shoot regeneration. Younger leaves cut into segments of at least 2 cm^2 yielded the highest number of shoots per explant.

When A. tumefaciens is used as a vector for introducing foreign genes into plant cells, it is convenient to engineer a dominant selectable marker into the T-DNA along with the gene of interest. A selectable marker allows direct selection of transformed tissue by conferring resistance to an antibiotic. The most commonly used selectable marker for plant transformation is the bacterial gene coding for neomycin phosphotransferase (NPTII') (5,14,17). This gene confers resistance to the antibiotic kanamycin and has proved to be an effective marker in crop species such as tobacco (18), tomato (10,12), and lettuce (19). To test whether selection for kanamycin resistance was appropriate for Populus NC5339, we plated leaf explants onto MS1BZ medium containing 0, 20, 40, 60, and 100 mg/l kanamycin. Concentrations of 60 mg/l or greater completely inhibited callus growth and shoot initiation; therefore, we used kanamycin at 60 mg/l in our co-cultivation experiments.

With a regeneration system in place and a suitable marker for selection, we began the Agrobacterium/explant co-cultivation experiments. Initially, a binary disarmed strain of A. tumefaciens (LBA4404/587/85) was used in the co-cultivation experiments. This strain contains two plasmids that function in trans to promote the transfer of T-DNA. One plasmid carries an intact vir region and lacks the T-DNA, while the other plasmid, pPMG587/85 (13), has T-DNA border sequences flanking the two NPTII' genes and the aroA gene (8) and no vir region. Expression of the NPTII' gene is promoted by regulatory sequences derived from the octopine synthetase gene of A. tumefaciens strain A6 (4). Expression of the other NPTII' gene and the aroA gene is regulated by sequences derived from the mannopine synthetase gene of A. tumefaciens strain A6 (4). A binary oncogenic strain of A. tumefaciens (C58/587/85) containing a wild-type C58 plasmid (15) and the plasmid pPMG587/85 (described above) was also used as a DNA vector.

A modified leaf explant co-cultivation system was used to transform poplar (10,18). Leaf explants from in vitro-grown Populus NC5339 plants were cut into approximately 2-cm^2 segments in H_2O and preincubated on tobacco feeder plates for 24 hr under low light conditions. The tobacco feeder plates were prepared two days prior to use by pipetting 0.5 ml of tobacco suspension culture onto petri dishes (100 x 25 mm) containing 50 ml of Murashige minimal organics medium, supplemented with 2,4-dichlorophenoxyacetic acid (2,4-D) (0.2 mg/l), kinetin (0.1 mg/l), thiamine hydrochloride (0.9 mg/l), potassium acid phosphate (200 mg/l), and Difco Bacto agar (0.8%). After two days a disc of sterile filter paper (Whatmann 3 mm) was placed on top of the tobacco cells. The leaf segments were incubated in a liquid broth culture of A. tumefaciens (LBA4404/587/85: 5 x 10^8 bacteria per ml; C58/587/85: 2 x 10^8 bacteria per ml) for 30 min, blotted, and placed back on the tobacco feeder plates for 48 hr. Following the co-cultivation period, the leaf segments were plated onto regeneration medium containing carbenicillin (500 mg/l) and kanamycin (60 mg/l). When leaf explants were co-cultivated with the binary disarmed vector LBA4404/587/85, transformed shoots were never recovered. However, when A. tumefaciens strain C58/587/85 was employed as a vector, kanamycin-resistant shoots with two distinct phenotypes developed (13).

Since a binary oncogenic vector was employed, it was necessary to distinguish between the three possible insertion events. By including kanamycin in the regeneration medium, we selected against the development of cells that obtained only a copy of the C58 wild-type T-DNA. The other two insertion events were easily distinguished by conducting nopaline assays. Kanamycin-resistant shoots that did not produce nopaline contained only the disarmed T-DNA, as evidenced by Southern blot analysis. Kanamycin-resistant shoots that also produced nopaline contained both the disarmed and wild-type T-DNA. The two distinct shoot phenotypes (normal and teratomic) which developed following co-cultivation also correlated with the absence or presence of the wild-type T-DNA in the poplar genome.

Thirty-eight of the 40 phenotypically normal shoots, which developed and subsequently rooted on medium containing kanamycin, exhibited NPTII' enzyme activity as assayed by the protocol of Reiss et al. (20). A western blot analysis was conducted on the kanamycin-resistant Populus NC5339 plants to demonstrate the presence of the bacterial aroA protein. With bacterial EPSP synthase antiserum we could detect a polypeptide with the correct molecular weight in only two of the 40 kanamycin-resistant shoots tested (see Fig. 1). This polypeptide was absent in untransformed plants.

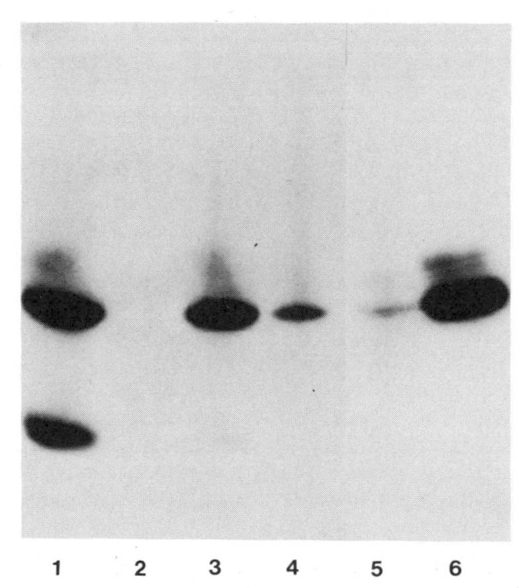

Fig. 1. Detection of bacterial EPSP synthase protein in transformed Populus NC5339 plants by western blot analysis. Lane 1, leaf extract from a tomato plant transformed using Agrobacterium tumefaciens strain LBA4404 pPMG587/85; lane 2, leaf extract from a control untransformed Populus NC5339 plant; lanes 3 and 4, leaf extracts from transformed Populus NC5339 plants; lane 5, 250 mg purified EPSP synthase protein added to untransformed Populus NC5339 leaf extract; lane 6, 250 mg purified EPSP synthase protein and no leaf extract. From Ref. 13, with permission.

To test whether the introduced aroA gene conferred tolerance to the herbicide glyphosate, kanamycin-resistant transformants derived from independent transformation events and untransformed control plants were sprayed with the equivalent of 0.07 or 0.28 kg/ha glyphosate (active ingredient) and compared to unsprayed control plants. The growth rate of five of the six transformants tested was significantly greater than that of untransformed control plants after treatment with 0.07 kg/ha glyphosate (Fig. 2). When plants were sprayed with 0.28 kg/ha glyphosate, transformants grew significantly taller than the untransformed controls; however, the height differences were less than at the lower dosage (D. Riemenschnieder, B.E. Haissig, J. Sellmer, and J.J. Fillatti, ms. in

Fig. 2. Glyphosate spray experiment (from left to right): (a) control untransformed poplar plant sprayed with the equivalent of 0.07 kg/ha of glyphosate; (b) poplar plant producing the aroA protein sprayed with 0.07 kg/ha of glyphosate; and (c) control untransformed poplar plant sprayed with H_2O only. The transformed poplar clone shown above (b, middle) exhibited the highest level of tolerance in preliminary greenhouse tests.

prep.). The aroA protein was not detected in three of the five poplar clones that exhibited tolerance to glyphosate when sprayed at 0.07 and 0.28 kg/ha. However, when a Southern blot analysis was conducted on one of these three glyphosate-tolerant clones, the results indicated that the aroA gene had integrated into the poplar genome. It is possible that the efficiency of protein recovery may be low due to a high concentration of phenolics present in crude leaf extracts. Western blot analysis conducted on poplar plants therefore may not accurately reflect the amount of aroA protein present in vitro.

CONCLUSIONS

We have demonstrated that expression of the mutant aroA gene confers tolerance to the herbicide glyphosate in *Populus*; however, the degree of tolerance is not yet at a commercial level. Through further modifications of the DNA sequences controlling expression and compartmentalization of the enzyme, we expect to obtain higher levels of glyphosate tolerance. This work represents the first demonstration of introduction and expression of foreign genes of agronomic value into a forest tree species. Gene transfer technology will continue to provide a powerful tool for improving forest tree species as well as a valuable system for studying the regulation of gene expression. Furthermore, the availability of a glyphosate-tolerant *Populus* cultivar may have a great impact on the economics of commercial wood production from *Populus* by allowing chemical control of weeds. This will result in substantially decreased losses due to weed competition during the establishment and subsequent cultivation of *Populus* plantations.

REFERENCES

1. Akinymiju, O.A., J.G. Isebrands, N.D. Nelson, and D.I. Dickmann (1982) Use of glyphosate in the establishment of *Populus* in short rotation intensive culture. In *Proceedings of the North American Poplar Council Meeting*, Rhinelander, Wisconsin, Kansas State University, Manhattan, Kansas, pp. 94-102.
2. Amrhein, N., B. Deus, P. Gehrke, and H.C. Steinrucken (1980) The site of the inhibition of the shikimate pathway by glyphosate. II. Interference of glyphosate with chlorimate formation *in vivo* and *in vitro*. *Plant Physiol.* 66:830-834.
3. Amrhein, N., D. Johaenning, J. Schab, and A. Schulz (1983) Biochemical basis for glyphosate tolerance in a bacterium and a plant tissue culture. *FEBS Lett.* 157:191-196.
4. Barker, R.F., K.B. Idler, D.V. Thompson, and J.D. Kemp (1983) Nucleotide sequence of the T-DNA region from the *Agrobacterium tumefaciens* octopine Ti plasmid pTi15955. *Plant Mol. Biol.* 2:335-350.
5. Bevan, M., R.B. Flavell, and M.-D. Chilton (1983) A chimeric antibiotic resistance gene as a selectable marker for plant cell transformation. *Nature* 304:184-187.
6. Chilton, M.-D., R.K. Saiki, N. Yadav, M.P. Gordon, and F. Quietier (1980) T-DNA from *Agrobacterium* Ti plasmid is in the nuclear DNA fraction of crown gall tumor cells. *Proc. Natl. Acad. Sci., USA* 77:4060-4064.
7. Comai, L., L. Sen, and D. Stalker (1983) An altered aroA gene product confers resistance to the herbicide glyphosate. *Science* 221:370-371.
8. Comai, L., D. Facciotti, W.R. Hiatt, G. Thompson, R. Rose, and D. Stalker (1985) Expression in plants of a mutant aroA gene from

Salmonella typhimurium confers tolerance to glyphosate. Nature 317:741-744.

9. DeGreve, H., J. Leemans, J.-P. Hernalsteens, L. Thia-Toong, M. DeBeuckeleer, L. Willmitzer, L. Otten, M. Van Montagu, and J. Schell (1982) Regeneration of normal and fertile plants that express octopine synthase, from tobacco crown galls after deletion of tumour controlling functions. Nature 300:752-755.
10. Fillatti, J.J., J. Kiser, R. Rose, and L. Comai (1986) Efficient transformation of tomato and the introduction and expression of a gene for herbicide tolerance. In Tomato Biotechnology Symposium, D.J. Nevins and R.A. Jones, eds. Alan R. Liss, Inc., New York, pp. 199-210.
11. Fillatti, J.J., B. McCown, J. Sellmer, and B. Haissig (1986) The introduction and expression of a gene conferring tolerance to the herbicide glyphosate in Populus NC5339. In TAPPI Research and Development Proceedings, TAPPI Press, Atlanta, Georgia, pp. 83-85.
12. Fillatti, J.J., J. Kiser, R. Rose, and L. Comai (1987) Efficient transfer of a glyphosate tolerance gene into tomato using a binary Agrobacterium tumefaciens vector. Bio/Technology 5:726-730.
13. Fillatti, J.J., J. Sellmer, B. McCown, B. Haissig, and L. Comai (1987) Agrobacterium mediated transformation and regeneration of Populus. Mol. Gen. Genet. 206:192-199.
14. Fraley, R.T., S.G. Rogers, R.B. Horsch, P. Sanders, J. Flick, S. Adams, M. Bittner, L. Brand, C. Fink, J. Fry, G. Galluppi, S. Goldberg, N.L. Hoffmann, and S. Woo (1983) Expression of bacterial genes in plant cells. Proc. Natl. Acad. Sci., USA 80:4803-4807.
15. Hamilton, R.H., and M.Z. Fall (1971) The loss of tumor initiating ability in Agrobacterium tumefaciens by inoculation at high temperature. Experientia 27:229-230.
16. Hansen, E.A., and D.A. Netzer (1985) Weed Control Using Herbicides in Short-Rotation Intensively Cultured Poplar Plantations, U.S. Department of Agriculture Forest Service Research Paper NC-260.
17. Herrera-Estrella, L., A. Depicker, M. Van Montagu, and J. Schell (1983) Expression of chimeric genes transferred into plant cells using Ti plasmid-derived vector. Nature (London) 303:209-213.
18. Horsch, R.B., J.B. Fry, N.L. Hoffmann, M. Wallroth, D. Eichholtz, S.G. Rogers, and R.T. Fraley (1985) A simple and general method for transferring genes into plants. Science 227:1229-1231.
19. Michelmore, R.W., E. Marsh, S. Seely, and B. Landry (1987) Transformation of lettuce (Lactuca sativa) mediated by Agrobacterium tumefaciens. Plant Cell Reports (in press).
20. Reiss, B., R. Sprengle, H. Will, and H. Schaller (1984) A new sensitive method for qualitative and quantitative assay of neomycin phosphotransferase in crude cell extracts. Gene 30:211-217.
21. Rogers, S.G., L. Brand, S. Holder, E.S. Sharp, and M.M. Bracken (1983) Amplification of the aroA gene from E. coli results in tolerance to the herbicide glyphosate. Appl. Environ. Microbiol. 46:37-43.
22. Shah, D.M., R.B. Horsch, H.J. Klee, G.M. Kishore, J.A. Winter, N.E. Tumer, C.M. Hironaka, P.R. Sanders, C.S. Gasser, S. Aykent, N.R. Siegel, S.G. Rogers, and R.T. Fraley (1986) Engineering herbicide tolerance in transgenic plants. Science 233:478-481.
23. Stalker, D.M., W.R. Hiatt, and L. Comai (1985) A single amino acid substitution in the enzyme 5-enolpyruvylshikimate 3-phosphate synthase confers resistance to the herbicide glyphosate. J. Biol. Chem. 260:4724-4728.
24. Steinrucken, H.C., and N. Amrhein (1980) The herbicide glyphosate is a potent inhibitor of 5-enolpyruvyl-shikimic acid-3-phosphate synthase. Biochem. Biophys. Res. Comm. 94:1207-1212.

INHERITANCE AND EVOLUTION OF CONIFER ORGANELLE GENOMES

David B. Neale and Ronald R. Sederoff

Institute of Forest Genetics
Pacific Southwest Forest and Range Experiment Station
U.S. Department of Agriculture Forest Service
Berkeley, California 94701

ABSTRACT

Inheritance of organelle genomes in conifers is different from that in angiosperms. Restriction fragment-length polymorphisms have been used to demonstrate the paternal inheritance of chloroplast DNA in Douglas fir and coast redwood and the maternal inheritance of mitochondrial DNA in loblolly pine. Conifers have evolved mechanisms to exclude the maternal transmission of plastids and the paternal transmission of mitochondria. The genetic data from intraspecific crosses are consistent with expectations based on ultrastructural observations of organelle transmission during fertilization in conifers. Conifers have followed the general evolutionary trend towards uniparental inheritance of organelles, but curiously have evolved mechanisms for the paternal transmission of chloroplasts. Strict paternal inheritance of chloroplasts is not found in other taxa of higher plants.

INTRODUCTION

It has long been known that genetic elements in the cytoplasm of higher plants have non-Mendelian modes of inheritance. Correns (23) showed that variegated phenotypes in the four-o'clock flower (Mirabilis jalapa) were inherited strictly through the maternal parent, whereas Baur (2) showed that variegated phenotypes in geranium (Pelargonium) were transmitted through both parents (biparental inheritance), but did not segregate in normal Mendelian ratios. Although it was not known at the time, the variegated phenotypes resulted from mutations of the DNA contained within chloroplasts. Chloroplast inheritance has since been studied in numerous higher plants. Most of these have strictly maternal inheritance; but there are exceptions, such as geranium, which have biparental inheritance.

Mitochondrial genomes appear to be strictly maternally inherited in higher plants just as they are in animals. Much of the evidence for maternal inheritance of mitochondria in higher plants is based on the inheritance of cytoplasmic male-sterility (CMS), a trait of considerable interest to plant breeders (25). This trait is always transmitted through the female parent, but it was not until 1976 that Levings and Pring (36) suggested that CMS is associated with mitochondrial DNA (mtDNA).

The inheritance of chloroplast and mitochondrial genomes is in striking contrast to the inheritance of nuclear genomes in sexually reproducing higher plants. Organelle genomes do not undergo meiosis or follow the basic rules of Mendelian inheritance. These fundamental differences in the inheritance and transmission of organelle versus nuclear genomes raise several important questions. Is there an adaptive advantage of uniparental inheritance of organelle genomes? What mechanisms have evolved to prohibit the transmission of these genomes through the pollen parent? Given that uniparental inheritance is the preferred state for organelle genomes, why have mechanisms evolved for exclusion of organelles from the pollen versus the egg? Why have exceptions to strict maternal inheritance of chloroplasts appeared in several taxa? Application of new techniques, such as restriction fragment-length polymorphisms (RFLPs) to organelle genetics may begin to provide answers to these basic evolutionary questions.

Among the seed plants, organelle inheritance and evolution has been studied almost exclusively within the angiosperms; the conifers and other gymnosperms have been almost completely ignored. Molecular genetic studies of conifer nuclear genomes now underway will provide the basic framework for the application of genetic engineering to forest trees. It may also be possible to genetically engineer conifer organelle genomes in the near future; however, the basic rules of organelle inheritance and transmission in conifers must first be established.

In this chapter we present data which confirm the paternal inheritance of chloroplast DNA (cpDNA) in Douglas fir [*Pseudotsuga menziesii* (Mirb.) Franco] and coast redwood (*Sequoia sempervirens* D. Don Endl.) and the maternal inheritance of mtDNA in loblolly pine (*Pinus taeda* L.). The molecular techniques used to study conifer organelle inheritance are presented. The nature and extent of organelle DNA variation in conifers, some of the mechanisms which determine the mode of inheritance for each organelle, and the evolution of organelle inheritance in higher plants are also discussed.

STRUCTURE, FUNCTION, AND INHERITANCE OF ORGANELLE GENOMES IN ANGIOSPERMS

Chloroplast Genomes

Chloroplast genomes have been studied in numerous angiosperms (for reviews, see Ref. 4, 24, 49, 50, 52, and 76). Chloroplast DNA usually exists as a homogeneous population of circular DNA molecules. Each molecule is approximately 120 to 160 kilobases (kb) in length, and up to 1,000 copies of the genome can be present per chloroplast (5). The cpDNA encodes a complete set of rRNA and tRNA genes and approximately 50 protein-encoding genes. The vast majority of chloroplast proteins are encoded in the nucleus and transported to the chloroplast (26).

The cpDNA molecule is organized into four major sections: (a) two inverted repeats that contain genes for the 23S, 16S, and 5S rRNA; (b) a large single-copy region that contains most of the protein-encoding genes;

and (c) a small single-copy region that separates the inverted repeats. A few species, such as pea and broad bean (53), lack the inverted repeat and have only one set of the ribosomal genes. Restriction and gene maps have been constructed for dozens of species (49), including one gymnosperm, Ginkgo biloba (56). Complete cpDNA nucleotide sequences have recently been determined for a liverwort (Marchantia polymorpha) (47) and tobacco (66). The tRNA, rRNA, and more than 50 protein-encoding genes have been identified, including several unidentified open reading frames for both plants.

The organization and nucleotide sequence of chloroplast genes is highly conserved across a broad array of taxa. This high degree of conservation has made cpDNA restriction site variation particularly useful for constructing molecular phylogenies (20,48,54,57,58). It also has led to the expectation that cpDNAs would be highly monomorphic at the intraspecific level. However, cpDNA polymorphisms have been detected in small surveys of several species of angiosperms (6,19,20,54,58,61,70) and in more extensive population genetic studies of Lupinus texensis (1) and Hordeum spontaneum (31).

Chloroplast DNA in angiosperms is usually inherited strictly through the maternal parent; the exceptions to this are species with biparental inheritance (for reviews, see Ref. 28, 33, and 62). Restriction fragment-length polymorphisms have recently been used to confirm the maternal or biparental inheritance of organelle DNAs in a variety of species (21,30,32, 42,72). There are no cases among angiosperms where chloroplast DNA is strictly paternally inherited, although there are examples of biparental inheritance (some species of Oenothera and Pelargonium, Phaseolus vulgaris, and Secale cereale). Recent evidence from Nicotiana (41) has shown that there may be low frequency transmission of paternal cpDNA in species which were commonly thought to have strict maternal inheritance.

Mitochondrial Genomes

Mitochondrial genomes of higher plants are larger and more variable in size than chloroplast genomes (for reviews, see Ref. 35, 38, 63, and 64). Estimates range from approximately 200 to 2,400 kb (3). The structural organization of mtDNA is extremely variable, existing as both linear and circular molecules of varying size classes. Higher plant mtDNAs code for three rRNAs, at least 22 tRNAs, and several subunits of inner membrane proteins such as cytochrome oxidase, apocytochrome b, and ATPase. These proteins are all involved in oxidative phosphorylation. As in chloroplasts, most proteins in the mitochondria are encoded in the nucleus. Restriction and gene maps have been constructed for mtDNA of two species of Brassica (17,55) and maize (39). Much less is known about gene organization and sequence diversity in mitochondria than in chloroplasts. In general, there is much variation in mtDNA gene organization, possibly due to recombination and rearrangement, but structural genes are at least as highly conserved as nuclear or chloroplast genes. Intraspecific restriction site polymorphisms have been detected in mtDNA of several species (59, 65,70), although the only extensive survey is that of H. spontaneum (31).

It is generally assumed that mitochondria are maternally inherited in higher plants. The only direct attempt to infer inheritance in sexual crosses using molecular markers is that for Zea (21) which showed maternal inheritance. There are no reports of paternal transmission of mitochondria in sexual crosses of higher plants; however, mitochondria are transmitted by the mt (paternal) parent in the green algae Chlamydomonas reinhardtii (7).

PURIFICATION OF CONIFER ORGANELLE DNA AND RESTRICTION FRAGMENT ANALYSIS

Two general procedures have been used to study organelle genomes in plants. One approach is to isolate pure cpDNAs or mtDNAs from plant tissues, cut these DNAs with restriction enzymes, and then visualize ethidium-bromide-stained DNA under UV light. Protocols have been developed for the isolation of high-molecular-weight cpDNAs and mtDNAs from conifer tissues (34,51,74,75). All of the cpDNA isolation procedures are based on the separation of intact chloroplasts in sucrose density gradients, followed by lysis of chloroplasts, and cesium chloride purification of cpDNA. Chloroplast and mtDNA isolated by these methods can easily be cloned for mapping and sequencing. We have chosen an approach which utilizes a rapid method of total DNA isolation from conifer tissues and which surveys conifer organelle polymorphisms by Southern hybridization with heterologous probes.

Total DNA is extracted from 10 g of fresh needle tissue following a modification of the CTAB procedure of Murray and Thompson (44) (see Ref. 45 and 72 for modifications). This procedure yields approximately 200 μg of DNA per gram fresh tissue in approximately a 90:10:1 ratio of nuclear DNA, cpDNA, and mtDNA, respectively. One person can complete 50 to 100 of such DNA preparations per week. The DNAs are then cut with restriction enzymes, fractionated on agarose gels, and transferred to nylon matrix membranes following standard procedures (40). The hexanucleotide restriction enzymes that we have found most useful for detecting conifer organelle DNA polymorphisms are HindIII, EcoRI, BamHI, BclI, and SmaI. The ^{32}P-labeled probes are then prepared by nick translation (40) and hybridizations are done for approximately 12 hr in 2X SSC and 0.1% sodium dodecyl sulfate (SDS) at 65°C. We have used the petunia PstI and SalI cpDNA clones of Palmer (67) and several mtDNA gene clones from maize [e.g., 18-5S rRNA (15), cytochrome oxidase II (27), and ATPase-alpha (8)] as hybridization probes. These clones have sufficient homology to conifer cpDNA and mtDNA sequences such that strong signals are obtained on autoradiograms after stringent hybridization conditions.

CHLOROPLAST AND MITOCHONDRIAL DNA REARRANGEMENTS IN CONIFERS

Inheritance studies of all types require genetic markers to determine patterns of segregation in progenies from experimental matings. There have been very few genetic markers of organelle genomes, aside from the chloroplast mutants in Chlamydomonas reinhardtii (28), until the widespread application of RFLPs. RFLPs are easily detected among closely related species within genera and have been used to infer chloroplast and mitochondrial inheritance in a number of species crosses (21,30,72). However, it may be misleading to infer the detailed modes of organelle inheritance in intraspecific crosses from data on interspecific crosses, particularly if the transmission of organelles is affected by fertilization abnormalities in species crosses. We have chosen to study organelle inheritance in intraspecific conifer crosses. This necessitates the identification of RFLPs within species.

Our investigations have been restricted to four conifers [Douglas fir, loblolly pine, coast redwood, and incense-cedar (Calocedrus decurrens Torr.)] which represent three major families of Coniferales (Pinaceae, Taxodiaceae, and Cupressaceae). Our approach has been to find RFLPs among trees that have been used in full-sib matings. The DNAs from parent trees are cut with a number of restriction enzymes and Southern blots are hybridized with petunia cpDNA probes. A number of RFLPs have been detected in this way (Tab. 1).

Chloroplast Rearrangements

The most thorough investigation of cpDNA variation in Douglas fir was conducted on 16 trees at the Institute of Forest Genetics (IFG), Placerville, California (Tab. 1). These included two trees from Arizona, eight from California, and six from British Columbia. Most restriction enzyme-probe combinations were invariant for these trees with the exception of three insertion/deletion variants in SmaI, BamHI, and EcoRI digests hybridized with petunia Pst6. Variable BamHI fragments range from approximately 2,500 to 3,080 bp. Insertions and deletions have also been detected among Douglas fir seed orchard (SO) clones and among rangewide (RW) sources for this same region of the chloroplast genome (Tab. 1).

An insertion/deletion polymorphism was also detected among 13 coast redwood trees in EcoRI, BclI, EcoRV, and HindIII digests hybridized with Pst6 (Tab. 1). The common genotype had a 3,890-bp EcoRI fragment, whereas one tree had a deletion of 134 bp and two others had insertions of 139 bp and 43 bp, respectively. Both Douglas fir and redwood have rearrangements in homologous sequences. This may be a "hot-spot" for cpDNA rearrangement in conifers.

Another insertion/deletion polymorphism was detected among just three incense-cedars in HindIII and EcoRI digests hybridized with Pst4 (Tab. 1). Unlike the polymorphisms in Douglas fir and redwood, the

Tab. 1. Chloroplast DNA polymorphisms in conifers detected by Southern hybridizations with cloned cpDNA probes from petunia.

Species*	Number of trees	Number of enzymes	Number of cpDNA clones	Percent genome surveyed	RFLP**	Type of RFLP	Number of variants
Douglas fir (SO)	30	4	8	84	Pst6	I/D†	6
Douglas fir (IFG)	16	14	6	54	Pst6	I/D	3
Douglas fir (RW)	64	2	1	11	Pst6	I/D	6
Loblolly pine	30	6	5	46	None	-	None
Redwood	15	6	2	18	Pst6	I/D	4
Incense-cedar	3	6	6	68	Pst4	I/D	2

*SO, seed orchard; IFG, Institute of Forest Genetics; RW, rangewide.
**RFLP refers to the petunia cpDNA clone homologous to conifer polymorphism.
†I/D, insertion/deletion.

polymorphism in incense-cedar is homologous to a petunia fragment which covers sections of the small single-copy region and the inverted repeat. Two trees had a 4,133-bp BamHI fragment whereas the third had a 3,928-bp fragment (deletion of 205 bp). These trees have been crossed in all combinations and seed collections exist for inheritance analysis.

The 30 loblolly pine trees were invariant for the regions of the chloroplast genome we have surveyed (Tab. 1). This suggests that cpDNA polymorphism in loblolly pine is lower than that in other conifers we have studied. However, our sampling was restricted to 30 trees from North Carolina and approximately one half of the chloroplast genome.

Mitochondrial DNA Rearrangements

We have just begun surveys to determine the extent and nature of mtDNA variation in conifers using the cloned gene probes from maize. RFLPs have been detected for loblolly pine, Douglas fir, and redwood with the 18-5S rRNA and the cytochrome oxidase II clones. Most rearrangements appear to be small insertions or deletions, but studies with additional enzymes are needed to verify which types of rearrangements produced the polymorphisms.

CHLOROPLAST DNA IS PATERNALLY INHERITED IN CONIFERS

We previously reported on the paternal inheritance of cpDNA in Douglas fir (45), which was the first report of strict paternal inheritance of chloroplasts in higher plants based on molecular data. This mode of inheritance was an unexpected result. The study was based on three full-sib crosses from the Weyerhaeuser Company and a total of 37 offspring. Subsequently, we analyzed an additional 60, 64, and 51 offspring from the same three Douglas fir crosses. Parent and offspring DNAs were cut with BamHI and hybridized with petunia Pst6 (Fig. 1). All offspring had the same restriction fragment as the paternal parent without exception. We conclude that Douglas fir has strict paternal inheritance of cpDNA or that the frequency of transmission of the maternal genome is rare.

Paternal inheritance of cpDNA has also been demonstrated in coast redwood. Full-sib crosses had been made by Simpson Timber Co. between redwoods, some of which had different cpDNA RFLP genotypes. We determined genotypes of offspring from two crosses; all had the same cpDNA genotype as the male parent (Fig. 2). This is the second conifer for which paternal inheritance has been confirmed from intraspecific crosses.

Additional evidence for the paternal inheritance of cpDNA has been obtained for several other conifers. Ohba et al. (46) demonstrated that a chlorophyll mutant marker in *Cryptomeria japonica* is transmitted through the paternal parent, although approximately 10% of the offspring in controlled crosses did not have the mutant phenotype. Paternal inheritance of cpDNA has also been shown in at least three interspecific crosses using RFLPs. The F_1's have the restriction fragment band pattern characteristic of the paternal parent in *Pinus contorta* x *P. banksiana* hybrids (73), *Larix* hybrids (68), and *Picea glauca* x *P. omorika* hybrids (W.M. Cheliak, pers. comm.). These data are consistent with the data we have obtained from intraspecific crosses. Paternal inheritance of cpDNA has now been shown from intraspecific crosses or inferred from interspecific crosses for four genera of Pinaceae (*Pinus*, *Pseudotsuga*, *Picea*, and *Larix*) and for two genera of Taxodiaceae (*Sequoia* and *Cryptomeria*).

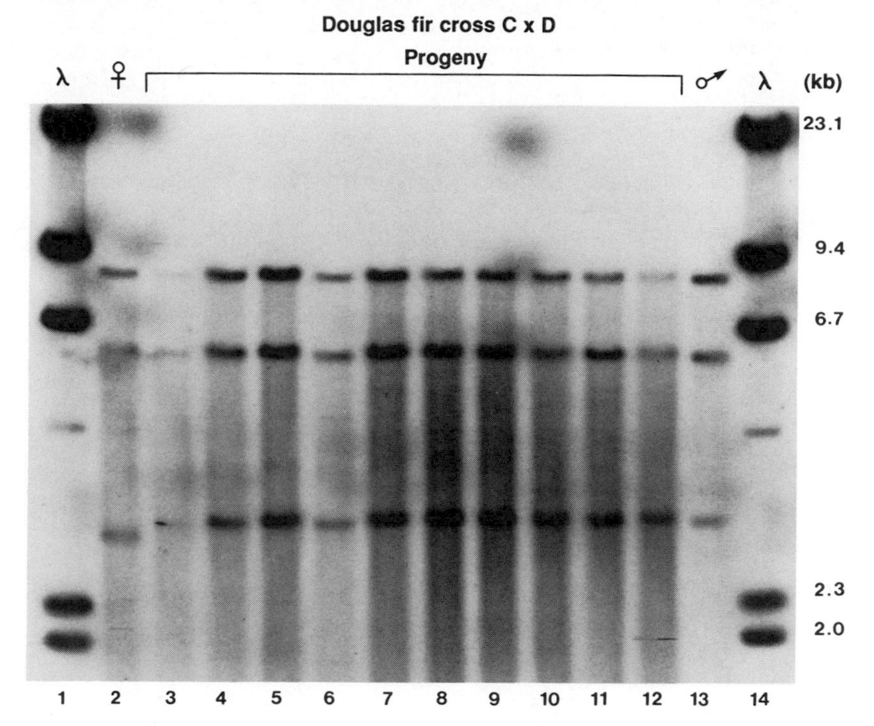

Fig. 1. Paternal inheritance of cpDNA in Douglas fir. Parents and offspring DNAs were cut with BamHI and hybridized with petunia Pst6. The male parent and all offspring have a 3,083-kb BamHI fragment whereas the female parent has a 2,902-kb fragment.

MITOCHONDRIAL DNA IS MATERNALLY INHERITED IN LOBLOLLY PINE

Mitochondrial DNA is presumed to be strictly maternally inherited in all higher plants. Much of the evidence for maternal inheritance of mtDNA comes from studies of the transmission of mitochondria-encoded CMS (9,18, 43,60). There is no evidence for CMS in conifers nor has the mode of mtDNA inheritance been shown. We assayed 20 full-sib progeny from each of two loblolly pine crosses of the Weyerhaeuser Company which had male and female parents with different 18-5S rRNA RFLP genotypes (Fig. 3). All 40 progeny had the same restriction fragments as the maternal parent. This is the first confirmation of maternal inheritance of mtDNA in conifers. Maternal transmission of mtDNA has also been shown by RFLPs in an F_1 from the cross Pinus strobus x P. griffithii (C.A. Loopstra, pers. comm.).

ULTRASTRUCTURAL OBSERVATIONS OF CONIFER FERTILIZATION SUPPORT GENETIC DATA ON ORGANELLE INHERITANCE

Genetic analyses can determine the ultimate fate of organelles but do not reveal the mechanisms by which organelle transmission is controlled. These questions are currently best addressed by ultrastructure analysis.

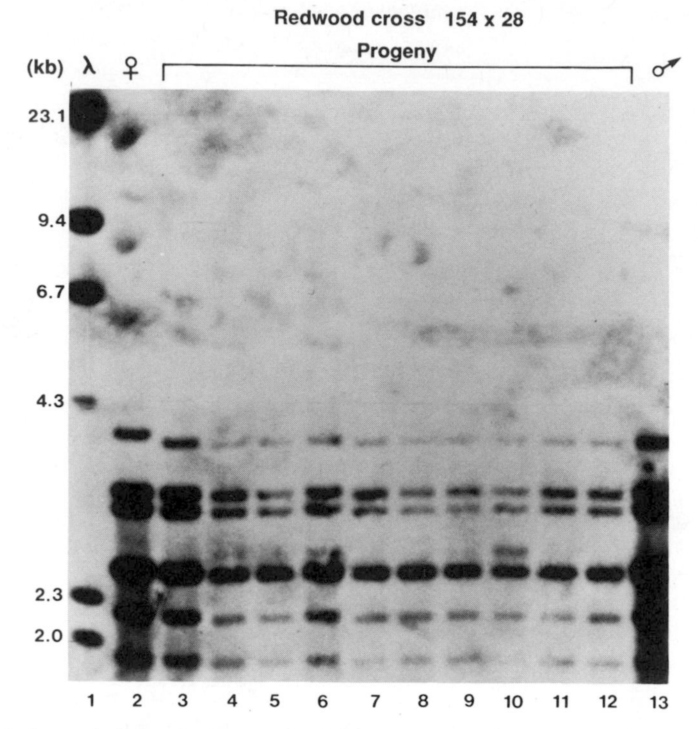

Fig. 2. Paternal inheritance of cpDNA in redwood. Parent and offspring DNAs were cut with EcoRI and hybridized with petunia Pst6. The male parent and all offspring have a 3,891-kb EcoRI fragment whereas the female parent has a 3,757-kb fragment.

The fate of chloroplasts and mitochondria in egg and pollen cells at fertilization has been studied in several conifers [reviewed by Chesnoy and Thomas (16)]. The presence of plastids and mitochondria in the pollen generative cell has been confirmed for Pinus sylvestris (77), Pinus nigra (12,13), Pseudotsuga menziesii (J.N. Owens, pers. comm.), and several Cupressaceae [Thuja orientalis, Calocedrus decurrens, and Chamaecyparis lawsonia (16)]. There appear to be no mechanisms for plastid or mitochondrial exclusion from the pollen generative cell such as those that exist in angiosperms (22,29,62). There are several examples, however, of angiosperms that contain organelles in the pollen generative cell that are not transmitted to offspring [e.g., Solanum (18) and Pelargonium (37,71)].

During early development of the archegonium, the egg cell contains both plastids and mitochondria (16). As development of the egg cell continues, the plastids and mitochondria are localized to distinct regions. The plastids are contained completely within large inclusions in the egg cell which become membrane bound and eventually the plastids within them degenerate. These inclusions have been shown in P. nigra (10,11), Larix decidua (14), and Pseudotsuga menziesii (69). The inclusions appear to be the mechanism whereby maternal plastids are eliminated. The mitochondria, on the other hand, migrate toward the egg nucleus to form what is called the perinuclear zone (69).

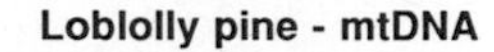

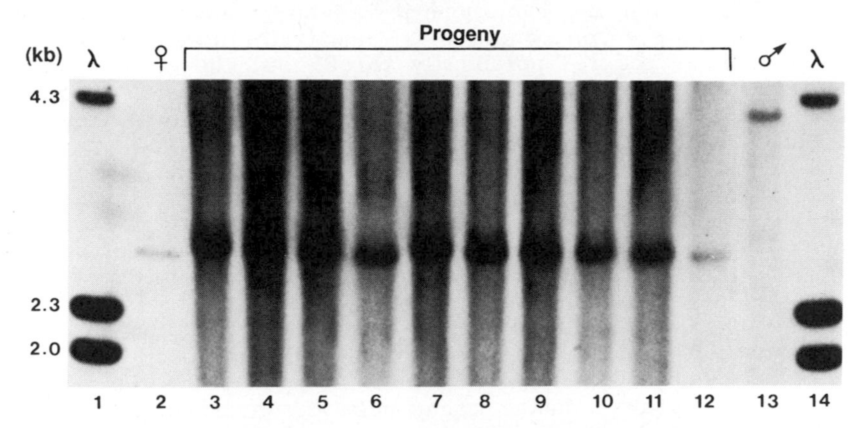

Fig. 3. Maternal inheritance of mtDNA in loblolly pine. Parent and offspring DNAs were cut with *EcoRI* and hybridized with the 18-5S rRNA mtDNA clone from *maize*. The female parent and all offspring have a 2.7-kb fragment whereas the male has a 2.8-kb fragment and a 4.0-kb fragment.

At fertilization in Pinaceae, the pollen tube penetrates the neck of the archegonium and discharges the sperm nuclei and cytoplasm containing plastids and mitochondria into the archegonium. The sperm and egg nuclei join, undergo a couple of mitotic divisions, and eventually migrate to the base of the archegonium. The neocytoplasm accompanying the zygote consists of plastids of paternal origin, an abundance of maternal mitochondria, and at least some mitochondria of paternal origin (16). In Cupressaceae, it is possible that most, if not all, of the mitochondria are of paternal origin (16).

The general picture that emerges from the literature on conifer fertilization ultrastructure is that there are two organelle exclusion mechanisms in conifers that appear to function independently. The first is the exclusion of maternal plastids in the egg cell prior to fertilization; this mechanism seems reasonably clear from ultrastructure. The second mechanism, whereby paternal mitochondria appear to be selectively eliminated, is not understood. Genetic data from large samples and additional species will be needed to establish whether or not mitochondria are strictly maternally inherited in all conifers. It will be of interest to study the inheritance of mtDNA in other conifers such as *Calocedrus decurrens* where ultrastructural observations suggest one might find paternal transmission.

EVOLUTION OF ORGANELLE INHERITANCE IN CONIFERS

It has been difficult to trace the evolution of the inheritance of organelles in plants due to the lack of information from most taxa aside from angiosperms. There seems to be little deviation from strict maternal inheritance of mtDNA in the plant kingdom; therefore, we will focus our discussion on the evolution of plastid inheritance. Sears (62) presents a very thorough discussion of the elimination of paternal plastids following an evolutionary path from algae to mosses and liverworts, to ferns, to gymnosperms, and finally to angiosperms. This survey reveals that although maternal inheritance is predominant, biparental transmission is found in both primitive and evolutionarily advanced taxa. It is therefore difficult to determine which condition is ancestral. However, because biparental inheritance occurs sporadically throughout the plant kingdom, it seems more probable that strict maternal inheritance was developed early and those extant taxa with biparental inheritance have lost mechanisms to exclude paternal plastids. Conifers appear to have evolved a unique exclusion mechanism by which maternal plastids are excluded.

Many theories have been advanced as to why maternal inheritance would have evolved, such as: (a) elimination of recombination between sexes at fertilization and hence preservation of optimal gene arrangements; (b) elimination of cytoplasm from the sperm to increase mobility and thus increase the likelihood of fertilization; and (c) elimination of the complexities associated with multiple forms of plastid gene products. In contrast to evolutionary advantages obtained from recombination of nuclear genomes in sexually reproducing organisms, the evolution of the inheritance of organelle genomes has been towards strict maternal inheritance, thus eliminating the potential for recombination.

The strict paternal inheritance of plastids in conifers is consistent with the trend toward uniparental inheritance; but, interestingly, the chosen route is through the pollen rather than the egg, while at the same time maternal transmission of mitochondria is maintained. Sears (62) suggests that the presence of plastids in pollen is to fulfill a need for starch in the pollen during the long period of time between pollination and fertilization (as much as 14 months in pines). However, even if there is a metabolic need for plastids in the pollen prior to fertilization, this is not sufficient to explain why paternal transmission has evolved. There are many angiosperms that have plastids in the pollen which are excluded or degraded just prior to or at fertilization. There are no apparent adaptive advantages to paternal versus maternal inheritance of chloroplasts; thus we propose that paternal inheritance of chloroplasts in conifers is the result of a random evolutionary event that has been maintained in present-day conifers.

SUMMARY

We have used RFLPs to demonstrate paternal inheritance of cpDNA in Douglas fir and coast redwood, and maternal inheritance of mtDNA in loblolly pine. These genetic data are consistent with expectations based on ultrastructural observations of the transmission of organelles at fertilization in conifers. This unusual mode of organelle inheritance is unique to higher plants; there are no other plants that are known to have strict paternal inheritance of cpDNA. There are several explanations for the evolution of uniparental inheritance of organelles, but it is difficult to reconcile why conifers evolved mechanisms for exclusion of maternal plastids.

Aside from the evolutionary questions raised by our data, it also has important applications to plant breeding and genetic engineering. The most obvious application will result from the fact that we now know from which parents organelle DNAs are transmitted. Genetic variances for mitochondria-encoded traits could be measured from open-pollinated families, whereas a mating design such as a tester would be needed for chloroplast-encoded traits. These considerations will become important as it becomes possible to engineer conifer organelle genomes for traits of commercial interest.

ACKNOWLEDGEMENTS

The authors thank Tom Adams (Oregon State University), David Harry (University of Illinois), Bill Libby (University of California, Berkeley), Kim Rodrigez (Simpson Timber Company), and Nick Wheeler (Weyerhaeuser Company) for providing materials used in this study. We also thank Bill Critchfield and Tom Ledig (Pacific Southwest Forest and Range Experiment Station, U.S. Department of Agriculture Forest Service) and John Owens (University of Victoria) for reviews of the manuscript.

REFERENCES

1. Banks, J.A., and C.W. Birky (1985) Chloroplast DNA diversity is low in a wild plant, Lupinus texensis. Proc. Natl. Acad. Sci., USA 82:6950-6954.
2. Baur, E. (1909) Das Wesen und die Erblichkeitsverhältrisse der 'Varietates albomarginatae hort.' von Pelargonium zonale. Z. Verebungslehre 1:330-351.
3. Bendich, A.J. (1985) Plant mitochondrial DNA: Unusual variation on a common theme. In Genetic Flux in Plants, B. Hohn and E.S. Dennis, eds. Springer-Verlag, New York, pp. 111-138.
4. Birky, C.W. (1987) Evolution and variation in plant chloroplast and mitochondrial genomes. In Plant Evolutionary Biology, L.D. Gottlieb and S.K. Jain, eds. Chapman and Hall, London.
5. Boffley, S.A., and R.M. Leech (1982) Plant Physiol. 69:1387-1391.
6. Bowman, C., G. Bonnard, and T. Dyer (1983) Chloroplast DNA variation between species of Triticum and Aegelops: Location of the variation on the chloroplast genome and its relevance to the inheritance and classification of the cytoplasm. Theor. Appl. Genet. 65: 247-262.
7. Boynton, J.E., E.H. Harris, B.D. Burkhart, P.M. Lamerson, and N.W. Gillham (1987) Transmission of mitochondrial and chloroplast genomes in crosses of Chlamydomonas. Proc. Natl. Acad. Sci., USA 84:2391-2395.
8. Braun, C.J., and C.S. Levings III (1985) Nucleotide sequence of the F_1-ATPase alpha subunit gene from maize mitochondria. Plant Physiol. 79:571-577.
9. Brown, G.G., H. Bussey, and L.J. DesRosiers (1986) Analysis of mitochondrial DNA, chloroplast DNA, and double-stranded RNA in fertile and cytoplasmic male-sterile sunflower (Helianthus annuus). Can. J. Genet. Cytol. 28:121-129.
10. Camefort, H. (1962) L'organization du cytoplasme dans l'oosphère et al cellule centrale du Pinus laricio Poir. (var. austriaca). Ann. Sci. Nat. Bot. (12e Ser.) 111:265-291.
11. Camefort, H. (1965) Une interprétation nouvelle de l'organization du protoplasme de l'oosphère des Pins. In Tavaux Dedies a Lucien Plantefol, Masson, ed. Paris, pp. 407-436.
12. Camefort, H. (1966) Observations sur les mitochondries et les plastes d'origine pollenique après leur entrée dans une oosphère chez le Pin noir (Pinus laricio Poir. var. austriaca = P. nigra Arn.). C.R. Hebd. Seanc. Acad. Sci., Paris 263:959-962.
13. Camefort, H. (1966) Etude en microscopie électronique de la dégénerescence du cytoplasme maternel dans les oosphères embryonnées du Pinus laricio Poir. var. austriaca (P. nigra Arn.). C.R. Hebd. Seanc. Acad. Sci., Paris 263:1371-1374.
14. Camefort, H. (1967) Observations sur les mitochondries l'oosphère du Larix decidua Mill. (L. europea D.Z.). C.R. Hebd. Seanc. Acad. Sci., Paris 265:1293-1296.

15. Chao, S., R.R. Sederoff, and C.S. Levings III (1984) Nucleotide sequence and evolution of the 18S ribosomal RNA gene in maize mitochondria. Nucl. Acids Res. 12:6629-6644.
16. Chesnoy, L., and M.J. Thomas (1971) Electron microscopy studies on gametogenesis and fertilization in gymnosperms. Phytomorphology 21:50-63.
17. Chetrit, P., C. Mathieu, J.P. Muller, and F. Vedel (1984) Physical and gene mapping of cauliflower (Brassica oleraceae) mitochondrial DNA. Curr. Genet. 8:413-421.
18. Clauhs, R.P., and P. Grun (1977) Changes in plastid and mitochondrion content during maturation of generative cells of Solanum (Solanaceae). Am. J. Bot. 64:377-383.
19. Clegg, M.T., A.H.D. Brown, and P.R. Whitfeld (1984) Chloroplast DNA diversity in wild and cultivated barley. Genet. Res. 43:339-343.
20. Clegg, M.T., J.R.Y. Rawson, and K. Thomas (1984) Chloroplast DNA evolution in pearl millet and related species. Genetics 106:449-461.
21. Conde, M.F., D.R. Pring, and C.S. Levings III (1979) Maternal inheritance of organelle DNA in Zea mays-Zea perennis reciprocal crosses. J. Hered. 70:2-4.
22. Connett, M.B. (1987) Mechanisms of maternal inheritance of plastids and mitochondria: Developmental and ultrastructural evidence. Plant Mol. Biol. Reporter 4:193-205.
23. Correns, C. (1909) Verekbungsversuche mit blaB(gelb) grünen und buntblättrigen Sippen bei Mirabilis jalapa, Urtica pilulifera und Lunaria annua. Z. Verebungslehre 2:331-340.
24. Curtis, S.E., and M.T. Clegg (1984) Molecular evolution of chloroplast DNA sequences. Mol. Biol. Evol. 1:291-301.
25. Edwardson, J.R. (1970) Cytoplasmic male sterility. Bot. Rev. 361: 341-420.
26. Ellis, R.J. (1981) Chloroplast proteins: Synthesis, transport and assembly. Ann. Rev. Plant Physiol. 32:111-137.
27. Fox, T.D., and C.J. Leaver (1981) The Zea mays mitochondrial gene coding cytochrome oxidase subunit II has an intervening sequence and does not contain TAG codons. Cell 26:315-323.
28. Gillham, N.W. (1978) Organelle Heredity, Raven Press, New York.
29. Hagemann, R. (1979) Genetics and molecular biology of plastids of higher plants. In Staedler Symposium, Vol. 11, G. Redei, ed. Missouri Agricultural Experiment Station, pp. 91-115.
30. Hatfield, P.M., R.C. Shoemaker, and R.G. Palmer (1985) Maternal inheritance of chloroplast DNA within the genus Glycine, subgenus soja. J. Hered. 76:373-374.
31. Holwerda, B.C., S. Jana, and W.L. Crosby (1986) Chloroplast and mitochondrial DNA variation in Hordeum vulgare and Hordeum spontaneum. Genetics 114:1271-1291.
32. Ichikawa, H., and A. Hirai (1983) Search for the female parent in the genesis of Brassica rapus by chloroplast DNA restriction patterns. Jap. J. Genet. 58:419-424.
33. Kirk, J.T.O., and R.A.E. Tilney-Bassett (1967) The Plastids, W.H. Freeman, San Francisco.
34. Kondo, T., T. Ishibashi, M. Shibata, and A. Hirai (1986) Isolation of chloroplast DNA from Pinus. Plant Cell Physiol. 27:741-744.
35. Leaver, C.J., and M.W. Gray (1982) Mitochondrial genome organization and expression in higher plants. Ann. Rev. Plant Physiol. 33:373-402.
36. Levings, C.S., III, and D.R. Pring (1976) Restriction endonuclease analysis of mitochondrial DNA from normal and Texas cytoplasmic male-sterile maize. Science 193:158-160.

37. Lombardo, G., and F.M. Gerloa (1968) Ultrastructure of the pollen grain and taxonomy. Planta 82:105-110.
38. Lonsdale, D.M. (1984) A review of the structure and organization of the mitochondrial genome of higher plants. Plant Mol. Biol. 3:201-206.
39. Lonsdale, D.M., T.P. Hodge, and C.M.-R. Fauron (1984) The physical map and organization of the mitochondrial genome from the fertile cytoplasm of maize. Nucl. Acids Res. 12:9249-9261.
40. Maniatis, T., E.F. Fritsch, and J. Sambrook (1982) Molecular Cloning: A Laboratory Manual, Cold Spring Harbor Laboratory, Cold Spring Harbor, New York.
41. Medgyesy, P., A. Páy, and L. Márton (1987) Transmission of paternal chloroplasts in Nicotiana. Mol. Gen. Genet. 204:195-198.
42. Metzlaff, M., T. Borner, and R. Hagemann (1981) Variations of chloroplast DNAs in the genus Pelargonium and their biparental inheritance. Theor. Appl. Genet. 60:37-41.
43. Mikami, T., S. Sugiura, and T. Kinoshita (1984) Molecular heterogeneity in mitochondrial and chloroplast DNAs from normal and male-sterile cytoplasms in sugar beets. Curr. Genet. 8:319-322.
44. Murray, M.G., and W.F. Thompson (1980) Rapid isolation of high molecular weight DNA. Nucl. Acids Res. 8:4321-4325.
45. Neale, D.B., N.C. Wheeler, and R.W. Allard (1986) Paternal inheritance of chloroplast DNA in Douglas-fir. Can. J. For. Res. 16:1152-1154.
46. Ohba, K., M. Iwakawa, Y. Okada, and M. Murai (1971) Paternal transmission of a plastid anomaly in some reciprocal crosses of Sugi, Cryptomeria japonica D. Don. Silvae Genet. 20:101-107.
47. Ohyama, K., H. Fukuzawa, T. Kohchi, H. Shirai, T. Sano, S. Sano, K. Umesono, Y. Shiki, M. Takeuchi, Z. Chang, S.-I. Aota, H. Inokuchi, and H. Ozeki (1986) Chloroplast gene organization deduced from complete sequence of liverwort Marchantia polymorpha chloroplast DNA. Nature 322:572-574.
48. Palmer, J.D. (1985) Chloroplast DNA and molecular phylogeny. Bio-Essays 2:263-267.
49. Palmer, J.D. (1985) Comparative organization of chloroplast genomes. Ann. Rev. Genet. 19:325-354.
50. Palmer, J.D. (1985) Evolution of chloroplast and mitochondrial DNA in plants and algae. In Monographs in Evolutionary Biology: Molecular and Evolutionary Genetics, R.J. MacIntyre, ed. Plenum Press, New York, pp. 131-240.
51. Palmer, J.D. (1985) Isolation and structural analysis of chloroplast DNA. In Methods in Enzymology, A. Weissbach and H. Weissbach, eds. Academic Press, Inc., Orlando, Florida.
52. Palmer, J.D. (1987) Chloroplast DNA evolution and biosystematic uses of chloroplast DNA variation. Am. Naturalist (in press).
53. Palmer, J.D., and W.F. Thompson (1982) Chloroplast DNA rearrangements are more frequent when a large inverted repeat is lost. Cell 29:573-550.
54. Palmer, J.D., and D. Zamir (1982) Chloroplast DNA evolution and phylogenetic relationships in Lycopersicon. Proc. Natl. Acad. Sci., USA 79:5006-5010.
55. Palmer, J.D., and C.R. Shields (1984) Tripartite structure of the Brassica campestris mitochondrial genome. Nature 307:437-440.
56. Palmer, J.D., and D.B. Stein (1986) Conservation of chloroplast genome structure among vascular plants. Curr. Genet. 10:823-833.
57. Palmer, J.D., G.P. Singh, and D.T.N. Pillay (1983) Structure and sequence evolution of three legume chloroplast DNAs. Mol. Gen. Genet. 190:13-19.

58. Palmer, J.D., R.A. Jorgensen, and W.F. Thompson (1985) Chloroplast DNA variation and evolution in Pisum. Patterns of change and phylogenetic analysis. Genetics 109:195-213.
59. Pring, D.R., and C.S. Levings III (1978) Heterogeneity of maize cytoplasmic genomes among male-sterile cytoplasms. Genetics 89:121-136.
60. Pring, D.R., M.F. Conde, and K.F. Schertz (1982) Organelle genome diversity in sorghum: Male sterile cytoplasms. Crop Sci. 22:414-421.
61. Scowcroft, W.R. (1979) Nucleotide polymorphism in chloroplast DNA of Nicotiana debneyi. Theor. Appl. Genet. 55:133-137.
62. Sears, B. (1980) Review: Elimination of plastids during spermatogenesis and fertilization in the plant kingdom. Plasmid 4:233-255.
63. Sederoff, R.R. (1984) Structural variation in mitochondrial DNA. Adv. Genet. 22:1-108.
64. Sederoff, R.R. (1987) Molecular mechanisms of mitochondrial genome evolution in plants. Am. Naturalist (in press).
65. Sederoff, R.R., C.S. Levings III, D.H. Timothy, and W.W.L. Hu (1981) Evolution of DNA sequence organization in mitochondrial genomes of Zea. Proc. Natl. Acad. Sci., USA 78:5953-5957.
66. Shinozaki, K., M. Ohme, M. Tanaka, T. Wakasugi, N. Hayashida, T. Matsubayashi, N. Zaita, J. Chunwongse, J. Obokata, K. Yamaguchi-Shinozaki, C. Ohto, K. Torazawa, B.Y. Meng, M. Sugita, H. Deno, T. Kamogashira, K. Yamada, J. Kusuda, F. Takaiwa, A. Kato, N. Tohdoh, H. Shimada, and M. Sugura (1986) The complete nucleotide sequence of tobacco chloroplast genome: Its organization and expression. EMBO J. 5:2043-2049.
67. Sytsma, K.J., and L.D. Gottlieb (1986) Chloroplast DNA evolution and phylogenetic relationships in Claexia sect. Peripetasma (Onagnaceae). Evolution 40:1248-1261.
68. Szmidt, A.E., T. Aldén, and J.-E. Hällgren (1987) Paternal inheritance of chloroplast DNA in Larix. Plant Mol. Biol. 9:59-64.
69. Thomas, M.J., and L. Chesnoy (1969) Observations relatives aux mitochondries Feulgen positives de la zone périnucléaire de l'oosphère du Pseudotsuga menziesii (Mirb.) Franco. Revue Cytol. Biol. Veg. 32:165-182.
70. Timothy, D.H., C.S. Levings III, D.R. Pring, M.F. Conde, and J.L. Kernicke (1979) Organelle DNA variation and systematic relationships in the genus Zea. Proc. Natl. Acad. Sci., USA 76:4220-4224.
71. Vaughn, K.C. (1981) Organelle transmission in higher plants: Organelle alteration vs. physical exclusion. J. Hered. 72:335-337.
72. Vedel, F., F. Quetier, Y. Cauderon, F. Dosba, and G. Doussinault (1981) Theor. Appl. Genet. 59:239-245.
73. Wagner, D.B., G.R. Furnier, M.A. Saghai-Maroof, S.M. Williams, B.P. Dancik, and R.W. Allard (1987) Chloroplast DNA polymorphisms in lodgepole and jack pines and their hybrids. Proc. Natl. Acad. Sci., USA 84:2097-2100.
74. White, E. (1986) A method for extraction of chloroplast DNA from conifers. Plant Mol. Biol. Reporter 4:98-101.
75. White, E.E., and J.-E. Hällgren (1983) Isolation of cytoplasmic DNA from Pinus sylvestris and Pinus contorta. Biochemical Genetics of Pinus contorta, E.E. White, Dissertation, The Swedish University of Agricultural Sciences, Alnarp, Sweden.
76. Whitfeld, P., and W. Bottomley (1983) Organization and structure of chloroplast genes. Ann. Rev. Plant Physiol. 34:279-326.
77. Willemse, M.Th.M., and H.F. Linskens (1969) Développement du microgamétophyte chez la Pinus silvestris entre la méiose et la fécondation. Revue Cytol. Biol. Veg. 32:121-128.

DNA ANALYSIS DURING GROWTH AND DEVELOPMENT

Sukhraj S. Dhillon

Department of Botany
North Carolina State University
Raleigh, North Carolina 27695-7612

ABSTRACT

Nuclear DNA changes in relation to growth and development were followed in *Populus deltoides*. These observations were supported with findings in other plant species. The genome size of *P. deltoides* was small and was comprised of a 2C DNA content of 1.07 pg. The genome showed an average increase of 27% under long-day photoperiod as compared to short-day photoperiod. Plants under long-day conditions displayed significantly higher growth. Furthermore, the young leaves possessed higher DNA values as compared to mature leaves. In view of our present as well as previous observations in several species, it appears that higher DNA values are related to increasing protein synthesis for maintaining growth. The senescing leaves displayed an average reduction in nuclear DNA content of 21% as compared to mature leaves. Purified DNAs from nuclei and chloroplasts of mature and senescing leaves did not display different thermal profiles. However, chloroplast DNA displayed a significantly higher %GC ratio than nuclear DNA in mature leaves. The %GC was 33.92 for nuclear and 35.95 for chloroplast DNA. Studies are in progress to understand the growth- and senescence-related DNA changes at the molecular level.

INTRODUCTION

The accumulation of literature on DNA variation has been continuously questioning the dogma of DNA constancy. It has been suggested that in plants there is a mechanism of phylogenetic increase of nuclear DNA through duplication, repeats, and generative polyploidy. In the absence of such a mechanism, an ontogenetic increase of nuclear DNA, as a prerequisite for differentiation, is adopted by the plants through endomitotic division (16). While some of the mechanisms of DNA variation, such as endopolyploidy, aneuploidy, and B-chromosomes, are now understood (but not their control), the possibility of differential DNA replication remains rather puzzling. However, more literature on DNA amplification and underreplication is appearing (3,9,13) due to the sophistication in techniques that enables one to detect small variations in DNA quantity (e.g., by scanning microspectrophotometry) as well as quality (e.g., by DNA cloning). Amplification is assumed to be a step of early differentiation and is a transitory character, while underreplication is confined to the end of the differentiation process and is irreversible. Underreplication of DNA and DNA amplification have been suggested to be of fundamental importance in the understanding of gene regulation, gene expression, and thus cell differentiation and morphogenesis (17).

The changes in DNA in relation to senescence of cotyledons (3,7,14) and leaf tissue (5,12,18,19) have been reported in various plant species. The systematic studies of DNA changes during growth, however, are less often reported except for a few recent studies in our laboratory. In this chapter, I will discuss patterns of DNA changes, including amplification and loss during growth and development and their significance.

PROCEDURES

The various tissues were collected at specific stages of growth as mentioned in the text. The following procedures were used for analyses of DNA.

Feulgen Cytophotometry

Tissues were fixed in Carnoy's No. 2 (6:3:1 ethanol:chloroform:glacial acetic acid) for 2 hr, then transferred to 70% ethanol, and kept in a refrigerator until used. The squashing of tissue, Feulgen staining, and internal standard were the same as reported in earlier studies (4,8,9). The determinations of DNA and hetero-euchromatin ratios were accomplished with a Vickers M86 scanning microdensitometer-interferometer, as described previously (5,15).

Flow Cytometry

The procedure was a modification of Galbraith et al. (11), as published by Dhillon (4). Fresh tissues were used to separate intact nuclei and the mithramycin-stained DNA was measured with a Becton Dickinson FACS (Fluorescence Activated Cell Sorting) system.

DNA Extraction

Plant DNA extraction requires special attention primarily due to large quantities of polysaccharides, pigments, and nucleases in plant cells; low nuclear/cellular volume ratio; and cell walls which are difficult to break. The method for DNA extraction was the urea-phosphate-hydroxyapatite (MUP) procedure of Britten et al. (2), as modified by Dhillon et al. (10).

For separation of chloroplasts and nuclei to study organelle and nuclear genomes, the procedures were followed from Dhillon (4). This method provided two layers of chloroplasts and pelleted nuclei that were used for DNA extraction, as mentioned above.

Thermal Denaturation

Thermal denaturation was obtained by using a Gilford system 2600 thermoprogrammer, as described in Dhillon (4). Thermal denaturation of

DNA samples in 0.1 sodium saline citrate (SSC) (1X SSC is 0.15 M NaCl, 0.015 M sodium citrate, pH 7.0) was carried out at a heat rate of 1°C/min, and the increase in absorbance at 260 nm was recorded. The system was equipped with a computer that corrected for expansion by subtracting the reference blank containing 0.1 SSC solution for each reading.

Fractionation of Repetitive DNA and Cloning

Repetitive DNA was fractionated according to DNA reassociation kinetics (2,10). The rest of the procedures were followed as described in Chang et al. (3).

FINDINGS AND SIGNIFICANCE

Both early- and late-senescing lines of *Populus deltoides* showed significantly higher growth under a long-day photoperiod as compared to a short-day photoperiod (Tab. 1). The cytophotometric determination of DNA amounts also showed an average of 26% higher DNA values under long-day conditions in both the early- and late-senescing lines (Tab. 2). Further studies with flow cytometry, using the DNA-specific fluorescence stain mithramycin, supported the above integrating microdensitometer data showing that plants under long days possessed higher DNA amounts (Tab. 3). The young leaves of plants growing under long days possessed more DNA than the leaves of those plants growing under short days; also, young leaves possessed more DNA than mature leaves (Tab. 4). Our studies with other plant species during growth and development have also shown that growing leaves invariably contained higher amounts of DNA as compared to mature leaves (Tab. 5).

The significance of higher DNA values during growth is difficult to explain. However, it appears that higher DNA values are related to increasing protein synthesis for maintaining growth. This view is supported by the increase in RNA and protein corresponding to increased DNA content during embryo development of soybean (7). Furthermore, the data obtained by Feulgen microspectrophotometry, as well as by flow cytometry,

Tab. 1. Growth data of late- and early-senescing *Populus deltoides* under different light conditions.

Photoperiod (fluorescence and incandescence)	Seed source	Plant height (cm)		Leaf number	
		66 days	99 days	66 days	99 days
9 hr (short day)	Late senescencing	55.98 ± 2.41	55.08 ± 2.53	19.53 ± 1.03	17.78 ± 0.84
	Early senescencing	33.93 ± 2.08	31.18 ± 2.62	14.00 ± 0.90	11.89 ± 0.71
9 hr + 3 hr dark interruption (long day)	Late senescencing	81.23 ± 7.51	159.44 ± 4.74	24.27 ± 1.55	35.17 ± 1.39
	Early senescencing	74.39 ± 6.14	147.23 ± 7.08	19.80 ± 1.42	22.78 ± 0.85

Note: Both early and late senescing showed significantly higher growth under long days as compared by the "t" test. The seed sources showed significant differences under short days; under long days, seed sources did not differ significantly except for leaf number at 99 days.

Temperatures were 26°C during light period and 22°C during dark period. Values are mean ± S.E. of 15 cuttings grown for 66 days and 99 days.

Tab. 2. Microspectrophotometric determination of DNA amounts (pg) in leaf nuclei of early- and late-senescing Populus deltoides under different light conditions.

Photoperiod	Seed source	Slide number 1	2	3	Mean + S.E.*
9 hr (short day)	Late senescencing	1.17 ± 0.06	1.09 ± 0.05	0.98 ± 0.05	1.08 ± 0.05 a
	Early senescencing	1.15 ± 0.04	1.04 ± 0.04	1.11 ± 0.04	1.10 ± 0.03 a
9 hr + 3 hr dark interruption (long day)	Late senescencing	1.49 ± 0.04	1.43 ± 0.03	1.44 ± 0.03	1.45 ± 0.02 b
	Early senescencing	1.48 ± 0.03	1.53 ± 0.03	1.51 ± 0.03	1.51 ± 0.02 b

*Mean ± S.E. followed by a common letter (a or b) does not differ significantly at the 1% level by Tukey's honestly significant difference test. Slides within treatment do not differ significantly.

Note: Each slide represents means ± S.E. of 35 nuclei. The average 2C value is 1.07 pg as determined from dividing root-tip nuclei.

also indicated a tendency for the early-senescing clones of Populus deltoides to possess more DNA than the late-senescing clones (Tab. 4). The significance of the differences in early- and late-senescing clones is not clear, but it is possible that higher DNA values in early-senescing seed sources help complete the growth of these sources earlier than do those in late-senescing seed sources.

Similar to the above suggestions, the higher amounts of DNA in cotyledons of legume species, as compared to other tissues, have been suggested to increase the protein-synthesizing capacity in order to meet the demand for high protein content in legume cotyledons. A legume cotyledon with a given number of cells has to synthesize, in a relatively short time, large quantities of reserve protein to ensure the growth of the embryo during germination. In the absence of cell divisions, the number of genes or cistrons bearing the information for the synthesis of the corresponding RNAs and proteins, might be the limiting factor in the processes of transcription and translation. In such cases, the nondividing cells can ensure higher rates of protein synthesis by formation of endopolyploid interphase

Tab. 3. Flow cytometric DNA measurements (pg) in leaf nuclei of early- and late-senescing Populus deltoides under different light conditions.

Photoperiod	Seed source	Sample number 1	2	3	Mean + S.E.
9 hr (short day)	Late senescencing	1.15	1.17	1.17	1.16 ± 0.01
	Early senescencing	1.20	1.23	1.20	1.21 ± 0.01
9 hr + 3 hr dark interruption (long day)	Late senescencing	1.23	1.23	1.23	1.23 ± 0.00
	Early senescencing	1.33	1.34	1.31	1.33 ± 0.01

Note: Each sample analyzed 10,240 nuclei, including chicken erythrocytes used as a standard.

Tab. 4. Nuclear DNA amounts (pg) from young and mature leaves of early- and late-senescing Populus deltoides under different light conditions.

Stage of leaf development	Photoperiod	Seed source	DNA amount (pg) Feulgen cytophotometry	Flow cytometry
Young leaves	Short day	Late senescencing	1.08 ± 0.09	1.16 ± 0.01
		Early senescencing	1.10 ± 0.06	1.21 ± 0.01
	Long day	Late senescencing	1.45 ± 0.03	1.23 ± 0.00
		Early senescencing	1.51 ± 0.03	1.33 ± 0.01
Mature leaves	Short day	Late senescencing	1.00 ± 0.02	1.12 ± 0.02
		Early senescencing	1.07 ± 0.01	1.20 ± 0.03
	Long day	Late senescencing	1.05 ± 0.01	1.06 ± 0.01
		Early senescencing	1.08 ± 0.04	1.20 ± 0.04

Note: Feulgen cytophotometry values are means ± S.E. of three slides, and flow cytometry values are mean ± S.E. of three samples.

nuclei, as noted in legume cotyledons (6,7). This view is supported by the increase in RNA and protein corresponding to increased DNA content (7). The comparison of RNA and protein contents between cotyledons and embryo axes indicated that RNA and protein values in cotyledon cells increased significantly, while cells of embryo axes showed no significant increase in either RNA or protein during embryogenesis (7).

In comparing the DNA contents in nuclei from various tissues of peanut, a higher DNA content in cotyledon nuclei as compared to root, shoot, or leaf tissue was noted (6). These observations, which are similar to the soybean DNA study, further support the role of high cotyledon DNA in enhancing protein-synthesizing capacity.

Tab. 5. Changes in nuclear DNA levels during leaf ontogeny. From Ref. 9, with permission of the American Association of Cereal Chemists.

Species	Growing leaf	Mature leaf	2C DNA amount
Arachis hypogaea	8.02 ± 0.02 (806)	7.48 ± 0.03 (610)	4.28
Glycine max	5.61 ± 0.08 (35)	3.10 ± 0.01 (35)	2.64
Nicotiana tabacum	11.37 ± 0.19 (75)	10.33 ± 0.18 (75)	9.28
Zea mays	---	5.75 ± 0.18 (45)	5.65
Populus deltoides	1.43 ± 0.03 (35)	1.06 ± 0.04 (35)	1.07

Note: Average 2C DNA amounts are determined from dividing meristem nuclei. Values represent mean ± S.E. (number of observations) in picograms.

Kowles and Phillips (13) followed DNA and nuclear volume changes in developing Zea mays endosperm, and found a rapid increase in DNA values and nuclear volume. After 16 to 18 days post-pollination, DNA had increased over 20-fold, when cell divisions were not apparent. They suggested polyteny, or preferential gene amplification, rather than polyploidy as a possibility for the increase in size of the Zea mays endosperm genome. These observations indicate that multiplication of the genome provides a mechanism for the nourishment of the embryo, and such multiplications occur preferentially in tissues involved in nourishment functions (e.g., cotyledons and endosperm).

The differentiated tissues of root, shoot, and leaves in peanut, soybean, and tobacco, and embryo axes of soybean showed DNA values intermediate between 2C and 4C amounts (6,7,9). These DNA variations indicate underreplication or amplification of certain genes or control sequences which are of fundamental importance for progress in understanding the gene regulation, gene expression, and thus cell differentiation and morphogenesis. These basic understandings are critical in improving yield and quality of economical plants.

Another kind of DNA change was in relation to senescence. In Populus deltoides, DNA changes were investigated in mature and senescing leaves by Feulgen cytophotometry and thermal denaturation of purified nuclear and chloroplast DNAs. Cytophotometric Feulgen-DNA measurements displayed an average of 1.75 pg DNA per nucleus in mature leaves and 1.38 pg per nucleus in senescing leaves, indicating a reduction of 21% in Feulgen DNA values (Tab. 6). The DNA loss was senescence-related and not due to compactness of chromatin, because the kinetics of hydrolysis curves for older nuclei were consistently lower than those of curves generated from mature leaf nuclei (5).

Purified DNAs from nuclei and chloroplasts of mature and senescing leaves did not display different thermal profiles. However, DNAs from both layers of chloroplasts separated on sucrose gradient displayed significantly higher %GC ratios than nuclear DNA in mature leaves. The %GC was 33.92% for nucleus and 35.95% for chloroplast DNA (Tab. 7). This study indicates that DNA changes during senescence involve a complex phenomenon which includes the possibility of small single-strand nicks undetectable by thermal denaturation.

Tab. 6. Microspectrophotometric determination of nuclear DNA amounts (pg) in mature and senescing leaves of Populus deltoides.

Stage of leaf development	Slide number			Mean ± S.E.*
	1	2	3	
Mature	1.91 ± 0.07	1.62 ± 0.08	1.71 ± 0.07	1.75 ± 0.09
Senescing	1.49 ± 0.06	1.35 ± 0.06	1.31 ± 0.05	1.38 ± 0.06

*Values between slides within a stage of development do not differ at the 5% level by Tukey's honestly significant difference test. Differences between mature and senescing leaves are significant.

Note: Each slide represents mean ± S.E. of 30 nuclei.

The studies on cotyledon senescence followed by recombinant DNA techniques demonstrated that repetitive DNA sequences were degraded selectively in the genome of senescing cotyledons (Fig. 1). Southern blots were made from four repetitive DNA inserts selected by colony hybridization. The inserts were those which would be disappearing during senescence. These four inserts were hybridized against ^{32}P-labeled DNAs isolated from various stages of cotyledon growth. The results of autoradiography showed that levels of two of the DNA sequences started to decrease at the 14th day of germination, while the other two DNA sequences remained at constant levels until the 17th day after germination.

Further observations during growth and development of tissue indicated increases in amounts of heterochromatin in senescing as well as dormant cotyledons as compared to healthy green cotyledons; the dormant roots possessed a higher proportion of heterochromatin as compared to germinating roots (Fig. 2). Age-related heterochromatin increase, similar to that in senescing cotyledon nuclei, was also noted in leaves of other plant species (9). Although the increase in heterochromatin is difficult to explain, heterochromatin is often associated with absence of gene expression (20) and would be expected to be found in higher amounts in the nonfunctioning dormant and senescing nuclei (6). In parallel with the above observations, heterochromatin observed in transmission electron microscope (TEM) micrographs was abundant in dormant, overwintering nuclei of pine, but decreased in the March collection (9). The visual estimate was quantified by using the point-counting technique (1,21), in which a field of regularly spaced points (dots) was overlaid on TEM micrographs to estimate the ratio of heterochromatin to euchromatin. In the winter the ratio was 1.97 (65:33) as compared to 0.58 (31:53) in the March collection. These observations by TEM support the heterochromatin findings by the microspectrophotometric method using Vickers M86 that the amount of heterochromatin is higher in inactive nuclei where absence of gene expression is expected.

The significance of age-related DNA loss as well as amplification during growth requires further investigations. Studies, however, are in progress to understand the growth and senescence-related changes in DNA at the molecular level.

Tab. 7. Tm values and base composition of leaf DNA in Populus deltoides as determined by thermal denaturation in 1 SSC.

DNA source	Tm* (°C)	Percent GC**
Total DNA	83.86 ± 0.14	35.52 ± 0.23 ab
Nuclear DNA	83.20 ± 0.12	33.92 ± 0.28 a
Lower chloroplast band	84.07 ± 0.13	36.03 ± 0.33 b
Upper chloroplast band	84.00 ± 0.12	35.87 ± 0.28 b

*Melting temperature for half-strand separation.
**Values followed by a common letter (a or b) do not differ significantly at the 1% level by Tukey's honestly significant difference test.

Note: Values represent mean ± S.E. of three observations.

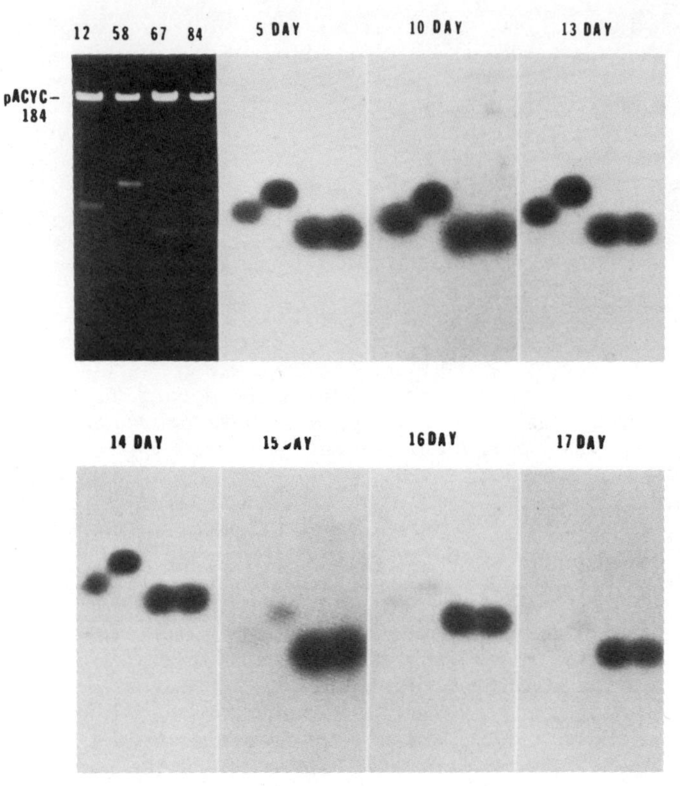

Fig. 1. Selective degradation of repetitive DNA sequences during soybean cotyledon senescence. Four sequences [12, 58, 67, 84 (selected by colony hybridization)] were hybridized with ^{32}P-labeled DNA from cotyledons of various ages. From Ref. 3, with permission.

CONCLUSIONS

The following general conclusions are reached with our present level of understanding:

(a) The current observations in Populus deltoides, as well as previous observations in several plant species (e.g., tobacco, peanut, soybean, and corn), have indicated that there is amplification of DNA during growth and decline during senescence.

(b) The increase in DNA values during growth appears to help in synthesizing more protein to maintain growth.

(c) The DNA loss during senescence involves selective degradation of repeated sequences.

(d) Heterochromatin proportion increased in dormant and senescing tissue, and appears to be related to the absence of gene expression.

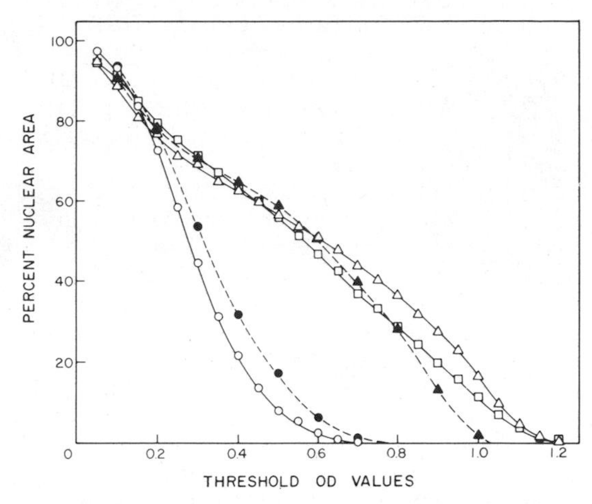

Fig. 2. The nuclear area scanned at different threshold optical densities in cotyledon and root nuclei of peanut in the following dormant and active metabolic states: dormant cotyledons (□); healthy and green cotyledons after two weeks of germination (o); senescing cotyledons after six weeks of germination (Δ); dormant root (▲); germinating root (●). From Ref. 6, with permission.

(e) More studies are required to understand the growth and senescence-related changes in DNA at the molecular level.

ACKNOWLEDGEMENTS

I wish to thank the U.S. Department of Agriculture (USDA) Forest Service, North Central Forest Experiment Station (Project No. 23-83.06) and the USDA/Science and Education Administration (Research Agreement 84-CRSR-2-2498) for partial support of this work.

REFERENCES

1. Berlyn, G.P., and J.P. Miksche (1976) Botanical Microtechnique and Cytochemistry, Iowa State University Press, Ames, Iowa.
2. Britten, R.J., D.E. Graham, and B.R. Neufeld (1974) Analysis of repeating DNA sequences by reassociation. Methods in Enzymology 29:363-418.
3. Chang, D.Y., J.P. Miksche, and S.S. Dhillon (1985) DNA changes involving repeated sequences in senescing soybean (Glycine max L.) cotyledon nuclei. Physiol. Plant. 64:409-417.
4. Dhillon, S.S. (1987) DNA in tree species. In Tissue Culture in Forestry, Vol. 1, J.M. Bonga and D.J. Durzan, eds. Martinus Nijhoff, Dordrecht, The Netherlands, pp. 298-313.
5. Dhillon, S.S., and J.P. Miksche (1981) DNA changes during sequential leaf senescence of tobacco (Nicotiana tabacum L.). Physiol. Plant. 51:291-298.
6. Dhillon, S.S., and J.P. Miksche (1982) DNA content and heterochromatin variations in various tissues of peanut (Arachis hypogaea). Am. J. Bot. 69:219-226.

7. Dhillon, S.S., and J.P. Miksche (1983) DNA, RNA, protein, and heterochromatin changes during embryo development and germination of soybean (Glycine max L.). Histochem. J. 15:21-37.
8. Dhillon, S.S., G.P. Berlyn, and J.P. Miksche (1977) Requirement of an internal standard for microspectrophotometric measurements of DNA. Am. J. Bot. 64:117-121.
9. Dhillon, S.S., J.P. Miksche, and R.A. Cecich (1983) Microspectrophotometric applications in plant science research. In New Frontiers in Food Microstructure, D.G. Bechtel, ed. American Association of Cereal Chemists, St. Paul, Minnesota, pp. 27-69.
10. Dhillon, S.S., A.V. Rake, and J.P. Miksche (1980) Reassociation kinetics and cytophotometric characterization of peanut (Arachis hypogaea L.) DNA. Plant Physiol. 65:1121-1127.
11. Galbraith, D.W., K.R. Harkins, J.M. Maddox, N.M. Ayers, D.P. Sharma, and E. Firoozabody (1983) Rapid flow cytometric analysis of the cell cycle in intact plant tissues. Science 220:1049-1051.
12. Harris, J.B., V.G. Schaeffer, S.S. Dhillon, and J.P. Miksche (1982) Differential decline in DNA of aging leaf tissue. Plant Cell Physiol. 23:1267-1273.
13. Kowles, R.V., and R.L. Phillips (1985) DNA amplification patterns in maize endosperm nuclei during kernel development. Proc. Natl. Acad. Sci., USA 82:7010-7014.
14. Krul, W.R. (1974) Nucleic acid and protein metabolism of senescing and regenerating soybean cotyledons. Plant Physiol. 54:36-40.
15. Miksche, J.P., S.S. Dhillon, G.P. Berlyn, and K.J. Landauer (1979) Nonspecific light loss and intrinsic DNA variation problems associated with Feulgen DNA cytophotometry. J. Histochem. Cytochem. 27:1377-1379.
16. Nagl, W. (1976) Endopolyploidy and polyteny understood as evolutionary strategies. Nature 261:614-615.
17. Nagl, W. (1979) Differential DNA replication in plants: A critical review. Z. Pflanzenphysiol. 95:283-314.
18. Spencer, P.W., and J.S. Titus (1972) Biochemical and enzymatic changes in apple leaf tissue during autumnal senescence. Plant Physiol. 49:746-750.
19. Udovenko, G.V., and L.A. Gogoleva (1974) Dynamics of DNA and RNA content in leaves of wheat during ontogenesis. Fisiol. Rast. 21:1076-1078.
20. Walbot, V., and R.B. Goldberg (1979) Plant genome organization and its relationship to classical plant genetics. In Nucleic Acids in Plants, T.C. Hall and J.W. Davies, eds. CRC Press, Inc., Boca Raton, Florida, pp. 3-40.
21. Weibel, E.R., and H. Elias (1967) Introduction to stereologic principles. In Quantitative Methods in Morphology, E.H. Weibel and H. Elias, eds. Springer-Verlag, New York, pp. 89-98.

THE ANAEROBIC STRESS RESPONSE AND ITS USE FOR STUDYING GENE EXPRESSION IN CONIFERS

David E. Harry,* Claire S. Kinlaw, and Ronald R. Sederoff**

Institute of Forest Genetics
U.S. Department of Agriculture Forest Service
Pacific Southwest Forest and Range Experiment Station
Berkeley, California 94701

ABSTRACT

For many angiosperms, a shift from an aerobic to an anaerobic environment elicits a stress response that involves increased transcription of specific genes. The best known of these genes are those encoding ADH. *Adh* genes have now been cloned from several angiosperm species and their regulation is being studied. Because of the available information for angiosperms, similar studies in conifers should provide insights into gene regulation in this evolutionarily distinct, commercially important group of woody plants. We have begun our studies using ADH from *Pinus radiata*. Induction of one of two ADH isozymes can be detected on starch gels within 12 hr after exposure of germinating seedlings to anaerobic conditions. We constructed a cDNA library in the phage vector λgt10 and screened this library using an *Adh* cDNA clone from corn as a probe. Several clones have been isolated and are being characterized by restriction analysis, filter hybridization, and DNA sequence analysis. Preliminary results show about 70% sequence homology between the pine cDNA clones and the corn *Adh* cDNA.

INTRODUCTION

Crop improvement by genetic engineering requires understanding gene expression in higher plants. This requirement stems from the need for engineered genes to be expressed in a regulated manner with respect to tissue, time, and the level of gene products.

---

*Present address: Department of Forestry, University of Illinois, Urbana, Illinois 61801.
**Present address: Department of Forestry, North Carolina State University, Raleigh, North Carolina 27695.

Several kinds of regulatory patterns are being studied, and each may serve a specific role in crop improvement. For genes that are constitutively expressed, levels of gene activity are determined by rates of synthesis, processing, and stability of the RNA and protein products. Such genes are likely to be involved with general cell metabolism. For genes that are developmentally regulated, levels of gene activity are also influenced by factors that are specific to developmental state. Some genes of this type have been characterized, such as those encoding seed storage proteins, but the role of most tissue-specific genes is not known. A third class of genes are induced or modulated by environmental stimuli. These genes encompass responses to such biotic stresses as fungal diseases or insect pests, or to such abiotic stresses as temperature extremes, water deficits or excess, or toxic chemicals (50). Although much work has been done, we are only beginning to define the molecular genetic mechanisms of these induced responses.

PLANT RESPONSES TO FLOODING

By reducing the availability of oxygen, flooding induces a variety of responses in plants (3,33,37,42,43,44,48). Although most responses are poorly understood, they are presumably adaptive solutions to the problems posed by flooding (Ref. 42 and references therein). For example, long-term exposure to waterlogged soils may cause development of root systems with altered morphologies or rooting depths (14,15,33,38,44,47). In general, plants that tolerate long-term flooding have adaptations that promote the transport of oxygen from aerated portions of the plant (36,37,66). Some flood-tolerant species, including a few forest trees, may shunt metabolites through alternative biochemical pathways to avoid build-up of toxic end-products (7,16,17,33). However, the significance of such alternative pathways remains unclear (36,66).

Short-term changes in plant growth processes begin rapidly. Hormonal fluctuations occur within hours after flooding and are thought to influence diverse functions (2,58). Decreased rates of shoot elongation, uptake of carbon dioxide, and transpiration were observed in *Pseudotsuga menziesii* within 4 to 5 hr after flooding (73). Rapid decreases in stomatal conductance and photosynthetic rates were also found in *Liquidambar styraciflua* (56) and in several other woody angiosperms (55).

One short-term response to flooding that is being actively studied in angiosperms is the response to anaerobic stress. Best studied in maize, this response results in characteristic changes in enzyme activity, protein synthesis, and RNA abundance (reviewed in Ref. 2 and 26). Of the genes and their products that are implicated in this response, alcohol dehydrogenase (ADH) was the first to be identified (62) and remains the best characterized (26). The biological role of ADH in the anaerobic response is not yet clear, but possibilities include the generation of energy through glycolysis (2), and the maintenance of intracellular pH (60). Several other proteins induced by anaerobiosis have now been identified, including sucrose synthase (67), phosphoglucose isomerase (40), aldolase (39), and pyruvate decarboxylase (46). All of these proteins are enzymes involved in converting sucrose to energy through glycolysis (Fig. 1).

The genes encoding ADH are ideal for studying the regulation of gene expression in gymnosperms. The induction of ADH is coordinated with other genes in the metabolic pathway from sucrose to ethanol (Fig. 1).

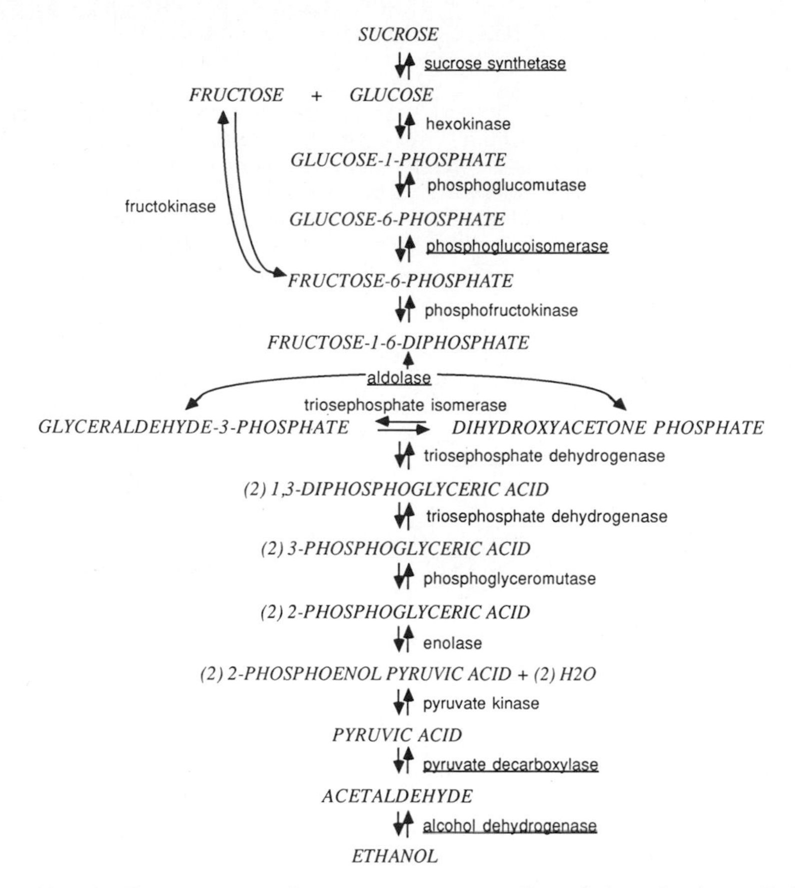

Fig. 1. Metabolic pathway from sucrose to ethanol in plants. Substrates and products are printed in upper-case letters; enzymes are printed in lower-case letters. Enzymes known to be induced during the anaerobic response in angiosperms are underlined.

Mechanisms that control the expression of ADH are being actively studied in angiosperms, and so are mechanisms that coordinately control expression of other anaerobic proteins. Hence, we can apply insights from such studies in angiosperms to learn more quickly about the regulation of gene expression in gymnosperms, including the commercially important conifers. Moreover, comparisons of regulatory mechanisms in diverse taxonomic groups will promote the formulation of generalizations applicable to all higher plants. In the following discussion, we focus on ADH and its involvement in the anaerobic response. First, we review and summarize information on ADH expression and regulation in angiosperms and gymnosperms. Next, we outline our progress in studying ADH and the anaerobic response in Monterey pine (*Pinus radiata*). Finally, we describe how results from current research in angiosperms may help us formulate specific questions leading towards an understanding of how specific DNA sequences regulate gene expression in conifers.

EXPRESSION AND REGULATION OF ALCOHOL DEHYDROGENASE IN ANGIOSPERMS

Alcohol dehydrogenase activity in angiosperms is highly regulated, showing developmental and tissue specificity and environmental induction. For example, ADH activity is typically high in seeds and germinating seedlings, and then declines following germination (5,22,70,71). In tomato (70), barley (32), and maize (26), expression of specific ADH genes changes during development. Expression of Adh1 in maize shows organ specificity that can be altered by mutation (9,26).

When maize roots are exposed to anaerobic conditions, activity of ADH in roots increases (30). This increase occurs within a few hours through differential expression of two genes (reviewed in Ref. 26 and 64). Increased ADH activity is due to increased rates of de novo protein synthesis (62); ADH1 and ADH2 are two of about 20 polypeptides that are selectively synthesized in anaerobically treated maize roots (24,63). With the exception of leaves, a similar pattern of anaerobic induction was seen for all maize organs tested (53). Anaerobic induction of ADH occurs in a number of angiosperms, including tomato (70), barley (32,51), Arabidopsis (22), rice (59), and sugarbeet (71). Likewise, shifts in protein synthesis following anaerobiosis are seen in Arabidopsis (22) and rice (59).

Several lines of evidence suggest that one mechanism regulating expression of the anaerobic response genes is at the level of gene transcription. When poly(A)+RNA is isolated from aerobic and anaerobic roots and then translated in vitro, differences in the protein products suggest that RNA populations are also different (24,63). Using cloned DNAs as molecular probes, the abundance of specific mRNAs was examined directly using northern blots. Evidence from these experiments also indicates that the abundance of particular mRNAs increases after anaerobiosis (31,69). Recent experiments using nuclear runoffs have confirmed the role of transcriptional regulation of ADH induction (72). Other regulatory mechanisms are also implicated for the anaerobic response genes. These include transcript stability (61) and translational controls (63).

Genomic clones of ADH isolated from several plant species are now being used to identify DNA sequences that are important in regulating ADH expression. In Arabidopsis, DNA sequence analysis identified a TATA box, a possible polyadenylation signal, and consensus sequences within each of the six introns that may be responsible for RNA splicing (8). Similar descriptive studies have been done for maize (19,20). But also in maize, ADH mutants and revertants are being studied to determine the DNA sequences responsible for changes in organ-specific expression of Adh1 (Ref. 9; M. Freeling, pers. comm.).

The regulatory roles of DNA sequences in the 5' flanking regions of the Adh1 gene are also being tested in maize. By itself, the Adh1 promoter functions poorly, if at all, in controlling gene function in dicots (23,27). But when a chimeric gene consisting of the Adh1 promoter and a reporter gene was introduced by electroporation into maize protoplasts, the chimeric gene was expressed and was also induced by anaerobic conditions (34). After sequences with enhancer-like properties in dicots were placed upstream of the Adh1 promoter, the maize promoter functioned efficiently in tobacco and was also responsive to anaerobic induction (23). This result has important implications for workers attempting to use DNA constructs in diverse species, because important aspects of gene regulation may be species- or taxon-specific (23). The implications of this observation are especially important for researchers attempting to genetically engineer conifers using DNA constructs designed for angiosperms.

The role of introns within eukaryotic genes remains intriguing. Intron locations may correspond to functional domains of the protein (4,26). All nine introns of maize Adh1 and Adh2 are in identical positions with respect to each other. Arabidopsis has only six introns, but this is apparently due to loss of the introns that correspond to maize introns 4, 5, and 7. The position of each remaining intron in Arabidopsis is identical with an intron in maize Adh (8). Introns are also implicated in regulating expression of maize Adh1. At first, circumstantial evidence suggested the involvement of intron 1: Adh1 mutants have been independently recovered from Robertson's mutator lines carrying active transposons, and several of the recovered mutants differ from their progenitor allele by an insertion within the first intron (26). There is now direct evidence for the involvement of introns in the expression of maize Adh1. DNA constructs containing the entire genomic sequence of maize Adh1 and DNA constructs with introns deleted were electroporated into maize protoplasts. Expression of ADH from these constructs was monitored using transient assays. Expression from constructs containing all nine introns was as much as 50 to 100 times as high as that in constructs lacking introns (Ref. 6 and 28; M. Fromm, pers. comm.).

EXPRESSION AND REGULATION OF ALCOHOL DEHYDROGENASE IN CONIFERS

Relatively little is known about ADH in conifers, but ADH expression appears to be developmentally regulated in conifers like it is in angiosperms. Alcohol dehydrogenase activity changes markedly during germination and early growth of Pinus palustris and P. elliottii seedlings (1). In embryos, ADH activity increases after sowing and reaches a peak at about the time of germination. Alcohol dehydrogenase activity then decreases in newly germinated seedlings. In megagametophytes, ADH activity is roughly constant until it begins to increase when the megagametophyte begins to senesce (1). Alcohol dehydrogenase isozymes are differentially expressed during germination and early seedling growth in P. attenuata (11). In pines, ADH is encoded by at least two unlinked loci (e.g., Ref. 10, 12, 52, 54, and 68), so it is possible that ADH genes in pine are also differentially regulated during development.

Alcohol dehydrogenase expression appears to be environmentally modulated in conifers. In Pseudotsuga menziesii, ADH activity cannot be detected in seeds unless they are soaked in water for about 24 hr (M.T. Conkle, pers. comm.). Alcohol dehydrogenase activity cannot be detected in freshly collected pine needles, but activity is observed in needles after they have been refrigerated and stored for one month (J. Hamrick and C. Niebling, pers. comm.). Pinus taeda roots are apparently capable of increased rates of glycolysis after long-term exposure to anaerobic conditions (18). We observed increased rates of ADH activity in roots after immersing P. taeda seedlings in water for 24 hr (data not shown).

EXPRESSION OF ALCOHOL DEHYDROGENASE IN MONTEREY PINE

In using ADH to study gene regulation in Monterey pine (P. radiata), one of our first goals was to determine how ADH activity is affected by exposure to anaerobic conditions. Wind-pollinated seeds were soaked overnight in chilled water, stored in the cold for one to six weeks, and then placed in an incubator for germination. After about one week, primary roots were about 2 to 5 cm long and cotyledons were beginning to emerge from their seed coats. These young seedlings were exposed to anaerobic conditions by immersing them in tap-distilled water for 6, 12, 24, 48, and 72 hr. Aerobically grown (control) seedlings were placed on moistened filter paper for comparable periods of time. All germinants were kept at room temperature (20° to 23°C) in the dark. Individual germinants were dissected following anaerobic treatments and extracts were made separately from roots and from shoots (including cotyledons and hypocotyls).

Total ADH activity was measured spectrophotometrically using methods modified from Denslow and Hook (21). Relative to aerobic controls, ADH activity in roots remained relatively constant up to 12 hr after immersion in water. After 24 hr of immersion, however, total ADH activity in roots increased substantially (Fig. 2). No additional increases in ADH activity were observed after 48 and 72 hr of anaerobic treatment (Fig. 2).

Extracts from roots and shoots were subjected to starch gel electrophoresis and gel slices were stained for ADH activity (13,68). Little or no difference in activity of ADH2 is apparent in extracts from aerobically and anaerobically treated roots (Fig. 3A) or shoots (data not shown). The mobility variants observed for ADH2 (Fig. 3A) are allozymes of the Adh2 locus (Ref. 57; S. Strauss, pers. comm.). In contrast to ADH2, ADH1 activity increases dramatically following anaerobic treatment (Fig. 3A, lanes a-f vs lanes g-l). Induction of ADH1 occurs in both roots and shoots. One difference in ADH activity between roots and shoots is that the background (i.e., uninduced) level of ADH2 activity is greater in shoots than in roots.

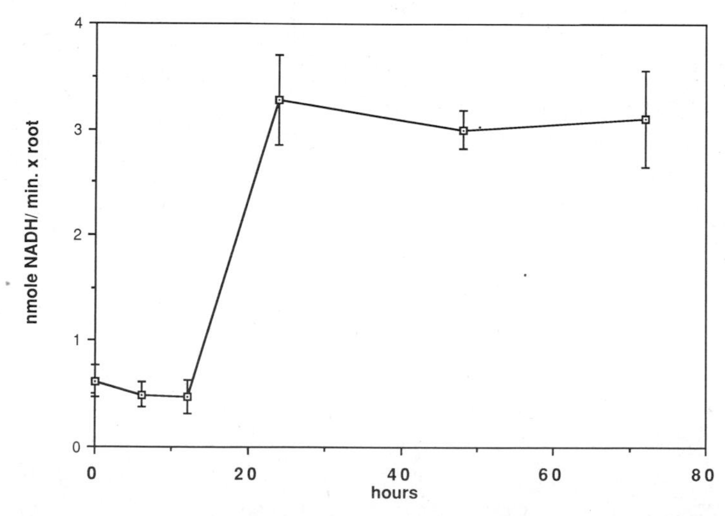

Fig. 2. Total ADH activity in roots of Monterey pine seedlings following a time-course of exposure to anaerobic conditions. Alcohol dehydrogenase activity is expressed as nmole NADH produced per min per root. Each sample was an extract made from five roots. Displayed are the means and standard errors estimated from four samples for each time point.

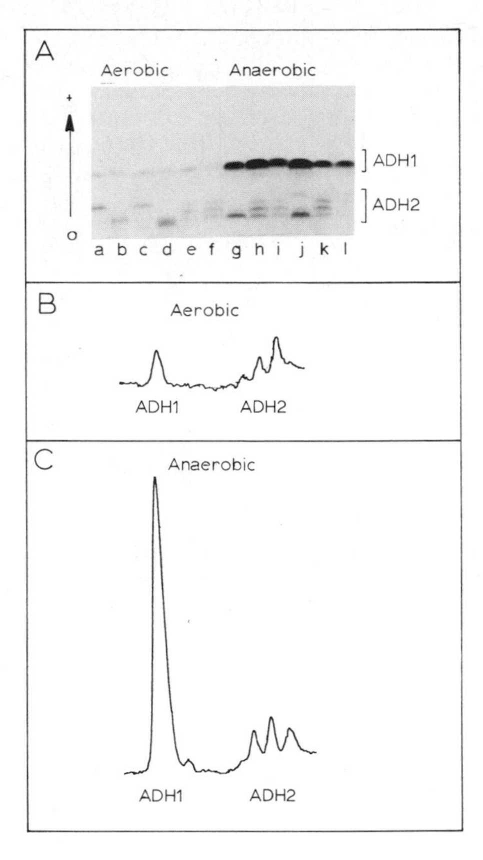

Fig. 3. Activity of ADH1 and ADH2 in roots of Monterey pine seedlings exposed to aerobic and anaerobic conditions. (A) Starch gel stained for ADH activity. Lanes a-f are from aerobically treated roots; lanes g-l are from anaerobically treated roots. Relative mobilities of ADH1 and ADH2 are shown. (B) Densitometer trace of one lane of a starch gel stained for ADH. The areas beneath the peaks labeled ADH1 and ADH2 are measures of the relative activities of these isozymes for a single aerobically treated seedling root. The trace depicted is from a root with an ADH phenotype such as in lane e or f. (C) Same as in (B), except the trace is from an anaerobically treated root with an ADH phenotype such as in lane h or k.

To quantify relative activities of ADH1 and ADH2 in shoots and roots, we scanned gels with a densitometer. Areas under each peak were integrated, and total ADH1 and ADH2 activities were estimated separately by pooling areas corresponding to appropriate peaks (Fig. 3B and C). In aerobic shoots, ADH1 activity is about 5% of ADH2 activity, while in anaerobic shoots ADH1 activity increases to about 17% of ADH2 activity (Fig. 4). By contrast, ADH1 activity in aerobic roots is about 30% of ADH2 activity, and after 24 hr of anaerobic treatment, ADH1 activity in roots is about two-fold greater than ADH2 activity (Fig. 4).

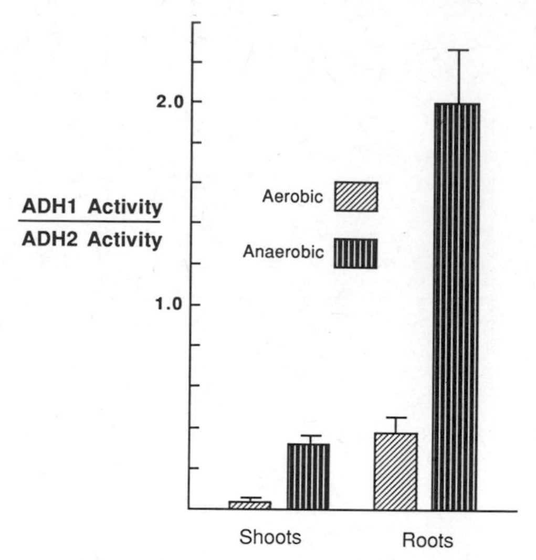

Fig. 4. Relative activity of ADH1 and ADH2 in shoots and roots of Monterey pine seedlings after 24-hr exposures to aerobic or anaerobic conditions. Ratios were calculated using ADH1 and ADH2 activities estimated from densitometer tracings such as those depicted in Fig. 3. Depicted are means and standard errors estimated from 16 samples for each organ and treatment combination.

These results clearly show that ADH activity affords an interesting opportunity for the study of gene regulation in conifers: ADH is induced in young seedlings of Monterey pine; most, if not all, of the increase in total ADH activity results from increased expression of one isozyme; and the relative levels of ADH1 and ADH2 activity and inducibility are tissue-specific.

MOLECULAR CLONING OF AN ALCOHOL DEHYDROGENASE cDNA FROM MONTEREY PINE

One strategy for cloning ADH from pine would be to use existing clones of ADH from angiosperms as molecular probes. Using such heterologous probes would greatly simplify our cloning efforts, but we first needed to test whether the DNA sequence homology between pine ADH and angiosperm ADHs would be sufficient for us to use standard DNA hybridization techniques. Alcohol dehydrogenase and other glycolytic enzymes are highly conserved evolutionarily (26,41), and there is a precedent for using DNA clones of ADH as heterologous probes in angiosperms. A maize cDNA clone corresponding to *Adh1* was used to isolate a maize *Adh2* cDNA clone (19), and a genomic fragment from maize *Adh1-S* was used to isolate a genomic clone from a dicot, *Arabidopsis thaliana* (8).

Our first attempts to identify ADH sequences from pine were done using genomic Southern blot hybridizations. The DNA isolated from Monterey pine was digested with several restriction enzymes, subjected to electrophoresis, and then blotted to nitrocellulose filters. These blots were then probed with radiolabeled DNA clones of ADH. However, even when we used conditions that would allow hybridization of DNA molecules with divergent sequences (i.e., reduced stringencies), we could not detect clear hybridization signals. We conducted similar experiments with other probes: an actin genomic clone from P. contorta (a gift from J. Kenny), and an aldolase cDNA clone from maize (pZM1154, Ref. 41). We observed better hybridization when the actin clone was used as a probe, but we could also detect hybridization using the aldolase clone as a probe (C.S. Kinlaw et al., ms. in prep.). Our failure to detect molecular hybridization using ADH probes is probably due to both the large genome size of P. radiata [about 13 pg per haploid nucleus (S. Dhillon, pers. comm.)] and the insufficient sequence homology.

Because the concentration of target sequences could be limiting, we reasoned that northern blots should provide a better test for a heterologous ADH probe. In a tissue known to express ADH, an RNA preparation from that tissue should contain a higher concentration of target molecules than a DNA preparation from that tissue. Using methods modified from Lizardi and Engelberg (49), we isolated total RNA from a rapidly growing suspension culture of Monterey pine cells known to express both ADH1 and ADH2 (data not shown). After electrophoresis, RNA was blotted to nitrocellulose and then, using reduced stringencies (C.S. Kinlaw et al., ms. in prep.), probed with a DNA insert purified from pZML793 (a nearly full-length cDNA clone of maize Adh1) (20). We observed hybridization to an RNA 100 to 200 nucleotides longer than the 1,600-nucleotide maize Adh1 RNA used as a positive control (Fig. 5). To determine that the hybridization we observed was specific to ADH, we probed similar blots using DNA clones for aldolase and sucrose synthase from maize. Different hybridization patterns were observed in both instances (data not shown). We concluded that we could use a cDNA clone from maize to identify ADH sequences from pine, and we based our cDNA cloning strategy on this assumption.

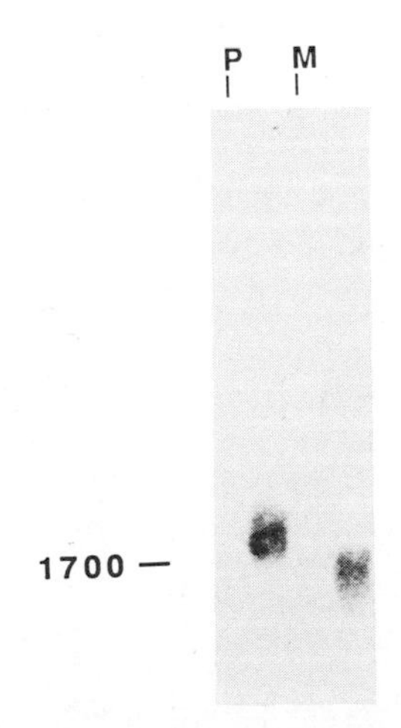

Fig. 5. Hybridization of a radiolabeled maize ADH cDNA probe to RNA isolated from maize (M) and Monterey pine (P). The 1,700-bp marker corresponds to the approximate mobility of ribosomal RNA.

The steps we used to construct a cDNA library are described in detail elsewhere (C.S. Kinlaw et al., ms. in prep.) and are summarized as follows. RNA was isolated from the Monterey pine suspension culture cells as described above. This RNA preparation was enriched for the poly(A)+-containing fraction before being used as a template for cDNA synthesis. The synthesized cDNA molecules were size-fractionated to enrich for fragments longer than 500 bp and then cloned into the EcoRI site of the cloning vector λgt10 (35). The resulting library contains about 500,000 independent recombinants, originally plated onto ten petri dishes (150 mm).

The library was screened using standard methods. Replicate lifts onto nylon filters were made from each petri plate. Both filters from each plate were subsequently probed with purified insert from pZML793 using the hybridization and washing conditions successfully used in our experiments with RNA blots. Filters were exposed to X-ray film and about 25 plaques with positive hybridization signals from both lifts were identified. Of these, we chose eight with the brightest hybridization signals to characterize further. These were subjected to a second round of plating and screening and individual plaques were picked. DNA was prepared from selected phage using plate lysates (35). Phage DNA digested with EcoRI and separated by electrophoresis shows two large fragments that correspond to the arms of the λgt10 vector. Cloned cDNA sequences are visualized as one or more insert fragments. Two of the clones we have examined so far show restriction patterns consistent with these expectations and contain insert fragments homologous to the maize probe (data not shown). These clones, designated RCS1019 and RCS1025, were subcloned into M13 for DNA sequencing.

The DNA sequence data from both clones were compared to the coding sequence of the ADH1-F allele from maize (20). Optimal sequence alignments and sequence homologies were determined using software available on the BIONET network (45). Homology matrix analyses were done using DNA Inspector II (29) on a Macintosh microcomputer. Sequence homology between RCS1025 and ADH1-F begins near the maize translational start site and extends downstream for about 400 bp (Fig. 6). Relative to RCS1025, RCS1019 shares homology with a downstream portion of maize ADH1-F, but RCS1025 and RCS1019 also share homology with each other (Fig. 6). We

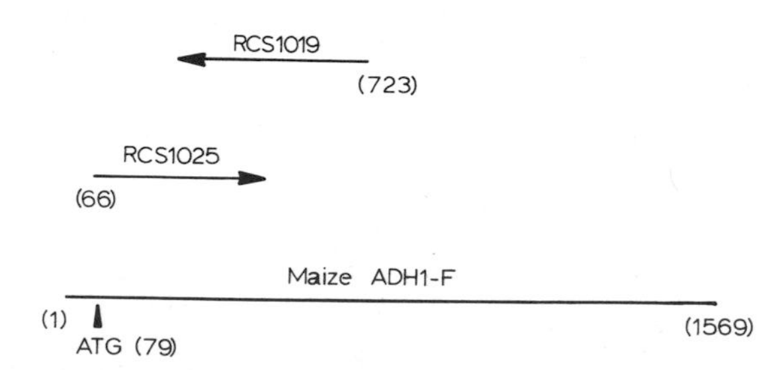

Fig. 6. Alignments of pine ADH cDNA clones RCS1019 and RCS1025 relative to the maize cDNA sequence ADH1-F. Arrows indicate the direction of sequencing for each pine cDNA. Numbers in parentheses refer to nucleotides in the maize sequence (5'-3'), e.g., the maize translational start site is at nucleotide 79, and clone RCS1025 begins at nucleotide 66 of maize.

do not yet know if RCS1019 and RCS1025 are cDNA clones from the same gene or from different genes. The diagonal pattern that emerges from the homology matrix analyses (Fig. 7A and B) assures us that both pine clones are from ADH.

The overall sequence homology between both pine cDNA clones and the maize ADH1-F sequence is about 70%. This homology seems reasonable given the homologies observed for other plant ADH genes: the coding sequence from Arabidopsis Adh shares 73% homology with maize Adh1 and 72% homology with maize Adh2 (8); the two maize ADH genes share 82% homology in their coding regions (19).

CONCLUSIONS AND FUTURE DIRECTIONS

Much work remains to understand the regulation of ADH expression in Monterey pine. The ADH clones isolated and those from our cDNA library will be useful to describe molecular events that occur after anaerobically treating pine seedlings. We can now ask whether specific mRNAs increase in abundance after anaerobiosis, which would suggest a similar regulatory mechanism as observed in angiosperms. However, a more thorough understanding of the mechanisms controlling the regulation of ADH in pines will require experiments using genomic clones. Hence, one important use of our cDNA clones will be the identification of ADH sequences from a genomic library.

Once genomic clones are in hand, we can begin to identify DNA sequences that are likely to be important in regulating ADH expression. We are intrigued by the fact that in Monterey pine, ADH1 activity is induced

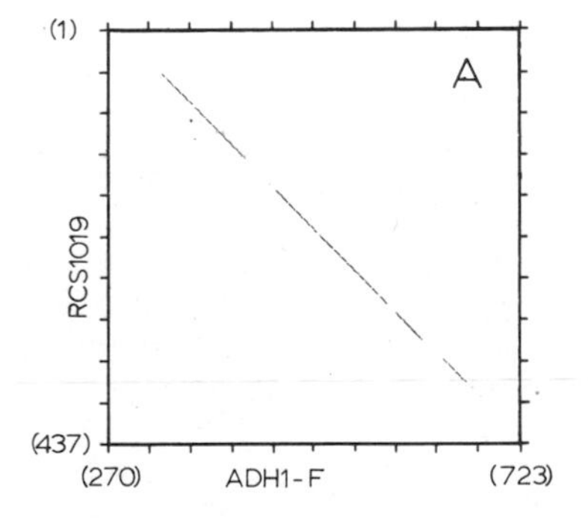

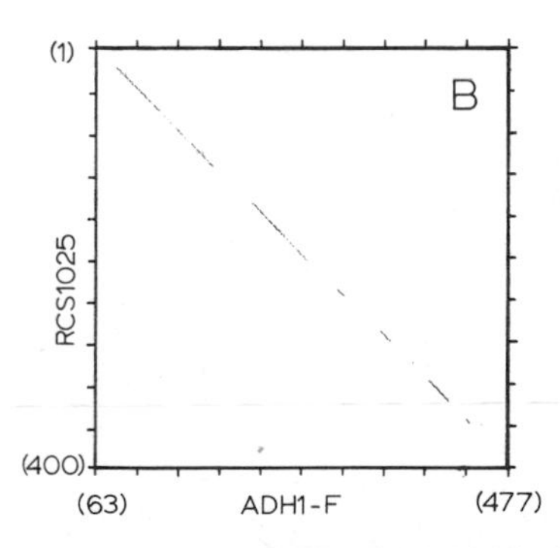

Fig. 7. Homology matrix analysis of ADH cDNA clones RCS1019 (A) and RCS1025 (B). For each clone, available sequence data (lengths are indicated by numbers in parentheses along vertical axes) were compared to sequence data from homologous regions of the maize cDNA ADH1-F (indicated by numbers in parentheses along the horizontal axes). Scales along both vertical and horizontal axes are relative and divide each sequence into tenths. Pairwise comparisons were made for subsets of 20 contiguous nucleotides from each sequence. For each comparison, a dot indicates that at least 14 of the 20 bases are identical. Hence, a diagonal line (adjacent dots) indicates regions of the pine and maize ADH sequences that share at least 70% homology. Gaps indicate regions of the two sequences that share less than 70% homology.

several fold by anaerobiosis, but ADH2 activity is relatively unaffected. Tentative identification of sequences with regulatory function will be made on the basis of sequence homology to regulatory sequences identified in angiosperms. Experimental manipulation, however, is necessary to confirm the importance of putative regulatory sequences. As discussed earlier, this work is currently underway for ADH in angiosperms (23,27,34). Such experiments require a reliable method for transferring and expressing foreign genes in pines. Recent progress has been made using *Agrobacterium tumefaciens* in pines (Ref. 65; Stomp et al., this Volume). However, these methods are slow and labor-intensive, and we feel that transient assays such as those using electroporation (27,34) would also be useful.

Another set of studies is to examine other genes that have been implicated in the anaerobic response. Such experiments would determine whether in Monterey pine other genes are also specifically induced, and whether these are the same genes as those identified in angiosperms. If indeed the induction of other genes also occurs, then experiments could be designed to determine how induction of these genes is coordinately regulated. Studies such as these also have their parallels in angiosperms: DNA sequence analysis of *Adh1* and *Sh1* (shrunken, a gene encoding sucrose synthase) shows a 13-bp sequence common to both genes that may play a role in their coordinate induction (67).

The long-term aim of experiments such as these is to provide a basis for understanding and manipulating genes in conifers. Because of the importance of ADH in understanding gene expression in angiosperms, and because of its involvement in the anaerobic response of angiosperms, ADH is ideal to begin understanding gene expression in gymnosperms, and in the commercially important conifers.

ACKNOWLEDGEMENTS

We thank Donald Sleeter for assistance with DNA sequencing; we also thank Marcella Barrios and Sally Aitken for technical assistance. Dr. Robert Teasdale provided us with the suspension culture line of Monterey pine used for RNA preparation. Dr. David Sharp provided assistance in constructing the cDNA library. Some of our initial experiments were done in the laboratory of Dr. Michael Freeling, and we thank him for advice and direction in the early phases of the project. This work was supported, in part, by U.S. Department of Agriculture Grant 86-FSTY-9-0190.

REFERENCES

1. Barnett, J.P., and A.W. Naylor (1969) Alcohol dehydrogenase activity and ethanol utilization in germinating longleaf and slash pine seeds. *For. Sci.* 15:400-403.
2. Bennett, D.C., and M. Freeling (1987) Flooding and the anaerobic stress response. In *Models in Plant Physiology and Biochemistry*, Vol. 3, D. Newman and K. Wilson, eds. CRC Press, Inc., Boca Raton, Florida.
3. Bradford, K.J., and S.F. Yang (1981) Physiological responses of plants to waterlogging. *HortScience* 16:25-30.
4. Brändén, C.-I., H. Eklund, C. Cambillau, and A.J. Pryor (1984) Correlation of exons with structural domains in alcohol dehydrogenase. *EMBO J.* 3:1307-1310.

5. Brzezinski, R.B., B.G. Talbot, D. Brown, D. Klimuszko, S.D. Blakeley, and J.-P. Thirion (1986) Characterization of alcohol dehydrogenase in young soybean seedlings. Biochem. Genet. 24:643-656.
6. Callis, J., M. Fromm, and V. Walbot (1987) Efficient expression of the maize ADH-1 gene: Requirement for intervening sequences (submitted for publication).
7. Carpenter, J.R., and C.A. Mitchell (1980) Flood-induced shift of electron flow between cyanide-sensitive and alternative respiratory pathways in roots of tolerant and intolerant tree species. J. Am. Hort. Sci. 105:688-690.
8. Chang, C., and E.M. Meyerowitz (1986) Molecular cloning and DNA sequence of the Arabidopsis thaliana alcohol dehydrogenase gene. Proc. Natl. Acad. Sci., USA 83:1408-1412.
9. Chen, C.-H., K.K. Oishi, B. Kloeckener-Gruissem, and M. Freeling (1987) Organ specific expression of maize Adh1 is altered after a Mu transposon insertion. Genetics (in press).
10. Conkle, M.T. (1971) Inheritance of alcohol dehydrogenase and leucine aminopeptidase isozymes in knobcone pine. For. Sci. 17:190-194.
11. Conkle, M.T. (1971) Isozyme specificity during germination and early growth of knobcone pine. For. Sci. 17:494-498.
12. Conkle, M.T. (1981) Isozyme variation and linkage in six conifer species. In Proceedings of the Symposium on Isozymes of North American Forest Trees and Forest Insects, M.T. Conkle, ed. General Technical Report PSW--48, Berkeley, California, Pacific Southwest Forest and Range Experiment Station, Forest Service, U.S. Department of Agriculture, pp. 11-17.
13. Conkle, M.T., P.D. Hodgskiss, L.B. Nunnally, and S.C. Hunter (1982) Starch Gel Electrophoresis of Conifer Species: A Laboratory Manual, General Technical Report PSW-64, Berkeley, California, Pacific Southwest Forest and Range Experiment Station, Forest Service, U.S. Department of Agriculture.
14. Coutts, M.P., and J.J. Philipson (1978) Tolerance of tree roots to waterlogging. I. Survival of sitka spruce and lodgepole pine. New Phytol. 80:63-69.
15. Coutts, M.P., and J.J. Philipson (1978) Tolerance of tree roots to waterlogging. II. Adaptation of sitka spruce and lodgepole pine to waterlogged soil. New Phytol. 80:71-77.
16. Crawford, R.M.M. (1978) Metabolic adaptations to anoxia. In Plant Life in Anaerobic Environments, D.D. Hook and R.M.M. Crawford, eds. Ann Arbor Science Publishers, Inc., pp. 119-136.
17. Crawford, R.M.M. (1982) Physiological responses to flooding. In Physiological Plant Ecology. II. Water Relations and Carbon Assimilation, Vol. 12b, O.L. Lange, P.S. Nobel, C.B. Osmond, and H. Ziegler, eds. Springer-Verlag, Berlin, pp. 453-477.
18. DeBell, D.S., D.D. Hook, W.H. McKee, Jr., and J.L. Askew (1984) Growth and physiology of loblolly pine roots under various water table level and phosphorus treatments. For. Sci. 3:705-714.
19. Dennis, E.S., M.M. Sachs, W.L. Gerlach, E.J. Finnegan, and W.J. Peacock (1985) Molecular analysis of the alcohol dehydrogenase 2 (Adh2) gene of maize. Nucl. Acids Res. 13:727-743.
20. Dennis, E.S., W.L. Gerlach, A.J. Pryor, J.L. Bennetzen, A. Inglis, D. Llewellyn, M.M. Sachs, R.J. Ferl, and W.J. Peacock (1984) Molecular analysis of the alcohol dehydrogenase (Adh1) gene of maize. Nucl. Acids Res. 12:3983-4000.
21. Denslow, S., and D.D. Hook (1986) Extraction of alcohol dehydrogenase from fresh root tips of loblolly pine. Can. J. For. Res. 16:146-148.

22. Dolferus, R., G. Marbaix, and M. Jacobs (1985) Alcohol dehydrogenase in Arabidopsis: Analysis of the induction phenomenon in plantlets and tissue cultures. Mol. Gen. Genet. 199:256-264.
23. Ellis, J.G., D.J. Llewellyn, E.S. Dennis, and W.J. Peacock (1987) Maize Adh-1 promoter sequences control anaerobic regulation: Addition of upstream promoter elements from constitutive genes is necessary for expression in tobacco. EMBO J. 6:11-16.
24. Ferl, R.J., M.D. Brennan, and D. Schwartz (1980) In vitro translation of maize ADH: Evidence for the anaerobic induction of mRNA. Biochem. Genet. 18:681-691.
25. Ferl, R.J., S.R. Dlouhy, and D. Schwartz (1979) Analysis of maize alcohol dehydrogenase by native-SDS two dimensional electrophoresis and autoradiography. Mol. Gen. Genet. 169:7-12.
26. Freeling, M., and D.C. Bennett (1985) Maize Adh1. Ann. Rev. Genet. 19:297-323.
27. Fromm, M., L.P. Taylor, and V. Walbot (1985) Expression of genes electroporated into monocot and dicot plant cells. Proc. Natl. Acad. Sci., USA 82:5824-5828.
28. Fromm, M., J. Callis, and V. Walbot (1987) Introns increase chimeric gene expression in maize (submitted for publication).
29. Gross, R.H. (1986) The DNA inspector II: A program for analyzing and manipulating DNA sequence on the Apple Macintosh. Gene Anal. Techn. 3:67-74.
30. Hageman, R.H., and D. Flesher (1960) The effect of anaerobic environment on the activity of alcohol dehydrogenase and other enzymes of corn seedlings. Arch. Biochem. Biophys. 87:203.
31. Hake, S., P.M. Kelley, W.C. Taylor, and M. Freeling (1985) Coordinate induction of alcohol dehydrogenase 1, aldolase, and other anaerobic RNAs in maize. J. Biol. Chem. 260:5050-5054.
32. Hanson, A.D., J.V. Jacobsen, and J.A. Zwar (1984) Regulated expression of three alcohol dehydrogenase genes in barley aleurone layers. Plant Physiol. 75:573-581.
33. Hook, D.D. (1984) Adaptations to flooding with fresh water. In Flooding and Plant Growth, T.T. Kozlowski, ed. Academic Press, Inc., Orlando, Florida, pp. 265-294.
34. Howard, E.A., J.C. Walker, E.S. Dennis, and W.J. Peacock (1987) Regulated expression of an alcohol dehydrogenase 1 chimeric gene introduced into maize protoplasts. Planta 170:535-540.
35. Huynh, T., R.A. Young, and R.W. Davis (1985) Constructing and screening cDNA libraries in lambda gt10 and lambda gt11. In DNA Cloning. Vol. 1. A Practical Approach, D.M. Glover, ed. IRL Press, Oxford, pp. 49-78.
36. Jackson, M.B., and M.C. Drew (1984) Effects of flooding on growth and metabolism of herbaceous plants. In Flooding and Plant Growth, T.T. Kozlowski, ed. Academic Press, Inc., Orlando, Florida, pp. 47-128.
37. Kawase, M. (1981) Anatomical and morphological adaptation of plants to waterlogging. HortScience 16:30-34.
38. Keeley, J.E. (1979) Population differentiation along a flood frequency gradient: Physiological adaptations to flooding in Nyssa sylvatica. Ecol. Monogr. 49:89-108.
39. Kelley, P.M., and M. Freeling (1984) Anaerobic expression of maize fructose-1,6-diphosphate aldolase. J. Biol. Chem. 259:14180-14183.
40. Kelley, P.M., and M. Freeling (1984) Anaerobic expression of maize glucose phosphate isomerase I. J. Biol. Chem. 259:673-677.
41. Kelley, P.M., and D.R. Tolan (1986) The complete amino acid sequence for the anaerobically induced aldolase from maize derived from cDNA clones. Plant Physiol. 82:1076-1080.

42. Kozlowski, T.T., ed. (1984) Flooding and Plant Growth, Academic Press, Inc., Orlando, Florida, 356 pp.
43. Kozlowski, T.T. (1984) Plant responses to flooding of soil. BioScience 34:162-167.
44. Kozlowski, T.T. (1984) Responses of woody plants to flooding. In Flooding and Plant Growth, T.T. Kozlowski, ed. Academic Press, Inc., Orlando, Florida, pp. 129-163.
45. Kristofferson, D. (1987) The BIONET electronic network. Nature 325:555-556.
46. Laszlo, A., and P. St. Lawrence (1983) Parallel induction and synthesis of PDC and ADH in anoxic maize roots. Mol. Gen. Genet. 192:110-117.
47. Levan, M.A., and S.J. Riha (1986) Response of root systems of northern conifer transplants to flooding. Can. J. For. Res. 16:42-46.
48. Levitt, J. (1980) Responses of Plants to Environmental Stresses. Vol. II. Water, Radiation, Salt, and Other Stresses, Academic Press, Inc., New York.
49. Lizardi, P.M., and A. Engelberg (1979) Rapid isolation of RNA using proteinase K and sodium perchlorate. Anal. Biochem. 98:112-116.
50. Matters, G.L., and J.G. Scandalios (1986) Changes in plant gene expression during stress. Devel. Genet. 7:167-175.
51. Mayne, R.G., and P.J. Lea (1984) Alcohol dehydrogenase in Hordeum vulgare: Changes in isoenzyme levels under hypoxia. Plant Sci. Lett. 37:73-78.
52. Millar, C.I. (1985) Inheritance of allozyme variants in bishop pine (Pinus muricata). Biochem. Genet. 23:933-946.
53. Okimoto, R., M.M. Sachs, E.K. Porter, and M. Freeling (1980) Patterns of polypeptide synthesis in various maize organs under anaerobiosis. Planta 150:89-94.
54. O'Malley, D.M., F.W. Allendorf, and G.M. Blake (1979) Inheritance of isozyme variation and heterozygosity in Pinus ponderosa. Biochem. Genet. 17:233-250.
55. Pereira, J.S., and T.T. Kozlowski (1977) Variations among woody angiosperms in response to flooding. Physiol. Plant. 41:184-192.
56. Pezeshki, S.R., and J.L. Chambers (1985) Stomatal and photosynthetic response of sweetgum (Liquidambar styraciflua L.) to flooding. Can. J. For. Res. 15:371-375.
57. Plessas, M.E., and S.H. Strauss (1986) Allozyme differentiation among populations, stands, and cohorts in Monterey pine. Can. J. For. Res. 16:1155-1164.
58. Reid, D.M., and K.J. Bradford (1984) Effects of flooding on hormone relations. In Flooding and Plant Growth, T.T. Kozlowski, ed. Academic Press, Inc., Orlando, Florida, pp. 195-219.
59. Ricard, B., B. Mocquot, A. Fournier, M. Delseny, and A. Pradet (1986) Expression of alcohol dehydrogenase in rice embryos under anoxia. Plant Mol. Biol. 7:321-329.
60. Roberts, J.K.M., F.H. Andrade, and I.C. Anderson (1985) Further evidence that cytoplasmic acidosis is a determinant of flooding intolerance in plants. Plant Physiol. 77:492-494.
61. Rowland, L.J., and J.N. Strommer (1986) Anaerobic treatment of maize roots affects transcription of Adh1 and transcript stability. Mol. Cell. Biol. 6:3368-3372.
62. Sachs, M.M., and M. Freeling (1978) Selective synthesis of alcohol dehydrogenase during anaerobic treatment of maize. Mol. Gen. Genet. 161:111-115.

63. Sachs, M.M., M. Freeling, and R. Okimoto (1980) The anerobic proteins of maize. Cell 20:761-767.
64. Sachs, M.M., E.S. Dennis, J. Ellis, E.J. Finnegan, W.L. Gerlach, D. Llewellyn, and W.J. Peacock (1985) Adh1 and Adh2: Two genes involved in the maize anaerobic response. In Cellular and Molecular Biology of Plant Stress, J.L. Key and T. Kosuge, eds. ARCO PCRI and UCLA Symposium, Alan R. Liss, Inc., New York, pp. 217-226.
65. Sederoff, R., A.-M. Stomp, W.C. Chilton, and L.W. Moore (1986) Gene transfer into loblolly pine by Agrobacterium tumefaciens. Bio/-Technology 4:647-649.
66. Smith, A.M., C.M. Hylton, L. Koch, and H.W. Woolhouse (1986) Alcohol dehydrogenase activity in the roots of marsh plants in naturally waterlogged soils. Planta 168:130-138.
67. Springer, B., W. Werr, P. Starlinger, D.C. Bennett, M. Zokolica, and M. Freeling (1986) The Shrunken gene on chromosome 9 of Zea mays L. is expressed in various plant tissues and encodes an anaerobic protein. Mol. Gen. Genet. 205:461-468.
68. Strauss, S.H., and M.T. Conkle (1986) Segregation, linkage, and diversity of allozymes in knobcone pine. Theor. Appl. Genet. 72: 483-493.
69. Strommer, J.N., S. Hake, J. Bennetzen, W.C. Taylor, and M. Freeling (1982) Regulatory mutants of the maize Adh1 gene caused by DNA insertions. Nature 300:542-544.
70. Tanksley, S.D., and R.A. Jones (1981) Effects of O_2 stress on tomato alcohol dehydrogenase activity: Description of a second ADH coding gene. Biochem. Genet. 19:397-409.
71. Van Geyt, J., and M. Jacobs (1986) Mode of inheritance and some general characteristics of sugarbeet alcohol dehydrogenase. Plant Sci. 46:143-149.
72. Vayda, M.E., and M. Freeling (1986) Insertion of the Mu1 transposable element into the first intron of maize Adh1 interferes with transcript elongation but does not disrupt chromatin structure. Plant Mol. Biol. 6:441-454.
73. Zaerr, J.B. (1983) Short-term flooding and net photosynthesis in seedlings of three conifers. For. Sci. 29:71-78.

REGULATION OF GENE EXPRESSION

METABOLIC PHENOTYPES IN SOMATIC EMBRYOGENESIS AND POLYEMBRYOGENESIS

Don J. Durzan

Department of Environmental Horticulture
University of California
Davis, California 95616

ABSTRACT

The recovery of somatic embryos from Prunus spp. and a variety of conifers using cell suspension cultures has created an opportunity to compare metabolic phenotypes among somatic and zygotic embryonic systems. Metabolic phenotypes are based on, but not limited to, the integration over time of families of metabolites involved in amino acid metabolism. Metabolic phenotypes are displayed as physiological state-networks by computer-assisted graphics. While the interpretation of such maps remains difficult, their potential value in revealing true-to-type phenotypes for clonal development and culture practices will be illustrated.

INTRODUCTION

This chapter reviews current thinking on metabolic phenotypes in somatic embryogenesis (SE) and polyembryogenesis (SPE) of the following conifers: Douglas fir (Pseudotsuga menziesii Mirb. Franco), sugar pine (Pinus lambertiana), loblolly pine (Pinus taeda), a tropical pine (Pinus merkusii), and Norway spruce (Picea abies Karst.).

The concept of metabolic phenotypes attempts to capture the dynamic behavior of gene expression for precise environmental adaptation in complex, long-lived woody perennials (9,11). It is also an extension of the subject of phenetics (e.g., Ref. 46) and biochronology (24). For conifers, the concept derives from our interest in nitrogen metabolism (16,39) and the control of cellular development in vitro and in situ (15,29). More recently, metabolic phenotypes for zygotic embryo development have been displayed as state-network maps (9,17) and interpreted in relation to the activity of plant growth regulators (10,11). This approach is now extended to include new observations with embryogenic cell suspensions and with protoplasts capable of SPE (20,21,22,23).

The term "somatic polyembryogenesis" is appropriate for those gymnosperms which are naturally polyembryonic in zygotic embryogenesis.

Zygotic embryos of Douglas fir are not normally polyembryogenic (2), but in suspension cultures, polyembryony is observed (13,14). In this case, the distinction between SE and SPE is useful.

Clonal products, aimed at capturing genetic gains (e.g., Ref. 41), should be true-to-type in their gene expression (cf. Ref. 25). Somatic embryos generated by SPE should pass through predicted stages of proembryony and early and late embryony, and should be converted to uniform seedlings in soil. For example, proembryony should involve free nuclei, neocytoplasm, and basal plans of development that lead to embryogenesis and polyembryogenesis. Many of these events are indeed observed in vitro.

In SPE, the rescued products of fertilization represent the new generation and not callus or the mother tree as in other woody perennials. In _Citrus_, SE originates from the nucellus (38), whereas in _Larix_, embryogenesis originates from the haploid female gametophyte (28). In our earlier published work with sugar pine (21), "callus" was used to describe the proliferating embryonal suspensor masses (ESMs) obtained from mature seeds. However, diagnostic cytochemical tests developed in our laboratory for _Prunus_ spp. (cf. Ref. 33) reaffirm that these ESMs are not calli. This conclusion is based on the cellular composition, developmental fate, and comparison of ESMs with nonembryogenic calli (e.g., Ref. 14, 22, and 23).

This chapter therefore will build upon the above considerations and on unpublished observations of a number of colleagues (P. Krogstrup, P.K. Gupta, A. Dandekar, R. Falk, M. Boulay, M. Mota, I. Umboh, and F. Ventimiglia). This synthesis is related speculatively to metabolic phenotypes for SPE. Unfortunately, we are still a long way off from judging the utility of metabolic phenotypes. We need more progress with cell suspension cultures, morphogenesis, protoplast regeneration of plants, and genetic engineering to help us take the first step in testing our hypotheses and in dealing with observations that until a year ago were never possible.

FACTORS IN SOMATIC POLYEMBRYOGENESIS

Properties and Origin of Embryonal Cells

We must distinguish initially between cells of embryogenic and nonembryogenic potential (Fig. 1). Embryogenic cell lines originate in the developing seed as a proliferating ESM. For conifers, the ESM is often best isolated from immature seeds three to four weeks after fertilization. Embryonal suspensor masses can also be induced in protodermal cells by a mixture of plant growth regulators (PGRs), specific factors that supplement commonly used media, and by physical contact with the mucilaginous substance associated with the ESM (Fig. 2). Responsive cells have thus far been found in hypocotyls, cotyledons, and on buds and leaves excised from mature trees but not in coleorhizae as in _Pinus merkusii_ (I. Umboh, P.K. Gupta, and D.J. Durzan, unpubl. results). In our experience, responsive cells are usually protodermal in origin and type.

Considering the long reproductive process in conifers, the fully totipotent embryonal cells and morphogenic protoplasts, derived from the ESM, are comparatively recent products of meiosis and fertilization. The ESM recapitulates patterns of conifer embryogeny that have been classified as one of several basal plans (37). At fertilization, the nucleus of the zygote contributes to a free-nuclear stage that launches the embryonic developmental process. Both the free-nuclear stage and the embryonic process have been observed in somatic cells and protoplasts derived from the ESMs of loblolly pine, Douglas fir, and Norway spruce. A free-nuclear stage also exists in Larix SE (J.M. Bonga, pers. comm.; J.M. Bonga et al., this Volume; Ref. 23).

We have observed associated with acetocarmine-reactive free nuclei, a dense cytoplasm that also stains red. This cytoplasm may be considered equivalent to the neocytoplasm described by Camefort (5) and by Willemse (44). The neocytoplasm originates at fertilization from the nucleoplasms of male and female nuclei. The concept of neocytoplasm represents a useful working hypothesis for the study of embryonic development, although its validity needs to be reaffirmed by more study (e.g., Ref. 23).

In SE and SPE, the neocytoplasm accounts for the reactivity, optical properties, and structure of embryonal cells at the first cell division. Not only does this cytoplasm stain with acetocarmine, it also has some natural refraction under polarized light (Fig. 1 and 3). Furthermore, this neocytoplasm reacts with calcofluor white and produces a bright fluorescence under UV light (excitation, 365 nm; emission, >418 nm). These properties have enabled the tracing of the origin and fate of protoplasts of embryonal cells and have created new opportunities for evaluating the origin of histogenic algorithms (e.g., Ref. 42) in embryonic development.

Biomimetic Properties of Extracellular Materials

The ESM is characterized by its origin from the seed, its cellular composition, and by the production of an extracellular mucilaginous material that is not found in calli (nonembryogenic) even from the same explant source. In this context, Romberger and Tabor (35) also found a polysaccharide-rich material, but this was deposited beneath excised cultured shoot tips of Norway spruce. Krogstrup (26) was probably the first to recognize the importance of the slimy mucilaginous material in somatic proembryogenesis in Norway spruce.

Production of the mucilage in SE characterizes the totipotent physiological state of embryogenic cells of carrot (cf. Ref. 36). In conifers, mucilage production is enhanced by 2,4-dichlorophenoxyacetic acid (2,4-D), kinetin, amino acids, and high levels of myoinositol (11). When the ESM is placed in suspension culture, cellular components of the mass are separated mechanically and the mucilage is dispersed into the medium. Separation facilitates the study of differentiation as cells pass through the free-nuclear proembryonic stage and through early embryogeny.

During pro- and early embryony in darkness, extracellular mucilaginous materials are continually produced to condition the medium. Under the influence of α-naphthaleneacetic acid (NAA), this material may be excreted and polymerized. Often the condensed and polymerized material takes the shape of different stages of SPE. To the unwary eye, these structures resemble somatic embryos (Fig. 3).

Through a series of multiple-stains (6) and fluorescent or polarized light microscopy, we have observed some unusual properties of the mucilaginous gel. Substances in the gel appear to originate from neocytoplasms associated with true-to-type embryonic development. This view is supported by the affinity of these substances for acetocarmine and by their general fluorescent properties under UV and polarized light. We have now developed an automated analytical method to measure the fluorescence of compounds that react with 9,10-phenanthrenequinone (43).

Most unexpectedly, even in the culture medium the extracellular material may become polymerized, and it continues to react with acetocarmine (Fig. 4). We do not know if the shapes produced are significantly biomimetic, i.e., does this material represent a substance responsible for SPE

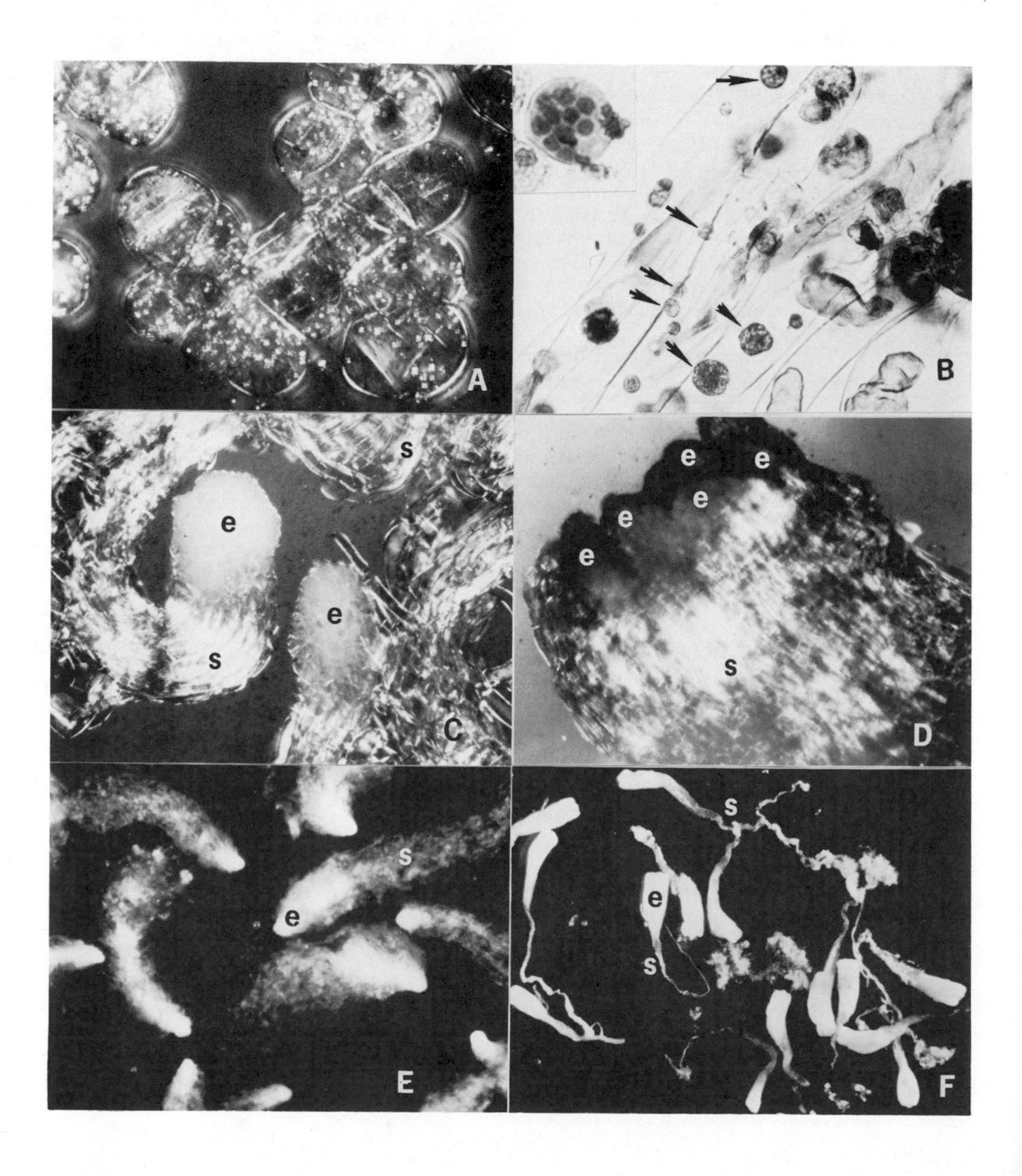

that has somehow escaped the containment of the cell or is it the diffusible morphogen or oligosaccharin (1) that conditions the medium and accounts for the transformation of cells as in contact studies of explants with the ESM? For the development of metabolic phenotypes, the properties of large molecules in the mucilage may be extended by diffusion beyond the properties of their chemical subgroups (32).

As for the adherence of mucilaginous or polymerized materials to somatic proembryos, we have observed a nonacetocarmine-reactive layer around embryos (cf. Fig. 3F for Douglas fir). The water solubility of the gel and the need to remove excess stain make the application of the acetocarmine staining to study the mucilaginous materials outside of cells very difficult.

Biochemical studies have been initiated to isolate externally produced polymerized materials and the glycoproteins and other fractions of the neocytoplasm that could account for acetocarmine reactivity (see below). These extracellular materials are considered to be components of the phenotypic expression of different physiological states as determined by specific cell lines in response to PGRs under specific culture conditions.

Fig. 1. Somatic embryogenesis and polyembryogenesis in conifer cell suspension cultures. (A) Nonembryonic callus derived from an immature Douglas-fir embryo three to four weeks after fertilization. Cells are photographed under polarized light to reveal wall thickening and star-like birefringence of starch grains (after P.K. Gupta and D.J. Durzan, unpubl. observations) (X40). (B) Free-nuclear stage in Douglas-fir ESM (X55). Arrows point to nuclei of varying sizes and affinity for acetocarmine and Evan's blue. Cells shown have already been stained. Inset at upper left shows multiple free nuclei in *Picea abies* (after M.H. Boulay and D.J. Durzan, unpubl. observations). (C) Somatic embryos (e) of Douglas fir in cell suspension cultures of the ESM under polarized light (X25). Note weak, diffuse refraction of light by embryonal cells bearing a neocytoplasm as opposed to suspensor cells (s) which have strongly highlighted wall thickening. (D) Polyembryonic mass of at least ten Douglas-fir double-stained somatic embryos (X75). Embryos were photographed under polarized light to outline cell walls and to show affinity of multiple embryos for acetocarmine and reaction of suspensor cells with Evan's blue. The acetocarmine reaction with the neocytoplasm subdues the refraction of light seen in (C). Polyembryony of this type does not normally occur in the zygotic seed and is a function of the level of ABA in the culture medium. (E) Effect of ABA (5 μM) on separating polyembryonic masses of loblolly pine to encourage the development of individual somatic embryos with suspensors (cf. Ref. 14) (X28). (F) Recovery of individual somatic embryos at the early stage of cotyledon development in Douglas fir under influence of ABA. Embryos shown have well-developed suspensors and are beginning to show cotyledonary development. Each elongated embryo is approximately 2 mm long.

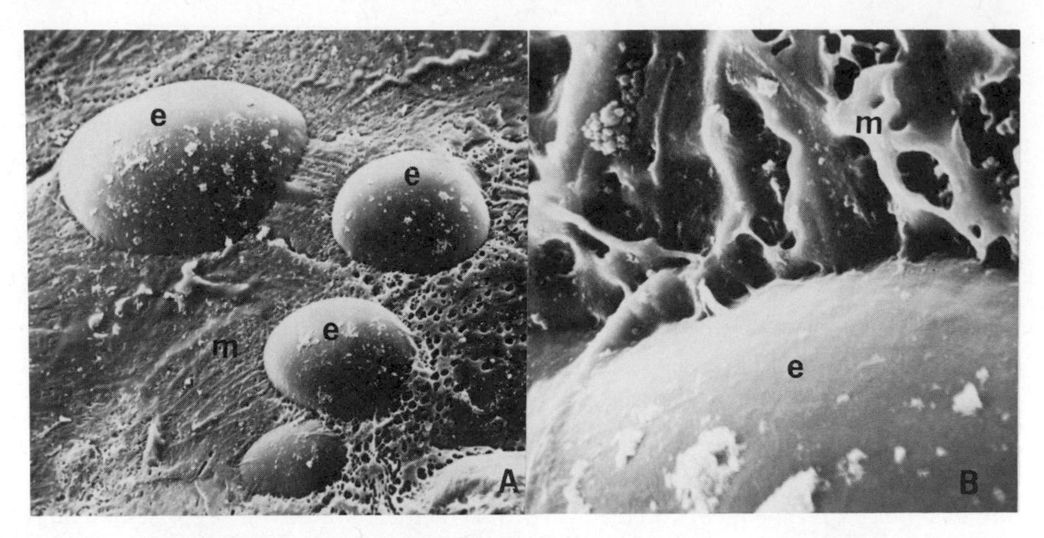

Fig. 2. Scanning electron micrographs of loblolly pine embryonal cells cryopreserved to reveal (A) interaction of mucilage (m) with embryonal cells (e) (X400), and (B) point of contact of a cell with mucilaginous material (X4000). 2,4-Dichlorophenoxyacetic acid could not be localized in this material by cathode luminescence induced by the scanning electron microscope (R. Falk and D.J. Durzan, unpubl. observations; cf. Ref. 18 and 30).

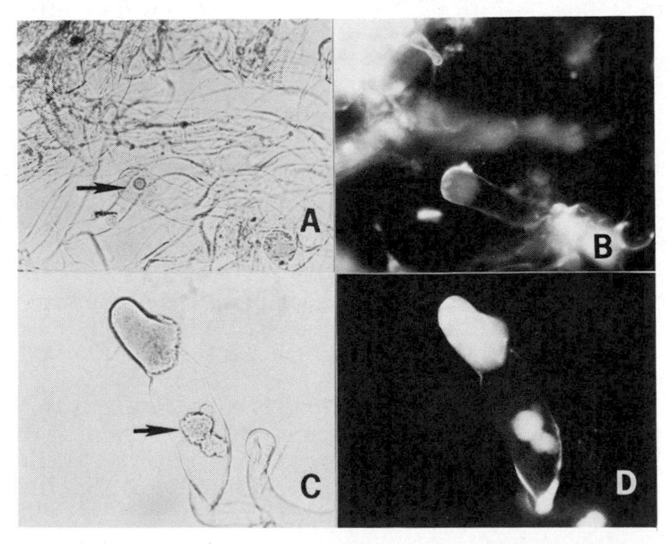

Fig. 3. Properties of the neocytoplasm in proembryonal Douglas-fir cells. (A) Cell suspension of ESM showing nucleus surrounded by neocytoplasm (arrow) under bright field (X65). (B) Same cells as in (A) but under polarized light (X65). (C) Proembryonal cell with several nuclei each surrounded by neocytoplasm (e.g., arrow) stained with calcofluor white and observed under bright field (X65). (D) Same cell as in (C) but under polarized light (X65).

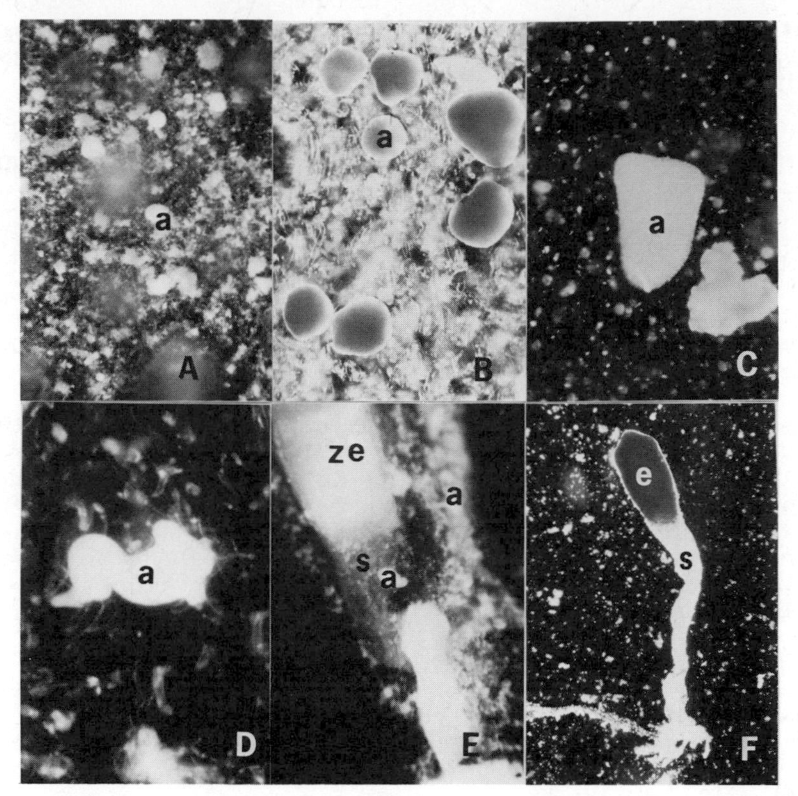

Fig. 4. Release of an acetocarmine-reactive material that simulates embryonic shapes when polymerized in culture media bearing ESMs. (A) Release into medium of acetocarmine-reactive (a) material by an auxin-stressed (cf. Ref. 40) embryonal cell (X65). (B) Release of unstained material (white) into medium after three weeks (X10). (C) Growth of clumps of extracellular material after four to five weeks (X12). (D) Biomimetic shape of extracellular materials after six weeks (X10). (E) Natural occurrence of unstained extracellular material in an ESM three weeks after removal of the ESM from a young seed (three to four weeks after fertilization) (X18). Results indicate that an acetocarmine-reactive material is released into culture medium by ESM (ze, zygotic embryo). (F) Acetocarmine-stained somatic embryo of Douglas fir to show non-stained layer surrounding the embryo (e) and contiguous with the suspensor(s) (Durzan and Gupta, unpubl. observ.) (X10).

Division of Labor in the Proembryo

The first indications of differentiation and segregation of an independent metabolic phenotype among cells of the ESM relate, as far as we have studied, to properties of nuclei at the free-nuclear stage. Some nuclei, at lower frequency, and usually associated with suspensor cells, only stain weakly red. These are more permeable to Evan's blue than nuclei from embryogenic cells. The application of the double-staining method with acetocarmine and Evan's blue to the ESM therefore reveals at least two types of nuclei.

At the free-nuclear stage of SE, there exists a nuclear-type linked allocation of neocytoplasm (Fig. 2) that carries through all daughter cells in early proembryony, i.e., the developmental fate of cells in the proembryo seems to be sequentially related to a "base address" as diagnosed by red staining nuclei (Fig. 3). It is important to reiterate that the neocytoplasm originates from the nucleoplasm of the male and female nucleus at fertilization (5,44).

We now have to specify somehow the "physiological states" derived from the types of nuclei in the ESM. Here two additional challenges emerge. One is to understand how somatic cells, bearing neocytoplasm, recapitulate the embryonal basal plans typical of zygotic embryogenesis (cf. Ref. 7 and 37), and, second, to understand how these metabolic phenotypes relate to the emergence of histogenic algorithms.

In terms of a simple model for gene expression, we can postulate that a set of embryonal genes are somehow activated to amplify the production of the acetocarmine-reactive substances (glycoproteins) in the nucleo- and neocytoplasm. Should these proteins have enzyme activity, then they would contribute to metabolic changes in the soluble and insoluble phase by sequential conversions of substrates into a variety of products. These reactions may be viewed as hierarchies of evolving intercalated series of metabolic networks (see Fig. 8).

The sequential developmental logic, which derives from the base address, namely, genes in nuclei of male and female gametes that generate a nucleoplasm, reflects a linked allocation and mass flow of enzymic information in gene expression of proembryonal cells. This linked allocation may be termed "precessing," a term which derives from topological notions (Fig. 5).

One interesting focus of developmental events is the proplastid. In Douglas fir, proplastids of proembryonal cells appear to have microtubules (M. Mota and D.J. Durzan, unpubl. results). The role of microtubules in precessing and the evolving developmental logic in embryonal cells remains unknown.

Precessing of macromolecules can have its metabolic basis in post-translation events and in membrane transformation and flow (Fig. 6). The structural polymorphism, based on the 140 post-translational modifications of protein amino acid and the processing of mRNA (base methylation, splicing, polyadenylation, and transport), endows a degree of determinism in subcellular structure and function in advance of many changes in the soluble pool of cells. Hence, enzymes that process and schedule macromolecular synthesis may allow the cell to quantify the effect of multiple external nutritional and environmental factors that impinge on daughter cells of the ESM. This biochemical determinism is reinforced when enzymes are formed at specific locations (e.g., membranes, vacuoles, organelles) in the cell and convert substrates into products through assimilation, redox reactions, ligases, hydrolases, and other processes. This further "locks in" the fate of cells, especially if products are hormonal or inhibitory. Hence enzymatic reactions that contribute to the formation of homologs, analogs, cyclization, ring-closure, conjugation, and methylation among others become important (3,8,16). These processes are postulated as occurring progressively and sequentially over time as cells pass from totipotent to mature states in the life cycle (10).

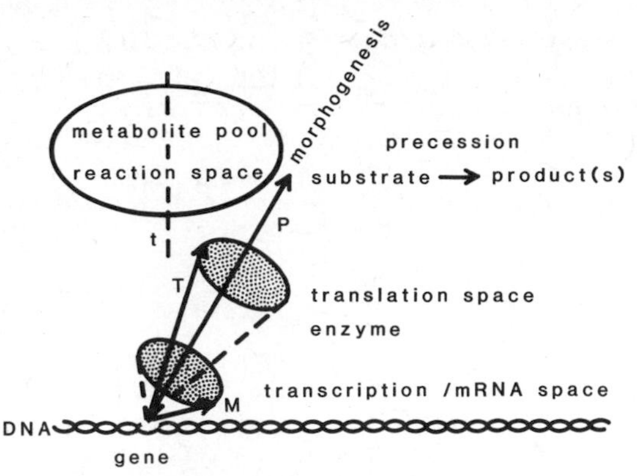

Fig. 5. Precessing in gene expression. Transcription and translation of information in certain genes contributes to a discrete developmental outcome (vector P in morphogenesis). This outcome can be described by vectors M, T, and P. Vector M defines transcription space topologically. Vector T describes translation space. In our notation, M and T vectors are precessed about the P vector. "Precess" denotes progressive expression with a directed flow whereby changes in P occur about the space related to M and T. This concept is projected as a cone. Similarly, the vector P can be precessed about the "reaction space" occupied by the metabolic pools where substrates are converted into products (e.g., Fig. 8). The reaction space may be located conceptually around an axis t. This axis may represent, for example, the orientation of a membrane microtubule or organelle (cf. Fig. 3). Dynamics of the overall metabolic pool can be depicted as a state-network map (Fig. 8). Topological notations of this type aim to integrate biological phenomena (Fig. 1 and 3) by formal mathematical terms with varying degrees of success.

When time-lapse studies are made on cells (cf. Ref. 12), we are faced with tracing and explaining the dynamic behavior of precessed, dissipative physiological states in a neocytoplasm engaged in self-organization with a discrete developmental outcome, namely, the somatic embryo with its metabolic phenotypes.

Polyembryogenesis

Conditions that lead to alternative mitotic states in embryonal cells, i.e., cleavage rather than differentiation of a daughter cell, appear to be related to the activity of plant growth regulators. In "cleavage division" of an embryonal cell to give another proembryo, the neocytoplasm is reset to a "basal state." We can postulate that the cleavage phenomenon is due to locally high levels of growth regulators. We are not yet sure what is the critical number of cleavage-type mitoses before the cleavage property is lost or becomes aberrant, nor why such a process is inhibited in seeds of Douglas fir.

One clue to the physiological basis of polyembryony comes from the use of abscisic acid (ABA) with ESMs of loblolly pine (13), Douglas fir (14), and Norway spruce (4) (Fig. 1). Abscisic acid inhibits the repetitive cleavage polyembryonic process and enables the recovery of individual

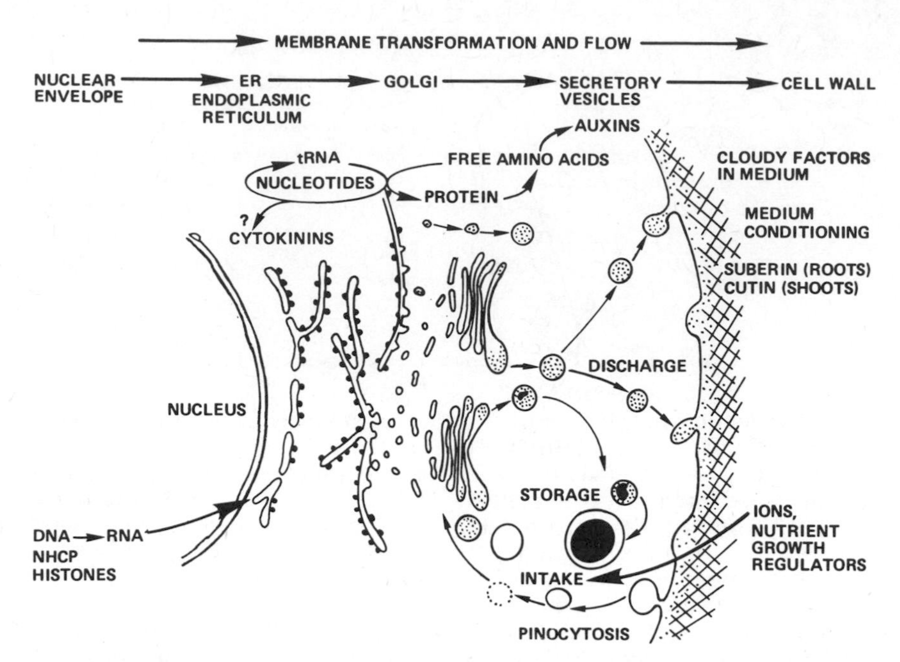

Fig. 6. Membrane transformation and flow in embryonal cells modified from Mühlethaler (27). In our model, the neocytoplasm originates from the nucleoplasm and the information coding for the acetocarmine-reactive materials is postulated to pass by membrane transformation and flow into the cytoplasm of an embryonal cell. Proteins and their conjugates are released inside the cell and outside the cell as diffusible factors. These contribute to the formation of mucilaginous materials which characterize ESMs and condition the culture medium. These materials have been called cloudy factors in the earlier literature. Auxins and cytokinins (endogenous and added to the medium) interact somehow with protein and nucleic acid metabolism to facilitate membrane transformation and flow characteristic of the embryonic state. Nonhistone chromosomal proteins (NHCPs) are believed to be components of the nucleoplasms that somehow generate the acetocarmine-reactive neocytoplasm. The addition of an amino acid supplement and high levels of myoinositol facilitates the culture of embryonal cells and contributes substrates to drive the postulated reactions and flows.

embryos from suspension cultures of the ESM. The propensity of the ESM for continual polyembryogenesis appears to be a function of ABA activity. Ultimately, this phenomenon may somehow relate to gene-controlled sequential over-production of a cleavage-type mitotic promoter or inhibitor in the neocytoplasm. That is, ABA assists in inhibiting the cleavage process. In Douglas fir seeds, ABA levels could be postulated as being naturally high (14). When cells are cultured and the ABA associated with the ESM is diluted, this could reestablish the propensity for polyembryony. Polyembryony can become aberrant particularly if rates of transfer and conditions are not maintained. The result is the appearance of incompletely separated embryos, organogenesis, and newly emergent senescent subsets of cells (e.g., Picea abies) (4).

The recovery of aberrant cells from the ESM may, in some cases, be worthwhile, especially if they represent transformations or mutations or even products of relic nuclei that are sometimes associated with fertilization (cf. Ref. 37). Recovery of aberrant or relic cells requires cultures of the entire ESM on a basal medium and the removal of normal cells in the proliferating ESM by screening. Aberrant and relic types may be recovered as small clusters of delayed or inhibited cells that pass through the screen. This population of cells could then be evaluated on enriched basal media according to the method of Woodward et al. (45) to develop mutant or relic phenoclonal variants.

The recovery of one embryo per seed has raised notions of lethality among polyembryonic masses in the reproductive process. Our results indicate that inhibitory conditions in the seed may account for some of this behavior. A lot more work needs to be done because of the various types of lethality that could emerge (cf. Ref. 34). Sorting out lethality represents a major task. In simplest terms, it points to the need to distinguish between mutational and inhibitional events associated with metabolic phenotypes presented by the cell population dynamics derived from the ESM.

Metabolic Phenotypes

Because of the limited progress and recalcitrance associated with the genetics of conifers, we have placed more emphasis on establishing useful phenotypic baseline information. This should contribute ultimately to the search for algorithms describing the expression of elite traits. Knowledge of the origin of physiological states and the performance of metabolic networks may someday improve our knowledge of gene expression and the interaction of the gene with the environment and with the maturity state. These phenotypes are also useful for understanding the precise environmental adaptation of our cloned germplasm.

A major barrier to progress is the complexity of phenotypes. Complexity relates to pleiotropy (i.e., one gene affects several traits) and polymery (i.e., several genes affect one trait). While we cannot yet fully sort out the "precedence" of any phenotype derived from the neocytoplasm, we still lack suitable methods of representing molecular phenotypes associated with growth and development (Fig. 5). Several attempts have been made to represent the flux of amino acids in somatic and zygotic embryogenesis (9), to account for the activity of PGRs (11) and to understand phase-directed totipotency in the life cycle (10).

Embryogenic and Nonembryogenic Models

Soluble N pool. Unfortunately, for conifers, we do not have the resources to obtain as complete a picture for the soluble N pool as with some fruit tree species (9,11,17). However, the results in Tab. 1 and 2 for Norway spruce are sufficiently detailed to indicate how soluble nitrogen may be networked in embryonal tissues and what amino acid families dominate seed development.

Increased levels of 2,4-D reduce total soluble N in cells presumably with an increase in embryonal protein associated with neocytoplasm (Fig. 7). Soluble N is dominated by alanine (embryogenic tissue) and arginine, dicarboxylic acids, and amides (nonembryogenic tissue) (Tab. 1). As somatic embryos develop (Tab. 2), the total soluble N accumulates and is dominated by arginine in advance of building up the embryonic proteinaceous reserves. This build-up is related to a decrease in alanine N.

Tab. 1. Effect of level of 2,4-dichlorophenoxyacetic acid (ppm) on percent distribution of free amino acid N in the soluble N of embryonal-suspensor masses (ESMs) of Picea abies on day 9 of a ten-day subculture schedule.

Free amino acid	Level of 2,4-D (ppm)				
	0	0.5*	5.0	50	NON*
Aspartic acid	2.1	1.7	1.5	5.3	1.0
Threonine	3.3	3.4	7.1	4.5	0.4
Serine	4.7	5.8	6.3	7.2	5.7
Asparagine	15.3	12.3	1.1	2.2	15.7
Glutamic acid	4.0	4.6	5.6	8.7	15.1
Glutamine	19.1	7.7	1.4	1.4	2.0
Proline	3.5	6.1	3.0	6.1	8.9
Glycine	2.9	2.7	6.9	4.1	0.6
Alanine	21.8	35.9	23.8	12.7	17.9
Citrulline	--	t	t	t	t
Valine	3.5	4.0	4.9	4.6	3.1
Cysteine	--	0.4	0.2	--	0.1
Methionine	2.3	1.3	4.0	8.3	--
Cystathione	3.0	1.8	1.9	7.7	--
Isoleucine	2.9	2.1	3.7	4.4	1.4
Leucine	1.8	1.7	3.7	6.9	1.3
Tyrosine	0.4	0.5	1.4	0.6	0.5
β-Alanine	0.4	0.7	0.9	t	0.8
Phenylalanine	0.9	0.6	0.7	1.2	0.2
γ-Aminobutyrate	4.9	4.4	5.3	1.2	8.2
Ethanolamine	2.4	1.3	t	t	1.5
Ornithine	--	0.2	0.4	0.7	--
Lysine	0.2	0.6	11.6	6.4	0.9
Histidine	t	t	3.4	2.3	--
Arginine	t	t	t	3.5	14.3
Total μg soluble N (g/FW)	49.1	52.1	19.4	34.2	373.4

*Single sample only.

Note: NON is a nonembryogenic callus on the 50 ppm 2,4-D medium. Nonembryogenic calli were derived from embryos that yielded the ESMs. Data are based on the mean of duplicate samples unless noted otherwise (D.J. Durzan, P.K. Gupta, and P. Krogstrup, unpubl. results). FW, Fresh weight.

From these preliminary results, we postulate that transaminases involving alanine and glutamate are pivitol in developing metabolic phenotypes in embryonal cells. For conifers, our goal is to construct a state-network map for gene expression in families of amino acids found in the neocytoplasm of embryonal and suspensor cells (i.e., related to alternative discrete outcomes).

Soluble proteins. For Douglas fir, the soluble protein from embryogenic cells in the ESM and from nonembryogenic calli are shown in Fig. 7. The separation of proteins on a 12% polyacrylamide gel reveals low molecular weight proteins that are dominant in embryonal cells and some that are absent in callus. More work needs to be done to isolate and purify proteins unique to SE and SPE, prepare antibodies to these, use antibodies to recover mRNA by precipitation of polysomes, and then copy the information

Tab. 2. Percent distribution of free amino acid N in the total soluble N of somatic embryos of Picea abies at different stages of development.

Free amino acid	Stages of somatic embryo development						
	ESM	PRO*	GLO	EM-1	EM-2	ZYG	ABER*
Aspartic acid	2.5	2.0	2.3	2.3	1.8	1.1	4.7
Threonine	4.9	4.1	5.6	2.8	1.7	1.5	4.2
Serine	5.5	7.2	11.6	7.3	6.8	3.5	18.9
Asparagine	2.1	5.2	9.6	8.2	11.6	2.0	15.5
Glutamic acid	3.6	4.0	7.3	7.4	6.3	3.4	5.7
Glutamine	0.3	0.2	1.1	0.7	0.3	0.1	0.3
Proline	t	3.2	5.3	9.9	4.5	1.7	4.0
Glycine	9.0	4.6	7.4	2.3	2.4	2.4	11.0
Alanine	33.3	53.0	40.8	30.2	9.6	3.4	16.2
Citrulline	--	--	t	t	--	--	--
Valine	5.7	3.3	4.1	2.3	1.3	1.9	3.1
Cysteine	--	--	--	--	--	--	--
Methionine	t	t	t	t	t	t	t
Cystathione	--	--	--	--	--	--	--
Isoleucine	t	0.8	t	t	1.1	1.4	1.9
Leucine	9.7	1.4	2.0	1.5	1.2	2.1	t
Tyrosine	1.6	0.3	0.6	0.4	0.4	0.7	0.9
β-Alanine	--	--	1.0	0.8	--	--	--
Phenylalanine	1.9	0.5	0.9	0.5	0.2	t	1.2
γ-Aminobutyrate	3.0	t	t	7.0	2.6	t	3.2
Ethanolamine	2.3	5.4	t	1.8	1.6	7.2	1.4
Ornithine	--	--	0.5	0.4	0.6	0.5	3.0
Lysine	8.9	2.9	t	1.5	1.2	2.1	1.9
Histidine	5.3	0.9	t	t	1.4	3.5	3.2
Arginine	t	t	t	12.7	43.6	61.3	--
Total μg soluble N (g/FW)	60.2	162.5	87.9	117.6	572.3	363.6	145.5

*Single sample only.

Note: PRO, proembryos of ESM; GLO, globular embryos in ESM; EM-1, embryo with cotyledons in light for two days; EM-2, embryo with cotyledons in light for two weeks; ZYG, zygotic embryo excised from ripe seed; ABER, aberrant development of globular somatic embryos due to premature exposure to white light. Data are based on the mean of duplicate samples unless noted otherwise (D.J. Durzan, P.K. Gupta, and P. Krogstrup, unpubl. results). FW, Fresh weight.

in mRNAs to produce cDNA probes. Unique proteins isolated in this way will help us to establish precedence in gene expression during SE and SPE as depicted by state-network maps (Fig. 8).

Physiological States and Metabolic Phenotypes

Every important analytical aspect of embryonic development leading to discrete metabolic phenotypes seems to arise somewhere in the context of sorting and searching. To get some insight into the utility of these processes, we hope to compare our state-network maps (e.g., Fig. 8) with one another (e.g., somatic vs zygotic, normal vs aberrant). A second approach is to use the numerical matrices representing maps to define and identify branching processes (bifurcations) in metabolic networks and physiological states. A third alternative is to use "hashing" or "scatter

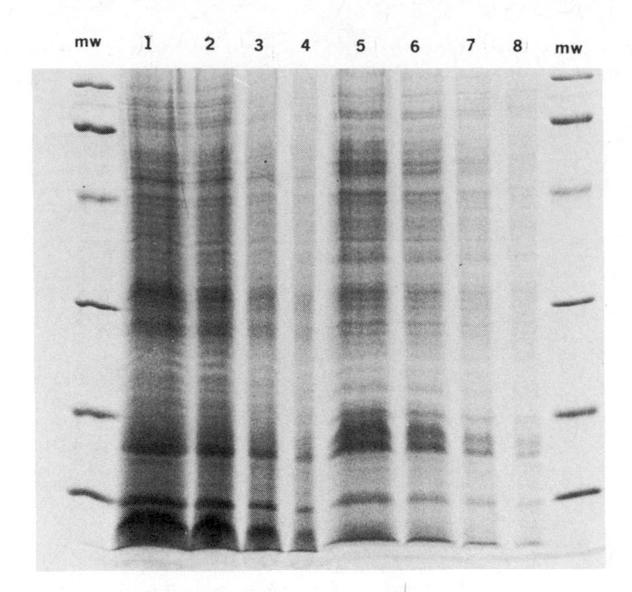

Fig. 7. Separation of soluble proteins (pH 8.3) on 12% polyacrylamide gel from (A) ESMs (lanes 1 to 4) and (B) nonembryogenic callus (lanes 5 to 8). Concentration of protein is reduced by 50% as lanes progress left to right. Molecular weights (MW) are in daltons as follows (top to bottom): 92.5, 66.2, 45.0, 31.0, 21.5, and 4.4. Note occurrence of enriched protein bands from ESMs (lanes 1 to 4). These bands are believed to represent in part some of the acetocarmine-reactive factors in the neocytoplasm. Work is in progress to separate these proteins by two-dimensional gel electrophoresis using different gel cross-link percentages (F. Ventimiglia and D.J. Durzan, unpubl. results).

and storage" techniques for analytical purposes. This involves computing mathematic functions of the metabolic fluxes, e.g., some metabolite or PGR-directed forcing function, and inserting these values as the address or map location where the search and pattern-match begins. This enormous task remains incomplete because of the need for voluminous data. Gaps in knowledge also limit our ability to construct meaningful maps and to assess and track developmental events.

SOME APPLICATIONS OF METABOLIC PHENOTYPES

Somatic Cell Genetics

Now that the origin of cells in the somatic and zygotic embryo can be traced to properties of the nucleus and neocytoplasm, new hypotheses relating to somatic cell variations (namely, recombination, mutation, segregation of nonchromosomal factors, and infectious agents) and their corresponding metabolic phenotypes can be explored in developmental terms. One approach involves the recovery of somatic embryos of conifers from morphogenic protoplasts that have been fused with protoplasts and organelles having useful markers or electroporated with foreign DNA (e.g., Ref. 19).

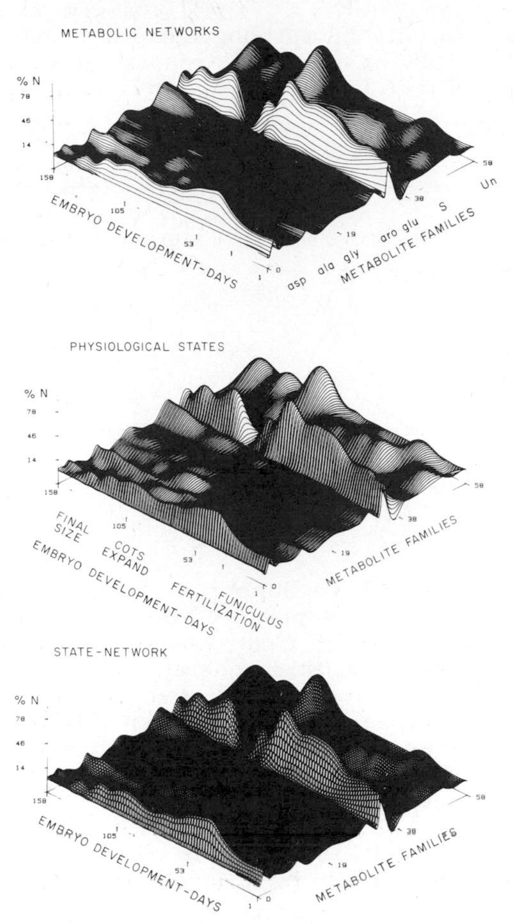

Fig. 8. Phenotype for amino acid metabolism in zygotic embryos (Pistacia vera) from fertilization to filling of the nut at 158 days (17). (A) Metabolites arranged arbitrarily into amino acid families, e.g., aspartate (lanes 1-5), alanine (10-12), glycine (17-19), aromatic amino acids (24-27), glutamate (32-35), methionine and sulfur amino acids (40-43), and total unidentified ninhydrin-reactive compounds (Un, lane 58). Amino acids in families arranged, wherever possible and appropriate, in the sequence of their biosynthesis. (B) Each day of embryonic development is integrated over composition to describe the physiological state of the zygotic embryo. Initially, the funiculus and ovule develop (1-15 days), followed by fertilization (days 15-18) and globular and heart-shaped stages (75 days) of embryonic development. Final size is reached at ∿112 days. (C) Combination of (A) and (B) to provide a combined state-network map. Through computer-assisted technology, the map can be rotated and viewed at various degrees of magnification to reveal information hidden by peaks. This inspection further reveals compounds contributing to a very low percentage of total soluble N. Also, the map can be compared with mutant expressions or aberrations to provide, by subtraction, patterns revealing differences between phenotypes. Maps of this type are constructed from data of the type found in Tab. 1 and 2 but taken at many more sampling times. From Ref. 17, with permission of A.R. Liss, Inc.

Tab. 3. Luciferase activity in morphogenic protoplasts of Douglas fir and loblolly pine 36 hr after electroporation (after P.K. Gupta, A. Dandekar, and D.J. Durzan, unpubl. results).

	Light units* per 10^6 protoplasts			
	Douglas fir		Loblolly pine	
	Without PEG**	With PEG	Without PEG	With PEG
No DNA	0	0	0	0
With DNA (luc)	52	105	103	196

*Based on light flashes produced by 1 pg of firefly tail enzyme extract in 1 min.
**Polyethylene glycol (3% w/v).

In unpublished work (P.K. Gupta, A. Dandekar, and D.J. Durzan), we have taken morphogenic protoplasts of Douglas fir and loblolly pine and introduced the luc genes (31) by electroporation. Forty-eight hours later, the recovered protoplasts start to regenerate cell walls, and when luciferin is added, protoplasts that have incorporated the luc genes can be identified by their bioluminescence (Tab. 3). Given this transient expression of a foreign gene, more work is needed to use luc as a reporter gene to evaluate the ability of stably transformed cells to produce daughter cells with metabolic phenotypes having the luc gene.

We hope someday to initiate studies to recover lost "phenes" and their genes from the extraordinarily high levels of DNA in the gymnosperm genome.

ACKNOWLEDGEMENT

The author thanks Prof. E.M. Gifford, Botany Department, University of California, Davis, for discussions relating to conifer embryology.

REFERENCES

1. Albersheim, P., and A.G. Darvill (1985) Oligosaccharins. Am. Scientist 253:58-64.
2. Allen, G.S., and J.N. Owens (1972) The Life History of Douglas-fir, Environment Canada, Forestry Service, Ottawa, 139 pp.
3. Bidwell, R.G.S., and D.J. Durzan (1975) Some recent aspects of nitrogen metabolism. In Historical and Current Aspects of Plant Physiology. A Symposium Honoring F.C. Steward, P. Davies, ed. New York State College of Agricultural Science, Cornell University, Ithaca, New York, pp. 162-227.
4. Boulay, M.H., P.K. Gupta, and D.J. Durzan (1988) Conversion of somatic embryos from cell suspension cultures of Norway spruce (Picea abies Karst.). Plant Cell Reports (submitted for publication).

5. Camefort, H. (1969) Fécondation et proembryogénèse chez les Abiétacées (notion de neocytoplasme). Rev. Cytol. Biol. Veg. 32:253-271.
6. Conn, H.J. (1961) Biological Stains, 7th ed., Williams and Williams Company, Philadelphia.
7. Dogra, P.D. (1967) Seed sterility and disturbances in embryony in conifers with particular reference to seed testing and breeding in Pinaceae. Stud. For. Suecica No. 45, pp. 1-97.
8. Durzan, D.J. (1982) Nitrogen metabolism and vegetative propagation of forest trees. In Tissue Culture in Forestry, J.M. Bonga and D.J. Durzan, eds. Martinus Nijhoff/Dr. W. Junk Publishers, The Hague, pp. 256-324.
9. Durzan, D.J. (1987) Physiological states and metabolic phenotypes in embryonic development. In Cell and Tissue Culture in Forestry, Vol. 2, J.M. Bonga and D.J. Durzan, eds. Martinus Nijhoff/Dr. W. Junk Publishers, Dordrecht, The Netherlands, pp. 405-439.
10. Durzan, D.J. (1987) Plant growth regulator-directed phase specificity in cell and tissue culture for tree improvement. In Proceedings of the Society American Foresters National Convention, Birmingham, Alabama, October 5-8, 1986, pp. 218-222.
11. Durzan, D.J. (1987) Plant growth regulators in cell and tissue culture of woody perennials. In Hormonal Control of Tree Growth, S.V. Kossuth and S. Ross, eds. Martinus Nijhoff Publishers, Dordrecht, The Netherlands (in press).
12. Durzan, D.J., and G. Bourgon (1976) Growth and metabolism of cells and tissue of jack pine. I. Observations on cytoplasmic streaming and effects of L-glutamine and its analogues on subcellular activities. Can. J. Bot. 54:507-517.
13. Durzan, D.J., and P.K. Gupta (1987) Somatic embryogenesis and polyembryogenesis in conifers. Adv. Biotech. Proc. Vol. 9 (in press).
14. Durzan, D.J., and P.K. Gupta (1987) Somatic embryogenesis and polyembryogenesis in Douglas-fir cell suspension cultures. Plant Sci. (in press).
15. Durzan, D.J., and F.C. Steward (1970) Morphogenesis in cell cultures of gymnosperms: Some growth patterns. (International Union Forest Research Organizations, Section 22, Workshop, May 28-June 5, Helsinki, Finland, 20 pp. plus 8 plates.) Abstr. Inst. For. Fenn. 74(6):16.
16. Durzan, D.J., and F.C. Steward (1983) Nitrogen metabolism. In Plant Physiology A Treatise, Vol. 8, F.C. Steward, ed. Academic Press, Inc., New York, pp. 55-265.
17. Durzan, D.J., and K. Uriu (1986) Metabolic networks in developing pistachio embryos (Pistachia vera cv. Kerman). In Progress in Developmental Biology, Part A, Harold C. Slavkin, ed. Alan R. Liss, Inc., New York, pp. 199-202.
18. Falk, R.H. (1972) Scanning electron microscope induced cathodeluminescence. Proceedings of the 30th Annual Meeting of the Electron Microscopy Society of America, Los Angeles, California, D.J. Arceneaux, ed.
19. Fromm, M., L.P. Taylor, and V. Walbot (1986) Stable transformation of maize after gene transfer by electroporation. Nature 319:791-793.
20. Gupta, P.K., and D.J. Durzan (1986) Plantlet regeneration via somatic embryogenesis from subcultured callus of mature embryos of Picea abies (Norway spruce). In Vitro Cell. Develop. Biol. 11:685-688.
21. Gupta, P.K., and D.J. Durzan (1986) Somatic polyembryogenesis from callus of mature sugar pine embryos. Bio/Technology 4:643-645.

22. Gupta, P.K., and D.J. Durzan (1987) Biotechnology of somatic embryogenesis in loblolly pine. Bio/Technology 5:147-151.
23. Gupta, P.K., and D.J. Durzan (1987) Somatic embryos from protoplasts of loblolly pine proembryonal cells. Bio/Technology 5:710-712.
24. Halberg, J., E. Halberg, F. Halberg, and L.C. Olson (1973) Internal circadian acrophases for plant physiologists in comparative biochemistry or photoperiodism. Intl. J. Chronobiol. 1:81-90.
25. Kester, D.E. (1983) The clone in horticulture. HortScience 18:831-837.
26. Krogstrup, P. (1984) Micropropagation of conifers. Ph.D. Thesis. Royal Veterinary and Agricultural University, Department of Horticulture, Copenhagen, July, 1984.
27. Mühlethaler, K. (1975) The ultrastructure of cells. In Historical and Current Aspects of Plant Physiology, P.J. Davies, ed. New York State College of Agriculture and Life Sciences, Cornell University, Ithaca, New York, pp. 226-242.
28. Nagmani, R., and J.M. Bonga (1986) Embryogenesis in subcultured callus of Larix decidua. Can. J. For. Res. 15:1088-1091.
29. Norstog, K. (1982) Experimental embryology of gymnosperms. In Experimental Embryology of Vascular Plants, B.M. Johri, ed. Springer-Verlag, Berlin, New York, pp. 25-48.
30. Ong, B.Y., R.H. Falk, and D.E. Bayer (1973) Scanning electron microscope observations of herbicide dispersal using cathode luminescence as the detector mode. Plant Physiol. 51:415-420.
31. Ow, D.W., K.V. Wood, M. DeLuca, J.R. DeWet, D.R. Helinski, and S.H. Howell (1986) Transient and stable expression of the firefly luciferase gene in plant cells and transgenic plants. Science 234:856-859.
32. Platt, J.R. (1961) Properties of large molecules that go beyond the properties of their chemical sub-groups. J. Theor. Biol. 1:342-358.
33. Powledge, T.M. (1984) Biotechnology touches the forest. Bio/Technology 2:763-772.
34. Reiger, R., D. Michaels, and M. Green (1976) Glossary of Genetics and Cytogenetics, Springer-Verlag, New York.
35. Romberger, J.A., and C.A. Tabor (1975) The Picea abies shoot apical meristem in culture. II. Deposition of polysaccharides and lignin-like substances beneath cultures. Am. J. Bot. 62:660-671.
36. Satoh, S., H. Kamada, H. Harada, and T. Fujii (1986) Auxin-controlled glycoprotein release into the medium of embryogenic carrot cells. Plant Physiol. 81:931-933.
37. Singh, H. (1978) Embryology of gymnosperms. In Encyclopedia of Plant Physiology, W. Zimmermann, S. Carlquist, P. Ozenda, and H.D. Wulff, eds. Gebrüder, Borntraeger, Berlin, pp. 192-198.
38. Speigel-Roy, P., and A. Vardi (1984) Citrus. In Handbook of Plant Cell Culture. Vol. 3. Crop Species, P.V. Ammirato, D.A. Evans, W.R. Sharp, and Y. Yamada, eds. Macmillan Publishing Company, New York, pp. 355-372.
39. Steward, F.C., and D.J. Durzan (1965) Metabolism of nitrogenous compounds. In Plant Physiology, an Advanced Treatise, Vol. 4, Part A, F.C. Steward, ed. Academic Press, Inc., New York, pp. 379-686.
40. Theologis, A. (1986) Rapid gene regulation by auxin. Ann. Rev. Plant Physiol. 37:407-438.
41. Timmis, R., M.M. Abo-El-Nil, and R.W. Stonecypher (1986) Potential gain through tissue culture. In Cell and Tissue Culture In Forestry. Vol. 1. General Principles and Biotechnology, J.M. Bonga and D.J. Durzan, eds. Martinus Nijhoff Publishers, Dordrecht, The Netherlands, pp. 198-215.

42. Veen, A.H., and A. Lindenmayer (1977) Diffusion mechanism for phyllotaxis. Theoretical physico-chemical and computer study. Plant Physiol. 60:127-139.
43. Ventimiglia, F., and D.J. Durzan (1986) The determination of mono-substituted guanidines using a dedicated amino acid analyzer. Liquid Chromatography/Gas Chromatography 4:1121-1124.
44. Willemse, M.T.M. (1974) Megagametogenesis and formation of neocytoplasm in Pinus sylvestris L. In Fertilization in Higher Plants, H.F. Linskins, ed. North-Holland, Amsterdam, pp. 97-102.
45. Woodward, V.M., J.R. DeZeeuw, and A.M. Srb (1954) The separation and isolation of particular biochemical mutants of Neurospora by differential germination of conidia followed by filtration and selected plating. Proc. Natl. Acad. Sci., USA 40:192-200.
46. Yablokov, A.V. (1986) Phenetics: Evolution, Population, Trait, Columbia University Press, New York, 171 pp.

ALKALOID PRODUCTION FROM CINCHONA CELL AND ORGAN SYSTEMS

E. John Staba

Department of Medicinal Chemistry and Pharmacognosy
University of Minnesota
Minneapolis, Minnesota 55455

ABSTRACT

Alkaloids from members of the genus *Cinchona* are important because they are used for antimalarial and antiarrhythmic therapy, and as a beverage bitter. In vitro cultures of species such as *Cinchona ledgeriana* and *C. pubescens* are being studied in various laboratories for their production of indole alkaloids, quinoline alkaloids, and anthraquinones, and for use as a micropropagation system. The extent to which cell and organ systems express the quinoline alkaloids quinine and quinidine and how their production is affected by growth regulators, differentiation and morphology, and alkaloid intermediates will be discussed. Examples of alkaloid intermediates to be considered are tryptophan, secologanin, strictosidine (isovincoside), 10-methoxystrictosidine, vincoside, and 10-methoxyvincoside.

INTRODUCTION

Cinchona plants are the most important source of the alkaloids quinine and quinidine. Quinine is most often used as a schizonticide against the erythrocytic stage of the malaria parasite and as a beverage bitter. Quinidine is used as a cardiac depressant or antiarrhythmic agent. Two other major quinoline alkaloids are cinchonidine, the nonmethoxylated form of quinine, and cinchonine, the nonmethoxylated form of quinidine. Cinchonidine and cinchonine are both schizonticides and toxic convulsants (38). It is estimated that the 1980 world production of *Cinchona* bark was greater than 5,000 tons, and that the value of the *quinidine* produced in 1982 (500 tons) was 50 million dollars (52). In 1985, India planned to produce 50 tons of quinine and five tons of quinidine (10).

BOTANY

Sixty-five evergreen *Cinchona* shrubs and trees have been identified in the Rubiaceae family (38), and these plants may grow in Africa, India, Indonesia, South America, and elsewhere. *Cinchona ledgeriana* Moens, commonly called Yellow Bark or Ledger Bark, is the species most often propagated. Cinchona ledgeriana is a fast-growing tree, up to 20 feet tall, with a poor root stock. Cinchona pubescens Vahl (C. succirubra Pavon ex. Klotzsch) or Red Bark may be grown as the root stock for C. ledgeriana. Although cinchona hybrids have been made (C. ledgeriana x C. pubescens; C. robusta), their alkaloid content is low.

High-density planting (4,000 trees/ha) of cinchona is most often done with seeds (20). Seeds germinate in three to six weeks, but have a rapid loss of viability and produce progeny that are low in alkaloids and high in heterogeneity (19). Cinchona plants are easily cross-fertilized. Seeds from C. ledgeriana plants containing 12 to 19% alkaloids yielded plants with an alkaloid content considerably lower than that of the parent plant (6).

Preharvest alkaloid analysis and selection has been done; however, the selected plants often do not propagate and root well (20). Cuttings that have a high alkaloid content are more difficult to root. Propagation may be done by layering or by treating two- to three-year-old branch cuttings (gootees) with 2% indolebutyric acid (IBA) in lanolin for 24 hr (38) or for 30 to 75 days with indoleacetic acid (IAA), naphthaleneacetic acid (NAA), or IBA (5).

If cinchona plants do not receive shade, an acidic soil (pH 4.2 to 5.6), and moisture, the plants will become stunted and the concentration of alkaloids in the root will be low. Plants should not be water-logged, and shading favors formation of superficial rootlets. The contribution to the total alkaloids by the root is greater at lower altitudes, since the amount of stem bark is smaller at lower than at higher altitudes (6). Plants are harvested from the fourth year to a maximum of about 10 to 12 years. Stumps may be dug out (grubbing) and the alkaloid-containing bark removed from roots. Coppicing may be used from 7 to 20 years (38).

CHEMISTRY

There are approximately 35 indole and quinoline alkaloids in the cinchona plant, and they are principally combined with quinic acid and cinchotannic acid (10). Indole alkaloids are precursors to quinoline alkaloids and are not of commercial importance; quinoline alkaloids of commercial importance are cinchonidine and its methoxylated derivative quinine, as well as cinchonine and its methoxylated derivative quinidine. The four asymmetric carbon atoms in cinchonine result in 16 optical isomers. Quinidine is the naturally occurring dextrorotary stereoisomer of quinine, and is with 8R,9S-stereochemistry. Quinine can be chemically isomerized to quinidine but at significant cost. The yield of quinidine produced synthetically from quinine varies from 10% to approximately 40%, although patents probably exist for more efficient processes (31). Various extraction (51), thin-layer chromatography (44), high-performance liquid chromatography (HPLC) (7,17), mass spectroscopic (22), and immunological (37) assay procedures have been developed for cinchona alkaloids and anthraquinones.

The major alkaloid of C. ledgeriana stem bark is quinine (typically 5 to 7%, but also higher than 14%) and that of C. pubescens stem bark is cinchonine (1 to 3%) (38). Cinchonine and cinchonidine may be present together in the bark in a concentration of 10% w/v (10). Quinidine may be as high as 3% in C. pubescens, but is typically about 0.2% and rarely above 1.0% (31).

Cinchona leaves contain about 1% total alkaloids. Young leaves, particularly the petioles, contain more alkaloids than old leaves. Leaves of _C. ledgeriana_ contain at least 12 monomeric or quasidimeric indole alkaloids (cinchophyllines). Indole and quinoline alkaloids may be present in _C. ledgeriana_ and _C. pubescens_ leaves but their content varies significantly for reasons that may be genetic or edaphic (17). The quinoline alkaloids are believed to be formed during the descent of the sap. Trace amounts of cinchonine and quinine are present in the heartwood of _C. ledgeriana_ (11). The alkaloid content is low in twigs but increases down the stem to a maximum in the root bark. In _C. ledgeriana_, the concentration of quinine in stem bark and root bark is 90% and 60%, respectively (38). Negligible amounts of alkaloids have been detected in _C. ledgeriana_ seeds (2).

A number of indole alkaloids are biosynthesized from a C^9-monoterpenoid and tryptophan (16). The early stage of cinchona alkaloid biosynthesis requires secologanin and tryptophan (21) to make strictosidine (isovincoside) (Fig. 1A and B). The middle stage utilizes strictosidine synthase (30,40) to form strictosidine. Strictosidine (30) and not vincoside (3) is the key intermediate to the indole alkaloids. Other enzymes and substrates form indole alkaloids such as quinamine, cinchonamine, and cinchophylline (Fig. 1C) (4,45). The late stage involves the transformation of the indole alkaloids into the cinchona quinoline alkaloids (Fig. 1D). The cinchona alkaloid biosynthetic pathway may compete with the cinchona anthraquinone biosynthetic pathway for the common intermediates mevalonic acid and chorismic acid (Fig. 1A) (35,45).

TISSUE CULTURE

Murashige and Skoog (MS) or Gamborg-B5 (B5) media with supplements or adsorbents to minimize browning are often used to establish cinchona tissue cultures for micropropagation or to study alkaloid metabolism (Tab. 1). Supplements such as cysteine (23), phloroglucinol (14,19,20), phlorizin (20), and polyvinylpyrrolidine (PVP) (2) are beneficial for growth and proliferation but are not essential. Root organ cultures have been established directly from seedlings (46) or from callus and shoot cultures (Fig. 2A, B, and C). Frequent medium replacement has proven beneficial for cinchona shoot growth (42). Cinchona plantlets should be transferred to a well-drained acidic soil and grown in a moist atmosphere. Activated charcoal and reduced inorganics do not favor root formation (19). Adventive embryoids from cell suspensions should be possible but have not yet been studied (20).

Cinchona callus and cell suspensions have been established to study alkaloid and anthraquinone production and biosynthesis (Tab. 2). Both MS and B5 media result in dense, highly aggregated cells or fine cell suspensions that grow slowly and may run down (Fig. 1D and E) (9,15, 42). The large, brown cell clumps of cinchona are physiologically active with respect to growth and membrane integrity and do not contain necrotic centers (29,34). Cell suspensions are often selected from the supernatant liquid about callus or organ cultures growing in liquid medium with appropriate growth regulators (2,9,15,18,32,40). Maintenance of the cell suspension is probably best in a medium containing an auxin [2,4-dichlorophenoxyacetic acid (2,4-D), NAA, IAA, or IBA] and a cytokinin [benzyladenine (BA), kinetin, zeatin riboside, or zeatin] which is then grown in the light. The concentration of sucrose is typically 2 to 4%; however, a 2% glucose concentration has also been recommended (1).

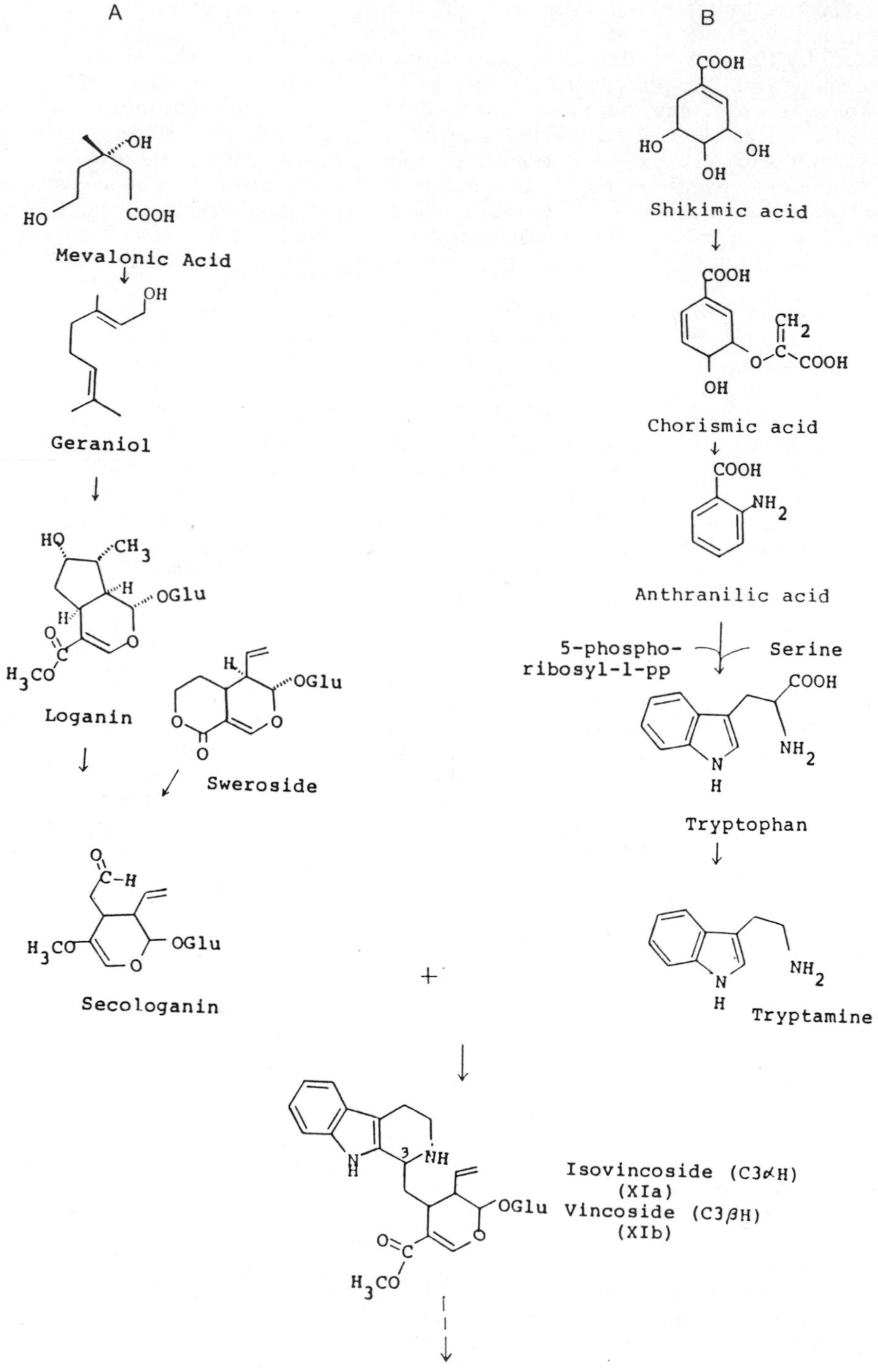

A

Mevalonic Acid

Geraniol

Loganin

Sweroside

Secologanin

B

Shikimic acid

Chorismic acid

Anthranilic acid

5-phospho-ribosyl-1-pp

Serine

Tryptophan

Tryptamine

Isovincoside (C3αH) (XIa)

Vincoside (C3βH) (XIb)

Fig. 1. Cinchona alkaloid biosynthetic pathways. (A, B) Early stages.

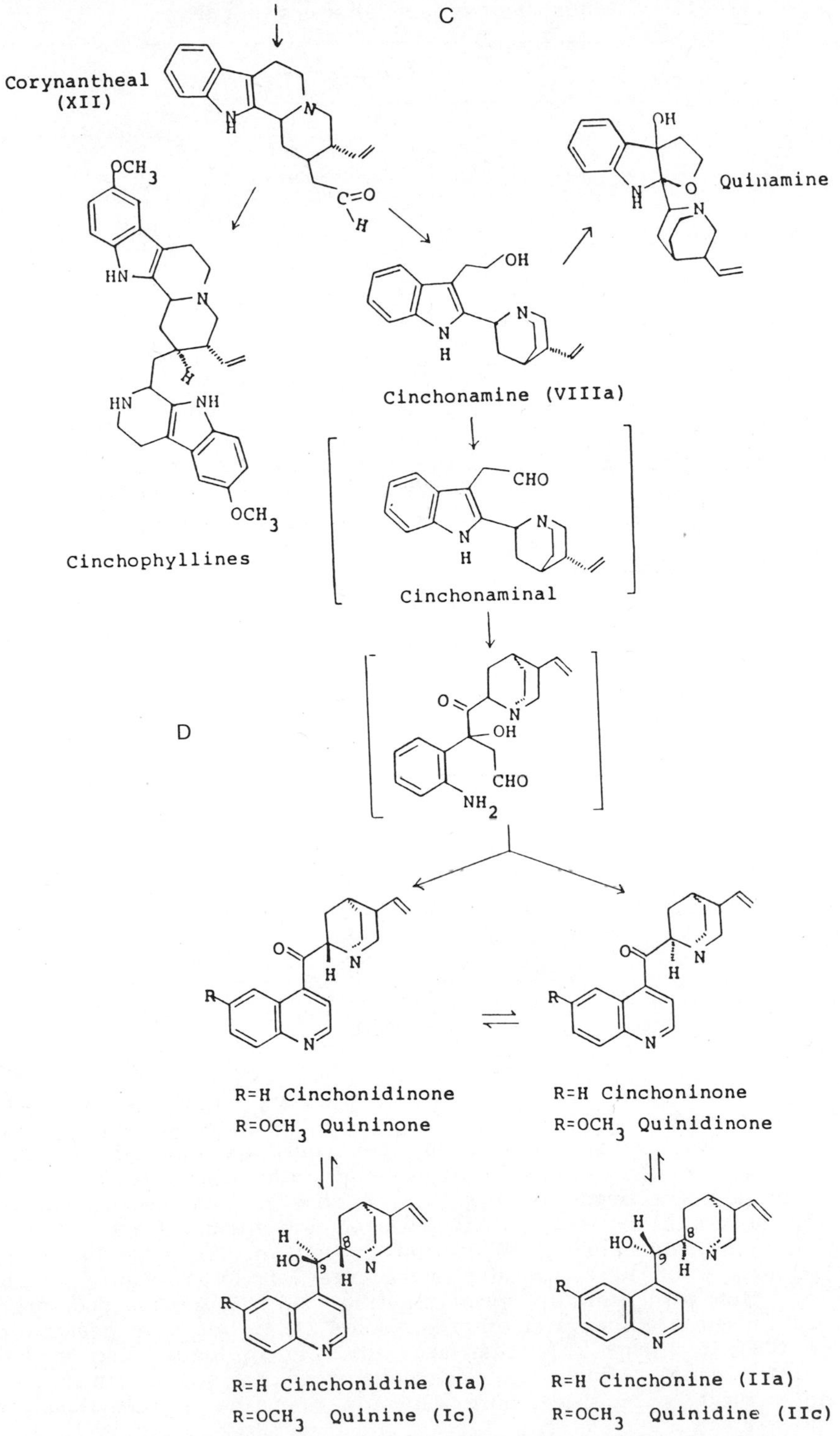

C

Corynantheal (XII)

Quinamine

Cinchophyllines

Cinchonamine (VIIIa)

Cinchonaminal

D

R=H Cinchonidinone
R=OCH_3 Quininone

R=H Cinchoninone
R=OCH_3 Quinidinone

R=H Cinchonidine (Ia)
R=OCH_3 Quinine (Ic)

R=H Cinchonine (IIa)
R=OCH_3 Quinidine (IIc)

Fig. 1 (continued). (C) Middle stage. (D) Late stage.

Tab. 1. Establishment of cinchona organ cultures.

Explant*	Medium additives**	Reference
1979: Hunter, UK		
CL seedling shoot tips	Rooting: phloroglucinol, 162 ppm; giberellic acid, 1 ppm; IBA, 1 ppm.	14
1981: Dougall et al., USA		
CL seedlings	Roots: NAA, 2.0 ppm; casein hydrolysate, 0.1%.	46
1981: Krikorian et al., USA		
CL shoot and node tips	Shoots: N^6-BAP, 10 ppm; IBA, 1 ppm.	20
	Rooting: increase IBA to ∿10 ppm.	
1981, 1987: Staba et al., USA		
CL/CP seedlings	Shoots: BA, 5.0 ppm.	42
	Roots: IBA, 3.0 ppm, or NAA, 3.0 ppm.	9
1982, 1986: Phillipson et al., UK		
CL seedlings and cotyledons	Shoots and root suspension cultures: 2,4-D, 1.0 ppm; kinetin, 0.1 ppm; sucrose, 5.0%.	13
	Above with polyvinylpyrrolidone, 1.0%.	2
1983: Koblitz et al., DDR		
CL/CP shoot tip and cotyledons	Shoots: IAA, 1.0 µM/1; BA, 0.2 µM/1.	19
	Roots: IBA, 0.1 mM/1; arginine, ∿50 µM/1.	

*CL, Cinchona ledgeriana; CP, C. pubescens (C. succirubra).
**Murashige and Skoog's medium was used in all cases except Dougall et al. (46), who used Gamborg's B5 medium. BA, benzyladenine; BAP, benzylaminopurine; 2,4-D, 2,4-dichlorophenoxyacetic acid; IAA, indoleacetic acid; IBA, indolebutyric acid; NAA, naphthaleneacetic acid.

Cinchona stem bark five or more years of age often contains about 5,000 to 7,000 mg% (mg/100 g dry weight) of total alkaloids. Juvenile cinchona shoots (approximately 4 to 40 weeks old) are composed of leaves and petioles that contain about 85 to 450 mg% total alkaloids (9,34), with quinine and quinidine predominating (2,9) (Tab. 3). This amount is similar to the alkaloids present in one-year-old greenhouse-grown cinchona plants and is approximately 2.0% of that present in the stem bark (2,9, 42). Cinchona root bark contains more total alkaloids than the plant stem bark, often with quinidine predominating (38). Cinchona root organs contain quinine, quinidine, and other alkaloids (46), but in concentrations [0.1 mg% (42); 2.9 mg% (2)] that are significantly lower than in shoot organs. Interestingly, fine suspension cells from root organ cultures contained approximately 15 times more alkaloids than the organs themselves (2).

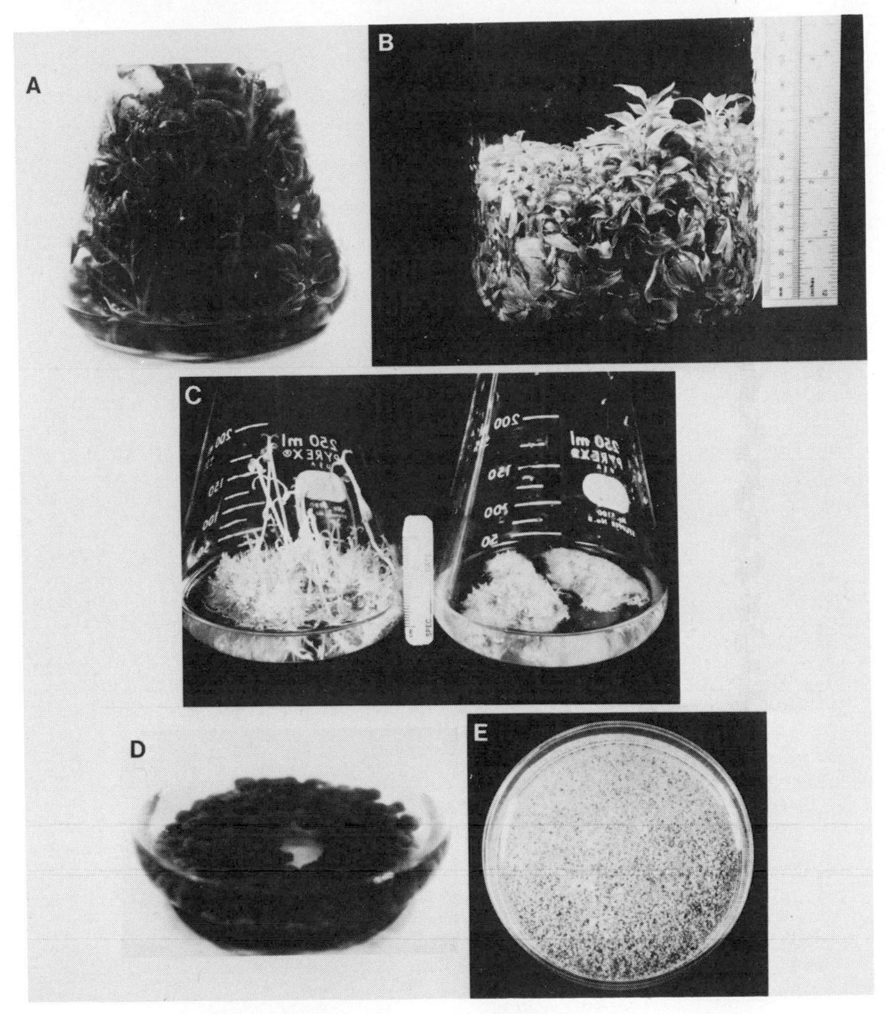

Fig. 2. *Cinchona ledgeriana* tissue cultures. (A) Shoot organ culture, six week olds, BA (5 ppm). (B) Shoot organ culture, 18 weeks old, BA (5 ppm). (C) Root organ culture from shoot organ culture. Left: first generation, NAA (3 ppm); right: second generation, four weeks old, NAA (3 ppm). (D) Large cell aggregate suspension culture in 500-ml flask, 2,4-D (0.1 ppm). (E) Small cell aggregate suspension culture, NAA (3.5 ppm)/BA (0.1 ppm).

It has been observed that high concentrations of NAA resulted in less fluorescent compounds in root organ cultures and that the alkaloids were absent from unorganized cells (46). Quinine production was enhanced in shoot organ cultures by BA (5 ppm) and/or shoot formation. Quinidine production was enhanced by IAA (5 ppm), the absence of BA, and/or root formation. High concentrations (25 to 50 ppm) of abscisic acid or mefluidide inhibited growth and alkaloid production (9).

Tab. 2. **Establishment of cinchona callus/cell suspension cultures.**

Explant*	Medium additives**	Reference(s)
1981, 1987: Staba et al., USA		
CL/CP seedlings	Suspensions grown at 80 rpm in 16-hr light cycle with weekly medium replacement.	
	2,4-D, 0.5 ppm; BA, 5 ppm (aggregated).	42
	NAA, 3.5 ppmp; BA, 0.1 ppm (friable).	9
1982: Hunter et al., UK		
CL internodal stems from shoot cultures.	2,4-D, 1.0 ppm; then hormone-free medium for 30 weeks. Or in 2,4-D, 1.0-2.0 ppm; BAP or K, 1.0 or 0.5 ppm; and coconut water (10%) for six weeks.	15
1982, 1986: Phillipson et al., UK		
CL root cultures	Supernatant from root organ cultures grown in liquid medium containing 2,4-D, 1 ppm; kinetin, 0.1 ppm; PVP, 1.0%; sucrose, 5.0%; and shaken at 100 rpm (fine cell suspension).	2
	Above medium without PVP.	13
1982, 1986: Verpoorte et al., The Netherlands		
CP stem segments and	Callus: zeatin, 1 µM; 2,4-D, 1 µM.	25, 26, 48
CL one-year-old plants	Callus: BA, 2 ppm; 2,4-D, 1 ppm.	23
CL seedling roots	Suspensions grown in B5 medium at 140 rpm	50
CL hypocotyls	and 12 hr light for ten days with half medium replaced every four days. 2,4-D, 1.0 ppm; K, 0.2 ppm.	12
1982: Wiryowidagdo et al., Indonesia		
CL shoot tip	Callus: Nitsch and Nitsch; NAA, 7.5 ppm; K, 0.5-1.0 ppm.	53
1983, 1985: Koblitz et al., DDR		
CL/CP leaf	Suspensions grown at 75 rpm in 12 hr light; subcultured every four weeks; medium supplemented every three to four days. 2,4-D, 4.5 µM/l; K, 1 µM/l; GA3, 0.3 µM/l.	18, 39
1986: Robins et al., UK		
CL internodes from shoot cultures	Suspensions grown in B5 medium at 100 rpm in 16 hr light; subcultured four weeks. Initially, 2,4-D, 2.0 ppm; BA, 0.5 ppm. Then 2,4-D, 0.5 ppm; BA, 0.1 ppm (green-brown aggregates)	32, 34, 35
1986: Scragg et al., UK		
CL seedlings	Callus: B5 medium with 2,4-D, 1.0 ppm; K, 0.1 ppm; coconut water (10%). Suspension: NAA, 1.0 ppm; BA; glucose, 2.0%; without coconut water; in light.	40
	Or 2,4-D, 1.0 ppm; K, 0.1 ppm; glucose, 2.0%; PVP, 1.0%; 150 rpm.	1

*CL, *Cinchona ledgeriana*; CP, *C. pubescens* (*C. succirubra*).
**Murashige and Skoog's medium was used in all cases except where noted. BA, benzyladenine; BAP, benzylaminopurine; 2,4-D, 2,4-dichlorophenoxyacetic acid; K, kinetin; NAA, naphthaleneacetic acid; PVP, polyvinylpyrrolidine.

Tab. 3. Alkaloid content of cinchona organ cultures.

Organ*	Total alkaloids/Comments	Reference
Dougall et al.		
CL roots	NR/Root cultures with quinine, quinidine, and dihydroquinidine.	46
Staba et al.		
CL roots	4.5 mg per 100 g dry weight/Quinidine higher than quinine.	
CL shoots	91 mg/Cinchonidine highest; quinine higher than quinidine.	
CP roots	0.1 mg.	42
CP shoots	80 mg/Cinchonidine highest.	9
	Tryptophan increased alkaloid yields of shoot cultures	8
Phillipson et al.		
CL roots	1.6-2.9 mg for quinine, quinidine, cinchonine, and cinchonidine per 100 g dry weight/Quinine and quinidine about equal.	
CL shoots	20.6-29.3 mg for quinine, quinidine, cinchonine, and cinchonidine/Quinine or cinchonidine the highest.	2
CL roots	Tryptophan increased alkaloid 5X but also inhibited growth.	13
Robins et al.		
CL leaves	12-82 μg quinine per 100 g fresh weight.	
CL stems	80-782 μg quinine per 100 g fresh weight.	37
CL leaves	4,200-6,300 μg total alkaloids per 100 g fresh weight.	
CL stems	7,900-51,400 μg total alkaloids per 100 g fresh weight.	34

*CL, *Cinchona ledgeriana*; CP, *C. pubescens* (*C. succirubra*).

NR, Not reported.
Nonmethoxylated: cinchonine; cinchonidine.
Methoxylated: quinidine, quinine.
CL, Stem bark contains principally 5-7% quinine, but may be higher than 14%.
CP, Stem bark contains principally cinchonine.
CL/CP, Root bark typically contains more alkaloids than stem bark (38).
CL Plant (one-year-old): Leaf-stem = 220 mg%
Root = 142 mg%
Mature stem bark = 7,100 mg% (Ref. 9)

Although the major cinchona alkaloids have been reported absent from unorganized cultures (42,46), most laboratories report that unorganized callus and suspension cells do produce quinoline alkaloids (Tab. 4). Trace amounts (15), 1 to 13 mg% yields (9,12,13,18,47), and alkaloid yields of about 50 mg% (2,40) are reported. Quinidine may be the dominant alkaloid in *C. ledgeriana* callus (40). In *C. ledgeriana* cell suspensions, the dominant alkaloid may be quinidine (2), quinine (1), or cinchonine (35,50). Alkaloid production is low during the growth phase and increases during the stationary phase (45).

The biomass yield of cinchona is reported to range from a high of 11.0 g/l in 14 days (39) to about 6.8 g/l in 14 days if grown in a 7-l airlift bioreactor (1). The biomass doubling time is 105 (50) to 115 hr (1).

Tab. 4. Alkaloid content of cinchona callus/cell suspension cultures.

Culture*	Total alkaloids/Comments	Reference(s)
Dougall et al.		
CL	Unorganized sections of tissue cultures with fluor compounds but not the alkaloids in root cultures.	46
Staba et al.		
CL/CP	No major alkaloids/Large aggregates.	42
CL	1.1 mg per 100 g dry weight/Fine aggregates; quinine, quinidine, cinchonine, and cinchonidine present.	9
Hunter et al.		
CL	9-17 mg per 100 g dry weight/Cell suspension, more than seven months old. Quinine highest, then cinchonine and cinchonidine. No quinidine. Tryptophan increased alkaloids ∿25%.	15
Phillipson et al.		
CL	22-43 mg per 100 g dry weight/Fine cell suspension, fourth generation, 14-day-old cultures. Quinidine or cinchonine highest. Very significant extra-cellular secretion of alkaloids.	2, 13
Verpoorte et al.		
CP	NR/Callus contains quinine, quinidine, cinchonine, and cinchonidine, which were isolated. Yields stimulated by light.	24, 25, 26
CL	1-13 mg per 100 g dry weight/Cell suspension. Tryptophan converted principally to β-carbolines.	12, 47
Wiryowidagdo et al.		
CL	NR/Callus with higher total nitrogen and alkaloids different than explant.	53
Koblitz et al.		
CP	10 mg per 100 g dry weight/Tryptophan increased alkaloids ∿100x, but growth ∿25% control, alkaloid into medium (1-6 μg/ml).	18
Robins et al.		
CL	2,000 μg total alkaloids per 100 g fresh weight/Up to 75% in medium, green-brown aggregates.	34
	14,000 μg total alkaloids per 100 g fresh weight.	32
Scragg et al.		
CL	53 mg quinidine per 100 g dry weight/Callus, quinine very low or absent.	40
Sejourne et al.		
CP	NR/Differences in alkaloid pattern found.	41

*CL, *Cinchona ledgeriana*; CP, *C. pubescens* (*C. succirubra*).

NR, Not reported.

The effect of various media on C. pubescens (24,25) and C. ledgeriana (40) callus induction and growth and C. ledgeriana cell suspension and growth (12,50) has been studied. High auxin concentrations and C/N ratios increased the production of the indole alkaloids cinchonamine and quinamine (12,50). Quinoline alkaloid production from cell suspensions is increased when low levels of auxins are combined with cytokinins (32,34) or when suspensions are grown in the dark with a low concentration of cytokinin and 4% sucrose (50). The cytokinin BA favored cinchonine production in cinchona cell suspensions and retarded the release of alkaloids into the medium (34). Significant amounts of cinchona alkaloids are released extracellularly (2,18,27), and this release is not suitably enhanced by dimethyl sulphoxide (28). Also studied has been the immobilization characteristics of cinchona dense cell aggregates that were releasing a mucilage onto polymeric surfaces (36).

It has been demonstrated that cinchona tissue cultures (49) and plants (52) produce anthraquinones in response to fungal pathogens which can be considered to be phytoalexins. The fungal pathogens tested did not stimulate alkaloid production (49,52). At least 12 to 15 anthraquinones are present in C. pubescens callus (24) and C. ledgeriana callus and cell suspensions (26,35). Approximately five or six new anthraquinones have been isolated from C. pubescens callus cultures (48) and C. ledgeriana cell suspensions (35). Anthraquinone production is increased and occurs earlier in the growth cycle if IAA and zeatin riboside are used (35). About 80% of the anthraquinone aglycones are released into the medium and inhibit culture growth. Sucrose (8%) (12,50) and polymeric adsorbents (33) stimulated C. ledgeriana cell suspensions to produce about 20 mg/l of anthraquinones. The production of anthraquinones does not appear to be correlated to either growth or alkaloid production (12,35).

Biosynthesis

The cinchona anthraquinone and alkaloid biosynthetic pathways are competing with each other for at least two common intermediates, chorismic acid and mevalonic acid (35,39,47).

Chorismic acid is a precursor both to tryptophan which is required for alkaloid biosynthesis, and to isochorismic acid which is required for anthraquinone biosynthesis. Chorismic acid is derived from shikimic acid. Mevalonic acid is required to form secologanin for alkaloid biosynthesis and to form ring C in the anthraquinone structure.

Enzymes considered to be critical in the anthraquinone and alkaloid pathways are tryptophan decarboxylase to form tryptamine (39), strictosidine synthase to form corynantheal (43,47), and NADPH oxidoreductase to reduce cinchoninone or quinidinone to cinchonine/cinchonidine or quinine/-quinidine, respectively (47).

Tryptophan (approximately 50 to 500 mg/l) has been shown to significantly stimulate cinchona alkaloid production in C. pubescens callus cultures (19), C. ledgeriana cell suspensions (15), C. ledgeriana root organ suspension cultures (13), and C. ledgeriana shoot cultures (8). The alkaloid increases reported were: C. pubescens callus cultures, 90X or 900 mg%; C. ledgeriana cell suspension, 34 mg% alkaloids; C. ledgeriana root organ suspension cultures, 5X; and C. ledgeriana shoot cultures, 66% increase or 143 mg%. Patents have been applied for by Koblitz et al. for the improved tissue culture production of cinchona alkaloids by feeding

L-tryptophan (18). Anthranilic acid (35) and tryptophan can inhibit cinchona cell and tissue growth. It has been suggested that tryptamine may be less toxic than tryptophan to culture systems (47). Tryptophan has been converted to β-carboline alkaloids, most probably as artifacts, by nonquinoline alkaloid-producing C. ledgeriana suspension cultures (47). Tryptophan may (13,18) or may not (35,47) stimulate anthraquinone production in cinchona tissue culture systems.

Approximately 0.25% of L-(methylene-^{14}C)-tryptophan was incorporated into the total alkaloids produced by C. ledgeriana root suspension cultures (13). The incorporation of L-[3'-^{14}C]tryptophan into C. ledgeriana shoot cultures was as follows: cinchonidine (0.04%), cinchonine (0.01%), quinidine (0.01%), and quinine (0.08%) (8). Plants of C. pubescens have incorporated 0.7 to 0.9% of labeled DL-tryptophan into quinine (21). Corynantheal (Fig. 1) did not result in quinoline alkaloid formation in C. ledgeriana suspension cultures (47); however, C. ledgeriana shoot cultures incorporated [5-^{14}C]strictosidine·HCl into cinchonidine (0.01%), cinchonine (0.003%), quinidine (0.003%), and quinine (0.005%), and [5-^{14}C]-10-methoxy-strictosidine·HCl equally (0.01%) into quinidine and quinine (8).

Cell suspensions of C. pubescens grown in air-lift fermenters containing 3.5 l medium were fed tryptophan and analyzed for the following enzymes: DAHP synthase, chorismate mutase, anthranilate synthase, glucose-6-phosphate dehydrogenase, and tryptophan decarboxylase. It was concluded that tryptophan decarboxylase enzyme is necessary to produce significant yields of cinchona alkaloids (39). Cell suspensions of C. ledgeriana that do not produce alkaloids are deficient in tryptophan decarboxylase (47).

An enzyme catalyzing the late stage of cinchona alkaloid biosynthesis, cinchoninone:NADPH oxidoreductase, has been characterized (47). The extent to which this enzyme and others control important hydroxylation and methoxylation reactions in cinchona plant tissue culture systems remains to be studied.

SUMMARY

Micropropagation systems have been developed to efficiently produce juvenile cinchona plantlets (14,19,20,42) that might prove to be more cost-effective than cinchona seed propagation. However, the micropropagation system is probably not as cost-effective as propagating cinchona plants by stem cuttings.

It remains to be determined if cinchona plant alkaloid races selected from either cell or organ cultures can be micropropagated without reintroducing chemical variability.

The cinchona plant species, or the tissue explant used, did not appear to affect tissue culture alkaloid patterns or quinine, quinidine, cinchonine, and cinchonidine concentrations.

Various chemical races of quinine, quinidine, cinchonine, and cinchonidine have been identified in both unorganized (40) and organized cinchona tissue cultures (37). Quinine production may be enhanced by conditions that enhance shoot formation and quinidine production, by conditions that enhance root formation (9).

Unorganized cinchona cultures generally produce less total alkaloid than root organ (2.9 mg%) (2) or shoot organ (91 mg%) (9) cultures. They also produce negligible (47) or significant (900 mg%) (47) amounts of alkaloids, both intracellularly and extracellularly.

A major enzyme for expressing high cinchona alkaloid production is tryptophan decarboxylase (39,47) and perhaps strictosidine synthase and NADPH oxidoreductase (47).

Unorganized and shoot cinchona cultures can incorporate labeled tryptophan (8,13) and labeled strictosidine•HCl into quinoline alkaloids (8).

Anthraquinone production does not seem to directly affect alkaloid production in cinchona tissue culture systems (35,47). Anthraquinones, and not alkaloids, appear to be phytoalexins in both cinchona plants and tissue cultures (49,52).

ACKNOWLEDGEMENTS

Support from NATO (grant number 599/83) for collaborative studies with R. Verpoorte and his staff, Center of Bio-Pharmaceutical Sciences, State University of Leiden, The Netherlands, is gratefully acknowledged. The author also acknowledges Dr. C.-T. Alex Chung for the technical data provided from his doctorate studies at the University of Minnesota, College of Pharmacy, Minneapolis.

REFERENCES

1. Allan, E.J., and A.H. Scragg (1986) Comparison of the growth of Cinchona ledgeriana Moens suspension cultures is shake flasks and 7 litre air-lift bioreactors. Biotechnol. Lett. 8:635-638.
2. Anderson, L.A., A.T. Keene, and J.D. Phillipson (1982) Alkaloid production by leaf organ, root organ and cell suspension cultures of Cinchona ledgeriana. Planta Med. 46:25-27.
3. Battersby, A.R., and R.J. Parry (1971) Biosynthesis of the cinchona alkaloids. Middle stages of the pathway. J. Chem. Soc. (Sect. D) Chem. Comm. 1:30-31.
4. Brillanceau, M.H., C. Kan-Fan, S.K. Kan, and H.-P. Husson (1984) Guettardine, a possible biogenetic intermediate in the formation of corynanthe-cinchona alkaloids. Tetrahedron Lett. 25:2767-2770.
5. Chatterjee, S.K. (1974) Vegetative propagation of high quinine yielding cinchona. Indian J. Hort. 31:174-177.
6. Chatterjee, S.K. (1982) Cultivation of cinchona in Darjeeling Hills, West Bengal. In Cultivation and Utilization of Medicinal Plants, Vol. II, C.K. Atal and B.M. Kapur, eds. CSIR, Jammu-Tawi, India, pp. 222-229.
7. Chung, C.-T.A., and E.J. Staba (1984) Separation and quantitation of cinchona major alkaloids by high-performance liquid chromatography. J. Chromatog. 295:276-281.
8. Chung, C.-T.A., and E.J. Staba (1987) Effect of precursors on growth and alkaloid production in Cinchona ledgeriana leafshoot organ cultures. Planta Med. (submitted for publication).
9. Chung, C.-T.A., and E.J. Staba (1987) Effects of age and growth regulators on growth and alkaloid production in Cinchona ledgeriana leaf-shoot organ cultures. Planta Med. 53:206-210.

10. Doraswamy, K., and K.P. Venkatratnam (1982) Chemistry, production and marketing of cinchona alkaloids. In Cultivation and Utilization of Medicinal Plants, Vol II, C.K. Atal and B.M. Kapur, eds. CSIR, Jammu-Tawi, India, pp. 230-243.
11. Dutta, N.L., and C. Quassim (1968) Isolation and characterization of glycosides and alkaloids from heartwood of Cinchona ledgeriana Linn. Indian J. Chem. 6:566-567.
12. Harkes, P.A.A., L. Krijbolder, K.R. Libbenga, R. Wijnsma, T. Nsengiyaremge, and R. Verpoorte (1985) Influence of various media constituents on the growth of Cinchona ledgeriana tissue cultures and the production of alkaloids and anthraquinones therein. Plant Cell Tissue Organ Culture 4:199-214.
13. Hay, C.A., L.A. Anderson, M.F. Roberts, and J.D. Phillipson (1986) In vitro cultures of cinchona species. Precursor feeding of C. ledgeriana root organ suspension cultures with L-tryptophan. Plant Cell Reports 5:1-4.
14. Hunter, C.S. (1979) In vitro culture of Cinchona ledgeriana L. J. Hort. Sci. 54:111-114.
15. Hunter, C.S., D.V. McCally, and A.J. Barraclough (1982) Alkaloids produced by cultures of Cinchona ledgeriana L. In Plant Tissue Culture. Proceedings Fifth International Congress of Plant Tissue Cultures, A. Fujiwara, ed. Tokyo, p. 317.
16. Hutchinson, C.R., A.H. Heckendorf, J.L. Straughn, P.E. Daddona, and D.E. Crane (1979) Biosynthesis of camptothecin. 3. Definition of strictosamide as the penultimate biosynthetic precursor. Assisted by ^{13}C and ^{2}H NMR spectroscopy. J. Am. Chem. Soc. 101:3358-3369.
17. Keene, A.T., L.A. Anderson, and J.D. Phillipson (1983) Investigation of cinchona leaf alkaloids by high-performance liquid chromatography. J. Chromatog. 260:123-128.
18. Koblitz, H., D. Koblitz, H.-P. Schmauder, and D. Groger (1983) Studies on tissue cultures of the genus Cinchona L. Alkaloid production in cell suspension cultures. Plant Cell Reports 2:122-125.
19. Koblitz, H., D. Koblitz, H.-P. Schmauder, and D. Groger (1983) Studies on tissue cultures of the genus Cinchona L. In vitro mass propagation through meristem-derived plants. Plant Cell Reports 2:95-97.
20. Krikorian, A.D., M. Singh, and C.E. Quinn (1981) Aseptic micropropagation of Cinchona: Prospects and problems. In Proceedings of the COSTED Symposium on Tissue Culture of Economically Important Plants, A. Rao, ed. National University of Singapore, pp. 167-174.
21. Leete, E., and J.N. Wemple (1966) Biosynthesis of the cinchona alkaloids. The incorporation of [3-^{14}C]-geraniol into quinine. J. Am. Chem. Soc. 88:4743.
22. Mellon, F.A., M.J.C. Rhodes, and R.J. Robins (1986) The potential of linked scanning at constant B/E for the rapid identification of cell culture products. Biomed. Env. Mass Spect. 13:155-158.
23. Mulder-Krieger, Th., R. Verpoorte, and A.B. Svendsen (1982) Tissue culture of Cinchona pubescens: Effects of media modifications on the growth. Planta Med. 44:237-240.
24. Mulder-Krieger, Th., R. Verpoorte, M. van der Kreek, and A.B. Svendsen (1984) Identification of alkaloids and anthraquinones in Cinchona pubescens callus cultures: The effect of plant growth regulators and light on the alkaloid content. Planta Med. 50:17-20.
25. Mulder-Krieger, Th., R. Verpoorte, Y.P. de Graaf, M. van der Kreek, and A. Baerheim-Svendsen (1982) The effects of plant growth regulators and culture conditions on the growth and the alkaloid content of callus cultures of Cinchona pubescens. Planta Med. 46:15-18.

26. Mulder-Krieger, Th., R. Verpoorte, A. de Water, M. van Gessel, B.C.J.A. van Oeveren, and A.B. Svendsen (1982) Identification of the alkaloids and anthraquinones in Cinchona ledgeriana callus cultures. Planta Med. 46:19-24.
27. Parr, A.J., R.J. Robins, and J.C. Rhodes (1986) Alkaloid transport in Cinchona spp. cell cultures. Physiol. Veg. 24:419-429.
28. Parr, A.J., R.J. Robins, and M.J.C. Rhodes (1986) Permeabilization of Cinchona ledgeriana cells by dimethyl-sulphoxide. Effects on alkaloid release and long-term membrane integrity. Plant Cell Reports 3:262-265.
29. Parr, A.J., J.I. Smith, R.J. Robins, and M.J.C. Rhodes (1984) Apparent free space and cell volume estimation: A non-destructive method for assessing the growth and membrane integrity/viability of immobilised plant cells. Plant Cell Reports 3:161-164.
30. Pfitzner, U., and M.H. Zenk (1982) Immobilization of strictosidine synthase from Catharanthus cell cultures and preparative synthesis of strictosidine. Planta Med. 46:10-14.
31. Rao, P.R. (1982) Conversion of quinine to quinidine. In Cultivation and Utilization of Medicinal Plants, Vol II., C.K. Atal and B.M. Kapur, eds. CSIR, Jammu-Tawi, India, pp. 244-250.
32. Rhodes, M.J.C., J. Payne, and R.J. Rhodes (1986) Cell suspension cultures of Cinchona ledgeriana. II. The effect of a range of auxins and cytokinins on the production of quinoline alkaloids. Planta Med. 52:226-229.
33. Robins, R.J., and M.J.C. Rhodes (1986) The stimulation of anthraquinone production by Cinchona ledgeriana cultures with polymeric adsorbents. Appl. Microbiol. Biotechnol. 24:35-41.
34. Robins, R.J., J. Payne, and M.J.C. Rhodes (1986) Cell suspension cultures of Cinchona ledgeriana. I. Growth and quinoline alkaloid production. Planta Med. 52:220-225.
35. Robins, R.J., J. Payne, and M.J.C. Rhodes (1986) The production of anthraquinones by cell suspension cultures of Cinchona ledgeriana. Phytochemistry 25:2327-2334.
36. Robins, R.J., D.O. Hall, D.-J. Shi, R.J. Turner, and M.J.C. Rhodes (1986) Mucilage acts to adhere cyanobacteria and cultures plant cells to biological and inert surfaces. FEMS Microbiol. Lett. 34:155-160.
37. Robins, R.J., A.J. Webb, M.J.C. Rhodes, J. Payne, and M.R.A. Morgan (1984) Radioimmunoassay for the quantitative determination of quinine in cultured plant tissues. Planta Med. 50:235-238.
38. Sastri, B.N. (1950) The Wealth of India, Vol. II, CSIR, Delhi, India, pp. 163-173.
39. Schmauder, H.-P., D. Groger, H. Koblitz, and D. Koblitz (1985) Shikimate pathway activity in shake and fermenter cultures of Cinchona succirubra. Plant Cell Reports 4:233-236.
40. Scragg, A.H., P. Morris, and E.J. Allan (1986) The effects of plant growth regulators on growth and alkaloid formation in Cinchona ledgeriana callus culture. J. Plant Physiol. 124:371-377.
41. Sejourne, M., G. Resplandy, C. Viel, J.C. Chenieux, and M. Rideau (1986) Bioproduction of quinoline alkaloids by Cinchona succirubra strains cultured in vitro. Fitoterapia 57:121-123 (cited in Chem. Abstr. 105:57949g).
42. Staba, E.J., and A.C. Chung (1981) Quinine and quinidine production by cinchona leaf, root and unorganized cultures. Phytochemistry 20:2495-2498.

43. Van der Heijden, R., I. Hegger, E.J.M. Pennings, R. Verpoorte, R. Wijnsma, and J.A. Duine (1986) An HPLC assay for strictosidine synthase activity. Annual Meeting of the American Society of Pharmacognosy, Ann Arbor, Michigan (presented poster).
44. Verpoorte, R., Th. Mulder-Krieger, J.J. Troost, and A.B. Svendsen (1980) Thin-layer chromatographic separation of cinchona alkaloids. J. Chromatog. 184:79-96.
45. Verpoorte, R., R. Wijnsma, Th. Mulder-Krieger, P.A.A. Harkes, and A.B. Svendsen (1985) Plant cell and tissue culture of Cinchona species. In Primary and Secondary Metabolism of Plant Cell Cultures, K.-H. Neumann, W. Barz, and E. Reinhard, eds. Springer-Verlag, New York, pp. 196-208.
46. Whitten, G.H., and D.K. Dougall (1981) Quinine and quinidine accumulation by root, callus and suspension cultures of C. ledgeriana. In Vitro 17:220.
47. Wijnsma, R. (1986) Anthraquinones and alkaloids in cell and tissue cultures of Cinchona species. R. Verpoorte, Copromotor, Doctorate Dissertation, State University of Leiden, Leiden, The Netherlands.
48. Wijnsma, R., J.T.K.A. Go, P.A.A. Harkes, R. Verpoorte, and A.B. Svendsen (1986) Anthraquinones in callus cultures of Cinchona pubescens. Phytochemistry 25:1123-1126.
49. Wijnsma, R., J.T.K.A. Go, I.N. van Weerden, P.A.A. Harkes, R. Verpoorte, and A.B. Svendsen (1985) Anthraquinones as phytoalexins in cell and tissue cultures of Cinchona species. Plant Cell Reports 4:241-244.
50. Wijnsma, R., R. Verpoorte, P.A.A. Harkes, T.B. van Vliet, H.J.G. ten Hoopen, and A.B. Svendsen (1986) The influence of initial sucrose and nitrate concentrations on the growth of Cinchona ledgeriana cell suspension cultures and the production of alkaloids and anthraquinones. Plant Cell Tissue Organ Culture 7:21-29.
51. Wijnsma, R., T.B. van Vliet, P.A.A. Harkes, H.J. van Groningen, R. van der Jeijden, R. Verpoorte, and A.B. Svendsen (1987) A method for the quantitative determination of anthraquinones and alkaloids in cell and tissue cultures of Cinchona sp. Planta Med. 53:78-84.
52. Wijnsma, R., I.N. van Weerden, R. Verpoorte, P.A.A. Harkes, Ch.B. Lugt, J.J.C. Scheffer, and A.B. Svendsen (1986) Anthraquinones in Cinchona ledgeriana bark infected with Phytophthora cinnamomi. Planta Med. 3:211-212.
53. Wiryowidagdo, S. (1982) Pembentukan alkaloid dalam kalus kina. Doctorate Dissertation, Institut Teknologi Bandung, Indonesia.

BIOSYNTHESIS OF LOWER TERPENOIDS: GENETIC AND PHYSIOLOGICAL CONTROLS IN WOODY PLANTS

C. Bernard-Dagan

Laboratoire de Physiologie Cellulaire Végétale
Unité associée au Centre National
de la Recherche Scientifique n° 568
Université de Bordeaux I
33405 Talence Cedex, France

ABSTRACT

Biosynthesis of lower terpenoids has been studied in maritime pine (Pinus pinaster) using three complementary approaches: (a) biochemistry, (b) electron microscopy, and (c) cell-free systems. Monoterpene hydrocarbon synthesis is performed by the transitory activity of a specific organelle--the leucoplast. Monogenic control of terpenes is expressed differently in primary and secondary structures. In needles, sesquiterpene production is continuous. The cyclization processes take place in endomembranes, presumably in the endoplasmic reticulum. The role of C_{10} and C_{15} acyclic compounds is demonstrated by in vivo and in vitro results. The existence of six major genes, each controlling the level of one terpene, has been demonstrated and some practical applications developed. The cell regulation of terpene biosynthesis is disturbed by various aggressions which induce either stimulation of quiescent specialized cells or delocalized synthesis.

INTRODUCTION

Lower terpenoids have been extensively studied in such oil-bearing plants as Mentha, Pelargonium, Tanacetum, Foeniculum, Salvia, Artemisia, and Rosa, in which the major isoprenoid compounds are oxygenated forms (e.g., mainly alcohols, aldehydes, esters, and glucosides). Croteau (26,27) has recently published comprehensive reviews of the biosynthesis and functions of these compounds in a large number of herbaceous aromatic species. But there are few data on the biosynthesis of lower terpenoids in evergreen woody plants, especially resinous species. In fact, the metabolism of resinous trees is not well documented although they are very important in problems of afforestation.

For more than 15 years at the University of Bordeaux (France), biosynthesis of oleoresin has been studied in Pinus pinaster, a tree which is widely spread over southwestern Europe. This review of the repartition, biosynthetic pathways, cell localization, and physiological controls of terpene elaboration in a woody plant is based upon cumulative biochemical data, ultrastructural observations, and subcellular fractionations.

NATURE AND REPARTITION OF TERPENES IN MARITIME PINE

Oleoresin is found throughout the entire tree. The main constituents of oleoresin are diterpene resin acids with volatile monoterpene and sesquiterpene hydrocarbons. Oxygenated compounds, monoterpenes, and, to a lesser extent, sesquiterpenes are also present, especially in the needles (Fig. 1) (56). As shown in Fig. 1, the degree of cyclization may be more or less complex, ranging from acyclic structures (e.g., myrcene, ocimene, geraniol, and farnesene) to bicyclic (e.g., α- and β-pinenes, caryophyllene, cadinenes, and anticopalic acid), or even tricyclic compounds (α-copaene, longifolene, and abietic and pimaric acids). Some of these compounds are found throughout the tree (e.g., α- and β-pinenes and abietic and pimaric acids, with minor amounts of caryophyllene, longifolene, and humulene), whereas others are found only within specific organs or tissues.

Sesquiterpenes occur mainly in needle oils. Their concentration increases from a level near 0 when these organs burst out of the buds to 40% at the end of their growing period (67). The concentration continues to increase during the aging of the organ. In contrast, the ratio of oxygenated compounds to hydrocarbons decreases during leaf development.

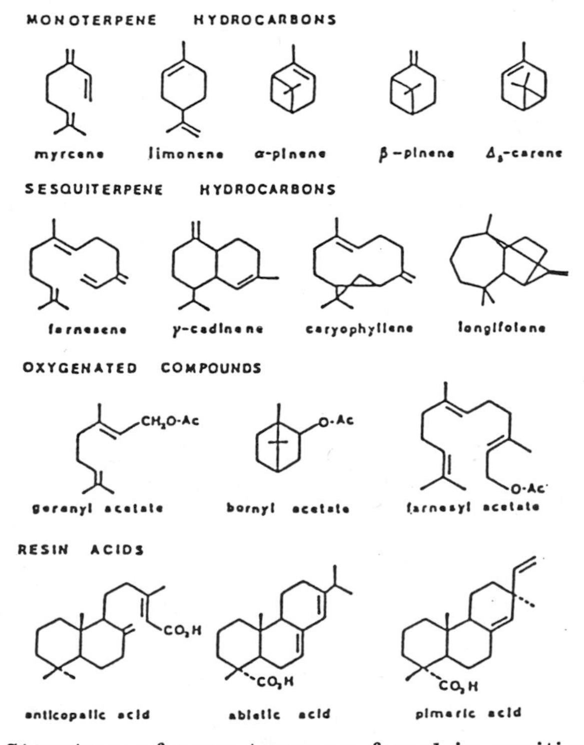

Fig. 1. Structure of some terpenes found in maritime pine.

This last fact may be explained by the sequential metabolism of these compounds (1,14,49), whereas the continuous accumulation of sesquiterpenes may be explained by a compartmentalized biosynthesis of the different classes of terpenes (11). In every case, major modifications occur in very young organs. Furthermore, large between-tree and within-tree variability is observed (12). Figure 2 presents the relative composition of volatile hydrocarbons within buds and stem tissues at different levels in three individual trees. The variations shown are the consequence of cumulative factors which control terpene biosynthesis.

GENETIC CONTROL OF TERPENE HYDROCARBON BIOSYNTHESIS

Because terpene composition is strongly modified during bud and needle aging, and relatively constant in woody tissues (Fig. 2), the studies on genetic control have been performed using the "cortical tissues" (cortex

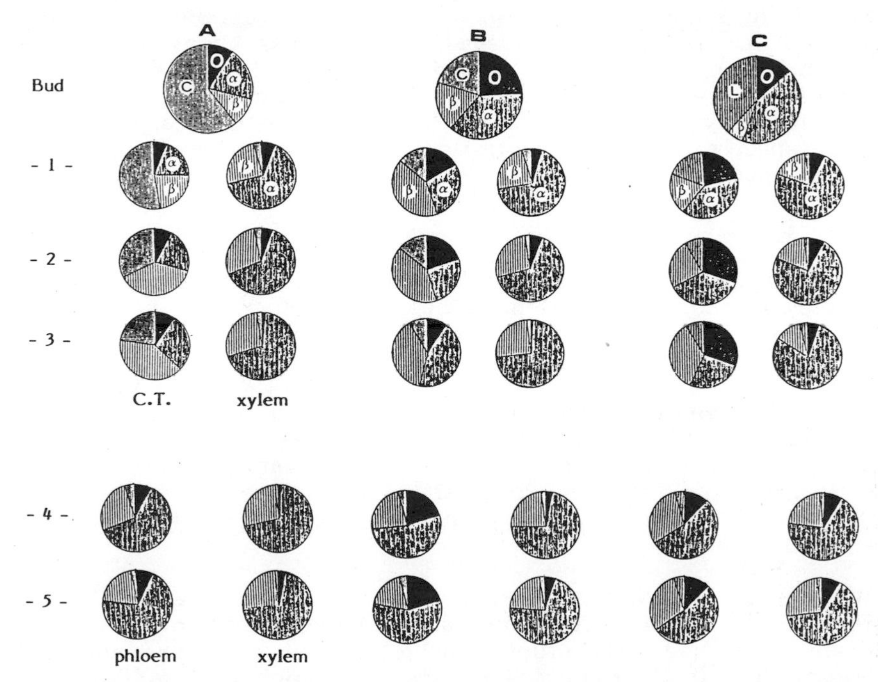

Fig. 2. Terpene hydrocarbon contents of three individual maritime pines (A, B, and C) in bud and five different levels of shoot and stem tissues (1 to 5). Three different phenotypes are represented, namely BmCl (tree A), BmCl (tree B), and BmcL (tree C). Trees A and B are carene-rich but the difference in carene amount is related to the genotypes: C^+/C^+ (A) and C^+/C^- (B). At levels 1, 2, and 3, xylem and cortical tissues (C.T.) have been collected separately. Cortical tissues include cambium plus phloem and chlorophyllous cortex. Shoots at level 1 are one-year old, whereas level 3 corresponds to the last chlorophyllous shoots. Levels 4 and 5 represent the middle and the base of the trunk, respectively. Two kinds of tissues have been collected at levels 4 and 5: xylem and phloem (plus cambium) without bark.

plus phloem) from one-year-old shoots. After elongation, the terpene composition of shoots is practically constant until the next growing period (67). Studies on the hereditary transmission mechanisms of terpene hydrocarbon biosynthesis in Pinus pinaster have been performed on selected clones (with known "terpene profiles") and full-sib families obtained by controlled pollination. These progeny tests have been established in the Landes area by the Laboratoire d'Amelioration des arbres forestiers, Institut National de Recherche Agronomique, Pierroton-Cestas. Comparisons between expected and observed frequencies of the different phenotypes in each kind of mating indicate that there is monogenic control of biosynthesis of four monoterpenes and two sesquiterpenes: Δ_3-carene, terpinolene (7), myrcene (6), limonene (55), caryophyllene, and longifolene (54). Such a simple transmission is also very likely for β-pinene. Results obtained by Walter et al. (66) suggest that anticopalic acid biosynthesis may also be genetically controlled.

Inheritance of monoterpene composition has been studied in various coniferous species. In the case of pines, Hanover (46,47,48) has clearly proved that in Pinus monticola the quantity of Δ_3-carene is controlled by two alleles at a single locus. A similar control has been demonstrated for myrcene and β-pinene in Pinus elliottii (62,64) and probably for limonene and β-phellandrene in Pinus taeda (63). But sesquiterpene inheritance has been studied only in Pinus pinaster (54). Except for caryophyllene, alleles that specify a high concentration are dominant or co-dominant over alleles that specify a low concentration.

Studies of Population Structures

In maritime pine, three geographical races--Atlantic coast, Mediterranean Europa, and Maghreb--have been determined on the basis of the six loci by using frequencies of richness alleles or by using the average relative concentration of each terpene (Fig. 3).

The Maghrebian populations are characterized by little genetic variation. Among the Mediterranean group, two types can be distinguished: the island populations (Corsica) and the European populations (southern France and eastern Spain). Among the Atlantic group, two types can be distinguished: the Landes populations and the northwestern Iberian peninsula populations.

More penetrating analyses provide interesting information about genetic differentiation of the species, characterization of meaningful isolates that should be preserved (4), studies on population structure and breeding mechanisms (5), and varietal tests. Practical application of varietal tests is now completely operational for identification of stands of unknown origin or provenances of seedlots in the Landes area. Consequently, an excellent tool is now available to avoid the introduction of poorly adapted varieties.

Biosynthetic Pathways

Another interesting consequence of the simple heredity of terpene hydrocarbons is the elucidation of biosynthetic pathways. When different precursors labeled with ^{14}C are applied to pine needles, the sequential labeling of hydrocarbons reveals the physiological role of acyclic hydrocarbons (34,35,36). As shown in Fig. 4, myrcene and trans-β-ocimene act as transitory metabolites prior to the biosynthesis of end-product monoterpenes such as α- and β-pinenes. The acyclic trans-β-farnesene also appears as a transitory compound prior to the further cyclizations towards

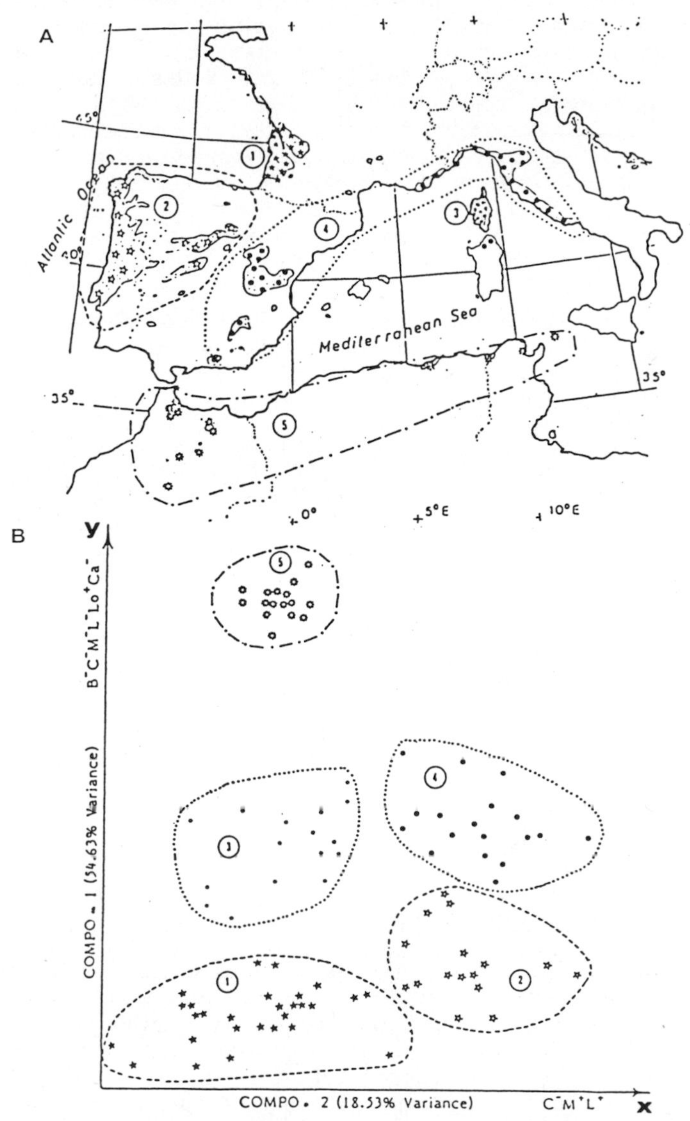

Fig. 3. (A) Natural geographical area of maritime pine: (1) Landes area in France; (2) western Iberian peninsula (Portugal and central Spain); (3) Corsica; (4) France, western Italy, and the Mediterranean shore of Spain; (5) Maghreb (Morocco, Algeria, and Tunisia). (B) Principal components analysis of 95 populations scattered throughout the entire natural area of maritime pine. The X axis (principal component 2 = 18.53% variance) is related to the increase of frequencies of the C^-, M^+, and L^+ alleles (see text for explanation). The Y axis (principal component 1 = 54.63% variance) is related to the increase of frequencies of the B^-, C^-, M^-, L^-, Lo^+, and Ca^- alleles.

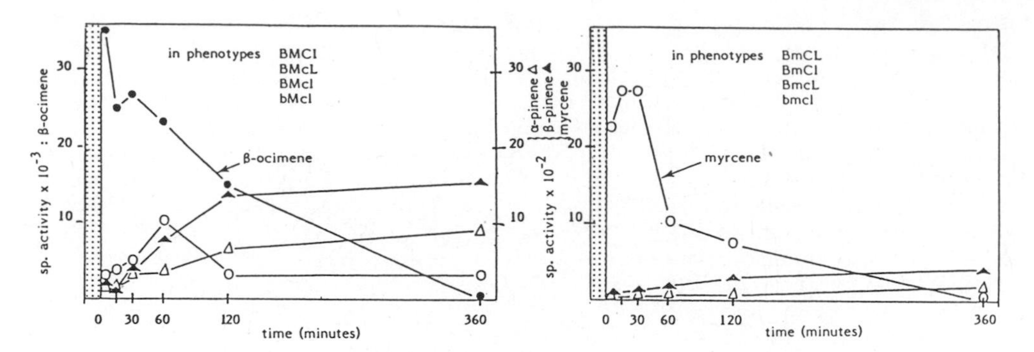

Fig. 4. Time course of the labeling of the main monoterpene hydrocarbons in the needles of different phenotypes of maritime pine after 15 min of exposure to $^{14}CO_2$.

the mono-, bi-, or tricyclic sesquiterpene hydrocarbons (43). Using different clones with known terpene profiles, it has been proven that the mechanisms of these biosynthetic pathways are genetically controlled (39). Myrcene appears as a transitory compound only in myrcene-poor (m) trees with the genotype M^-/M^-.* In this case, it is an intermediate only or is associated with trans-β-ocimene, which is always found in very small amounts in maritime pine. In myrcene-rich (M) trees, trans-β-ocimene is initially formed and then myrcene and α- and β-pinenes, with both the pinenes increasing in leaves. The gene action performed by the allele M^+ appears to be related to the elimination of a proton occurring preferentially on a specific form of carbonium ion (41). This hypothesis is consistent with the scheme proposed by Banthorpe (3) for biosynthesis of α- and β-pinenes in Pinus species. However, the expression of the genes controlling the terpene hydrocarbon enzymes is subject to various factors, the most important of which is the nature of the cell systems in which these compounds have been synthesized.

Expression of Terpene-Controlling Genes

The alleles M^+, L^+, and C^+, which control the ability to synthesize myrcene, limonene, and the couple Δ_3-carene plus terpinolene, respectively, only operate in the tissues that are differentiated from primary meristems (see Ref. 4 and 6). In other tissues, these compounds are not expressed and pinenes are predominately formed (A. Marpeau, unpubl. results). In other words, myrcene, limonene, and Δ_3-carene will only be found within buds, needles, and cortical chlorophyllous tissues of the shoots (Fig. 2). On the other hand, tissues originating from cambium activity (secondary conducting tissues) are unable to elaborate these specific compounds, and only the 1,5-bicyclic forms are accumulated in secondary

*In this review, for α-pinene, myrcene, Δ_3-carene, limonene, caryophyllene, and longifolene, the rich phenotypes will be designated by the upper-case letters B, M, C, L, Ca, and Lo, respectively, and the poor phenotypes by the lower-case letters b, m, c, l, ca, and lo. The alleles transmitting the ability (or the inability) to synthesize these terpenes will be designated B^+, M^+, C^+, L^+, Ca^+, Lo^+ (or B^-, M^-, C^-, L^-, Ca^-, Lo^-). The allele C^+ controls both Δ_3-carene and terpinolene.

resin ducts. The ratio of α-pinene to β-pinene is rather constant in the xylem of an individual tree. In contrast, the amounts of myrcene, limonene, and Δ_3-carene vary within the chlorophyllous tissues according to their age. The decrease from the youngest tissues (buds and growing shoots; level 1 in Fig. 2) to the oldest chlorophyllous cortical tissues (level 2, and then level 3 in Fig. 2) may be explained by the formation of new secondary structures which are differentiated every year and which induce the apparent decrease in first-year-elaborated compounds. These within-tree variations have been observed in pine species (33,45,60,61), explaining why chemotaxonomic or genetic studies have to be performed on cortical tissues from fully grown one-year-old shoots or buds.

The "other compounds" gathered into sector O (see Fig. 2) include trace monoterpenes (camphene, ocimene, and β-phellandrene) and sesquiterpene hydrocarbons (germacrene D, copaene, cubebene, caryophyllene, and humulene), the amounts of which may rise by up to 30% in the longifolene- and caryophyllene-rich phenotypes. The elaboration of these C_{15} compounds depends upon the cell and the physiological conditions, which are very different from those controlling the monoterpene synthesis.

COMPARISON OF THE CONDITIONS FOR THE BIOSYNTHESIS OF MONOTERPENE AND SESQUITERPENE COMPOUNDS

When the needles are fed with $^{14}CO_2$ in physiological conditions, the photosynthetic carbon is immediately incorporated into carbohydrates and then translocated or metabolized. A rapid immobilization of the carbon as insoluble food substances may occur within the needles. The nature of these accumulated nutrients varies according to the season: during spring, starch (as well as cell wall polysaccharides) is intensively accumulated as are lipids during fall and winter (Tab. 1). The behavior one-, two-, and three-year-old needles shows similar balance in food reserves (9,10). Under the same conditions, monoterpene synthesis only occurs in the growing needles from May to July.

Tab. 1. Distribution of radioactivity into carbohydrates, polysaccharides, lipids, and monoterpenes after application of $^{14}CO_2$ to one-, two-, and three-year-old needles (N_1, N_2, N_3) in June, November, and February.

Month	Needle age	Soluble carbohydrates	Polysaccharides	Lipids	Monoterpenes
June	N_1	7,000	8,195	40	288
	N_2	43,145	3,875	146	0
	N_3	38,950	6,950	99	0
November	N_1	29,600	1,780	660	0
	N_2	16,220	1,300	710	0
February	N_1	35,140	--	2,200	0
	N_2	24,100	--	1,380	0

Note: Results are expressed in $dpm \cdot g^{-1}$ fresh weight.

If different sections of the needles are considered (Fig. 5), it clearly appears that monoterpenes are mainly elaborated at the bases of the needle. Differentiating or newly differentiated cells are located in this zone, close to the basal meristem. In contrast, sesquiterpenes are formed all along the needle and accumulate in the course of needle aging (11). It is of interest to note that the de novo formations of tricyclic resin acids and monoterpenes are concomitant.

Furthermore, the different precursors are not incorporated into monoterpenes and sesquiterpenes in the same way (30,31). In Fig. 6, it appears that light is absolutely necessary for in vivo biosynthesis of monoterpenes both from carbonate and from acetate applied to intact leaves. In contrast, sesquiterpenes are formed under light or dark conditions from acetate, mevalonate, and isopentenyl pyrophosphate (IPP), with a higher incorporation when leaves were previously cut. Croteau and Loomis (28) obtained comparable results with Mentha piperita cuttings and explained the low incorporation of [^{14}C]mevalonic acid (MVA) as being due to compartmentalization effects (52,53).

Increases in monoterpene synthesis were previously observed after illumination of Mentha piperita plants (14) or Pinus radiata callus cultures (2). With Pinus pinaster seedlings, studies of different irradiances show that light activation is particularly striking for monoterpenes (39). The spectral dependency is photosynthetic-like when [^{14}C]carbonate is applied. With [1-^{14}C]- or [2-^{14}C]acetate, the most efficient radiations are located at 480 and 630 nm, and, to a lesser degree, at 685 nm. A phytochrome effect is probably not involved. Light activation is less efficient for sesquiterpene compounds.

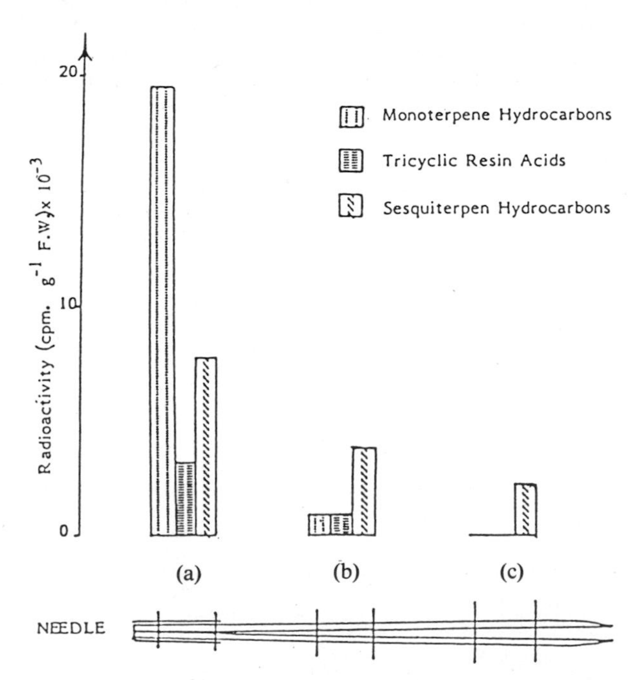

Fig. 5. Incorporation of $^{14}CO_2$ into the different classes of terpenes at the base (a), middle (b), and top (c) of maritime pine needle.

These data, on the whole, prove that C_{10} and C_{15} hydrocarbon biosynthesis is performed by independent systems probably located at different cell sites. Ultrastructural observations suggest that the activity level of a specific organelle--the leucoplast--may be a determinant in these compartmental phenomena (11,13).

CELL COMPARTMENTALIZATION OF TERPENE BIOSYNTHESIS

Monoterpene-Producing Systems

Because of chemical and anatomical parameters, as discussed by Gleizes (36), isolation of pure, active plastids from maritime pine tissues is extremely difficult. However, the development of specific plastids--the leucoplasts--in young secretory cells of pine needles (15), and a correlative study between these leucoplasts and the presence of monoterpenes in the essential oils (23), suggest a relationship between leucoplasts and monoterpene biosynthesis.

The leucoplasts involved in terpene-producing systems present a number of permanent features. These nongreen plastids are characterized by the absence of thylakoids and plastoribosomes (17,24). They possess a dense stroma and an envelope with a double membrane. The composition of plastid membranes resembles that of other plastids, especially as regards the nature and content of galactolipids (42). During the active secretory stage, leucoplasts take up a large part of the cell volume and their

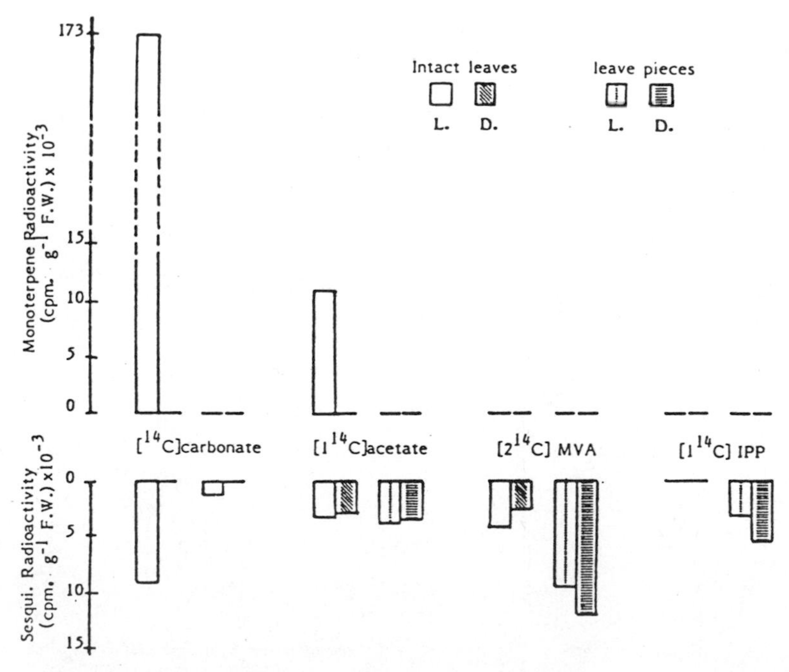

Fig. 6. Rates of incorporation into monoterpene and sesquiterpene hydrocarbons of the different radioactive precursors in intact leaves and leaf pieces under light (L) and darkness (D).

complex (often ameboid) shape enhances their exchange surface (21). Closely surrounding endoplasmic reticulum collects the synthesized terpenes and transfers them to the accumulation sites (16,18). A detailed description of this system will be given below.

Since isolation of leucoplasts from pine has been, until now, unsuccessful, another plant material--the peel of young fruits of Citrofortunella mitis (40,44)--was chosen which contains many secretory cavities replete with a monoterpene-rich oil. Factors controlling monoterpene biosynthesis at the cellular level were studied in vitro using subcellular fractions that are highly enriched in leucoplast vesicles. When incubated with radioactive IPP, the leucoplast preparation is able to synthesize monoterpene hydrocarbons, mainly limonene (44) and also the sequential terpenyl pyrophosphates up to the C_{20}: geranyl pyrophosphate (GPP), farnesyl pyrophosphate (FPP), and geranyl geranyl pyrophosphate (GGPP) (37). But neither FPP nor GGPP are metabolized towards the corresponding sesquiterpene hydrocarbons and diterpene (phytoene) or tetraterpene (carotenoids). These data demonstrate the high degree of specialization of leucoplast cyclases. Furthermore, among other expected precursors, acetate is rapidly used towards lipid metabolism and mevalonate is not metabolized.

As is true for other plastids, the leucoplasts depend on cytoplasmic activity for their IPP supply, which is assumed to be exclusively elaborated in this compartment (50,51). The utilization of IPP by leucoplasts may be somewhat modified by various exogenous factors (58). The addition of divalent cations (Mn^{2+} and to a lesser extent Mg^{2+}) and dimethylallyl pyrophosphate (DMAPP) increases the rate of monoterpene hydrocarbon biosynthesis. By demonstrating the low IPP-isomerase activity of the leucoplasts, this last point confirms the in vivo results comprehensively discussed by Poulter and Rilling (59). The main properties of the isolated leucoplasts are summarized in Fig. 7, illustrating once more the compartmentalization of IPP synthesis and the specificity of its utilization by plastids. Promising experiments using specific inhibitors are in progress.

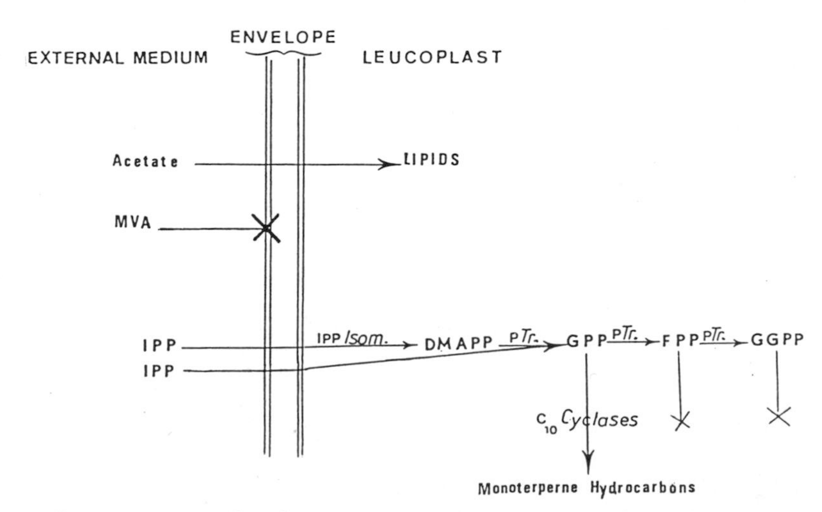

Fig. 7. Compartmentalization scheme of leucoplast functions. The localization of melavonic (MVA)-selective processes on internal or external membranes is unknown. IPPisom = Isopentenyl pyrophosphate isomerase; PTr = prenyltransferases.

Sesquiterpene-Producing Systems

The activity of light membranes in synthesizing cyclic sesquiterpene hydrocarbons is demonstrated by both in vivo and in vitro experiments. When light membrane fractions are isolated from an homogenate of primary leaves from Pinus pinaster seedlings after feeding with [^{14}C]acetate, highly enriched, labeled sesquiterpene hydrocarbons are found in two membrane fractions enriched in endoplasmic reticulum (Fig. 8). In these fractions, the radioactivity (radiochromatogram A in Fig. 8) is mainly distributed in the cyclic hydrocarbons caryophyllene, humulene, and cadinene (38).

Furthermore, active fractions of light membranes were purified from the peel of the fruits of Citrofortunella mitis. In the presence of [1-^{14}C]-IPP, such fractions are able to induce the elongation processes up to FPP (without GGPP formation) and convert FPP into the cyclic sesquiterpene hydrocarbons β-selinene and germacrene D (8). These fractions have been characterized as endoplasmic reticulum. This result demonstrates that endoplasmic reticulum is involved in the sesquiterpene cyclization processes.

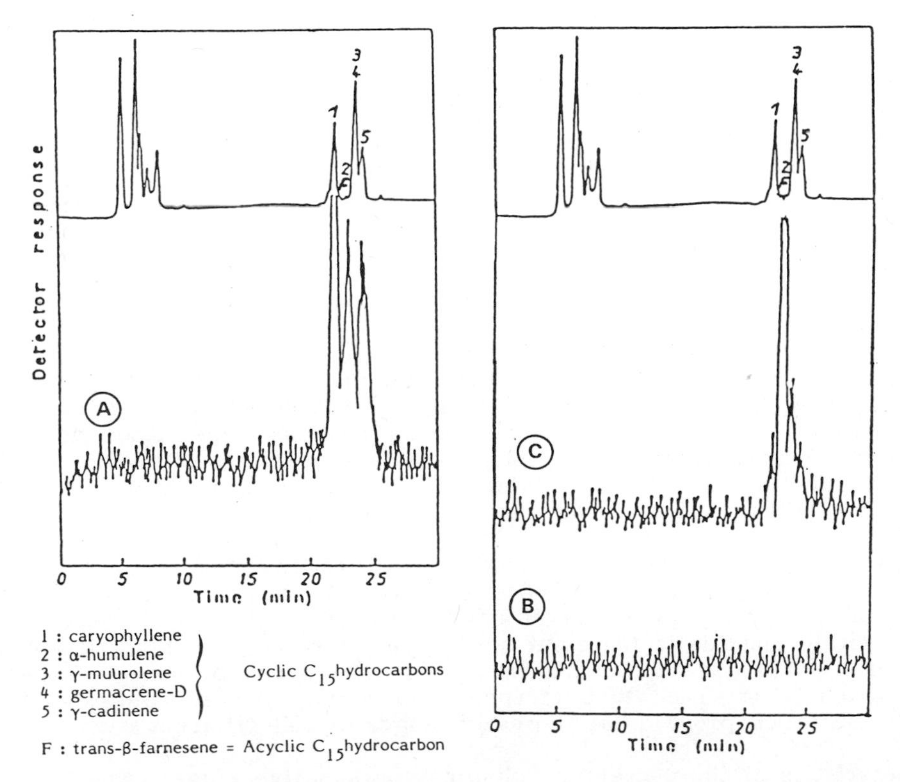

Fig. 8. Radio-gas liquid chromatograms of sesquiterpene hydrocarbons after incorporation of radioactive precursors in maritime pine leaves. (A) Endoplasmic reticulum, in vivo experiment ([^{14}C]-acetate). (B) Supernatant, in vivo experiment ([^{14}C]acetate). (C) Supernatant, in vitro experiment ([1-^{14}C]-IPP).

When supernatants are considered, the results are somewhat different, depending on the experimental conditions. When the radioactive precursor is applied before cell fractionation (in vivo experiments), no radioactive sesquiterpene is found in the supernatant (Fig. 8B). When the radioactive precursor is applied after cell fractionation (in vitro assay), the acyclic form is mainly found (Fig. 8C) with a lower ratio of cyclic hydrocarbons. Consequently, sesquiterpene synthetases appear as membrane enzymes more or less tightly linked to the endoplasmic reticulum and easily solubilized during homogenate preparation. The association between the C_{15}-cyclase and the endomembranes is often so loose that the membrane attachment of the enzyme cannot be detected. Such "soluble enzymes" with a sesquiterpene cyclization activity have been isolated from _Salvia officinalis_ (29) and _Citrofortunella mitis_ (8).

SECRETORY SYSTEMS: STRUCTURAL AND ULTRASTRUCTURAL DATA

Most of the biosynthetic processes leading to C_{10}, C_{15}, or C_{20} terpene compounds in maritime pine are assumed to take place in the resin ducts. These secretory systems are found in the parenchymas of the whole plant: buds, needles, young shoot chlorophyllous cortex, horizontal and vertical xylem parenchymas, and horizontal phloem parenchymas. The resin ducts are usually lined by one layer of secretory cells.

Differentiation

The differentiation of resin ducts proceeds very early from primary and secondary meristems. Primary meristems are located at the bud apex and at the base of the growing needles (Fig. 9, types I and II, respectively). Secondary meristems represent the cambial zone of the stem (Fig. 9, type III). Each year, the formation of additional resin ducts takes place from these meristems according to some general sequential events (15,19). The future secretory cells are first arranged concentrically (Fig. 10), then successive radial divisions lead to the formation of the secretory layer and the resin duct lumen, which enlarges by a lysogenous process (21).

The formation of new resin ducts in the outer part of elongating buds, from April to July, proceeds from the assembly of two cell lines issued from two bud scales (20). In growing needles, specialized areas of the basal meristem give rise to resin ducts located in mesophyll or close to the vascular bundles. In stem xylem or phloem, new resin ducts differentiate from cambial initials within parenchymal rays.

In every case, the most striking modifications concern the plastid compartment as well as the endoplasmic reticulum membranes. The initial proplastids (Fig. 10) give rise to amyloplasts (Fig. 11). The presence of starch within plastids is transient and, soon after, most of the cell volume is filled by large ameboid leucoplasts (Fig. 12). The presence of osmiophilic droplets, assumed to be terpenes, in the proplastids indicates that terpene secretion proceeds at the early stages of cell differentiation.

Characteristics of the Secretory Cell During Its Active Stage

Secretory cells of pine in the active secretory phase are characterized by large leucoplasts wrapped in a sheath of fenestrated endoplasmic reticulum (Fig. 12). These plastids contain few inner membranes, and their

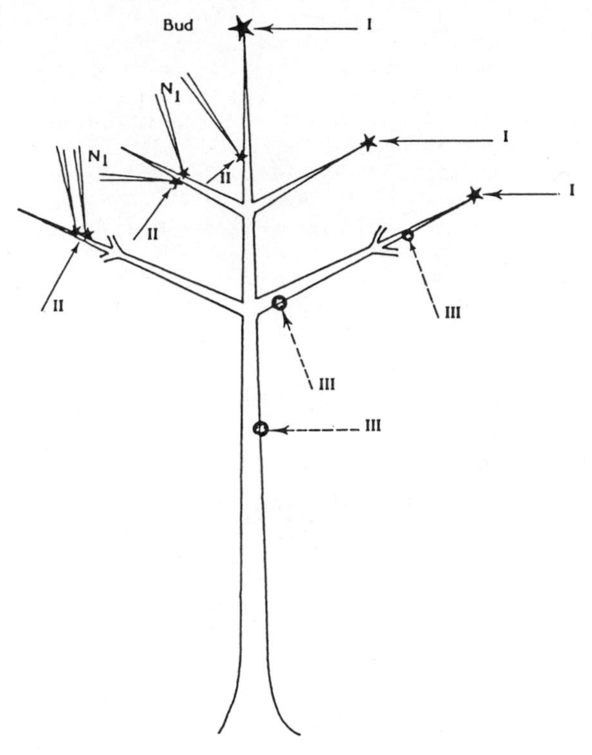

Fig. 9. Localization in a pine tree of the differentiation of type I (from primary meristem of buds), type II (from primary meristem of needles), and type III [from secondary meristem (cambium)] secretory systems.

very dense stroma is devoid of ribosomes (17,24). The envelope and some invaginations of the inner membrane contain osmiophilic droplets. The plastid outer membrane and endoplasmic reticulum are connected by luminal continuities which allow the translocation of terpenes from their elaboration site to their accumulation site. The extracellular outflow of terpenes is carried out by endoplasmic tubules budding with plasma membrane in the periplasmic space, giving rise to terpene vesicles bounded by two membranes (18). The terpene material then flows across the loosened wall into the lumen of the resin duct (Fig. 13). These typical structures are located in young secretory cells which are functional during an extremely short time (a few days) just after their differentiation from the meristem.

As previously shown, the monoterpenes elaborated by leucoplasts may be quite different according to the primary or secondary origin of the secretory systems. Leucoplasts from types I and II express the genes controlling synthesis of Δ_3-carene, myrcene, and limonene, whereas the type III leucoplasts do not. Furthermore, the endoplasmic reticulum, which is always implicated in terpene translocation, may also be involved in sesquiterpene biosynthesis, with the highest level found in cells from type II leucoplasts (38).

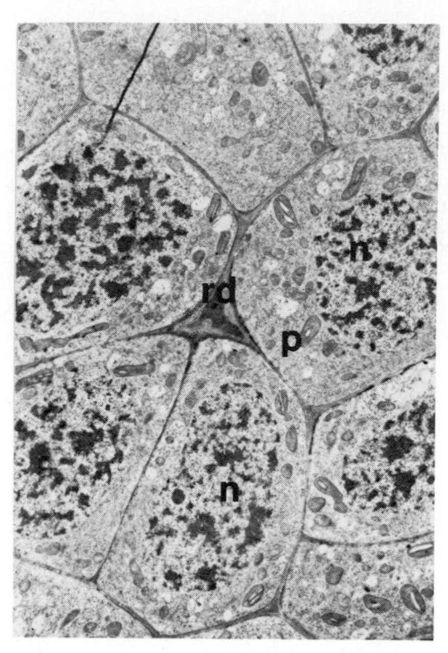

Fig. 10. Very young resin duct cells near the basal meristem of a growing needle. The cells contain prominent nuclei and small proplastids (X2,000). p, Plastid.

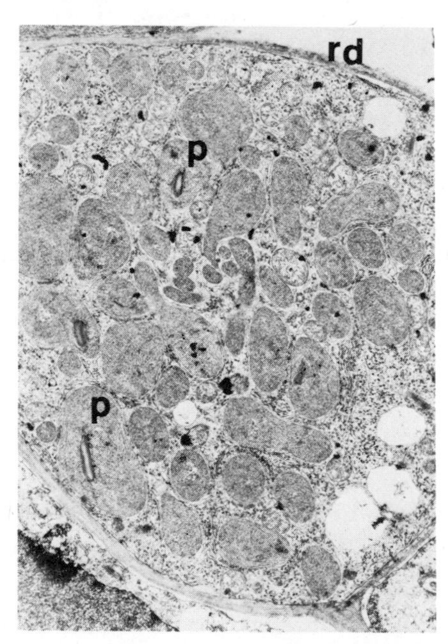

Fig. 11. Differentiating secretory cell with numerous plastids containing small starch grains (X12,000). p, Plastid; rd, resin duct lumen.

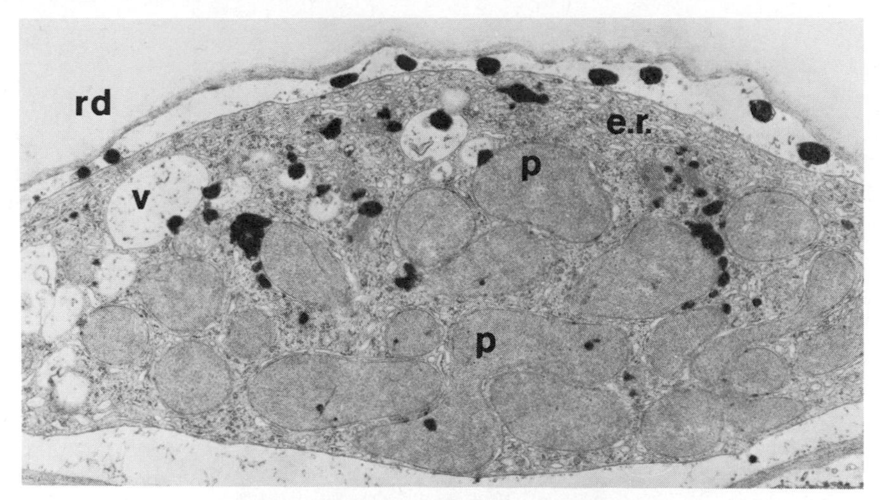

Fig. 12. Active secretory cell with large ameboid leucoplasts and osmiophilic terpene droplets close to plastids and in the periplasmic space (X16,000). The outer cell wall is loosened. er, Endoplasmic reticulum; p, plastid; rd, resin duct lumen; v, vacuole.

Decrease of the Secretory Activity

The cell wall lining the resin duct rapidly becomes more compact and prevents the diffusion of elaborated terpenes which accumulate or disappear by metabolization towards lipids. This results in dramatic cell reshaping, leading to quiescent or fully degenerated forms.

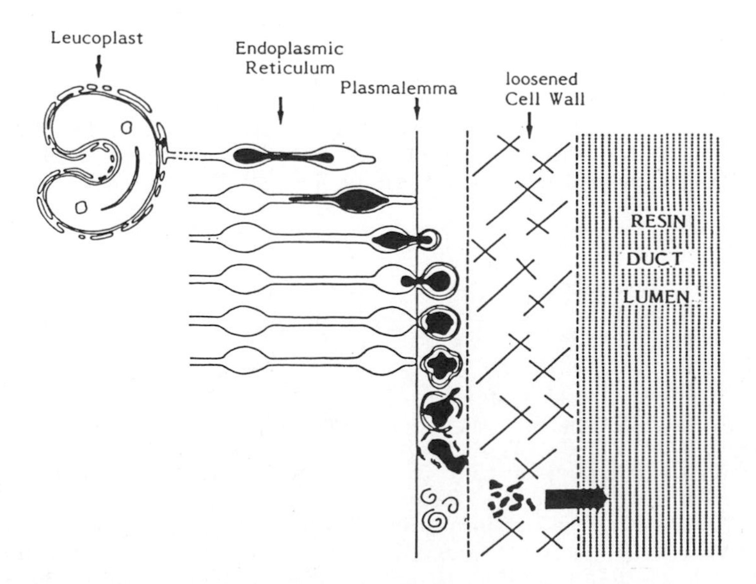

Fig. 13. Scheme of terpene translocation in the secretory cells from an active leucoplast to resin duct lumen.

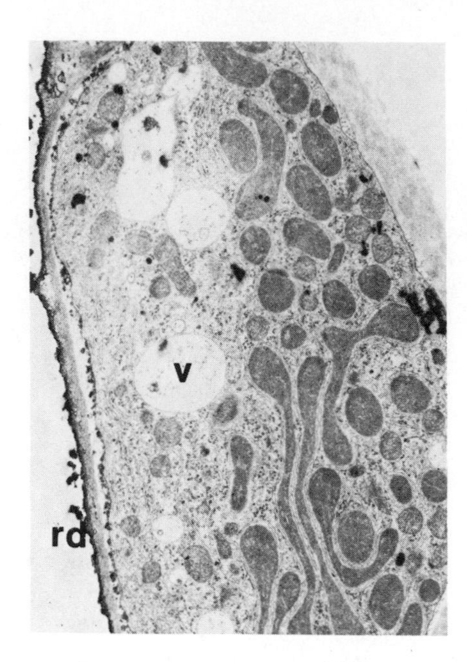

Fig. 14. Resin duct cell in a stem cortical resin duct after the secretory stage (X9,200). The plastids are smaller, the vacuole volume increases, and the outer cell wall is thickened. rd, Resin duct lumen; v, vacuole.

In bud, shoot, and cortical stem tissues, secretory cells (Fig. 9, types I and II) undergo a reversible regression (Fig. 14) and terpenes are progressively lysed as lipids. At the end of winter, numerous large lipid drops are gathered in a clear cytoplasm containing scarce proplastids and organelles (10,25). The secretory cells are then quiescent (Fig. 15). In needles, the secretory cells (type II) undergo a more drastic regression. The apical cell wall becomes thicker, the reticulum and cytoplasm density increases, and the leucoplast volume drops resulting in total disorganization characterized by a few organelles in a clear cytoplasm and large vacuoles. The degenerated cells are then no longer functional (Fig. 16).

Effects of Various Aggressions on Terpene Biosynthesis

Responses of secretory systems are different according to whether they are from type I, II, or III; in addition, the responses depend on the aggression form. As far as we know, the secretory systems of needles (type II) are not activated by wounding as are those of stem resin ducts (65), but they may be affected by atmospheric agents in the form of pollutants (C. Bernard-Dagan, unpubl. data). When quiescent tissues containing resin ducts (type I or III) are wounded or invaded with fungi, terpene accumulation occurs that is correlated with structural events involving the secretory cells and often the surrounding parenchyma cells of xylem or phloem (22,25). Wounding alone induces terpene accumulation. The synthesis of terpenes takes place in reactivated or supernumerary resin ducts which display structural features similar to those of secretory cells under normal conditions (Fig. 17).

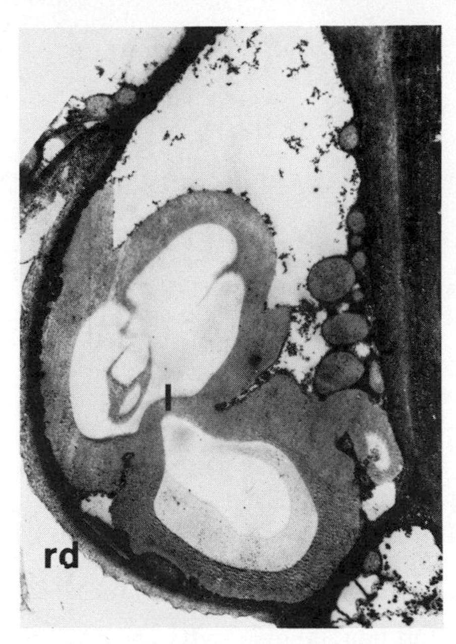

Fig. 15. Quiescent secretory cell in a stem cortical resin duct after the secretory stage (X6,500). Cell volume is nearly filled with lipid drops and vacuoles. l, Lipid droplet; rd, resin duct lumen.

Mycelium inoculation results in an exponential enrichment (X60 after 28 days in proximal phloem) as the enriched zone enlarges, indicating a diffusive process. Secretory ducts are first reactivated but soon undergo lysis and are no longer functional, whereas the rate of terpene enrichment is quickly increasing. The formation of terpene compounds is delocalized and occurs in nonspecific intra- or extracellular spaces formed by the lysis of cell contents and/or cell walls. The synthesis likely results from the random meeting of enzymes which can be synthesized in active parenchyma cells or released by lysis of structures that are involved in physiological terpene secretion (leucoplasts and endoplasmic reticulum) with the corresponding substrates which may be provided by degradation of cell contents (polyphenols, lipids) or by sieve flow (22,25). In all cases, newly synthesized terpenes are mainly α- and β-pinenes plus tricyclic resin acids.

GENERAL SCHEME OF TERPENOID BIOSYNTHESIS IN PINES

Resin constituents are elaborated close to the meristems when secretory systems are differentiating or when they have just differentiated. Specialized organelles (i.e., the leucoplasts) are involved in the biosynthesis of monoterpene hydrocarbons and probably tricyclic resin acids. In maritime pine, the biosynthesis of monoterpenes (Δ_3-carene plus terpinolene, myrcene, limonene, and probably β-pinene) and sesquiterpenes (caryophyllene and longifolene) has been demonstrated to be controlled by two alleles at the same locus. The genes are clearly expressed in primary secretory structures: bud, cortex of the elongating shoots, and growing needles (types I and II, Fig. 9). In the other tissues, only α- and β-pinenes with tricyclic resin acids are found (type III).

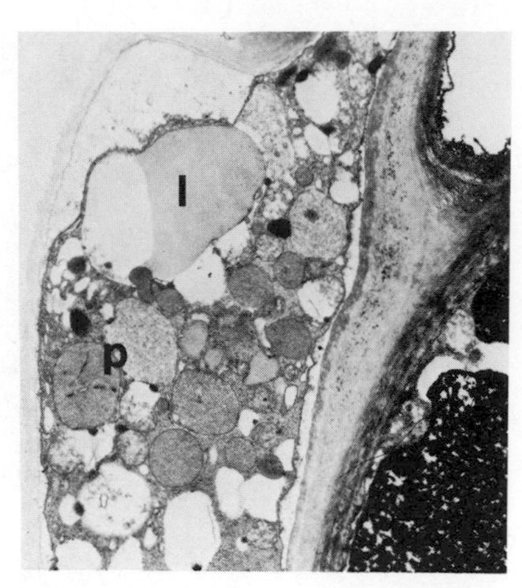

Fig. 16. Necrotic resin duct cell in a fully grown needle where the degenerative process is nonreversible (X9,800). l, Lipid droplet; p, plastid.

Under normal conditions, the activity period of the secretory cells is always very short (a few days). Rapidly their specific structures disappear and they become either quiescent (types I and III) or fully degenerated (type II). If quiescent duct cells are wounded, a local reactivation is observed with a correlative accumulation of pinenes (plus abietane and pimarane resin acids). When mycelia are inoculated or chemicals injected, a diffusing process induces the destruction of the parenchyma structures (secretory and all surrounding cells) and the metabolization of the cell components (wall polysaccharides, phenolics, membrane lipids, cytoplasm proteins, etc.) towards a delocalized intensive synthesis of terpenoids. Rapidly, large spaces are filled with resin and the nutrition of the pine is highly compromised. How and why these processes act as a defense against attack by fungus (and, indirectly, by insects) remains up to now an unsolved question.

At the cellular level, the biosynthesis of the different terpenoid classes is strictly compartmentalized. Monoterpene hydrocarbons are elaborated during the transitory activity of secretory cells by specialized leucoplasts. The enzymic properties of these plastids are highly specific: weak IPP-isomerase and strong prenyltransferase (until GGPP) and C_{10}-cyclase activities. The plastids depend on the cytoplasm for their IPP supply. Other precursors are not incorporated in the plastids (MVA) or used in other metabolic pathways (acetate → lipids). The secretory stage (during normal differentiation or after some aggressions) always involves the same sequential events: metabolism of accumulated lipids into transitory starch, and then into terpenes, with a correlated evolution of transient amyloplasts to leucoplasts. Reverse features are observed when secretory cells become quiescent. This observation may suggest a model for the metabolization of terpenes and their use as a carbon nutrient. Studies on *Mentha* and *Salvia* (27) suggest that monoterpene catabolism is highly probable. In *Pinus pinaster* three types of observation could support this assumption: (a) a rapid turnover of monoterpenes in needles (36); (b) an important decrease in heartwood resin just before the winter accumulation of lipids throughout the tree (9); and (c) the reversible conversion of lipids in differentiating or aging secretory cells.

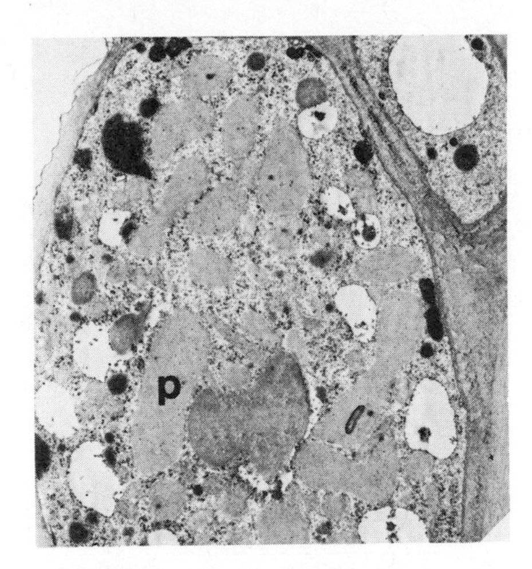

Fig. 17. Quiescent secretory cell reactivated by wounding which induces a strong synthesis of cytoplasm, membranes, and organelles (X7,200). p, Plastid.

Under normal conditions, terpene synthesis, translocation, and accumulation are achieved by specialized cell compartments, thus avoiding any diffusion of terpenes throughout the cytoplasm. Endoplasmic reticulum is implicated in cell compartmentalization both as a sesquiterpene hydrocarbon synthesizing system and as a membrane support allowing the translocation of terpenes from their elaboration site to their accumulation site. Any factor inducing disorganization of this cell compartmentalization initiates an irreversible disturbance, e.g., inhibition of mitochondria or chloroplast electron transfer (32,57) or "autophagic" processes leading to an uncontrolled terpenoid accumulation following inductions by chemical or biological elicitors.

REFERENCES

1. Attaway, J.A., A.P. Pieringer, and L.J. Barabas (1967) The origin of Citrus flavor components. III. A study of the percentage variations in peel and leaf oil terpenes during one season. Phytochemistry 6:25-32.
2. Banthorpe, D.V., and V.C.O. Njar (1984) Light dependent monoterpene synthesis in Pinus radiata cultures. Phytochemistry 23:295-299.
3. Banthorpe, D.V., O. Ekundayo, and V.C.O. Njar (1984) Biosynthesis of chiral α- and β-pinenes in Pinus species. Phytochemistry 23:291-294.
4. Baradat, P., C. Bernard-Dagan, and A. Marpeau (1978) Variations of terpenes within and between populations of maritime pine. In Biochemical Genetics of Forest Trees, D. Rudin, ed. Sveriges Lantbruksuniversitat, Umea, pp. 151-169.
5. Baradat, P., A. Marpeau, and C. Bernard-Dagan (1984) Les terpènes du Pin maritime: Aspects biologiques et génétiques. VI. Estimation du taux moyen d'autofécondation et mise en évidence d'écarts à la panmixie dans un verger à graines de semis. Ann. Sci. Forest. 41: 107-134.

6. Baradat, P., C. Bernard-Dagan, G. Pauly, and C. Zimmermann-Fillon (1975) Les terpènes du Pin maritime. III. Hérédité de la teneur en myrcène. Ann. Sci. Forest. 32:29-54.
7. Baradat, P., C. Bernard-Dagan, C. Fillon, A. Marpeau, and G. Pauly (1972) Les terpènes du Pin maritime: Aspects biologiques et génétiques. II. Hérédité de la teneur en monoterpènes. Ann. Sci. Forest. 29:307-334.
8. Belingheri, L., G. Pauly, M. Gleizes, and A. Marpeau (1987) Isolation by an aqueous two polymer phase system and identification of endomembranes from Citrofortunella mitis fruits for sesquiterpene hydrocarbon synthesis. J. Plant Phys. (in press).
9. Bernard-Dagan, C. (1987) Les substances de réserve du Pin maritime: Rôle éventuel des métabolites secondaires. Bull. Soc. Bot. Fr. (in press).
10. Bernard-Dagan, C. (1987) Variations of energy giving substances and biosynthesis of terpenes in maritime pine. In Mechanisms of Woody Plant Resistance to Insect and Pathogens, W.J. Mattson, ed. Springer-Verlag, New York (in press).
11. Bernard-Dagan, C., J.P. Carde, and M. Gleizes (1979) Étude des composés terpéniques au cours de la croissance des aiguilles de Pin maritime: Comparison de données biochimiques et ultrastructurales. Can. J. Bot. 57:255-263.
12. Bernard-Dagan, C., C. Fillon, G. Pauly, P. Baradat, and G. Illy (1971) Les terpènes du pin maritime: Aspects biologiques et génétiques. I. Variabilité de la composition monoterpénique dans un individu, entre individus et entre provenances. Ann. Sci. Forest. 28:223-258.
13. Bernard-Dagan, C., G. Pauly, A. Marpeau, M. Gleizes, J.P. Carde, and P. Baradat (1982) Control and compartmentation of terpene biosynthesis in leaves of Pinus pinaster. Physiol. Veg. 20:775-795.
14. Burbott, A.J., and W.D. Loomis (1967) Effects of light and temperature on the monoterpenes of Peppermint. Plant Physiol. 42:20-28.
15. Carde, J.P. (1976) Evolution infrastructurale du système sécréteur dans les aiguilles du Pin maritime. Bull. Soc. Bot. Fr. 123:181-189.
16. Carde, J.P. (1979) Le fonctionnement des cellules sécrétrices des canaux chez le Pin maritime: Données de la microscopie électronique. C.R. 104e Congr. Soc. Sav. II:275-286.
17. Carde, J.P. (1984) Leucoplasts: A distinct kind of organelle lacking typical 70S ribosomes and free thylakoids. Eur. J. Cell Biol. 34:18-26.
18. Carde, J.P., C. Bernard-Dagan, and M. Gleizes (1980) Membrane systems involved in synthesis and transport of monoterpene hydrocarbons in pine leaves. In Biogenesis and Function of Plant Lipids, P. Mazliak, P. Benveniste, C. Costes, and R. Douce, eds. Elsevier/-North-Holland Biomedical Press, Amsterdam, pp. 441-444.
19. Charon, J. (1983) Differenciation et fonctionnement des canaux sécréteurs dans les bourgeons de Pin maritime (Pinus pinaster Ait.). Études structurale et ultrastructurale. Thèse 3e cycle Cytologie et Morphogénèse Veg., Paris VI, France.
20. Charon, J., J. Launay, and E. Vindt-Balguerie (1986) Ontogénèse des canaux sécréteurs d'origine primaire dans le bourgeon de Pin maritime. Can. J. Bot. 64:2955-2964.
21. Charon, J., J. Launauy, and J.P. Carde (1987) Spatial organization and volume density of leucoplasts in pine secretory cells. Protoplasma 138:45-53.

22. Cheniclet, C. (1987) Effects of wounding and fungus inoculation on terpene producing systems of maritime pine. J. Exp. Bot. (in press).
23. Cheniclet, C., and J.P. Carde (1985) Presence of leucoplasts in secretory cells and of monoterpenes in the essential oils: A correlative study. Isr. J. Bot. 34:219-238.
24. Cheniclet, C., and J.P. Carde (1987) Ultrastructural localization of RNAs in plastids with an RNAse-gold method. Biol. Cell 59:79-88.
25. Cheniclet, C., C. Bernard-Dagan, and G. Pauly (1987) Terpene biosynthesis in pathological conditions. In Mechanisms of Woody Plant Resistance to Insects and Pathogens, W.J. Mattson, ed. Springer-Verlag, New York (in press).
26. Croteau, R. (1981) Biosynthesis of terpenes. In Biosynthesis of Isoprenoid Compounds, Vol. I, J.W. Porter and S.L. Spurgeon, eds. Wiley Interscience, New York, pp. 225-282.
27. Croteau, R. (1984) Biosynthesis and catabolism of monoterpenes. In Isoprenoids in Plants: Biochemistry and Function, W.D. Nes, G. Fuller, and L.S. Tsai, eds. Marcel Dekker, New York, pp. 31-64.
28. Croteau, R., and W.D. Loomis (1972) Biosynthesis of mono- and sesquiterpenes in peppermint from mevalonate-$2^{14}C$. Phytochemistry 11: 1055-1066.
29. Croteau, R., and A. Gundy (1984) Cyclization of farnesyl pyrophosphate to the sesquiterpene olefins humulene and caryophyllene by an enzyme system from sage (Salvia officinalis). Arch. Biochem. Biophys. 233:838-841.
30. Croteau, R., A.J. Burbott, and W.D. Loomis (1972) Apparent energy deficiency in mono- and sesquiterpene biosynthesis in peppermint. Phytochemistry 11:2937-2948.
31. Croteau, R., A.J. Burbott, and W.D. Loomis (1972) Biosynthesis of mono- and sesquiterpenes in peppermint from glucose-^{14}C and $^{14}CO_2$. Phytochemistry 11:2459-2467.
32. Douce, R., M. Neuburger, R. Bligny, and G. Pauly (1978) Effects of β-pinene on the oxidative properties of purified intact mitochondria. In Plant Mitochondria, G. Ducet and C. Lance, eds. Elsevier/North-Holland Biomedical Press, Amsterdam, pp. 207-214.
33. Forrest, G.I. (1978) Variation in monoterpene composition of the shoot cortical oleoresin within and between trees of Pinus contorta. Biochem. Syst. Ecol. 8:337-341.
34. Gleizes, M. (1976) Biosynthèse des carbures monoterpéniques du Pin maritime: Rôle du trans-β-ocimene. C.R. Acad. Sci. (Series D) 283:97-100.
35. Gleizes, M. (1978) Biosynthèse des carbures terpéniques du Pin maritime: Essai de localisation. C.R. Acad. Sci. (Series D) 286:543-546.
36. Gleizes, M. (1979) Biosynthèse et localisation cellulaire des hydrocarbures terpéniques dans le Pin maritime (Pinus pinaster Ait.). Thèse Doct. Etat, Université de Bordeaux I, Talence, France.
37. Gleizes, M., B. Camara, and J. Walter (1987) Some characteristics of terpenoid biosynthesis by leucoplasts of Citrofortunella mitis. Planta 170:138-140.
38. Gleizes, M., J.P. Carde, G. Pauly, and C. Bernard-Dagan (1980) In vivo formation of sesquiterpene hydrocarbons in the endoplasmic reticulum of pine. Plant Sci. Lett. 20:79-90.
39. Gleizes, M., G. Pauly, C. Bernard-Dagan, and R. Jacques (1980) Effects of light on terpene hydrocarbon synthesis in Pinus pinaster. Physiol. Plant. 50:16-20.

40. Gleizes, M., C. Bernard-Dagan, J.P. Carde, and A. Marpeau (1982) Function of plastids in terpene biosynthesis. Biochemistry and Metabolism of Plant Lipids, J.F.G.M. Wintermans and P.J.C. Kuiper, eds. Elsevier/North-Holland Biomedical Press, Amsterdam, pp. 511-514.
41. Gleizes, M., A. Marpeau, G. Pauly, and C. Bernard-Dagan (1982) Monoterpene hydrocarbon biosynthesis in Pinus pinaster needles: Behaviour of noncyclic compounds. Phytochemistry 21:2641-2644.
42. Gleizes, M., A. Marpeau, G. Pauly, and L. Belingheri (1984) Polar lipid composition and galactolipid biosynthesis in isolated leucoplasts of Citrofortunella mitis. In Structure, Function and Metabolism of Plant Lipids, P.A. Siegenthaler and Eichenberger, eds. Elsevier Science Publishers, pp. 343-346.
43. Gleizes, M., A. Marpeau, G. Pauly, and C. Bernard-Dagan (1984) Sesquiterpene biosynthesis in maritime pine needles. Phytochemistry 23:1257-1259.
44. Gleizes, M., G. Pauly, J.P. Carde, A. Marpeau, and C. Bernard-Dagan (1983) Monoterpene hydrocarbon biosynthesis by isolated leucoplasts of Citrofortunella mitis. Planta 159:373-381.
45. Hanover, J.W. (1966) Environmental variation of monoterpenes of Pinus monticola Dougl. Phytochemistry 5:713-717.
46. Hanover, J.W. (1966) Genetics of terpenes. I. Gene control of monoterpene levels in Pinus monticola Dougl. Heredity 21:73-84.
47. Hanover, J.W. (1966) Inheritance of 3-carene concentration in Pinus monticola. For. Sci. 12:447-450.
48. Hanover, J.W. (1971) Genetics of terpenes. II. Genetic variances and interrelationships of monoterpene concentrations in Pinus monticola. Heredity 27:237-245.
49. Königs, R., and P.G. Gülz (1974) Die Monoterpene in ätherischen Ol der Blätter von Cistus ladaniferus L. Z. Pflanzenphysiol. 72:237-248.
50. Kreuz, K., and H. Kleinig (1981) On the compartmentation of isoprenyl diphosphate synthesis and utilisation in plant cells. Planta 153:578-581.
51. Liedvogel, B. (1986) Acetyl co-enzyme A and isopentenyl pyrophosphate as lipid precursors in plant cells. Biosynthesis and compartmentation. J. Plant Physiol. 124:211-222.
52. Loomis, W.D., and R. Croteau (1973) Biochemistry and physiology of lower terpenoids. Recent Adv. Phytochem. 6:147-182.
53. Loomis, W.D., and R. Croteau (1980) Biochemistry of terpenoids. In The Biochemistry of Plants, Vol. 4, P.K. Stumpf and E.E. Conn, eds. Academic Press, Inc., New York, pp. 363-418.
54. Marpeau, A., P. Baradat, and C. Bernard-Dagan (1975) Les terpènes du Pin maritime: Aspects biologiques et génétiques. IV. Hérédité de la teneur en deux sesquiterpènes: Le longifolène et le caryophyllène. Ann. Sci. Forest. 32:185-203.
55. Marpeau-Bezard, A., P. Baradat, and C. Bernard-Dagan (1983) Les terpènes du Pin maritime: Aspects biologiques et génétiques. V. Hérédité de la teneur en limonène. Ann. Sci. Forest. 40:197-216.
56. Pauly, G., M. Gleizes, and C. Bernard-Dagan (1973) Identification des constituants de l'essence des aiguilles de Pinus pinaster. Phytochemistry 12:1395-1398.
57. Pauly, G., R. Douce, and J.P. Carde (1981) Effects of β-pinene on spinach chloroplast photosynthesis. Z. Pflanzenphysiol. 104:199-281.
58. Pauly, G., L. Belingheri, A. Marpeau, and M. Gleizes (1986) Monoterpene formation by leucoplasts of Citrofortunella mitis and Citrus unshiu. Plant Cell Reports 5:19-22.

59. Poulter, C.D., and H.C. Rilling (1981) Prenyl transferases and isomerase. In Biosynthesis of Isoprenoid Compounds, Vol. I, J.N. Porter and S.L. Spurgeon, eds. John Wiley and Sons, New York, pp. 161-224.
60. Rockwood, D.L. (1973) Variation in the monoterpene composition of two oleoresin systems of loblolly pine. For. Sci. 19:147-152.
61. Smith, R.H. (1966) Intratree measurements of the monoterpene composition of Ponderosa pine xylem resin. For. Sci. 14:418-419.
62. Squillace, A.E. (1971) Inheritance of monoterpene composition in cortical oleoresin of slash pine. For. Sci. 17:381-387.
63. Squillace, A.E., O.O. Wells, and D.L. Rockwood (1980) Inheritance of monoterpene composition in cortical oleoresin of loblolly pine. Silvae Genet. 29:141-152.
64. Strauss, S.H., and W.D. Critchfield (1982) Inheritance of β-pinene in xylem oleoresin of Knobcone X Monterey pine hybrids. For. Sci. 28:687-696.
65. Vassilyev, A.E., and J.P. Carde (1976) Effets du gemmage sur l'ultrastructure des cellules sécrétrices des canaux de l'écorce des tiges de Pinus sylvestris L. et Picea abies (L.) Karst. Protoplasma 89:41-48.
66. Walter, J., B. Delmond, and G. Pauly (1985) Les acides résiniques des aiguilles et des tissus corticaux de Pins maritimes (Pinus pinaster Ait.) des Landes et de Corse: Présence de l'acide anticopalique dans les aiguilles des Pins d'origine corse. C.R. Acad. Sci. (Series D) 301:539-542.
67. Zimmermann-Fillon, C., and C. Bernard-Dagan (1977) Variations qualitatives et quantitatives des carbures terpéniques au cours de la croissance des rameaux et des aiguilles du Pin maritime: Étude comparée de huit phénotypes. Can. J. Bot. 55:1009-1018.

APPROACHES TO ALTERING REGULATORY CONTROLS OF SECONDARY PATHWAYS IN CULTURED CELLS

Jochen Berlin

Gesellschaft für Biotechnologische Forschung m.b.H.
D-3300 Braunschweig, Federal Republic of Germany

ABSTRACT

Plant cell cultures are regarded as a potential source of commercially used compounds. However, without introduction of new approaches to this field of biotechnology, realization of this goal seems to be limited to just a few cases. Therefore, it is believed that with increasing biochemical and molecular knowledge, this technology may be applied to more systems as desired manipulations become possible. A short overview of our present knowledge of the enzymology and the regulation of secondary pathways as well as of the cloning of corresponding genes will be given. Using examples from our own laboratory, the chances and limits of using selection techniques for the recovery of variants with altered secondary metabolism will be described. Genetic transformation of cells seems to be the most attractive approach to influencing secondary pathways. The actual state of our aim of manipulating serotonin or quinolizidine alkaloid biosynthesis by genetic engineering will be outlined.

INTRODUCTION

Plant cell cultures are regarded as a potential source of commercially useful compounds (26,42), and the product levels of several culture systems seem to be high enough to warrant industrial production. Indeed, shikonins are the first natural compounds to be produced in large fermenters in Japan (16). The levels of rosmarinic acid in *Coleus* have been optimized to such an extent that cell culture technology may provide the most convenient procedure of producing this antiphlogistic compound (43). Berberines are produced by many culture systems at levels higher than those produced by the corresponding plants (51). The few rapidly growing and highly productive cell culture systems described in the literature, however, must not hide the fact that the many economically important secondary metabolites are found only in traces or not at all in rapidly growing cultured cells. Thus, cardiac glycosides are not formed by suspension cultures of *Digitalis*, and morphinan, tropane, and quinoline alkaloids are at best only present at extremely low levels in biotechnologically relevant cell culture systems (3).

For many years it was expected that the low productivities of many cell culture systems could be overcome by empirical approaches similar to those used for the optimization of antibiotic production in microorganisms. Analytical screening for highly productive and stable variants was only successful in a few cases (4). The use of production media undoubtedly helped to increase the production rate and to shorten the production time. However, technologically relevant increases were only seen for products that were also found at reasonable levels under other culture conditions. Today, the use of abiotic and biotic elicitors is believed to be a promising tool for the improvement of the yields of low-level or absent products in culture systems (12). However, a critical analysis of the reported cases clearly shows that convincing improvements are only found for true phytoalexins and for products that are usually present in cultured systems or that can be easily induced by other techniques. Thus, it is predicted that without finding very specific elicitors, the levels of morphinan, tropane, or quinoline alkaloids will not be improved to biotechnologically relevant levels by this technique.

Past results and the failure to achieve significant breakthroughs in improving product levels of very important secondary metabolites have changed the attitude of many researchers in the field towards acknowledging the need of greater biochemical and molecular knowledge of secondary metabolism (3,26,42). In many reviews it is clearly stated and emphasized that such knowledge may increase the chances of exploiting plant cell cultures for the production of useful plants. However, it is rarely stated how this can be approached and which steps are required. The purpose of this chapter is therefore to analyze in some detail the steps involved and the requirements needed for achieving the difficult goal of manipulating secondary metabolism. Approaches to altering secondary metabolism in cultured cells, as discussed here, include the selection of stable variants with altered secondary metabolism or genetic transformation with the aim of manipulating expression or regulation of a pathway. Altered productivities due to physiological changes caused by altered culture conditions are not considered.

At present, only in a very few cases has sufficient knowledge been collected or made available for starting a particular manipulation of a secondary pathway. Thus this review can only be a short description of our actual knowledge. But it will provide some suggestions as to how one may proceed to express more pathways or branches of pathways, or how one may establish more highly productive cell lines by such manipulations. Although I will mainly describe our own approaches, the conclusions and suggestions would be undoubtedly transferable to other secondary pathways.

PREREQUISITES FOR ALTERING CONTROLS OF SECONDARY PATHWAYS

It is evident that one needs detailed knowledge about a secondary pathway if one wants to influence its expression or its regulation. First of all, the decisive biosynthetic enzymes of this secondary pathway should be known. Though recently quite a number of novel enzymatic steps of several secondary pathways have been described (Tab. 1), it is clear that for the majority of secondary pathways the proposed biosynthesis (deduced from chemical considerations and feeding experiments) must await verification by the detection of the corresponding enzyme reactions. Thus, for many pathways, directed manipulations are not possible due to lack of the

Tab. 1. Some randomly chosen examples of recently detected novel enzyme reactions in the biosynthesis of various secondary pathways.

Enzyme*	Pathway	Source**	Reference
Acridone synthetase	Acridone alkaloids	Ruta graveolens (S)	Baumert and Gröger (2)
Cinchoninone:NADPH oxidoreductases	Quinoline alkaloids	Cinchona ledgeriana (S)	Isaac et al. (22)
Hyoscyamine - 6β-hydroxylase	Tropane alkaloids	Hyoscyamus niger (RC)	Hashimoto and Yamada (19)
Vinorine synthase	Ajmaline	Rauwolfia serpentina (S)	Pfitzner et al. (31)
Deacetylvindoline acetyltransferase	Vindoline	Catharanthus roseus (P)	DeLuca et al. (11)
SAM:Norreticuline N-methyltransferases	Benzylisoquinolines	Berberis vulgaris (S)	Wat et al. (47)
o-Succinylbenzoyl-CoA ligase	Anthraquinones	Galium mollugo (S)	Heide et al. (20)
Pinene cyclases	Monoterpenes	Salvia officinalis (P)	Gambliel and Croteau (17)

*These examples can only be an indication for progress in the field. A great number of enzymatic steps may be found in the literature especially for monoterpene indole alkaloids (52), isoquinoline alkaloids (53), nicotine (45), flavonoids (18), and monoterpenoids.
**S, Suspension culture; RC, root culture; P, plant.

enzymological background. However, recent progress indicates that with more groups focusing on the enzymology of secondary pathways, the gap in our present biochemical knowledge may be closed. Though the most impressive advances in the enzymology of secondary pathways were achieved with cell suspension culture systems (18,52,53), one should not forget that organ cultures or even the intact plants can be excellent sources for such enzymes. In cases where a pathway is not expressed in suspension cultures or its product is present at extremely low levels, one has no other choice than to use differentiated material if one wants to learn more about the biochemistry and regulation of that pathway (17,19).

The next question is of course how does the description and/or detection of an enzymatic step of a biosynthetic sequence help in altering its regulatory control? Clearly the "roles" of the enzymes within the sequence have to be realized, e.g., which enzymes are regulatory or rate-limiting. This requires detailed physiological studies on the behavior of the enzymes of the pathway under various culture conditions. The appearance and changes in enzyme activities in productive or nonproductive culture states may provide hints about which of the enzymes exert a control function and which are regulatory and thus candidates for alterations. At present, one would assume that enzymes at the beginning of a sequence, or at branching points of metabolism or of secondary pathways, are probably regulatory, especially when manifold increases of the activities of these enzymes are observed under conditions promoting dramatic increases of product levels or when their absence prevents de novo formation of the compounds. It is of course necessary to know whether the other biosynthetic enzymes are co-induced with the proposed regulatory enzyme or are permanently present even in nonproducing cultures. Such analyses may help one to evaluate the chances of a specific manipulation for increasing levels of the desired product.

Both the characterization of the enzymatic steps and the kinetic behavior of the enzyme activities of a complex pathway have only been presented for the light-induced flavonoid biosynthesis in parsley (18,21). Similar kinetics have recently been published for the joint enzymes of the closely related isoflavonoid-derived phytoalexin biosynthesis in biologically stressed bean cells (10). Interestingly, osmotically stressed soybean cells convert naringenin to apigenin (flavonoid branch), while elicitor-treated cells metabolize naringenin to genistein (isoflavonoid branch) (24). The induction of one or the other NADPH-dependent flavone synthase thus regulates the branch into which the common precursor is diverted (24). The branched pathway of flavonoid biosynthesis represents a suitable system for studying manipulations of a secondary pathway. This is the only secondary pathway where the enzymology has already led to detailed studies at the molecular level. For example, the coordinated synthesis of the enzymes could be related to coordinated inductions of the corresponding mRNAs (9,25,33).

The next step would be to clone the genes of the desired enzyme or at least the corresponding cDNA. To the best of our knowledge, the only gene of any secondary pathway which has been cloned, so that it might be used for transformations, is the chalcone synthase (CHS) gene. Chalcone synthase is a key enzyme in flavonoid biosynthesis, since it diverts simple phenylpropanoids into the flavonoid branch. Thus according to the above assumption (enzyme at a branching point) and based on experimental evidence, CHS may be regarded as a regulatory enzyme. The results reported so far indicate the complexity of gene regulation of secondary pathways which we had expected. Indeed, the CHS genes comprise a multigene family in *Petunia*, since at least six distinct, complete CHS genes were found (44). In the flowers, only one of these genes is expressed. Evidently genes of secondary metabolism also underlie tissue-specific expression, as was expected. In white flower segments of *Petunia* mutants, the CHS gene is not transcribed (44). The great interest in this enzyme may result from the fact that transposable elements affect the expression of anthocyanin biosynthesis and thus the pigmentation of flowers and kernels. Chalcone synthase genes have thus been isolated by the transposon tagging method (48). While complete cDNAs of other genes of the flavonoid biosynthesis that may be used for transformation experiments have not yet been reported, cDNAs of CHS are available from *Antirrhinum majus*, *Petunia hybrida*, *Petroselinum hortense*, and *Zea mays* (see Ref. 48). However, complete DNA sequences coding for chalcone flavanone isomerase (9,44), phenylalanine ammonia lyase (9), p-coumarate:CoA ligase (25), or stilbene synthase (a likely regulatory enzyme of another branch leaving the general phenylpropanoid pathway) (40), may soon be reported.

As cloning of genes of secondary metabolism has just begun, we do not know much about the overall organization of the genes of a secondary pathway or branch within the genome. It is likely that the genes are not clustered but distributed within the chromosome. One may assume that most genes, or at least the regulatory ones, will occur as gene families, and that the various genes of one family are each separately controlled and are thus expressed by different signals. This complexity indicates that directed alterations of such controls may be difficult to achieve. The situation may be even more complicated, since haploid plant cells for such manipulations would rarely be available.

WAYS OF ALTERING REGULATORY CONTROLS

Despite our limited knowledge of the biochemistry and molecular biology of secondary pathways, one may in certain cases attempt manipulations of these pathways. First we have to define the reason for altering the control. Assuming we want to increase or delete a secondary metabolite of an intact plant, we have to define where we want to see the alteration, since a pathway may underlie organ-specific expression. An alteration of that pathway may be achieved by conventional breeding programs, perhaps including cell culture steps for increasing the variability. However, as long as no selection pressure for the desired trait can be placed, the approach is not directed. It is questionable whether positive selections for the desired alterations at the cell culture level would be successful.

In the case of specific alterations of a secondary pathway in the intact plant, genetic engineering techniques seem to be required. Since suitable vectors have been developed (15,23,32), it is possible today to introduce and express a desired foreign gene in all parts of the intact plant; in some cases it might be possible to express the introduced gene in a specific organ such as the chloroplasts (1,8,13,39,41). Resistances and novel genes, unrelated to any pathway of the recipient plant, have been introduced by this technique. However, it is not presently known how the cell reacts to an interference of its own pathways. It will be very interesting to see the responses of the intact plant to genetic transformations involving its own pathways. The foreign gene should be under the control of a promoter that is permanently expressed or that can be induced by external signals. The CHS gene could be the first plant gene used for such transformations, with the aim of interfering with a secondary pathway. Experiments of repairing anthocyanin biosynthesis in mutants deficient in CHS by genetic transformation are planned (44); perhaps such transformed plants would also show an overall enhancement of anthocyanin biosynthesis.

In the case of altering regulatory controls of secondary pathways at the cell culture level, the chances of achieving this goal seem to be generally better, since selection may be applicable as an alternative technique in certain cases. The advantage would be that cloning of the genes or purification of the enzymes is not necessarily required. It might be sufficient to "understand" the pathway, that is, to know the quantitative behavior of enzyme activities of low- and high-productive culture states. However, as was concluded from our own selection experiences for desired traits (see below), successfully directed selection systems for secondary pathways remain the exception rather than the rule. Thus, in general, time consuming purification of enzymes, their characterization, and cloning of the corresponding genes will be required for manipulations of secondary metabolism in cell culture systems. The introduction of the desired gene of a pathway, but under an altered control, by using suitable vector constructions seems to be the most promising approach at present. Data supporting this assumption are, however, not yet available. The following will therefore describe some of our results on selecting for variants with altered secondary metabolism to show the value and limits of selection systems. I will outline two systems in which we are presently trying to affect secondary pathways by genetic transformation.

Selection for Enzymes Linking Primary and Secondary Metabolism

As pointed out above, the biosynthesis of secondary products is strictly regulated during plant development and in cultured cells. The presence or absence of any secondary metabolite may be dependent on the presence or absence of a single enzyme of the corresponding biosynthetic sequence or on the induction or repression of some or all enzymes of that pathway. Knowledge of whether one signal induces or increases one regulatory or all enzymes may have consequences for the strategy of influencing the pathway. During our studies on various secondary pathways, we and others have presented evidence that enzymes linking primary and secondary pathways may exert a regulatory function (30,35). Phenylalanine ammonia lyase, tryptophan decarboxylase (TDC), and lysine decarboxylase are typical enzymes for this. As we realized that high activities of these enzymes are required for high production of corresponding secondary metabolites, we devised a selection system for cell lines having greatly increased activities of these enzymes. However, product levels were not always increased in lines where these enzymes were present with higher activities, as will be shown by the examples given below.

Cinnamoyl putrescine formation in tobacco cells. Palmer and Widholm (29) selected a p-fluorophenylalanine (PFP)-tolerant tobacco cell line that produced ten times more cinnamoyl putrescines than did the wild-type cell. We were interested not only in the extremely high product levels which accounted for up to 10% of the dry mass, but also in what biochemical alterations occurred to make the cell produce the increased levels of cinnamoyl putrescines. The answer was clear since the high-producing cell line had ten- to 20-fold increased activities of the biosynthetic enzymes (5). We demonstrated that PFP selects for cell lines with increased phenylalanine ammonia lyase activity, since this enzyme detoxifies the amino acid analog (6). As the other biosynthetic enzymes of cinnamoyl putrescines are evidently co-induced in these selected cells, cinnamoyl putrescine biosynthesis and product levels are thus greatly increased in this tobacco line. Coinduction of the general phenylpropanoid pathway seems to be a general phenomenon and was first described by Hahlbrock's group while studying flavonoid biosynthesis in parsley (18,21). Without knowing which molecular changes were necessary to allow the better expression of the biosynthetic sequences of the pathway, we selected a cell type with altered regulatory control of that pathway. This trait has now been stably maintained for more than 12 years. The positive experience with the selection of PFP-tolerant phenolics overproducing cell lines enabled us to apply this technique to other pathways.

Restoration of serotonin biosynthesis in Peganum harmala. Freshly initiated cell cultures of Peganum harmala, showing some morphological differentiation at the end of the growth cycle, produce reasonable amounts of serotonin and simple β-carboline alkaloids, such as harmine and harmalol (34). However, over the years the initially high-producing cell lines changed into rapidly growing fine suspension cultures. The consequence was that neither serotonin nor β-carboline alkaloids were produced by these cells. We had shown that increased production of these compounds was accompanied by increased activities of TDC (35). In contrast to cinnamoyl putrescine biosynthesis, the biosynthetic enzymes after TDC are at present unknown. However, from feeding experiments with tryptamine it was at least known that the second and final step of serotonin biosynthesis--the hydroxylation of tryptamine at the 5-position--was also expressed in the rapidly growing fine cell suspension cultures that were unable to produce serotonin de novo (37). Here we have the unique case of the absence of serotonin in the fine suspension cultures being directly related to the absence of TDC. Consequently, in analogy to the tobacco system, we selected for 4-methyltryptophan-tolerant cell lines, since this analog can be detoxified by TDC. Indeed, we selected by this technique cell lines having sufficient TDC activity, such that serotonin levels of up to 2% were established (7). These producing cell lines show growth rates similar to those of wild-type cells. Thus by our selection systems we not only restored biosynthesis, which had been lost when the cultures changed from slowly growing aggregated cultures to rapidly growing fine suspension cultures, but also established this biosynthesis in fine suspensions. Again, we do not know which molecular changes were necessary to express this biosynthesis in the suspended cells.

Another observation indicates the limits of our selection systems. In none of the 4-methyltryptophan-tolerant *Peganum* cell lines was β-carboline alkaloid biosynthesis restored. Though we know that good TDC activity is required for high levels of β-carboline alkaloids, expression of this enzyme was evidently not sufficient to make the cells produce these alkaloids. As long as the subsequent biosynthetic enzymes of the β-carbolines are not co-induced, the desired products will remain absent even in cells with high TDC activity. Thus there seems to be another regulatory control between tryptamine and the first intermediate of β-carboline biosynthesis. Serotonin biosynthesis is only restored because the second enzyme, the tryptamine-5-hydroxylase, is always expressed in producing and nonproducing cells. Otherwise one would have expected increased accumulation of tryptamine.

Monoterpene indole alkaloid biosynthesis in Catharanthus. Results comparable to those above were obtained when we selected for 4-methyltryptophan-tolerant *Catharanthus* cell lines. While TDC and tryptamine levels distinctly increased, the levels of indole alkaloids remained low (36). Here the negative outcome may be not only a question of the absence or presence of the subsequent biosynthetic enzymes of the alkaloid branch but also a question of control in the biosynthesis of the monoterpenoid secologanin. Geraniol-10-hydroxylase seems to be an enzyme controlling the flow of terpenoids into the indole alkaloids (38). For this enzyme a positive selection system is not available at present.

In general, we have to conclude that positive selection for altered regulatory controls of secondary pathways can only be applied in special systems where co-induction occurs or where the other enzymes are present. If other regulatory controls are not affected by the selection system, the chances of finding high-producing cell lines may be rather low. This is due to the lack of specific selection systems for most enzymes of secondary pathways. On the other hand, even selections that are not aimed at specific targets of a secondary pathway may be useful. Selection of cell lines eliminates the majority of, if not all, wild-type cells and may yield new cell lines with new traits or traits previously overlooked. For example, the PFP-tolerant cells (29) are altered not only with respect to cinnamoyl putrescine biosynthesis but also with respect to uptake, cell structure, and tyrosine decarboxylase (46). Indeed, a completely new cell type was found. Thus, since the desired genetic manipulations will take time, the alternative of selecting for new cell types by killing the normal cells may be a useful approach.

Genetic Transformation of Secondary Pathways

The characterization of the above culture systems gave hints as to which of the systems might or might not be applicable for genetic transformations. The simplicity of serotonin biosynthesis and the fact that the lack of just one enzymatic step (TDC) prevents this biosynthesis make this pathway most interesting for such initial studies. Indeed, serotonin biosynthesis in nonproducing Peganum cells may be overcome not only by selections but also by genetic engineering. One could introduce a foreign gene coding for TDC into Peganum cells in order to restore this pathway. Consequently, we purified the TDC of Catharanthus roseus to homogeneity, prepared specific antibodies, and detected mRNAs which were able to form TDC in vitro (27,28). However, we are still screening for a cDNA clone containing the complete coding sequence for this enzyme. Even under optimal conditions, the mRNA for TDC represents a very small portion of the total mRNA of Catharanthus cultures. Though we have restored serotonin biosynthesis by selection techniques, we will continue on the transformation path, since we expect some rather clear answers from this very simple biosynthetic system. For example, although the foreign enzyme may be expressed, will it be supplied with substrate? Or, if tryptophan is decarboxylated to tryptamine, will the product be available for the tryptamine-5-hydroxylase, and, if so, will it be used? We expect that many of these and similar basic questions may be answered with this system.

When we realized how time consuming cloning of genes of secondary pathways for transformation purposes can be, naturally the question arose whether cloned plant genes are necessary for the transformations that we want to perform. If only the lack of one enzyme activity is the reason for the failure of cells to produce a desired compound de novo, one could imagine that a bacterial enzyme could also perform the missing reaction. Since a suitable bacterial source for a TDC gene was not available, another simple secondary pathway was chosen to answer this question. The tetracyclic ring structure of quinolizidine alkaloids in Lupinus is formed from lysine by two enzymatic steps (49). First, lysine is decarboxylated to cadaverine, then four cadaverine molecules are converted by oxosparteine synthase to the basic alkaloid structure. As in other systems, lysine decarboxylase activity seems at least partially to control the formation of the alkaloids in plants and cell cultures (30,50). Therefore we cloned the gene coding for lysine decarboxylase of the enterobacterium Hafnia alvei (14). Recently, we removed the sequence of the cloned DNA before the start codon which might impair expression of the gene when Agrobacterium-derived vectors are used for transformation (S. Herminghaus, ms. in prep.). Thus this bacterial gene is now ready for integration into suitable vectors for transformation experiments. We have recently transformed sterile-grown Lupinus hartwegii and L. polyphyllus plants with wild-type strains of Agrobacterium tumefaciens (DSM 30150) and A. rhizogenes (15834) (J. Berlin and C. Sator, ms. in prep.) so that successful transformations with strains containing vectors with the cloned lysine decarboxylase gene should also be possible. Again, we expect some general answers regarding the use of bacterial genes for interfering with secondary pathways of plants.

CONCLUSIONS

Unfortunately, no example could be presented that genetic manipulations of regulatory controls of secondary pathways will indeed help to improve the production rates of cell cultures for a desired secondary metabolite or will help to express more pathways in cultured cells. The interest in applying such techniques is recent, and the number of basic steps necessary before results are achieved could be frightening.

However, it is also evident that we have to apply such techniques to secondary metabolism if we want to alter and to manipulate secondary pathways. While the scientific value of this approach is undisputed, it is not clear today whether these manipulations of secondary pathways will improve the biotechnological impact of plant cell cultures on the production of useful compounds. It is most important that more genes of those biosynthetic enzymes that have a regulatory function become available for such experiments.

Although predictions of the outcome of improving production rates in cultured cells via genetic manipulations cannot be made as yet, product-oriented research should go in this direction. For such research it might not be so important to alter regulatory controls of the plant cell but rather to use the "side effects" of this research which might be biotechnologically even more relevant. Expression of cloned plant genes in bacteria and their use for stereospecifically difficult biotransformations could, for example, be a very attractive by-product. Indeed, many chemically difficult, but plant-specific reactions might be better performed by the corresponding enzymes than by organic synthesis. Whether for such purposes transformed plant cells, transformed microorganisms, or immobilized enzymes are best suited, is not decisive. In all cases, the genes of the enzymes have to be cloned and characterized first.

REFERENCES

1. Abel, P.P., R.S. Nelson, B. De, N. Hoffmann, S.G. Rogers, R.T. Fraley, and R.N. Beachy (1986) Delay of disease development in transgenic plants that express the tobacco mosaic virus coat protein. Science 232:738-743.
2. Baumert, A., and D. Gröger (1985) Synthesis of 1,3-dihydroxy-N-methylacridone by cell free extracts of Ruta graveolens cell suspension cultures. FEBS Lett. 187:311-313.
3. Berlin, J. (1986) Secondary products from plant cell cultures. In Biotechnology, Vol. 4, H.J. Rehm and G. Reed, eds. VCH Verlagsgesellschaft, Weinheim, pp. 630-658.
4. Berlin, J., and F. Sasse (1985) Selection and screening techniques for plant cell cultures. Adv. Biochem. Eng./Biotech. 31:99-132.
5. Berlin, J., K.H. Knobloch, G. Höfle, and L. Witte (1982) Biochemical characterization of two tobacco cell lines with different levels of cinnamoyl putrescines. J. Nat. Prod. 45:83-87.
6. Berlin, J., L. Witte, J. Hammer, K.G. Kukoschke, A. Zimmer, and D. Pape (1982) Metabolism of p-fluorophenylalanine in p-fluorophenylalanine sensitive and resistant tobacco cell cultures. Planta 155:244-250.
7. Berlin, J., C. Mollenschott, F. Sasse, L. Witte, G.W. Piehl, and H. Büntemeyer (1987) Restoration of serotonin biosynthesis in cell suspension cultures of Peganum harmala by selection for 4-methyltryptophan tolerant cell lines. J. Plant Physiol. (in press).
8. Broeck, G. van den, M.P. Timko, A.P. Kausch, A.R. Cashmore, M. von Montagu, and L. Herrera-Estrella (1985) Targeting of a foreign protein to chloroplasts by fusion of the trans peptide from the small subunit of ribulose 1,5-biphosphate carboxylase. Nature 313:358-363.
9. Cramer, C.L., T.B. Ryder, J.N. Bell, and C.J. Lamb (1985) Rapid switching of plant gene expression induced by fungal elicitors. Science 227:1240-1243.

10. Cramer, C.L., J.N. Bell, T.B. Ryder, J.A. Bailey, W. Schuch, G.P. Bolwell, M.P. Robbins, R.A. Dixon, and C.J. Lamb (1985) Co-ordinated synthesis of phytoalexin biosynthetic enzymes in biologically-stressed cells of bean (Phaseolus vulgaris). EMBO J. 4:285-289.
11. DeLuca, V., J. Balsevich, R.T. Tyler, U. Eilert, B.D. Panchuk, and W.G.W. Kurz (1986) Biosynthesis of indole alkaloids: Developmental regulation of the biosynthetic pathway from tabersonine to vindoline in Catharanthus roseus. J. Plant Physiol. 125:147-156.
12. DiCosmo, F., and M. Misawa (1985) Eliciting secondary metabolism in plant cell cultures. Trends in Biotech. 3:318-322.
13. Eckes, P., S. Rosahl, J. Schell, and L. Willmitzer (1986) Isolation and characterization of a light-inducible, organ-specific gene from potato and analysis of its expression after tagging and transfer into tobacco and tomato shoots. Mol. Gen. Genet. 205:14-22.
14. Fecker, L., H. Beier, and J. Berlin (1986) Cloning and characterization of a lysine decarboxylase gene from Hafnia alvei. Mol. Gen. Genet. 203:177-184.
15. Fraley, R.T., G.S. Rogers, R.B. Horsch, D.A. Eichholtz, J.S. Flick, C.L. Fink, N.L. Hoffmann, and P.R. Sanders (1985) The SEV System: A new disarmed Ti plasmid vector system for plant transformation. Bio/Technology 3:629-635.
16. Fujita, Y., M. Tabata, A. Nishi, and Y. Yamada (1982) New medium and production of secondary compounds with the two-staged culture method. In Plant Tissue Culture 1982, A. Fujiwara, ed. Maruzen Press, Tokyo, pp. 399-400.
17. Gambliel, H., and R. Croteau (1984) Pinene cyclases I and II. J. Biol. Chem. 259:740-748.
18. Hahlbrock, K., and H. Grisebach (1979) Enzymic controls in the biosynthesis of lignin and flavonoids. Ann. Rev. Plant Physiol. 30:105-130.
19. Hashimoto, T., and Y. Yamada (1986) Hyoscyamine 6β-hydroxylase, a 2-oxoglutarate dependent dioxygenase, in alkaloid producing root cultures. Plant Physiol. 81:619-625.
20. Heide, L., R. Kolkmann, S. Arendt, and E. Leistner (1982) Enzymic synthesis of o-succinylbenzoyl-CoA in cell free extracts of anthraquinone producing Galium mollugo cell suspension cultures. Plant Cell Reports 1:180-182.
21. Heller, W., B. Egin-Bühler, S.E. Gardiner, K.H. Knobloch, U. Matern, J. Ebel, and K. Hahlbrock (1979) Enzymes of general phenylpropanoid metabolism and of flavonoid biosynthesis in parsley. Plant Physiol. 64:371-373.
22. Isaac, J.E., R.J. Robins, and M.J.C. Rhodes (1987) Cinchoninone: NADPH oxidoreductases I and II--Novel enzymes in the biosynthesis of quinoline alkaloids in Cinchona ledgeriana. Phytochemistry 26:393-399.
23. Klee, H.J., M.F. Yanofsky, and E. Nester (1985) Vectors for transformation of higher plants. Bio/Technology 3:637-642.
24. Kochs, G., and H. Grisebach (1987) Induction and characterization of a NADPH-dependent flavone synthase from cell cultures of soybean. Z. Naturforsch. 42c:343-348.
25. Kuhn, D.N., J. Chapell, A. Boudet, and K. Hahlbrock (1984) Induction of phenylalanine ammonia lyase and 4-coumarate:CoA ligase mRNAs in cultured plant cells by UV-light and fungal elicitor. Proc. Natl. Acad. Sci., USA 81:1102-1106.
26. Misawa, M. (1985) Production of useful metabolites. Adv. Biochem. Eng./Biotech. 31:59-88.

27. Noé, W., and J. Berlin (1985) Induction of de novo synthesis of tryptophan decarboxylase in cell suspension cultures of Catharanthus roseus. Planta 166:500-504.
28. Noé, W., C. Mollenschott, and J. Berlin (1984) Tryptophan decarboxylase from Catharanthus roseus: Purification, molecular and kinetic data of the homogeneous protein. Plant Mol. Biol. 3:281-288.
29. Palmer, J.E., and J.M. Widholm (1975) Characterization of carrot and tobacco cell cultures resistant to p-fluorophenylalanine. Plant Physiol. 56:233-238.
30. Pelosi, L.A., A. Rother, and J.M. Edwards (1986) Lysine decarboxylase activity and alkaloid production in Heimia salicifolia cultures. Phytochemistry 25:2315-2319.
31. Pfitzner, A., L. Polz, and J. Stöckigt (1986) Properties of the vinorine synthase--The Rauwolfia enzyme involved in the formation of the ajmaline skeleton. Z. Naturforsch. 41c:103-114.
32. Potrykus, I., M.W. Saul, J. Petruska, J. Paszkowski, and R. Shilito (1985) Direct gene transfer to cells of a graminaceous monocot. Mol. Gen. Genet. 199:183-188.
33. Ragg, H., D.N. Kuhn, and K. Hahlbrock (1981) Coordinated regulation of 4-coumarate:CoA ligase and phenylalanine ammonia lyase mRNAs in cultured plant cells. J. Biol. Chem. 256:10061-10065.
34. Sasse, F., U. Heckenberg, and J. Berlin (1982) Accumulation of β-carboline alkaloids and serotonin by cell cultures of Peganum harmala. Plant Physiol. 69:400-404.
35. Sasse, F., K.H. Knobloch, and J. Berlin (1982) Induction of secondary metabolism of Catharanthus roseus, Nicotiana tabacum and Peganum harmala. In Plant Tissue Culture 1982, A. Fujiwara, ed. Maruzen Press, Tokyo, pp. 343-344.
36. Sasse, F., M. Buchholz, and J. Berlin (1983) Selection of cell lines of Catharanthus roseus with increased tryptophan decarboxylase activity. Z. Naturforsch. 38c:916-922.
37. Sasse, F., L. Witte, and J. Berlin (1987) Biotransformation of tryptamine to serotonin by cell suspension cultures of Peganum harmala. Planta Med. 53:354-359.
38. Schiel, O., L. Witte, and J. Berlin (1987) Geraniol-10-hydroxylase activity and its relation to monoterpene indole alkaloid accumulation in cell suspension cultures of Catharanthus roseus. Z. Naturforsch. Biosci. Vol. 42c (in press).
39. Schocher, R.J., R.D. Shilito, M.W. Saul, J. Paszkowski, and I. Potrykus (1986) Co-transformation of unlinked foreign genes into plants by direct gene transfer. Bio/Technology 4:1093-1096.
40. Schöppner, A., and H. Kindl (1984) Purification and properties of a stilbene synthase from induced cell cultures of peanut. J. Biol. Chem. 259:6806-6811.
41. Shah, D.M., R. Horsch, H.J. Klee et al. (1986) Engineering herbicide tolerance in transgenic plants. Science 233:478-481.
42. Stafford, A., P. Morris, and M.W. Fowler (1986) Plant cell biotechnology: A perspective. Enzyme Microb. Technol. 8:578-587.
43. Ulbrich, B., W. Wiesner, and H. Arens (1985) Large-scale production of rosmarinic acid from plant cell cultures of Coleus blumei. In Primary and Secondary Metabolism of Plant Cell Cultures, K.H. Neumann, W. Barz, and E. Reinhard, eds. Springer-Verlag, Berlin, Heidelberg, New York, pp. 293-303.
44. Veltkamp, E., and J.N.M. Mol (1986) Improved production of secondary metabolites in cultures of plant cells and microorganisms: The biosynthesis of flavonoids/anthocyanins in Petunia hybrida as model system. In Biomolecular Engineering Programme--Final Report, E.

Magnien, ed. Martinus Nijhoff Publishers, Dordrecht, The Netherlands, pp. 1071-1080.

45. Wagner, R., F. Feth, and K.G. Wagner (1986) Regulation of enzyme activities of the nicotine pathway in tobacco. Physiol. Plant. 68:667-672.
46. Walker, M.A., B.E. Ellis, E.B. Dumbroff, R.G. Downer, and R.J. Martin (1986) Changes in amines and biosynthetic enzyme activities in p-fluorophenylalanine resistant and wild type tobacco cell cultures. Plant Physiol. 80:825-828.
47. Wat, C.K., P. Steffens, and M.H. Zenk (1986) Partial purification of SAM:norreticuline N-methyltransferases from Berberis cell suspension cultures. Z. Naturforsch. 41c:126-134.
48. Wienand, U., U. Weydemann, U. Niesbach-Klösgen, P.A. Peterson, and H. Saedler (1986) Molecular cloning of the c2 locus of Zea mays, the gene coding for chalcone synthase. Mol. Gen. Genet. 203:202-207.
49. Wink, M., and T. Hartmann (1982) Localization of the enzymes of quinolizidine alkaloid biosynthesis in leaf chloroplasts of Lupinus polyphyllus. Plant Physiol. 70:74-77.
50. Wink, M., T. Hartmann, and L. Witte (1980) Biotransformation of cadaverine and potential biogenetic intermediates of lupanine biosynthesis by plant cell suspension cultures. Planta Med. 40:31-39.
51. Yamada, Y., and N. Okada (1985) Biotransformation of tetrahydroberberine to berberine by enzymes prepared from cultured Coptis japonica cells. Phytochemistry 24:63-65.
52. Zenk, M.H. (1980) Enzymatic synthesis of ajmalicine and related indole alkaloids. J. Nat. Prod. 43:438-451.
53. Zenk, M.H., M. Rueffer, M. Amann, and B. Deus-Neumann (1985) Benzylisoquinoline biosynthesis by cultivated plant cells and isolated enzymes. J. Nat. Prod. 48:725-738.

TISSUE CULTURES OF *EUPHORBIA* SPECIES

Yoshikazu Yamamoto

Technical Center
Nippon Paint Company, Ltd.
Neyagawa, Osaka 572, Japan

ABSTRACT

We induced calluses from three *Euphorbia* species and investigated their chemical constituents. We established, by clonal selection, a strain of cultured *Euphorbia millii* cells that produce a high and stable level of anthocyanin. This strain can be used to produce anthocyanin on a large scale for industrial application. The lipid constituents of *Euphorbia tirucalli*, *E. millii*, and *E. lathyris* calluses were sitosterol, stigmasterol, campesterol, palmitic acid, and linoleic acid. Callus from *E. millii* produced an anthocyanin, cyanidin glycoside. By cell-aggregate cloning, we selected a strain of cultured *E. millii* cells that produced a level of pigment seven times higher than that in the original cells. Statistical and cell-pedigree analyses using a computerized system proved that this cell strain has stable productivity for the red pigment. We determined the optimum composition of medium for anthocyanin production in suspension culture. Production of anthocyanin in suspension culture in modified Gamborg's medium is 4.5 times higher than that in MS medium, which we used in solid culture. A paddle with holes for agitation contributed to higher anthocyanin production in jar-fermenter culture than did other rotators.

INTRODUCTION

The genus *Euphorbia* consists of 2,000 to 3,000 species which grow in semi-arid lands world-wide. The latex from these species is highly vesicating; it irritates the skin and mucous membranes, and is used as a fish poison and an insecticide. Its biological functions have been reported to be caused by the fatty acid esters of diterpene alcohols (3). Calvin (1) has spearheaded a drive to develop *Euphorbia* species as future sources of gasoline, and has propagated *E. tirucalli* (African milk bush) and *E. lathyris* (gopherwood) on a large scale. His group has reported that chemical constituents of *Euphorbia* species are mainly triterpene alcohols (8).

We were interested in production of fuel compounds and biological substances of the genus *Euphorbia* by plant cell cultures. Therefore, we induced calluses from *E. tirucalli*, *E. lathyris*, and *E. millii* (13).

Usually plant cell cultures produce only small amounts of secondary metabolites. Recently, however, cell strains containing amounts of secondary metabolites greater than those found in intact plants have been isolated by clonal selection. There are two questions in respect to clonal selection: (a) How can we determine whether the high productivity for secondary metabolites in selected cell strains is stable? (b) How long must we continue clonal selection in order to obtain stable cell strains that produce secondary metabolites?

Pedigree analysis has been used in breeding programs of garden plants. The knowledge of parentage obtained by pedigree tracing can serve as an aid when we assign genotypes, and can give an indication of the degree of genetic diversity in breeding populations. Pedigree analysis performed manually is unwieldy, but when a computer system is used we can accommodate a large breeding program that involves the crossing of many genotypes and the collection of data on numerous progeny.

We continuously selected a cell strain that showed high content of a red pigment, an anthocyanin, from cultured E. millii cells, and proved by pedigree and statistical analyses that this strain had stable production of this pigment (14,15).

Anthocyanins, i.e., natural colorants, are widely distributed in plant species. They are mainly contained in flower petals and fruit peels. They have been used in foodstuffs for many years. However, it is difficult to maintain a stable and abundant supply of anthocyanins from the natural habitat.

There are many reports on anthocyanin production in cultured plant cells (e.g., carrot and grape). Two groups (2,12) have reported the use of suspension cultures for anthocyanin production. In this chapter, we report on the large amount of anthocyanin produced by jar-fermenter culture of E. millii cells.

MATERIALS AND METHODS

Callus Induction

Sterilized samples of young shoots of E. tirucalli Desf., leaves of E. millii Ch. des Moulins, or stems of E. lathyris L. were placed on Murashige and Skoog's (MS) basal medium (7) with 0.8% agar, 2% sucrose, 0.1-10 ppm 2,4-dichlorophenoxyacetic acid (2,4-D) or naphthaleneacetic acid (NAA), and 0.2% natural supplement [malt extract (ME), yeast extract (YE), or casein hydrolysate (CH)]. The cultures were kept under fluorescent light (3,000 lux) at 28°C. Calluses of E. tirucalli, E. millii, and E. lathyris appeared after two weeks of inoculation.

Extraction and Isolation of Chemical Constituents

Fresh calluses of E. tirucalli, E. millii, and E. lathyris (145 g, 140 g, and 644 g, respectively) were treated separately with acetone for 24 hr. Their acetone extracts were concentrated under reduced pressure, then flushed with ethyl acetate (EtOAc) after the addition of water. The EtOAc solution for each species was evaporated to give the EtOAc extracts (1.2 g from E. tirucalli callus, 0.8 g from E. millii callus, and 4.3 g from E. lathyris callus). The EtOAc extracts were separated into two fractions by silica gel column chromatography with hexane-EtOAc. These fractions were run through preparative thin-layer silica gel chromatography with hexane-EtOAc-acetic acid (AcOH) (80:20:1) as the solvent, which resulted in phytosterols and fatty acids.

Mottled red and white callus of E. millii (120 g) was combined with methanol containing 0.1% HCl and kept in the refrigerator overnight, after which it was treated one more time. The combined methanol solution was concentrated in vacuo and washed with ether after the addition of water. The acidic aqueous solution was evaporated in vacuo. The reddish residue was chromatographed on a cellulose column with H_2O-methanol (19:1) containing 2% HCl as the eluent, and red pigment (65 mg) was obtained.

Analyses of the Phytosterols and Fatty Acids

Phytosterols were analyzed by gas chromatography-mass spectrometry on a 3 mm x 1 m glass column packed with 2% OV-1 on Chromosorb W® (AW; DMCS), 80-100 mesh. Operating conditions were as follows: column temperature, 250°C; ionization voltage, 20 eV; accelerating voltage, 3 KV; chamber temperature, 180°C. The phytosterols were separated into four compounds. Retention times of these compounds were 5.32, 6.93, 7.23, and 8.45 min, respectively, and their molecular ions appeared at m/z 386, 400, 412, and 414. These values were consistent with those for authentic sterols.

The fatty acids were boiled in BF_3-methanol. The methyl esters were identified by comparing their GLCs with those of authentic samples. Gas-liquid chromatography analysis was made on a 3 mm x 2 m glass column packed with 15% diethylene glycol succinate on Chromosorb W® (AW; DMCS), 80-100 mesh, increasing the temperature from 150°C to 220°C at 6°C/min, with an N_2 flow rate of 32 ml/min.

Analysis of the Aglycon of Anthocyanin

The red pigment (10 mg) was dissolved in 20% HCl and boiled to hydrolyze it. Its solution was cooled immediately after the addition of water, and then washed with benzene and dried in vacuo to obtain the aglycon (5.6 mg).

The UV and visible absorption spectra of the aglycon were measured in methanol containing 0.01% HCl. The $AlCl_3$ shift was measured after the addition of a few drops of 5% $AlCl_3$ in EtOH to the sample in a cuvette.

Paper and thin-layer chromatographies were made on Toyo No. 51 paper and an Avicel thin-layer plate with the following solvent system (v/v): n-BuOH-AcOH-$H_2$0, 4:1:5, upper layer; AcOH-H_2O-HCl, 15:82:3; AcOH-HCl-H_2O, 5:1:5; AcOH-HCl-H_2O, 30:3:10; n-BuOH-HCl-H_2O, 5:1:4, upper layer; HCl-H_2O, 3:97; HCO_2H-HCl-$H_2$0, 1:2:1.

Selection and Culture

The outline of the selection method, cell-aggregate cloning, is shown in Fig. 1. The color of the original calluses of E. millii was mottled red and white. These calluses were cut into 128 segments (∿3 mm each) with a scalpel. Each segment was coded and placed on agar medium (25 ml) in a sectioned Petri dish 9 cm in diameter. The agar medium consisted of MS solution, 10^{-6} M 2,4-D, 0.2% ME, 2% sucrose, and 0.8% agar.

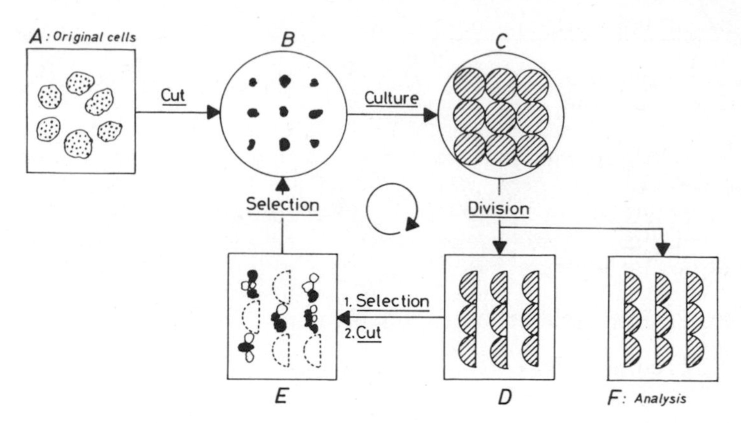

Fig. 1. Outline of the selection method (cell-aggregate cloning). ○ = unselected segment; ● = selected segment; ⁚ = unselected aggregate; ▨ = grown aggregate. From Ref. 11, with permission of Academic Press, Inc.

The segments were cultured at 28°C under fluorescent light (6,000 lux) for ten days. Each of the nine segments on a Petri dish was cut into two cell-aggregates: one (D_1) for subculture and the other (D_2) for quantitative analysis of the pigment. From the analysis of D_2, we selected the reddest D_1 cell-aggregate from each Petri dish. These selected cell-aggregates were cut into several segments (∿3 mm), and the segments were coded and transplanted onto fresh medium in a nine-section Petri dish. The code of each segment was put into the data file of a computer together with the code of the mother cell-aggregate. Segments selected at each transplantation were cultured on the same medium under the above conditions. This selection procedure was repeated 28 times. All codes for the 1,588 cell-aggregates from the 7th to 28th passages were put into the data file.

Computer System

The computer system (CELSEL) consists of four subsystems, MASTER, LIST, TREE, and ROOT, each of which is composed of a program and a file. Programs contained in these subsystems were written in DSM-11 as implemented for the PDP 11/34 system because of the convenience of using this language for handling data files that show a tree-structure. Input data for individual cell-aggregates consist of individual codes, anthocyanin content, fresh weight (mg), and daughter codes.

The records for 1,588 cell-aggregates, from the 7th to 28th passages, were put into the MASTER file with the MASTER program. The TREE and ROOT routine programs read the individual code, found the corresponding daughter or mother codes in the MASTER file, read the daughter or mother codes, and then traced the pedigrees through daughters or mothers by repeated use of the same sequence. When the TREE program did not find the corresponding daughter code, it stopped searching the parentage code and read the next individual code. The TREE and ROOT programs transferred the parentage codes to each file and printed them as a hard copy or on a visual display.

Determination of Pigment Contents

A random sample of 30 from the D_2 population was prepared for quantitative analysis of the red pigment. Each segment of the sample was soaked in methanol (4 ml) containing 2% HCl for 1 hr. The absorbance of the methanol solution was measured at 530 nm, then the pigment content [absorbance at 530 nm/fresh weight (mg) of the cell-aggregate] was calculated. A frequency distribution of cell-aggregates with various pigment contents was made (class-width was fixed at 0.5). Two characteristic terms, the mean ($\bar{C}$) and maximum (C^{max}) values, were calculated.

Determination of the Distribution Rate

The pedigree of the cell-aggregates was graphed, and a cell line (a group of cell-aggregates derived from one ancestor) was tabled by the computer program. The distribution rate (%) [(the number of cell-aggregates of a specific cell line/the number of total cell-aggregates) x 100] for each line was calculated for each generation. A frequency distribution then was made, and its mean value was calculated. This is marked in the histogram of the total cell-aggregates.

Suspension Culture

Cultured E. millii cells (3.5 g fresh weight) growing on agar medium were transferred to liquid medium (100 ml) in a 300-ml conical flask. These cells were maintained at 26°C on the rotatory shaker with 120 rpm under fluorescent light (∿6,000 lux) for ten days. Culture conditions were as follows: MS, Linsmaier and Skoog's (LS) (6), Gamborg's (GA) (4), Nitsch and Nitsch's (NN) (9), Heller's (HE) (5), or White's (WH) (10) solution was employed as basal inorganic solution with 10^{-6} M 2,4-D, 0.2% ME, and 2% sucrose. Sucrose at 1, 3, 5, 7, or 9% was added to GA medium with 10^{-6} M 2,4-D and 0.2% ME. Benzyladenine (BA) at 10^{-10}, 10^{-9}, 10^{-8}, or 10^{-7} M and 2,4-D at 10^{-7}, 10^{-6}, or 10^{-5} M were added to GA medium with 5% sucrose. Naphthaleneacetic acid at 10^{-6}, 10^{-5}, or 10^{-4} M and BA at 10^{-10}, 10^{-9}, 10^{-8}, or 10^{-7} M were added to GA medium with 5% sucrose. Nitrogen sources [NH_4NO_3 and KNO_3 (total amount: 15, 30, or 60 mM; NH_4^+/NO_3^- ratio: 0, 1/64, 1/32, 1/16, 1/8, 1/4, 1/2, or 1/1)] were added to GA medium with 5% sucrose, 10^{-6} M 2,4-D, 10^{-8} M BA, and no nitrogen source. The K^+ concentration was 30 mM in all media after controlling the additional amount of KCl. Cation (Fe^{2+}, Ca^{2+}, Mg^{2+}, or Cu^{2+}) and anion (PO_4^{3-} or SO_4^{2-}) sources were added to GA medium with 5% sucrose, 10^{-6} M 2,4-D, 10^{-8} M BA, and no corresponding ion source. The level of the concentration of these ions was half of, equal to, or two-fold that in GA medium.

Jar-Fermenter Culture

Cultured E. millii cells (150 g fresh weight) growing in liquid medium on the rotatory shaker were transferred to liquid medium (5 l) in a 14-l glass jar. These cells were maintained at 26°C under fluorescent light (∿6,000 lux) for seven days. Culture conditions were as follows: modified Gamborg's solution (EM) was employed as basal inorganic solution with 5% sucrose, 10^{-6} M 2,4-D, and 10^{-8} M BA. Two distinct paddles with holes, or a screw, were employed for agitation, and their rotation was 30, 50, 75, or 100 rpm. The dissolved oxygen value was controlled between 68 and 70%.

Measurement of Cell Growth in Suspension Culture

Fresh cultured E. millii cells were harvested after removing liquid medium and washing with water by suction filtration, and then dried for 24 hr in the freeze-dry apparatus. Cell growth was evaluated with dry weight.

Determination of Anthocyanin Content and Production in Suspension Culture

Dried cells (200 mg) were soaked in methanol (20 ml) containing 0.35% HCl for 1 hr. The methanol solution (1 ml) was diluted to 100-fold volume with the above acidic methanol. The absorbance of the methanol solution was measured at 530 nm; then the anthocyanin content [absorbance at 530 nm/dried cell weight (g)] and anthocyanin production (absorbance at 530 nm/flask) were calculated.

RESULTS AND DISCUSSION

Callus Induction

We obtained four calluses that were growing well from three Euphorbia species: yellowish-white callus on medium with ME and 5 x 10^{-7} 2,4-D from E. tirucalli; yellow callus on medium with ME and 5 x 10^{-6} 2,4-D and mottled red and white callus on medium with ME and 5 x 10^{-7} 2,4-D from E. millii; and yellow callus on medium with ME and 5 x 10^{-6} 2,4-D from E. lathyris.

Chemical Constituents

Gas chromatography-mass spectrometry showed that the sterol fraction consisted of sitosterol, stigmasterol, campesterol, and cholesterol. The fatty acid fraction was shown by GLC to contain palmitic and linoleic acids. We obtained a red pigment from a mottled red callus of E. millii. Using spectroscopic and chromatographic data, we identified cyanidin as the aglycon of this pigment. Its Fehling reaction was positive. These results showed that this pigment is a cyanidin glycoside. This is the first isolation of a cyanidin glycoside from cultured cells of E. millii; we have found no report of the existence of anthocyanins in this intact plant.

Selection of a Variant Cell Strain

The reddest cell-aggregates were selected continuously from each of the 28 subcultures of the original mottled red callus of E. millii. Two characteristic terms ($\bar{C}$ and C^{max}) in the frequency distribution of cell-aggregates from the 16th to 28th passages were plotted against the pigment content (Fig. 2). The mean value ($\bar{C}$) for the pigment content increased nearly three-fold from the 16th to 22nd passages. It came within the limit of 7.46 ± 0.56 after the 23rd passage, which was seven times higher than that of the original callus (1.05).

The 9A and 9F cell lines originated from the 9A and 9F cell-aggregates of the 9th subculture. In this subculture, the 9A and 9F lines produced pigment-rich descendants at high frequencies. The distribution rates of these two cell lines in the population of cell-aggregates from the 16th to 24th passages are shown in Fig. 3. The distribution rate of the 9F cell line was lower than that of the 9A cell line at the 16th subculture,

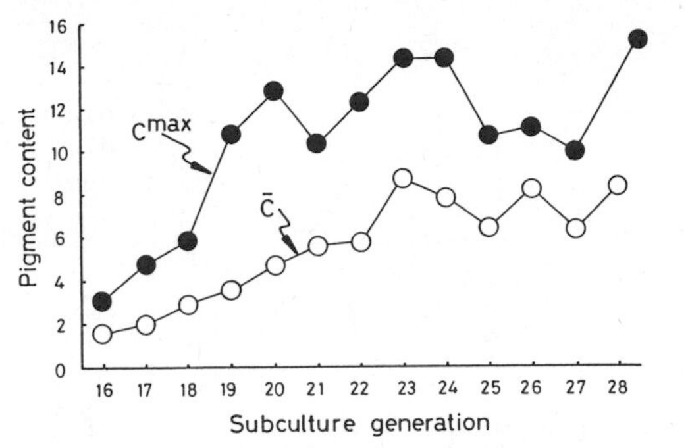

Fig. 2. Trends of the two characteristic terms, the mean value ($\bar{C}$) and maximum value (C^{max}), in the frequency distribution of cell-aggregates with various pigment contents from the 16th to 28th passages. From Ref. 16, with permission of the Japanese Association for Plant Tissue Culture.

but equaled it at the 20th subculture. This rate gradually increased and reached 100% at the 24th subculture. Figure 4 shows the histogram for frequency distributions of cell-aggregates from the 16th to 22nd passages.

If the mean value for the content of secondary metabolites is stable in a population of cell-aggregates chosen by successive clonal selection, and if all the cell-aggregates are derived from one cell-aggregate, they should consist of cells with high and stable production for secondary metabolites. These results show that we could isolate and culture a cell strain containing a high and stable level of pigment from E. millii callus after 24 successive clonal selections. We believe that during early passages cultured cells are heterogeneous in their production of pigments and that they contain variant cells that produce a large amount of pigment. Results show that we were able to select more cell-aggregates of the 9F cell line than of the 9A cell line at each subculture, and that the 9F cell-aggregate had variant cells that produced a high level of pigment. In preliminary experiments involving single-cell cloning, we failed to obtain a strain of cultured E. millii cells that produced a high level of pigment because culture of a single cell was too difficult.

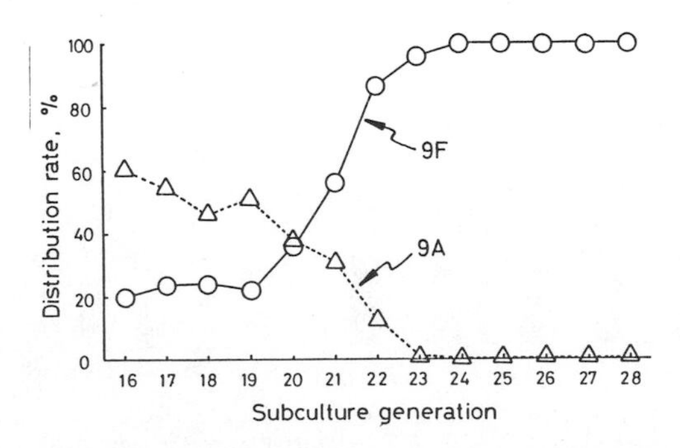

Fig. 3. Distribution rates of the 9A (Δ) and 9F (o) cell lines in the population of cell-aggregates from the 16th to 22nd passages. From Ref. 16, with permission of the Japanese Association for Plant Tissue Culture.

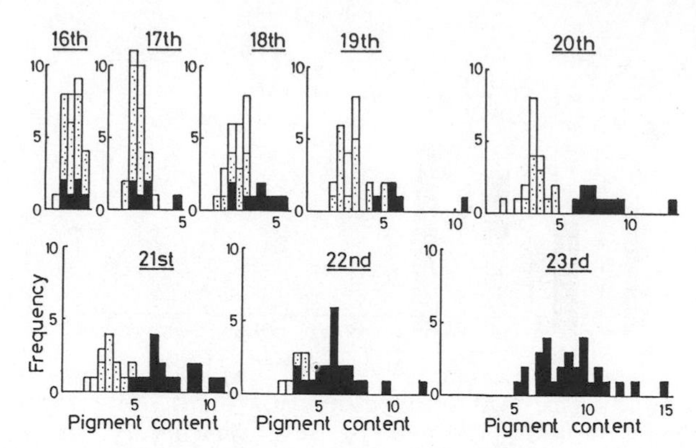

Fig. 4. Frequency distributions of cell-aggregates of the 9A ( ) and 9F ( ) cell lines in histograms of the population of cell-aggregates with various pigment contents from the 16th to 22nd passages. Cell-aggregates other than the 9A and 9F cell lines are designated by ☐. From Ref. 16, with permission of the Japanese Association for Plant Tissue Culture.

Effects of Medium Components on Anthocyanin Production in Suspension Culture

We investigated whether the medium composition of solid culture was also suitable for anthocyanin production in liquid culture. Cultured E. millii cells that produced a high level of anthocyanin on solid medium were cultured in liquid medium containing an inorganic solution of LS, MS, GA, NN, HE, or WH (Fig. 5). The anthocyanin production in GA medium is 1.3-fold that in MS medium used in the case of solid culture.

Figure 6 shows the effect of sucrose on anthocyanin production. The optimum concentration of sucrose for anthocyanin production was 5 to 7%. Sucrose at a concentration of 9% diminished cell growth. The absence of sucrose caused death of cells. The anthocyanin production in liquid culture with 5% sucrose is three-fold that in solid culture with 2% sucrose.

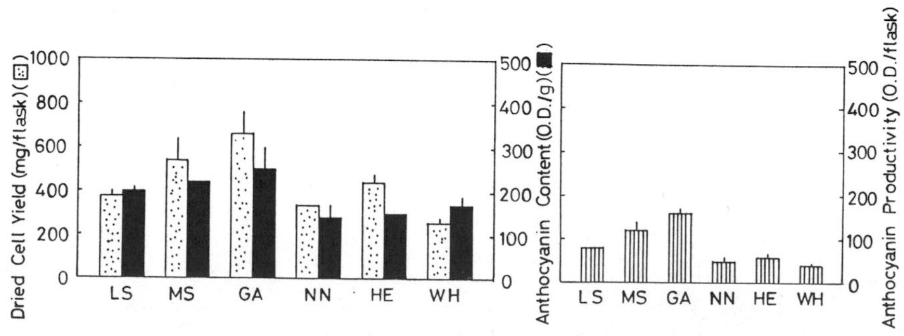

Fig. 5. Effect of basal medium on anthocyanin production in suspension cultures of E. millii cells. Cultures were maintained in various basal media with 2% sucrose, 0.2% ME, and 10^{-6} M 2,4-D on the rotatory shaker under fluorescent light (6,000 lux) at 26°C for ten days.

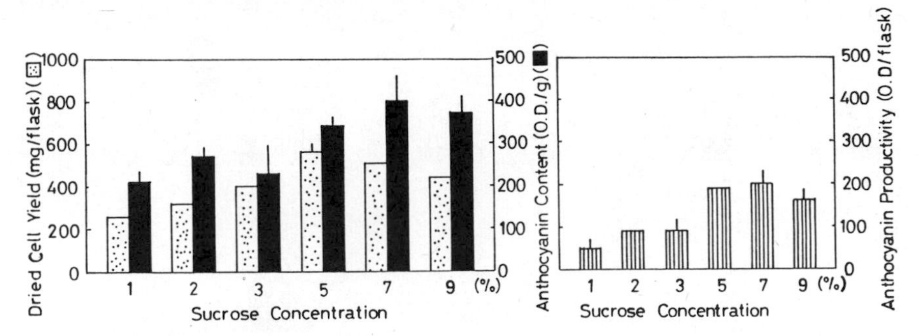

Fig. 6. Effect of sucrose on anthocyanin production in suspension cultures of E. millii cells. Cultures were maintained in GA medium with various concentrations of sucrose, 0.2% ME, and 10^{-6} M 2,4-D.

A combination of BA and either 2,4-D or NAA influenced anthocyanin production (Fig. 7). Anthocyanin production in 2,4-D medium is higher than that in NAA medium. The optimum combination is 10^{-6} M 2,4-D and 10^{-8} M BA. A concentration of BA higher than 10^{-6} M resulted in an extreme decrease in cell growth.

Figure 8 shows that the total amount of nitrogen in the medium and the NH_4^+/NO_3^- ratio (a/b ratio) also affected anthocyanin production. Anthocyanin production was promoted more at 30 mM total amount than at 15 or 60 mM. This amount is equal to the total amount of nitrogen in standard GA medium. Cell growth decreased in the medium that did not contained NH_4^+.

The effects of main ions on anthocyanin production are shown in Fig. 9. Higher concentrations of Fe^{2+} and SO_4^{2-} reduced anthocyanin production. Within the limit of the concentrations used here, Mg^{2+}, Cu^{2+}, and PO_4^{3-} did not influence this production.

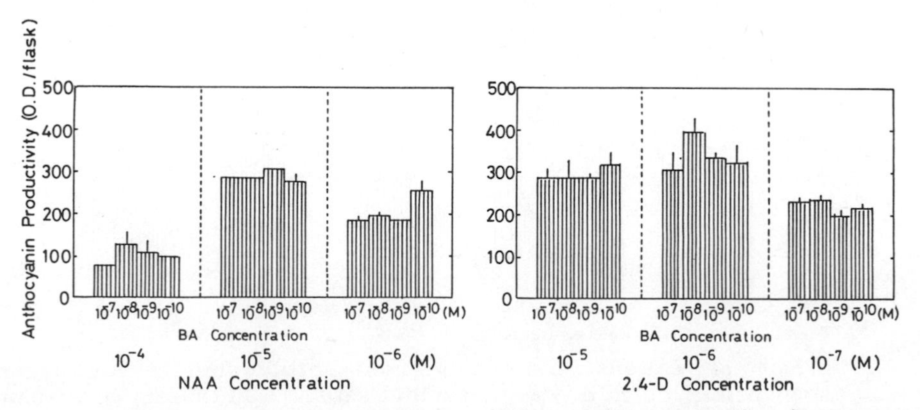

Fig. 7. Effect of phytohormones on anthocyanin production in suspension cultures of E. millii cells. Cultures were maintained in GA medium with various concentrations of BA and 2,4-D or NAA.

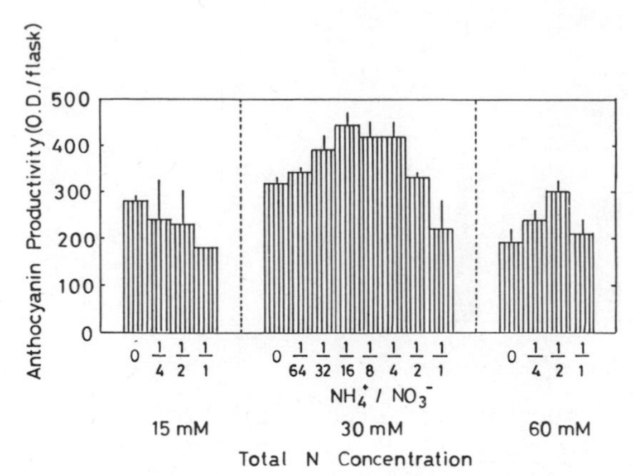

Fig. 8. Effect of total nitrogen amount and NH_4^+/NO_3^- ratio on anthocyanin production in suspension culture of E. millii cells. Cultures were maintained in GA medium of modified NH_4^+ and NO_3^- concentrations with 10^{-6} M 2,4-D, 10^{-8} M BA, and 5% sucrose.

We compared medium components and anthocyanin production in MS and EM media (Tab. 1). Dried cell yield in EM medium is 2.5-fold that in MS medium. The anthocyanin content of cells cultured in EM medium is higher than that in MS medium. Productivity of anthocyanin (O.D./flask) in EM medium is 4.5 times higher than that in MS medium.

Effects of Agitation on Anthocyanin Production in Jar-Fermenter Culture

The effects of agitation on anthocyanin production are shown in Tab. 2. A small paddle with holes used to produce agitation resulted in greater anthocyanin production in jar-fermenter culture than did other rotators. Optimum rotation was 50 to 100 rpm. Anthocyanin production in jar-fermenter culture was a quarter of that in rotatory culture. Euphorbia millii cells were too fragile to undergo agitation.

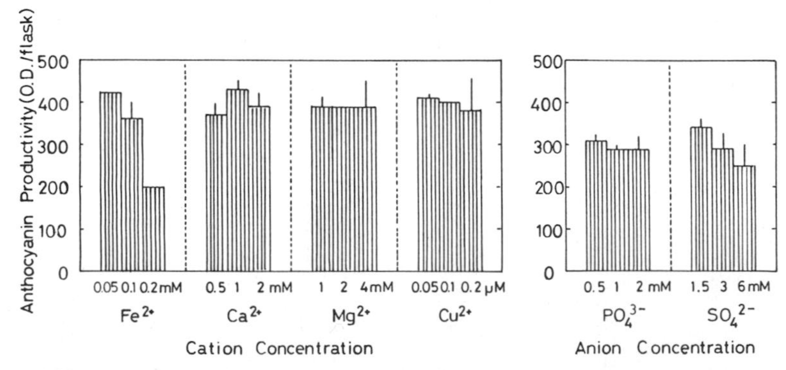

Fig. 9. Effect of cations and anions on anthocyanin production in suspension culture of E. millii cells. Cultures were maintained in GA medium of modified concentrations of cations (Fe^{2+}, Ca^{2+}, Mg^{2+}, and Cu^{2+}) and anions (PO_4^{3-} and SO_4^{2-}) containing 10^{-6} M 2,4-D, 10^{-8} M BA, and 5% sucrose.

Tab. 1. **Comparison of medium components and anthocyanin production between MS and EM media.**

	MS Medium	EM Medium
KNO_3	1,900 mg/l	2,648 mg/l
NH_4NO_3	1,650	141
$MgSO_4$ $7H_2O$	370	500
$CaCl_2$ $2H_2O$	440	150
KH_2PO_4	170	0
KCl	0	264
NaH_2PO_4 $2H_2O$	0	150
$FeSO_4$ $7H_2O$	27.8	13.9
Na_2EDTA	37.3	18.7
$MnSO_4$ $4H_2O$	22.3	10
$ZnSO_4$ $7H_2O$	8.6	2
$CuSO_4$ $5H_2O$	0.025	0.025
$CoCl_2$ $6H_2O$	0.025	0.025
KI	0.83	0.75
H_3BO_3	6.2	3
Na_2MoO_4 $2H_2O$	0.25	0.25
Inositol	100	100
Nicotinic acid	0.5	0.5
Pyridoxine HCl	0.5	0.5
Thiamine HCl	1	1
Glycine	2	2
Sucrose	20,000	50,000
Malt extract	2,000	0
2,4-D	10^{-6} M	10^{-6} M
BA	0	10^{-8} M
Dried cell yield (mg)	595	1,580
Content (O.D./g)	166	282
Productivity (O.D./flask)	99	445
EM/MS Ratio	1	4.5

Cultures were maintained in MS or EM medium at 26°C under fluorescent light (6,000 lux) for ten days on the rotatory shaker (120 rpm).

Tab. 2. **Effects of agitation on anthocyanin production in jar-fermenter culture.**

Rotator	Speed (rpm)	Final pH	Harvested cells (g)	Dried cells (g)	Anthocyanin content (O.D./g)	Anthocyanin productivity (O.D./l)
Paddle*	100	5.1	273	17.8	206	733
Paddle*	75	5.1	308	19.1	217	829
Paddle*	50	5.1	275	20.2	194	784
Paddle*	30	4.8	66	5.9	115	136
Paddle**	100	4.7	117	8.0	128	204
Screw	100	5.0	148	12.7	89	226

*100 mm width and 50 mm height with two holes (20 mm width and 30 mm height).
**120 mm width and 80 mm height with two holes (30 mm width and 40 mm height).

Note: Cultured Euphorbia millii cells (150 g fresh weight) were maintained in liquid EM medium (5 l) in a 14-l jar with 68 to 72% D.O. at 26°C under fluorescent light (6,000 lux) for seven days.

ACKNOWLEDGEMENTS

I wish to thank Prof. Y. Yamada of Kyoto University and Mr. R. Mizuguchi of Nippon Paint Company, Ltd., for their advice and encouragement, and Mr. N. Kadota of Nippon Paint Company, Ltd., for his help with the computer programming.

REFERENCES

1. Calvin, M. (1978) Chem. Eng. 56:30-36.
2. Dougall, D.K., and K.W. Weyrauch (1980) Growth and anthocyanin production by carrot suspension cultures growth under chemostat conditions with phosphate as the limiting nutrient. Biotech. Bioeng. 22:337-352.
3. Furstenberger, G., and E. Hecker (1977) New highly irritant Euphorbia factors from latex of Euphorbia tirucalli. Experientia 33:986-988.
4. Gamborg, O.L., R.A. Miller, and K. Ojima (1968) Nutrient requirements of suspension cultures of soybean root cells. Exp. Cell Res. 50:151-158.
5. Heller, R. (1953) Thèse Paris et Ann. Sci. Nat. (Bot. Biol. Veg.) 14:1-223.
6. Linsmaier, E.M., and F. Skoog (1965) Organic growth factor requirements of tobacco tissue cultures. Physiol. Plant. 18:100-127.
7. Murashige, T., and F. Skoog (1962) A revised medium for rapid growth and bioassays with tobacco tissue culture. Physiol. Plant. 15:473-497.
8. Nishimura, H., R.P. Philip, and M. Calvin (1977) Lipids of Hevea brasiliensis and Euphorbia coerulescens. Phytochemistry 16:1048-1049.
9. Nitsch, C., and J.P. Nitsch (1967) Induction of flowering in vitro in stem segments of Plumbago indica. I. The production of vegetative buds. Planta 72:355-370.
10. White, P.R. (1963) The Cultivation of Animal and Plant Cells, Ronald Press, New York, pp. 74-75.
11. Yamada, Y. (1984) Cell Culture and Somatic Cell Genetics of Plants, Vol. 1, Academic Press, Inc., New York, 631 pp.
12. Yamakawa, T., S. Kato, K. Ishida, T. Kodama, and Y. Minoda (1983) Production of anthocyanins by Vitis cells in suspension culture. Agric. Biol. Chem. 47:2185-2191.
13. Yamamoto, Y., R. Mizuguchi, and Y. Yamada (1981) Chemical constituents of cultured cells of Euphorbia tirucalli and E. millii. Plant Cell Reports 1:29-30.
14. Yamamoto, Y., R. Mizuguchi, and Y. Yamada (1982) Selection of a high and stable pigment-producing strain in cultured Euphorbia millii cells. Theor. Appl. Genet. 61:113-116.
15. Yamamoto, Y., N. Kadota, R. Mizuguchi, and Y. Yamada (1983) Computer tracing of the pedigree of cultured Euphorbia millii cells that produce high levels of anthocyanin. Agric. Biol. Chem. 47:1021-1026.
16. Yoshikazu, Y., R. Mizuguchi, and Y. Yamada (1982) Plant Tissue Culture 1982, The Japanese Association for Plant Tissue Culture, 284 pp.

CONTROL OF MORPHOGENESIS IN CALOCEDRUS DECURRENS TISSUE CULTURE

S. Jelaska[1] and W.J. Libby[2]

[1]Department of Biology
Faculty of Science
University of Zagreb
YU-41001 Zagreb, Yugoslavia

[2]Departments of Genetics and of
Forestry and Resource Management
University of California
Berkeley, California 94720

ABSTRACT

Adventitious and axillary buds were formed by incense-cedar [Calocedrus decurrens (Torr.) Florin] juvenile tissue explants on agar-solidified WPM medium with BA (7 to 10 μM). Buds were also induced on shoot-tip explants from five-year-old lateral branches after a pulse treatment with BA (125 μM); however, these had suppressed elongation growth. During elongation, shoots of both juvenile and fifth-year origins showed dimorphism of leaves. While shoots rarely rooted spontaneously, rooted shoots were reliably obtained by using auxins (IBA and NAA) as a supplement in the medium or as a 24-hr shoot treatment. Callus cultures were induced and maintained on 0.5 MS media plus varying concentrations of 2,4-D, NAA, and cytokinins, or the callus was modified when subcultured on DCR medium in the dark.

INTRODUCTION

Incense-cedar (Calocedrus decurrens) is of interest for at least three reasons. First, as this report indicates, it seems to perform about as well in cell and tissue culture as the most permissive conifers, and better than most. Thus, we and others (D. Harry, pers. comm.) are using this species to explore techniques of cell manipulation, tissue culture, and micropropagation.

Second, it is a species widely and successfully used in central, southern, and western Europe in parks and urban plantings. Limited trial plantings indicate that it may have a wider role as a production forest species in some European settings (13).

Third, it is an important member of the conifer forests of Oregon and California, successfully growing on a great variety of sites (10,16). Recent increases in the perceived importance of such species as Alaska yellow cedar (Chamaecyparis nootkatenesis) and western redcedar (Thuja plicata), also members of the Cupressaceae, may mean that the relative value of incense-cedar will increase as well and for the same reasons (Ref. 14; see Tab. 1).

Work on incense-cedar tissue culture was begun in Zagreb in 1981. It was, and to our knowledge remains, the only project investigating incense-cedar tissue culture. The first two years were characterized by media trials, and by problems of embryo survival and culture contamination. By 1983, the early results indicated that incense-cedar is a promising species for research on conifer tissue culture, and that tissue culture is a promising tool for incense-cedar research and perhaps for its management as well (11). This chapter reports mostly those trials and experiments conducted during the period 1983 to 1987 that produced successful or interesting results.

MATERIALS AND METHODS

Plant Materials

Stratified incense-cedar seeds of known provenances coded "LA County 77-33," "Booneville 31" and "32," and lateral branches from Zagreb clones "S1436," "S1439," and "S1573" (originally from the Greenhorn Mountains) were used in the experiments reported below.

Disinfestation of seeds was achieved by soaking in 3% Halamid (a chlorine product, Pliva, Zagreb) supplemented with 0.3% Tween 20® for 20 min, followed by a thorough rinse three times with sterile water. Following this treatment, the seeds used for embryo culture were kept at 4°C in moist conditions for 70 hr, were disinfested anew by soaking in 6% H_2O_2 for 10 min, and then were rinsed once in sterile water. These same two disinfestation steps were performed sequentially on intact seedlings immediately before explants were taken, omitting the 70-hr storage.

Branch tips, collected from five-year-old trees, were disinfested by soaking for 10 min in 1% $HgCl_2$ (prepared in 50% ethanol and 0.3% Tween 20®) after their cut surfaces had been sealed with hot wax. They were then rinsed once with sterile 5% $CaCl_2$ and three times in sterile water. The waxed surfaces were then cut off. This disinfestation protocol gave good results.

The following types of explants were used: an excised whole embryo; an excised embryo with the radicle removed; the intact cotyledonary node with a short hypocotyl stub from a 14-day-old seedling; an epicotyl from a six- to eight-week-old seedling; and a 1-cm-long shoot tip from a lateral branch of a five-year-old tree, the latter with scale-like leaves. (Callus induction was performed only on the tissue explants of the two-week-old seedlings.)

Culture Media

Several solid (0.8% agar) basal media containing macro- and micronutrients were used: MS, Murashige and Skoog (19); MCM, Bornman (1);

WPM, Lloyd and McCown (15); GD, Gresshoff and Doy (6); and DCR, Gupta and Durzan (7). In the shoot elongation phase of culture, MS and WPM media were supplemented with: 50 mg/l myoinositol, 1 mg/l thiamine:HCl, 0.5 mg/l pyridoxine:HCl, 0.5 mg/l nicotinic acid, 2.0 mg/l glycine, and 2% sucrose. The growth regulators N^6-benzyladenine (BA), indole- 3-butyric acid (IBA), 1-naphthaleneacetic acid (NAA), and 2,4-dichlorophenoxyacetic acid (2,4-D) were used either singly or in combination. The BA was incorporated in the agar media or applied as a high-concentration (125 μM) pulse for 3 hr to primary explants or to inocula, with intermittent shaking.

Culture Conditions

Cultures were incubated at 26°C under light or, for callus on DCR, in the dark. Illumination was from fluorescent lamps (warm white, 40 W, with a spectral range of 400 to 700 nm, 17 $W \cdot m^{-2}$) and a light-dark cycle of 16-8 hr. During each treatment cycle, 24 to 48 explants per replication were cultured, and often the more promising combinations were repeated two or more times. Results were quantified by the frequency of explants forming buds, by the mean number of buds per reactive explant, and by the mean number of buds formed per explant attempted.

Callus Induction

Callus induction was attempted with explants from 14-day-old seedlings. Explants used were the upper 1 to 2 cm of hypocotyl, a detached cotyledon, or the apical tip of the epicotyl. These were placed on one of two agar media (0.8%). Both media were 0.5 MS macronutrients, 1 MS micronutrients, supplemented with: 1,725 mg/l L-proline, 100 mg/l myoinositol, 100 mg/l L-glutamine, 2 mg/l glycine, 1 mg/l thiamine:HCl, 0.5 mg/l pyroxidine:HCl, 0.5 mg/l nicotinic acid, and 6,000 mg/l sucrose. Medium "a" (MSa) had 22.6 μM 2,4-D plus 2.3 μM kinetin; medium "b" (MSb) had 16.1 μM NAA plus 2.2 μM BA. Physical conditions were as described in "Culture Conditions," above.

Following establishment of calli, subculturing was done onto MSa from MSa or onto MSb from MSb; some subcultures of each were done onto DCR medium. The DCR medium was agarified (0.6%) and was supplemented with 450 mg/l L-glutamine, 500 mg/l casein hydrolysate, 100 mg/l myoinositol, and 3,000 mg/l sucrose, plus 226 μM 2,4-D, 2.3 μM BA, and 2.3 μM IAA. Physical conditions were as described in "Culture Conditions," above, except that subcultures on DCR were incubated in the dark.

All media discussed in "Materials and Methods" section were adjusted to pH 6.0 before autoclaving.

RESULTS

Primary Cultures

Excised embryos with or without radicle. Seeds from the "LA County 77-33" collection were used for these experiments. Embryos were excised zero to three days after the seeds were removed from 4°C and set to germinate. No consistent or significant differences were associated with removal of the radicle, and thus results for these two kinds of explant are reported together.

The embryos were small and proved difficult to handle. Results were variable, with from zero to more than 50 adventitious buds being observed per explant under similar conditions. When adventitious buds (Fig. 1) were formed, they were almost always formed on cotyledons (occasionally but rarely being formed on hypocotyls). While the embryo epicotyls sometimes elongated, they never became large enough to produce either adventitious or axillary buds before the primary culture was discontinued (six to eight weeks).

On the three induction media used, embryos on WPM survived in culture at a significantly higher percentage (95% vs 30% and 15%), and a higher percentage of the surviving explants on WPM produced buds (65% vs 25% and 22%) than did explants on full- and half-strength MS media, respectively.

The addition of NAA to WPM negatively affected both the percentage of embryos producing buds, and the average number of buds produced by the explants with buds (Fig. 2).

The addition of BA to either full- or half-strength WPM maximally affected an index of adventitious bud induction at concentrations between 5 and 10 μM, with a peak at 7 μM among the concentrations tested (Fig. 3).

Intact cotyledonary nodes from 14-day-old seedlings. Seedlings from "Booneville Open-Pollinated Families 31 and 32" mostly had two cotyledons, but about 20% had three. The cotyledons were laid flat on the surface of the medium, with the hypocotyl stub stuck into the medium. Intact nodes not only were easy to work with, but they produced more buds per cotyledon than did detached cotyledons.

Concentrations of BA in the neighborhood of 7 to 10 μM in full-strength WPM gave the best results of the combinations tried (Tab. 1).

Fig. 1. An early stage of adventitious bud induction on a cotyledon from a 14-day-old seedling cultured on WPM plus 7 μM BA.

The 14-day explants were easier to manipulate than the embryo explants, and gave more buds on average per explant (Fig. 2; Tab. 2A).

In all cases, the epicotyls of these explants actively grew. The lower primary needles typically thickened and often produced adventitious buds. These epicotyls produced only a few axillary buds (Tab. 2A).

Epicotyls from six- to eight-week-old seedlings. Seedlings from the two Booneville families ("31" and "32") donated their entire epicotyls as explants at 42 and 60 days after germination. Following a 3-hr immersion (pulse) in a 125-μM BA solution, they were cultured on cytokinin-free, half-strength WPM. No adventitious buds were formed (Tab. 2B), even when cotyledons were included as a part of the explant. (In these cases, the cotyledons soon died; on many donor seedlings of this age, the cotyledons were already dead before the explants were taken.)

These epicotyl explants grew vigorously, producing small numbers of axillary buds in cytokinin-free culture (and on intact seedlings), and much larger numbers of axillary buds following a BA pulse (Tab. 2B). These axillary buds appeared to be more vigorous, and on average they grew faster than the adventitious buds (see above).

Tips of branches on five-year-old trees. The distal 10 to 15 mm of lateral branches of clones "S1436," "S1439," and "S1573" were collected in March 1985, at a time of late dormancy. The trees were ∿1 m tall and the shoots had scale-like decussate needles in flat sprays, typical of mature

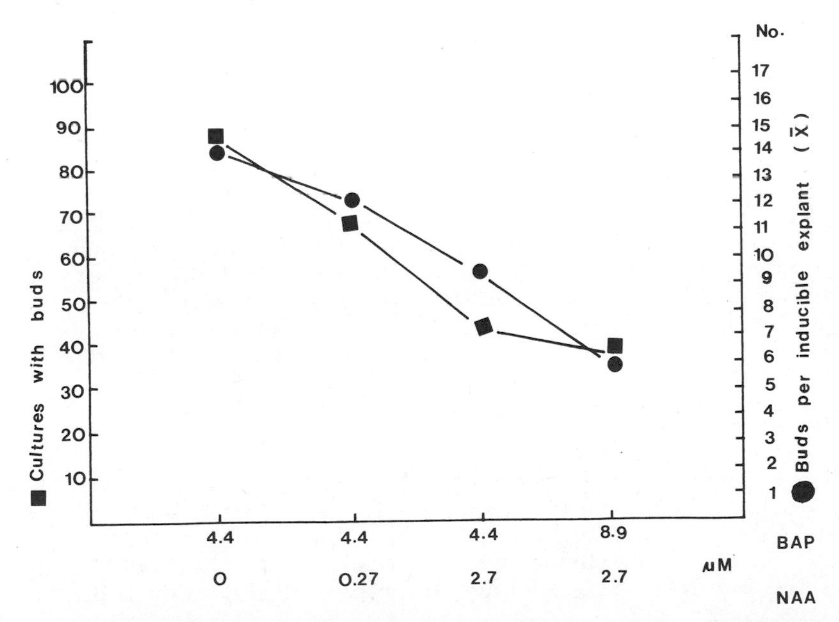

Fig. 2. The percentage of embryo explants that developed buds (squares, left axis) and the average number of buds on explants with buds (circles, right axis) on various primary culture media. The explants were cultured on agarified WPM plus sucrose (3%), supplemented with combinations of BA (lower axis, top) and NAA (lower axis, bottom).

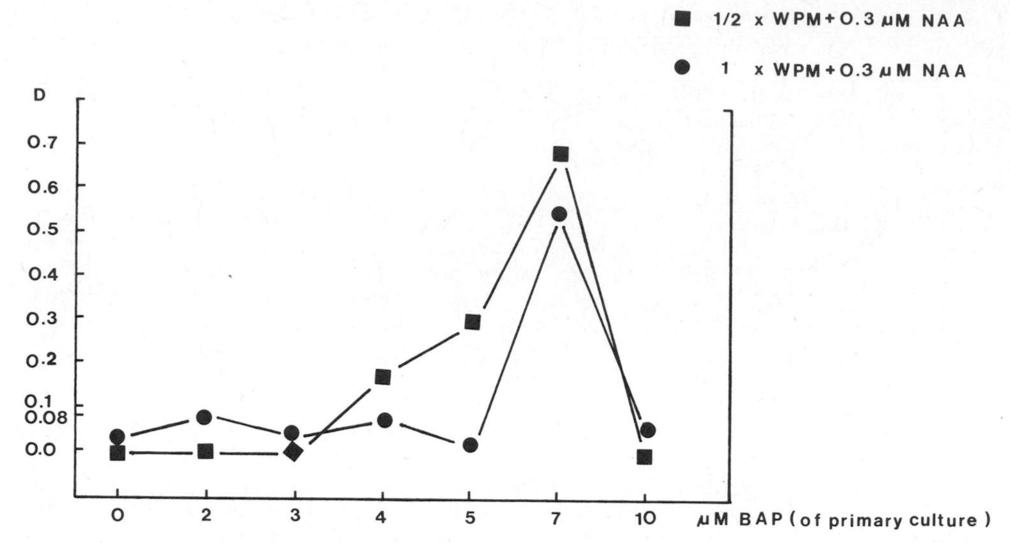

Fig. 3. Bud induction, expressed as an index "D" combining the number of reactive explants (of a constant number attempted) and the total number of buds produced [D=(number of cultures x number of buds)/100]. Data are from embryo explants cultured on half-strength (squares) or full-strength (circles) WPM supplemented with BA (0-10 μM, lower axis); observations taken after a month on the induction media and three weeks on the first subculture.

incense-cedar foliage. The branch tips were given the disinfestation treatment described in "Materials and Methods," above, and then immediately soaked for 3 hr in 125 μM BA solution and basally inserted into WPM supplemented with sucrose (3%), 50 mg/l myoinositol, and 0.03 μM NAA.

This severe disinfestation procedure caused the explants to turn brown shortly after being cultured, with most surface tissue apparently killed. About 80% of the explants were effectively disinfested. Of these, 55% soon (within a month) produced green elongating axillary buds emerging from the brown tissue of the explant. These axillary shoots all developed needles that were more like primary than secondary needles, although they were generally thicker than the typical primary needle.

Shoot Remultiplication

Subculturing. Elongation and multiplication were achieved by subculturing separated shoots on either full- or half-strength WPM supplemented with sucrose (2%), 50 mg/l myoinositol, and either 0.0054 or 0.054 μM NAA, the latter choice being made in response to the health and appearance of the culture. Subculturing was repeated at 21-day intervals.

Elongation. Elongation of the shoots was typically erratic and asynchronous. Elongation of some shoots to a length of 2 cm occurred in only two subcultures; apparently similar shoots of the same clone and history sometimes required four or more subcultures to reach a length of 2 cm. Shoots of adventitious origin commonly required longer to reach 2-cm elongation than did shoots of axillary origin. However, axillary shoots from

Tab. 1. **Effects of BA on percentage of explants from 14-day-old seedlings producing adventitious buds on cotyledons.**

Medium	BA concentration (μM)	Percent explants with buds*
WPM	0	0.0
0.5 WPM	7	35.0
WPM	7	87.5
WPM	10	55.5
WPM	125 (3-hr pulse)	50.0
MCM	7	12.5 (callus)
WPM + 1.4 mM myoinositol	7	0.0

*Minimum 24 explants per reported percentage.

shoots of adventitious origin, upon subculturing, elongated at approximately the same rates as did subcultured axillary shoots from shoots of axillary origin. Near the end of the current set of experiments, some of these shoots began producing scale-like needles in flat sprays. These have not yet been further subcultured.

Remultiplication. A 3-hr soak in 125 μM BA between subcultures resulted in a 10- to 20-fold increase in axillary bud production (Fig. 4). Without exception, multiplication was by axillary shoots. However, the

Tab. 2. **Adventitious and axillary bud formation following various methods of BA application.**

Medium	BA concentration (μM)	Average buds per explant*	
		Adventitious	Axillary
		$\bar{x}$ (min-max)	$\bar{x}$ (min-max)
A. In nutrient agar (explants from 14-day-old seedlings)			
WPM	0	0 (0-0)	0.0 (0-0)
WPM	7	11 (0-26)	0.6 (0-8)
0.5 WPM	7	17 (0-57)	0.1 (0-1)
WPM	10	22 (0-31)	Not observed
B. As a 3-hr pulse (explants from 42- and 60-day-old seedlings)			
0.5 WPM	0	0.0 (0-0)	2.0 (0-3)
0.5 WPM (42-day)	125	0.0 (0-0)	17.0 (10-22)
0.5 WPM (60-day)	125	0.0 (0-0)	17.0 (7-29)

*Minimum 24 explants per reported average.

Fig. 4. A cluster of axillary buds developing near the tip of an elongating shoot following pulse treatment with 125 μM BA.

subcultured shoots from the five-year-old trees elongated very little and did not proliferate axillary shoots following a BA pulse, but sometimes did produce abnormal-appearing "pseudobuds" (3) of adventitious origin on lower needles. The "pseudobuds" failed to develop further.

Rooting

Spontaneous rooting. Some shoots rooted spontaneously on the elongation media. Most of these spontaneously rooted shoots were of adventitious origin; only rarely did a shoot of axillary origin root spontaneously. Spontaneously rooted shoots typically initiated only one root, and rarely did they initiate more than two. Spontaneous rooting did not occur on any of the primary (bud induction) media.

Induced rooting. Addition of auxin greatly increased the rooting percentage and moderately increased the number of roots per rooted shoot, compared to spontaneous rooting. The rooting medium used was GD supplemented with 9.8 μM IBA and 2.7 μM NAA. Alternatively, shoot bases were soaked for 24 hr in a water solution containing 134.3 μM NAA and 122.9 μM IBA, and then subcultured on WPM plus sucrose (3%). The latter (dip) treatment appeared to produce somewhat better results than did adding auxin to the culture medium (Fig. 5). Clonal mixtures of shoots were used during these trials; the effects of genotype, explant origin, and previous treatment were therefore not assessed. Rooting occurred in four to eight weeks, with the average number of roots per rooted shoot varying from 1.4 to 3.1, and the rooting percentage varying in the range 30 to 50%.

Callus Induction

Induction and proliferation. With the exception of MCM (Tab. 1), callus was not consistently induced nor did it proliferate on the media used for primary cultures. In both of the callus-induction media [MSa and

Fig. 5. Three roots developing from an axillary shoot six weeks after treatment of the shoot base with an NAA/IBA solution for 24 hr.

MSb (see "Materials and Methods," above)], substantial callus induction and proliferation occurred on two of the three kinds of explants attempted (detached cotyledons and apical tips), but only negligible induction and proliferation was observed on the third (hypocotyl segments).

Maintenance. Four callus lines have been maintained for over 18 months, subculturing onto fresh MSa or MSb at approximately four-week intervals (Fig. 6). Each of the four lines had essentially the same growth rate, surface and internal morphology, and pattern of necrotic tissue. These remained consistent when the media were modified, e.g., by replacing glutamine with asparagine (100 mg/l), or by eliminating L-proline, or by decreasing auxin concentrations.

Modification of cells. Microscopic examination of callus tissue grown on MSa or MSb showed that fresh cells had various shapes (irregular, elongate, round), and that these variously shaped cells all had thin walls and a high accumulation of starch grains. Transferring these callus lines onto DCR and growing them in the dark did not alter their pattern of necrotic tissue. However, fresh tissue was changed in several respects. On DCR in the dark, it was white, translucent, and sometimes slimy, while previously it had been light-to-dark green, opaque, and never slimy. Individual cells from such DCR cultures were uniformly round and had fewer starch grains and even thinner walls than did cells grown on MSa or MSb.

Embryogenesis. No evidence of embryoid or proembryo formation was noted macroscopically, nor was evidence of such structures seen microscopically using acetocarmine-stained tissue.

Fig. 6. Morphology of established subcultures of line "N6" growing on 0.5 MS plus 16.1 μM NAA + 2.2 μM BA (MSb medium).

DISCUSSION

Performance of Incense-cedar

With respect to the development of adventitious and axillary buds and shoots and the rooting of these shoots in in-vitro culture, incense-cedar reacts rather like the other members of the Cupressaceae thus far investigated (4,5,12,21). However, the hypocotyls of *Biota orientalis* were rich locales for adventitious buds (20), while very few such buds formed on our incense-cedar hypocotyls, with young cotyledons being much more productive of adventitious buds. Furthermore, incense-cedar hypocotyls were not productive sites for callus induction and proliferation.

Effects of N^6-Benzyladenine

N^6-Benzyladenine was effective in inducing adventitious buds and in increasing the number of developing axillary buds of incense-cedar. High concentrations of BA (10 μM) in WPM had a strongly inhibitory effect on bud induction on embryo explants, but were suboptimally stimulative for explants from 14-day-old seedlings. Bornman (2) observed a similar effect with cultures from embryos and seedlings of *Picea abies*.

Adventitious buds were also induced on the lower primary leaves of epicotyls elongating from the 14-day explants when BA was included in the medium. "Nodules" and other incompletely formed structures [see description in Bornman (3)] also developed on higher needles, and particularly on the uppermost needles, of elongating epicotyls from 14-day explants in such conditions. A BA pulse induced axillary bud development and limited elongation in explants from five-year-old trees; many of the needles on such induced axillary shoots then produced adventitious "pseudobuds."

Maturation-State Markers

Recently, in vitro shoots with primary needles have begun to develop secondary needles. Conversely, more mature explants with secondary needles have produced buds and shoots with needles that have the appearance of thickened primary needles. This leaf dimorphism may serve as a marker for more fundamental changes in tissue and/or meristem maturation state, which may have been influenced by the conditions in the culture media and environments. Such changes are similar to those produced by serial micrografting of ten-year-old Thuja plicata shoot tips (Ref. 17; see also Ref. 18). The recovery as independent plants and the further characterization (22) of such shoots with changed leaf morphology seem attractive future steps for incense-cedar tissue culture. A better medium for subculturing and elongation of the buds from the five-year-old trees is a necessary prerequisite.

Callus Culture

The work with incense-cedar callus culture is in a very early stage. The fact that callus morphology, growth, and cell structure remained robustly consistent over several culture modifications but were substantially changed in another modification offers an additional route to fundamental studies with incense-cedar callus cultures (see, for example, Ref. 8 and 9).

Production of Plantlets

Although we have not yet grown incense-cedar plantlings in field conditions, our results with the 14-day and six- to eight-week explants indicate that at least modest numbers of incense-cedar clonal plantlets could be produced by developing the techniques reported above.

ACKNOWLEDGEMENTS

This research was conceived during WJL's Fulbright fellowship in Zagreb and was supported by the Science Research Council of SR Croatia and by a joint Yugoslav-U.S. research grant (FG-YU-258-YO-FS-104). We thank D. Harry, M. Guinon, and Z. Borzan for supplying the seeds and branch cuttings, and D. Rogers for a useful review of the manuscript.

REFERENCES

1. Bornman, C.H. (1981) In-vitro regeneration potential of the conifer phyllomorph. In Symposium on Clonal Forestry, G. Eriksson and K. Lundkvist, eds. Department of Forest Genetics, Swedish University of Agricultural Sciences, Uppsala (Swedish Research Note 32), pp. 43-53.
2. Bornman, C.H. (1983) Possibilities and constraints in the regeneration of trees from cotyledonary needles of Picea abies in-vitro. Physiol. Plant. 57:5-16.
3. Bornman, C.H. (1985) Hormonal control of growth and differentiation in conifer tissues in vitro. Biol. Plant. 27:249-256.
4. Coleman, W.K., and T.A. Thorpe (1977) In-vitro culture of western redcedar (Thuja plicata Donn). I. Plantlet formation. Bot. Gaz. 138:298-304.

5. Franco, E.O., and O.J. Schwarz (1985) Micropropagation of two conifers: Pinus oocarpa Schiede and Cupressus lusitanica Miller. In Tissue Culture in Forestry and Agriculture, R.R. Henke, K.W. Hughes, M.J. Constantin, and A. Hollaender, eds. Plenum Press, New York, pp. 195-213.
6. Gresshoff, P.M., and C.H. Doy (1972) Development and differentiation of haploid Lycopersicon esculentum (tomato). Planta 107:161-170.
7. Gupta, P.K., and D.J. Durzan (1985) Shoot multiplication from mature trees of douglas-fir (Pseudotsuga menziesii) and sugar pine (Pinus lambertiana). Plant Cell Reports 4:177-179.
8. Gupta, P.K., and D.J. Durzan (1986) Somatic polyembryogenesis from callus of mature sugar pine embryos. Bio/Technology 4:643-645.
9. Hakman, J., L.C. Fowke, S. von Arnold, and T. Eriksson (1985) The development of somatic embryos in tissue cultures initiated from immature embryos of Picea abies (Norway spruce). Plant Sci. 38:53-59.
10. Harry, D.E. (1984) Genetic structure of incense-cedar populations. Ph.D. dissertation. University of California, Berkeley, 163 pp.
11. Jelaska, S. (1987) Micropropagation of juvenile Calocedrus decurrens. Acta Hortic. (in press).
12. Konar, R.N., and Y.P. Oberoi (1985) In-vitro development of embryoids on cotyledons of Biota orientalis. Phytomorphology 15:137-140.
13. Libby, W.J. (1981) Some observations on Sequoiadendron and Calocedrus in Europe. Cal. For. For. Products 49:1-12.
14. Libby, W.J. (1987) Do we really want taller trees? Adaptation and allocation as tree-improvement strategies. The 1987 H.R. MacMillan Lecture in Forestry, University of British Columbia, Vancouver, 17 pp.
15. Lloyd, D.G., and B.H. McCown (1981) Commercially feasible micropropagation of mountain laurel (Kalmia latifolia) by use of shoot tip culture. Proc. Int. Plant Prop. Soc. 30:421-427.
16. McDonald, P.M. (1973) Incense-cedar--An American wood. USDA Forest Service FS-226, 7 pp.
17. Misson, J.P., and P. Giot-Wirgot (1985) Rejeunissement d'un clone de Thuja en vue de sa multiplication in vitro. AFOCEL Annales de Recherches Sylvicoles 1984:187-197.
18. Monteuuis, O. (1985) La multiplication végétative du sequoia géant en vue du clonage. AFOCEL Annales de Recherches Sylvicoles 1984:139-171.
19. Murashige, T., and F. Skoog (1962) A revised medium for rapid growth and bioassays with tobacco tissue cultures. Physiol. Plant. 15:473-497.
20. Thomas, M.J., and H. Tranvan (1982) Influence relative de la BAP et de l'IBA sur la neoformation de bourgeons et de racines sur les plantules du Biota orientalis (Cupressaceae). Plant Physiol. 56:118-122.
21. Thomas, M.J., E. Duhoux, and J. Vazart (1977) In-vitro organ initiation in tissue cultures of Biota orientalis and other species of the Cupressaceae. Plant Sci. Lett. 8:395-400.
22. Tran Thanh Van, K.M. (1981) Control of morphogenesis in in-vitro cultures. Ann. Rev. Plant Physiol. 32:291-311.

INTEGRATION OF GENETIC MANIPULATION INTO BREEDING PROGRAMS

POTENTIAL OF CELL CULTURE IN PLANTATION FORESTRY PROGRAMS*

A.F. Mascarenhas, S.S. Khuspe, R.S. Nadgauda,
P.K. Gupta, and B.M. Khan

Biochemical Sciences Division
National Chemical Laboratory
Poona 411 008, India

ABSTRACT

Micropropagation of mature, commercially important trees using in vitro methods is still difficult although interest is steadily increasing. One of the main reasons for this difficulty is that foresters and companies are not fully confident about the genetic gains that could accrue at harvest from clones derived from elites, and whether use of these clones will cover the additional costs involved in producing plants raised from tissue culture. We discuss preliminary field evaluation data obtained with tissue culture-raised plants from five important tree species: *Eucalyptus citriodora*, *E. tereticornis*, *E. torelliana*, *Salvadora persica*, and *Tectona grandis* (teak). The discussion will focus on the genetic gains, costs:benefits, and other advantages whereby tissue culture could have a potential for introduction into plantation and forestry improvement programs.

INTRODUCTION

A review of the progress of micropropagation of forest trees during the past decade shows a dramatic increase in the number of successful reports (5,21,35,68). The majority of these still describe methods from juvenile specimens such as embryos or young seedlings. More recently, many species have been rejuvenated prior to culture by special treatments such as spraying with benzylaminopurine (BAP) (1,60) or gibberellins, which has been done in Ivy (17) and *Eucalyptus ficifolia* (16,40), or by serial rooting or grafting on juvenile rootstocks (21). The problems of adult and juvenile explants have been reviewed (18). Serial subculture of explants from mature trees in vitro also results in a gradual increase in juvenility (6,27,50). One of the main limitations to the use of rejuvenating treatments is that these methods are not applicable to many tree species.

---

*National Chemical Laboratory Communication No. 4248.

In the cases where these methods are applicable in inducing juvenile sprouts, they often cannot be grown in culture (7). Moreover, in some cases where the process has been successful, considerable variation has been observed even in individual clones raised from these shoots (4). Reports on sexually mature tree species that have attained half their rotation age are still rare (21,47).

This report will be restricted mainly to the results of field evaluation studies on tissue culture propagules obtained from mature trees of the following species: Eucalyptus tereticornis, E. torelliana, E. citriodora, Tectona grandis (teak), and Salvadora persica. Economic and other benefits offered by these procedures will also be discussed. Greater emphasis has been laid on Eucalyptus. Based on preliminary information from field plantings, there are strong indications of the potential of tissue culture for use with these species. As a result, a second-phase program has been initiated between our laboratory and forest corporations from different states in India to effect wider application of processes that are being developed as part of a comprehensive tree improvement program. Attempts have been made to use tissue culture-raised plantlets in a breeding program. Success in application of the methods developed requires extensive research on refinements and improvements of the processes to the stage of out-planting, careful analysis of cost calculations, costs:benefits, and extensive field trials in different locations so that the best-suited elites can be used to maximum advantage in the best locations. This research is essential before implementation of methods for large-scale production (65,66).

RESULTS IN EUCALYPTUS

The propagation of Eucalyptus by rooting of cuttings has been performed for the last few years and has now resulted in production of genetically improved plantations with better quality trees and higher biomass yields (10). This method has also been used for establishment of germplasm banks and seed orchards. Studies using in vitro methods for propagation of some Eucalyptus species have also been reported (8,14,25,26, 27,45), although reports on a systematic evaluation of these plantlets by field testing, particularly those involving plantlets of mature tree origin, are rare. As early as 1980, we felt that field trials were essential before a large-scale production program could be undertaken. The case of oil palm, where abnormal flower development in clones has been reported (13), demonstrates the importance of field evaluation for woody perennials. The possibility of abnormalities can cast doubt on the future of tissue culture. Application of tissue culture in tree plantations is still in its infancy.

Source of Explants

In these studies, axillary and apical buds of E. tereticornis, E. torelliana, and E. citriodora were collected between March and October from a single elite tree each. The first two species were growing at Marakkhanam and the latter at Ootacamangalam in southern India. The selection of the elite trees was made on the basis of the physical appearance of the trees, the main trait being its high growth rate, which was over ten times greater than the average for E. tereticornis and E. torelliana, and on the basis of the higher oil and citronellal contents in E. citriodora. The selection was done at an intensity of one to two trees per hectare. Some characteristics of the plus trees are given in Tab. 1. The buds were collected from the upper branches of 20-year-old trees and cultured according to

Tab. 1. Data on elite trees: Eucalyptus.

	E. citriodora	*E. tereticornis*	*E. torelliana*
Year of planting	1967	1969	1969
Region	Ootacamangalam	Marakkhanam	Marakkhanam
Age of tree (years)	20	18	18
Seed source	Australia	Australia	Australia
Species (plus tree number)	DB-1	TC-1	TC-1
Age of forests (years)	20	18	18
Source of explants (height in m)	15-18	18-20	15-18
Number of plus trees per ha	1 to 2	1 to 2	1 to 2
Number of trees coppiced	Not felled	Not felled	Not felled
Height (m)	--	26.5	22.0
Girth (cm)	--	104	118
Oil content (%)	3.5	--	--
Normal (%)	0.5-2.0	--	--
Citronellal (%)	90	--	--
Normal (%)	60-70	--	--
Disease	----------- No infection since planting -----------		

the procedures described earlier (25,26,27,42). Plantlets produced were field-tested at Poona in June 1983, in a nonreplicated trial described elsewhere (36). In a 100-m^2 plot, 36 plants were grown at a spacing of 2 m by 2 m. Comparisons were made with seedlings raised from seed as control of the same elite trees from which the tissue cultures had been isolated. Preliminary roguing was carried out so that all plants were of similar height and morphological appearance at the start of the trial. In multilocational trials now being planned, comparisons will be made using controls from rooted cuttings. Characteristics (e.g., soil types, temperature, rainfall, and altitude, of Marakkhanam and Ootacamangalam (the regions where the elite trees were growing) and Poona, are given in Tab. 2 (22).

Tab. 2. Some characteristics of the Ootacamangalam, Marakkhanam, and Pune regions.

	Ootacamangalam	Marakkhanam	Pune
Latitude (N)	11° 24'	12° 28'	18° 31'
Longitude (E)	76° 44'	79° 94'	73° 55'
Altitude (m)	2,245	200	750
Mean annual temperature (°C)	14.2	27.5	26
Mean maximum temperature (°C)	18.9	35.2	38
Mean minimum temperature (°C)	9.4	22.0	10
Annual rainfall (mm)	4,500	1,500	650
Soil	Red sandy	Coastal alluvial	Medium black

Soil Analysis

Results of an analysis of the organic carbon and available phosphorus, potassium, and nitrogen levels of soil samples collected from zones 40 to 80 cm below the surface around the elite trees, from random locations in the same plantation, and also from different regions in the experimental plot at Poona, are given in Tab. 3. The methods of estimation were as follows: organic carbon and available nitrogen (67); available phosphorus (56); soil pH and electrical conductivity (34); and available potassium (58). The values reported are the average of five samples in each location. The composition of the soil samples both around the elite tree and at random locations was more or less similar. However, a more detailed analysis of the microelements and mycorrhizal associations, if any, is necessary to narrow down the possible explanations for the high growth rates of elite trees. The soil at Poona had higher pH and carbon, potassium, and nitrogen levels than the soil collected from locations around elite trees.

Growth Analysis and Biomass Yields

Field planting of E. tereticornis and E. torelliana was carried out in June 1983. Measurements of the height and girth were predominately made in the months of May and October. Figures 1 and 2 show the average measurements of the tissue culture and control plants with regard to height, diameter, and biomass. Biomass was estimated according to Chaturvedi and Venkataraman (12) for a 34-month period and the standard error was calculated using the method of Snedecor and Cochran (64). A preliminary communication of the biomass yields has been reported (36). The variation in height and diameter of tissue culture-raised plants was low when compared to the controls.

The data for height, diameter, and biomass in E. tereticornis are given in Fig. 1. The growth curves for these parameters in tissue culture-raised and control plants are similar and show hyperbolic patterns. Tissue culture-raised plants show greater height, diameter, and biomass values than control plants. At 12 months, the biomass of tissue culture-raised plants is 200% more than that of the control. On the other hand, at 34 months the biomass is only 38% more than that of the control.

Tab. 3. Soil analysis.

	pH (1:2.5) (soil:water)	EC* m mhos/cm (1:2.5) (soil:water)	Organic carbon (%)	Available phosphorus (kg/ha)	Available K_2O (kg/ha)	Total N (%) computed from organic C value
Marakkhanam (random)	5.1	0.05	0.19	11.10	50.0	0.018
Marakkhanam (around elite tree)	5.1	0.05	0.20	8.34	60.0	0.020
NCL,** Poona, India	7.6	0.25	0.50	10.42	444.0	0.050

*Electrical conductivity
**National Chemical Laboratory

The growth curve of tissue culture-raised E. torelliana (Fig. 2) shows a normal hyperbolic pattern, with the initial rate of increase in height and diameter being logarithmic up to 14 months and then gradually decreasing. The control plants of E. torelliana, on the other hand, show a sigmoidal growth curve, with the rate of increase in height and diameter being very slow up to 14 months and then steadily increasing up to 24 months. Thereafter, the growth rates decline.

The biomass growth curve also shows a hyperbolic pattern for tissue culture-raised plants and a sigmoidal pattern for control plants. The biomass of tissue culture-raised plants was 14 times more than that of the control plants after 12 months, 3.5 times more after 24 months, and two times more after 34 months.

The results both with E. torelliana and E. tereticornis indicate very large early increases in growth rates of tissue culture-raised plants up to 24 months compared to controls, the overall increases being dramatically larger with E. torelliana. This increased rate gradually slows down after 24 months, although the total biomass yields of tissue culture-raised plants of E. torelliana and E. tereticornis even at 34 months are 100% and 38%, respectively, higher than those of the control plants.

Observations on growth are being continued. In the Aracruz Forestry Programme (9), average yields over a seven-year period were 70 m^3/ha/year from the improved forests as compared to 33 m^3/ha/year from the unimproved forests, the gain being 112%. Similar gains were observed in the average density and pulp yields. In comparative greenhouse and field trials with tissue culture-raised plantlets of loblolly pine (Pinus taeda) and seedlings, numerous differences were observed (46). Plantlets were less efficient in nutrient uptake, but utilized the nutrients that were

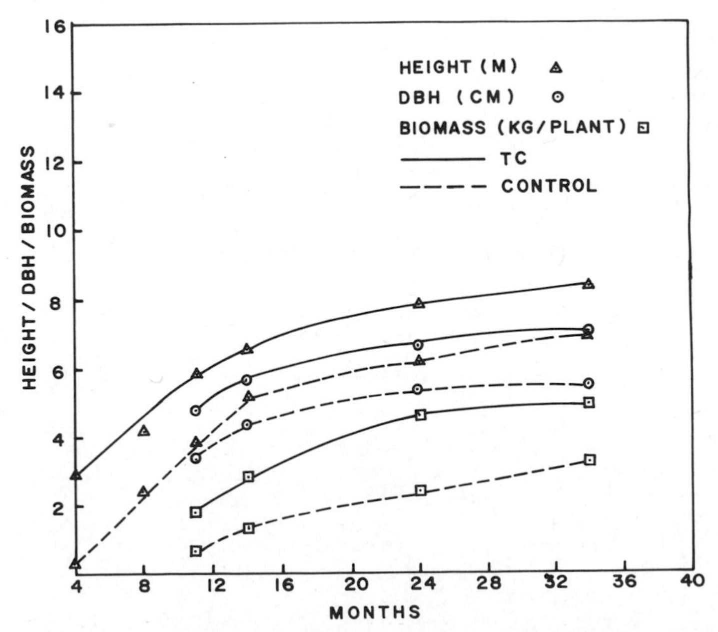

Fig. 1. Growth analysis and biomass yield of Eucalyptus tereticornis.

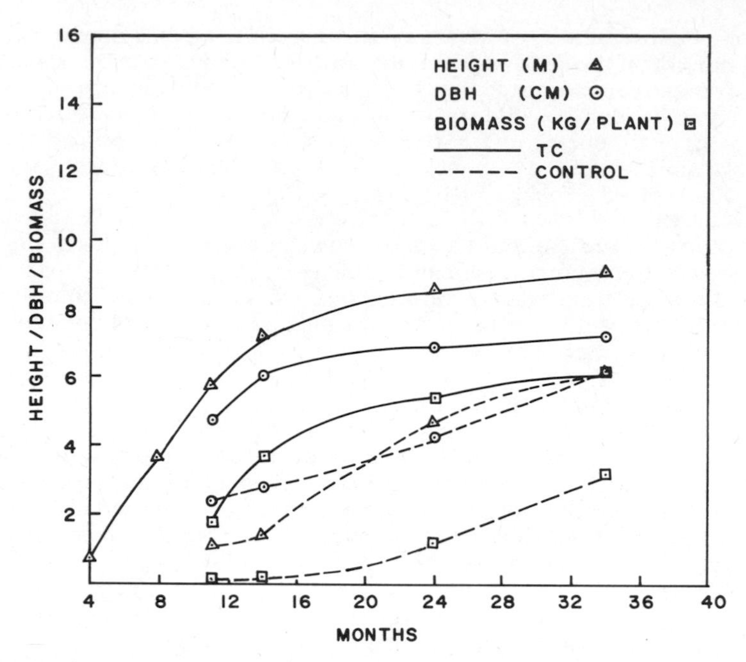

Fig. 2. Growth analysis and biomass yield of Eucalyptus torelliana.

absorbed quite effectively. Early growth in the field was slower for plantlets than for seedlings, which was in contrast to the present studies where early growth was strikingly higher.

Specific Gravity

The specific gravity of the wood was tested with samples selected randomly from 25% of the plants (nine plants per treatment) in both control and tissue culture-raised stands. These data are presented in Tab. 4. Wood samples were collected at breast height from 40-month-old trees using a cork borer and the specific gravity determined by the method of Smith

Tab. 4. Specific gravity measurements of Eucalyptus.

	E. torelliana		E. tereticornis	
	TC	Control	TC	Control
Maximum	720	720	790	630
Minimum	565	505	455	480
Average	655	645	640	585

Note: Specific gravity (oven-dried mass to green volume ratio) expressed as $kg \cdot m^{-3}$. The data has been computed based on measurements from nine plants for each treatment.

(62). The variation in specific gravity between the maximum and minimum values obtained with E. tereticornis and E. torelliana in tissue culture-raised and control plants was high. Similar observations were noted in the Aracruz experiments using rooted cuttings (9).

Oil Analysis

The analysis of the oil and citronellal contents in E. citriodora was carried out according to the procedure described earlier (24) with leaves collected separately from ten tissue culture-raised plants. The concentration of total oil and citronellal contents in leaves of tissue culture-raised plantlets was identical to that of the parent elite tree. This was observed in plants from the age of six months onwards. Projections as a result of this observation are given in Tab. 5. Cultivation of tissue-cultured plantlets could result in three times higher oil yields in the same planting area, thereby saving labor costs. The leaves could also be collected at an earlier age, since in seedling-raised plants at least three years are required before the oil concentration in the leaves attains a stable level. The results also indicate that tissue culture could play a role in the multiplication of other tree species screened for increased levels of high-value chemicals (29).

Cost:Benefit Analysis

The laboratory costs of the process for Eucalyptus were calculated on the basis of actual running costs, which included recurring expenditures for manpower, chemicals, power, and other consumables at all stages of the process, from collection of bud material to out-planting of propagules. The cost per 100 plantlets, with a 90% survival rate, works out to approximately $7.00 as compared to the cost per 100 seedlings, which is one-half the price of the plantlets (Tab. 6). Franclet and Boulay (20) also observed that the cost of frost-resistant Eucalyptus clones raised by tissue culture was twice as much as the cost of seed-raised plants. These costs could be strikingly lower if the plantlets were to be produced on a commercial scale and with a higher survival rate.

Tab. 5. Projected benefits of tissue culture for Eucalyptus citriodora.

Year	Weight of leaves (tons)	Oil content/ha (liters)	
		Trees (oil content 0.5-2%)	Plus trees (oil content 3.5%)
1	7	50	150
2	30	150	450
3	40	200	600

Benefits

1. Saving in plant area.
2. Saving in labor costs.
3. Saving in time. (It generally takes three years before the oil content is stabilized in seed-raised plants. In tissue culture-raised plants from mature trees, the oil levels are stable from six months.)

Tab. 6. Cost of production of one million Eucalyptus plants at an annual survival rate of 90%.

Recurring cost	Dollars (U.S.)
Manpower	27,048
Chemicals	3,091
Power	4,637
Nursery*	28,594
Breakages and unforeseen expenses	3,091
	66,461

*Includes transport from laboratory.

Note: Cost per plant, $.07 (based on recurring costs).

The costs:benefits have been calculated (Tab. 7) based on the growth analysis and biomass yields at the end of 34 months of E. tereticornis and E. torelliana planted at a spacing of 2 m by 2 m (Fig. 1 and 2). Gains with E. tereticornis and with E. torelliana plantlets raised from tissue culture are greater than those with seedlings, with the profits from E. torelliana being nearly double those from seedlings. The trials were conducted at Poona where the climatic and soil conditions are very different from those at Marakkhanam. Yields could be higher if tested at the original site. If capital costs (e.g., land, laboratory, and a greenhouse with a capacity of one million plants) are also taken into account, the price of 100 plants would increase to $14.00 (Tab. 8). Assuming this cost of plants, one could still derive a gain, though less from using tissue culture.

Tab. 7. Costs:benefits based on NCL trials: Eucalyptus.

Species	Cost/plant (recurring)	Yield (tons/ha) (air-dried)	Planting cost	Value of wood	Profit
Control					
E. tereticornis	$.03	7.56	$87	$233	$147
E. torelliana	$.03	7.59	$87	$235	$148
Tissue Culture (34 months)					
E tereticornis	$.07	11.93	$174	$369	$195
E. torelliana	$.07	15.00	$174	$464	$290

Note: Cost:benefit values based on cost of plantlets determined by recurring costs. Value of wood: $31 per ton (air-dried); plants per ha, 2,500.

Tab. 8. Annual cost of production of one million *Eucalyptus* plants.

Capital cost:		
Laboratory		$ 23,184
Equipment, greenhouse, generator, etc.		$170,015
Land		$ 7,728
	Total:	$200,927
Return cost on capital:		
20% Return on capital		$40,185
14% Interest on total capital		$28,130
	Total:	$68,315
Cost per plant on capital expenditure:	$.07	
Cost per plant on recurring expenditure:	$.07	
Total (capital and recurring costs):	$.14	

Hasnain et al. (30,31) have compared the economics of tissue culture-raised forest plants produced by the Forest Research Institute in New Zealand (63), by Weyerhaeuser in the United States (66), and in Canada (30,31) where the seedling costs are higher. As a result of the smaller differential between the costs in New Zealand and the United States as compared with those in Canada, micropropagation should be more feasible in Canada. Very recently, a simplified procedure for growing radiata pine in in vitro hedges has been reported (2). This process has greatly reduced the cost of plants. If applied to *Eucalyptus*, the entire process could become even more effective and result in large-scale production and raising of plantations. The main obstacles to mass propagation of forest trees by tissue culture have been the high production costs and the uncertainty of the benefits. A simplified, cost-effective procedure can change the picture dramatically.

Progeny Trials

The field plantings of tissue culture-raised *E. tereticornis* and *E. torelliana* were made in June 1983. Flowering was observed in August 1985 in the tissue culture-raised plants. Control plants have not yet flowered. As a result of the early reproductive cycles, tissue culture-raised plants can be used to raise seed orchards and in breeding programs to capture the best genetic traits. This can reduce the duration of a breeding program. For progeny testing, seeds were collected in February 1986 from flowers which were (a) bagged, (b) open-pollinated, and (c) from the parent elite tree. Seeds were germinated and the seedlings planted in July 1986. This progeny test was conducted from clones raised from a single individual. The main objective was to study the performance of the progeny of tissue culture-raised plants produced in a nutritionally rich and complex medium, using a rapid process differing from other vegetative methods. To our knowledge, there are no similar studies using forest trees raised from tissue culture.

The average numbers of seeds in selfed and open-pollinated capsules of 25 tissue culture-raised E. tereticornis and E. torelliana plants are given in Tab. 9. The variation in the number of seeds per capsule in selfed and open-pollinated plants was greater in E. tereticornis than in E. torelliana.

Eldridge (19) compared the average number of viable seed set from controlled selfing and crossing in three species of Eucalyptus and observed wide differences. In Eucalyptus, self-fertilization was between 20% and 40%, with indications that selfed zygotes and seedlings have higher chances of survival than those of conifers. A similar difference was observed in seed number in selfed and out-crossed progeny in E. tereticornis and E. torelliana. Eldridge suggested that since selfs appear to be more viable in Eucalyptus, the effects of selfing should not be ignored in the genetic improvement of Eucalyptus. In E. torelliana there were seeds of two sizes, termed big and small, which differed in weight (Tab. 10). The percent germination values in the big selfed and open-pollinated seed were more or less identical; however, in the small seed the percent germination was lower in the open-pollinated seed as compared to the selfed seed. The growth in height of plants from the big seed was significantly greater than that of plants from the small seed.

Field Planting

A replicate field trial was also carried out with seeds collected from E. tereticornis from three sources--the parent elite trees, flowers of the tissue-cultured plants which were bagged, and open-pollinated plants. The seedlings and plantlets at the beginning of the trial were of identical age and height (15 cm), and were selected after roguing out those of other sizes.

From September 1986, nitrogen fertilizer in the form of urea was given four times (8 g per plant) and the plot was irrigated at intervals of ten days. The data on height of plants as collected at nine months are presented in Tab. 11.

Tab. 9. Seed number per capsule of Eucalyptus tereticornis and E. torelliana--progeny trial.

	Number of seeds per capsule
E. tereticornis -- TC (selfed)	8-20
E. tereticornis -- (open)	4-40
E. torelliana -- TC (selfed)	3-12
E. torelliana -- (open)	6- 9

Note: Twenty-five-tree basis; five capsules per tree. The selfed conditions refer to inflorescences which were covered by paper bags; seeds collected from uncovered inflorescences are described as open.

Tab. 10. Seed germination of Eucalyptus torelliana.

	From tissue culture		From parent tree	
Seed size:	Big	Small	Big	Small
Days required for germination	5	5	7	7
100 Seed weight (mg)	332	177	526	182
Percent germination	38.0	34.5	37.5	22.5
Average maximum height (mm) ten days after germination	19.0	12.9	20.0	11.6

Note: Total number of seeds sown: 200 of each type.

There was a higher degree of uniformity in selfed and open-pollinated seedlings compared to controls. The average growth of control plants was more or less similar to that of the open-pollinated plants. Selfed seedlings were 54% taller than controls and more uniform.

The objective of the programs on Eucalyptus improvement throughout the world is to increase the yield and quality of wood from plantations. This wood is grown mainly for fuel and pulpwood. Recent research (33) on the proportion of selfing and out-crossing in Eucalyptus, while supporting the general predominance of out-crossing, has revealed that the proportion might differ from one Eucalyptus species to another, Eucalyptus having a much higher degree of selfing than pines (19). The height reached by deliberately selfed progeny of nearly all clones of E. grandis at 11-18 months was 8 to 40% less than that of out-crossed progeny. Moreover, selfed progeny were usually more crooked and included about

Tab. 11. First progeny trial of Eucalyptus tereticornis.

	Height (m)
Control (seed from elite tree)	1.25 ± 0.56
Tissue culture	
Open-pollinated	1.17 ± 0.44
Selfed	1.68 ± 0.45

Design:	RBD* (2 m x 2 m)
Number of plants per treatment:	40
Replications:	3
Number of plants per replication:	120

*Random block design.

Note: Planting data: July, 1986; observation data: May, 1987.

30% abnormal seedlings. However, in the case of E. regnans F. Muell (9), selfed seedlings did not differ in appearance and were not consistently smaller than out-crossed or open-pollinated seedlings up to the age of one year. Inbreeding depression was not obvious until six years of planting, and even then many individuals within selfed families were vigorous and of good form.

In our studies on observations taken up to nine months, a 54% height increase has been observed in selfed seedlings. This observation requires a longer period to determine inbreeding depression and also growth in successive generations. It will be necessary to carry out identical studies with propagules raised from a wider genetic base and with different species. The problems and dangers of inbreeding for short-term goals are too serious to be taken lightly (37,52,53). If this advantage is maintained over a longer period, and if an inexpensive technique for pollination is developed that will yield full-sib rather than half-sib seed from production orchards, the benefits will increase from using tissue culture to initially produce plantings in combination with conventional breeding.

RESULTS IN TECTONA GRANDIS LINN (TEAK)

The cultural conditions developed both for seedlings and for mature trees of teak, as well as the results of field trials carried out on four-year-old tissue culture-raised plants, have been reported elsewhere (28,43). More recent data on growth and wood properties with regard to specific gravity of these trees at seven years are presented in Tab. 12. The variation in specific gravity was higher in controls as compared to tissue culture-raised plants. This information over a long period is essential for timber-bearing trees where the hardness of the wood and other physical and mechanical qualities, such as durability and resistance to disease, are important.

For large-scale plantation of teak, one-year-old stumps from 30- to 40-cm-long seedlings are used. The stumps of 1 to 2 cm diameter include about 2 to 4 cm of the stem portion above the first pair of buds with the

Tab. 12. Specific gravity measurements of teak.

	Control (T1)	TC from seedling (T2)	TC from elite trees (T3)
Maximum	735	690	730
Minimum	520	540	630
Average	650	570	675

Note: Specific gravity (oven-dried mass to green volume ratio) expressed as $kg \cdot m^{-3}$ (61). The data has been computed based on measurements from nine plants for each treatment. T1, Seed raised (+ tree); T2, tissue culture from seedlings (+ tree); T3, tissue culture from elite tree (100 years old).

taproot cut to a length of 20 to 25 cm. These stumps are planted at an intensity of 2,500 plants per ha and at a spacing of 2 m by 2 m, leaving only a 1- to 2-cm portion of the stem above the ground (55). The cost per plants by this method is approximately $.08 whereas the cost of a tissue culture-raised plantlet based on the existing method is approximately $.24 to $.32 (43). In the first three years teak plantations are intercropped with short-duration crops. Following this period, intercropping is discontinued until the first thinning, which occurs after seven to ten years, at which time 50% of the plants are felled. The average height of plants at this age is approximately 3.6 to 4.5 m and the average diameter is around 10 cm. Poles of these specifications are used as posts or supports in construction and command a price of approximately $.96 to $1.20 each. A net profit of $1,040 after eight to ten years can be expected. If the diameter is less or if the plants are harvested at a younger age, they are used as rafters for construction or as handles for agricultural implements and thus have a lower market value.

Field Observations

The results of observations made with tissue culture-raised plantlets were that these plantlets reach a height of 3.6 to 4.5 m within four years so that returns can be expected much earlier. This is based on the initial fast growth of plantlets produced from over 100-year-old elite trees that were selected for their height, absence of side branches as a result of self pruning up to two-thirds of the total height, and good wood quality.

The second thinning in a teak plantation is made at 17 to 20 years, at which time the poles command a price three to four times higher than that at the seventh or tenth year. It is yet to be observed at what age tissue culture-raised plants will reach the correct height and girth for a second thinning.

Table 13 gives the growth increments of tissue culture-raised plantlets in the replicated field trials conducted in 1980 (43). The early growth increments in height and diameter observed up to 1984 both in plantlets raised from the mature elites and in seedlings have been reduced by the seventh year, at which time they are more or less identical to those of control plants although the overall measurements are still higher.

Teak is a highly cross-pollinating species and self incompatibility is high. In India and Thailand, it is reported that flowering normally starts at the age of eight to ten years but with considerable variation between individuals (e.g., some plus trees have not flowered even at the age of 27 years) (32). The age at which flowering occurs is under genetic control (39). In teak, the first inflorescence usually appears from the terminal bud and causes stem forking. In the present studies, tissue culture-raised plants flowered in the third year. A few seedling-raised plants flowered in the fourth year, and all flowered by the fifth year. This early flowering effect can be used to raise seed orchards for the production of superior seed. As observed by Larsen (39), the first flowering in some of the plants was from the terminal bud. Utilization of tissue culture for mass-scale production of plantlets will depend on the physico-chemical properties of the wood at the end of 15 to 20 years, since teak is mainly used as timber. The high cost factors for plantlet production will also have to be taken into account.

Tab. 13. Growth measurements of teak.

	Height increment (cm)				DBH* increment (cm)		
	1982	1983	1984	1987	1983	1984	1987
T1	166.9	90.1	115.9	229.5	5.30	1.93	2.32
T2	198.3	105.4	139.3	198.4	5.80	2.01	2.25
T3	216.3	130.9	140.6	187.7	6.97	2.15	2.42
S.E. (±)	7.39	4.33	9.75	18.37	0.84	0.078	3.05
C.D.	22.77	13.32	N.S.**	56.39	2.58	N.S.	N.S.

Number of plants per treatment: 2
Number of replications: 7
Number of treatments: 3
Spacing: 3.5 m x 3 m
Statistical design: Random block design
Date of flowering: TC, September, 1983; C, 1984, 1985.

*Diameter breast height.
**Not significant.

Note: T1, T2, and T3 are described in Tab. 12. C, Control.

RESULTS IN SALVADORA PERSICA LINN

Salvadora persica Linn. (tooth-brush tree) is a large, much branched evergreen tree found in the dry and arid regions of India, on saline soils and in coastal regions just above the high tide water mark. These trees have been identified in dry and arid areas and on saline and coastal regions, having a branching habit and a large crown area. These trees yield 20% more seed than the average from tooth-brush trees in other regions. They readily regenerate from seeds and coppice well although growth is slow. Three-year-old seedlings from the nursery are used for planting purposes.

The wood is soft and white, takes polish well, and because it is fairly termite-resistant, was used by the Egyptians for coffins. The leaves and tender shoots are eaten as salad and are also used as camel fodder. They possess antiscorbutic and astringent properties. Seeds yield a pale-yellow solid fat, the fatty acid composition of which is similar to coconut oil, having a high lauric acid content. It is used in the soap industry as a substitute for coconut oil (11).

Tissue Culture and Field Evaluation

A program on tissue culture was initiated in our laboratory to multiply high-yielding trees. Initially, a process was developed for mature trees. This method is now being utilized to propagate the elite trees (59). Small-scale plantings were made of these plantlets in June 1984, in a nonreplicated trial comprised of 15 plants each from seedlings and plantlets (Tab. 14). The first flowering was observed in control plants after 20 months, 11 months earlier than in tissue culture-raised plants. In

Tab. 14. Field evaluation of Salvadora persica.

Observations	Tissue culture	Control
Area of crown (m^2)	6.72 ± 0.183	5.90 ± 0.349
Diameter of the crown (cm)	286.17 ± 4.26	245.52 ± 8.78
Number of branches	7.0 ± 0.047	6 ± 0.097

Planting date: June 14, 1984

Flowering date: Tissue culture, December, 1986; Control, February, 1986

Spacing: 2.5 m x 5 m
Date of observation: April 2, 1987
Number of plants: 30
Number of plants per treatment: 15
Number of irrigations (annual): 14 (from October to June)

general, plantlets raised by tissue culture were more uniform and had a greater crown diameter. This could be an advantage in plants where seed yields are the final product. However, these data require confirmation using the elite tree material. Based on these data, projected benefits from elite plants--if the yields are 20% higher--are given in Tab. 15. An additional profit of $464 can be expected annually after the eighth year. Salvadora trees are known to grow for over a 100 years. Consequently, the initial cost of plants is not important since the end product is the seed and the annual gains will cover the initial cost.

Isolation of Salt-Tolerant Plants

The capacity of Salvadora cultures to withstand high levels of salt was also confirmed by exposing the cultures to enhanced levels of sodium

Tab. 15. Projected benefits of tissue culture for Salvadora persica.

	Number of trees per ha	Average seed yield per tree (age 8 years) (kg)	Seed yield per ha (tons)	Cost per ha ($.12/kg)	Oil yield per ha (kg)
Control trees	400	40	16	$1,855	5,600
Plus trees	400	50	20	$2,318	7,000

Advantages

1. Capacity to grow in degraded saline soils.
2. Initial cost of plants not important since end product is seed. Annual gains can pay for initial costs. (Additional $464 per ha after eighth year.)
3. Trees grow for over 100 years.
4. Cost of plants: TC, $.07; normal, $.02-.04

chloride and seawater. Interestingly, the cultures grew and survived on medium containing 50% sea water alone or on Murashige and Skoog medium (49) containing organic supplements and 3% sodium chloride.

India annually imports edible oils valued at over $640,000,000. One of the reasons for this is that edible oils are diverted towards the non-edible oil industry. There are over 100 million ha of wasteland in India, 7.5 million ha of which consist of saline and alkaline soils. If a portion of these tracts can be covered with elite Salvadora trees it will be a major benefit both to wasteland development and to the nonedible oil industry.

DISCUSSION AND CONCLUSIONS

When tissue culture methods were developed in our laboratory in the late 1970s for propagation of mature trees, we were interested in obtaining answers to three basic questions: (a) how will the plantlets behave as compared to seedlings?; (b) will all the ramets of a clone be uniform?; and (c) will there be any advantages to using this method? To answer these questions we established small-scale greenhouse and field plantings. As an outcome of these studies with different tree species, we are convinced that micropropagation can be profitable (41). The most striking observation is the accelerated growth rates resulting in a decrease in the rotation age. This factor alone can compensate for higher plant costs. A second factor is the high degree of uniformity. This has been clearly observed in Eucalyptus where the biomass yields at the end of 34 months have been higher than those of seed-raised controls. The net gains are also higher than in controls even after deducting the higher costs of tissue culture-raised plants. These results suggest that tissue culture has immediate commercial application. A gain, however small, in an applied forest tree improvement program can be of enormous economic benefit when distributed over thousands of hectares of plantations established annually.

Very recently, scientists at the Herty Foundation (Savannah, Georgia) have reported on the development of a process that produces very good quality newsprint and bond paper from one- to five-year-old sycamore trees. This process utilizes the entire tree--trunk, stem, and twigs. Young trees have usually been considered inferior for papermaking, but using such trees could double output of forest acreage, promising rotation every three to five years. Eucalyptus wood is the raw material for the paper pulp and rayon industries. The very early accelerated growth rates of tissue culture-raised plants in the first 18 months could be fully exploited by these industries if a similar process can be developed. This could reduce the rotation cycles even further. An additional advantage to using Eucalyptus is its coppicing ability.

There has been widespread interest in selective breeding of commercial forest trees, by which dramatic gains in value have been achieved. Our results indicate that tissue culture can also be incorporated into similar breeding programs since flowering, in most cases, is earlier than normal. For instance, Eucalyptus flowered in two years as opposed to four to five years. This early flowering could reduce the breeding cycle. A similar phenomenon was observed in teak and also in tamarind (38,44).

In general, the advantages of tissue culture will be known only after out-planting of the propagules. The benefits could vary for different tree species. For instance, in Dendrocalamus strictus (bamboo) plants raised from seedlings (51), early culm formation was observed in the second year as compared to being observed in the fourth year in plants raised from seed (A.F. Mascarenhas, unpubl. data). This factor can be exploited by the paper and pulp industry since bamboo culms are harvested two years after development.

Based on our results with *Eucalyptus* and on the net gains resulting from early growth and short reproductive rotation cycles, we have extended our program to cover different forest corporations and paper industries in different states of India by implementing a second phase in which more intensive trials will be conducted. The main objective of this project will be to cover a large number of elites from three different groups: *Eucalyptus*, bamboo, and *Salvadora*. These will be selected from different regions, and the plantlets produced will be tested in different provenances so that the best available trees can be used in the best regions. A second objective as a result of the early sexual maturity will be to raise seed orchards as part of a breeding program in which a large genetic base is maintained to avoid the deleterious effects of inbreeding. This could also provide superior seed with the capacity for improved growth rates and other useful properties, such as oil content, specific gravity, crown measurements, and wood quality. Breeding by mating of selected trees to combine the most favorable genes can result in even higher returns (54). To carry out the second phase of this program, tissue culture laboratories will be set up by forest corporations and other industries for production of plantlets based on the processes developed in our laboratory, before finally mass-producing plants. It is our view that this cautious approach is essential, particularly in perennials.

From our studies on *Eucalyptus*, appreciable gains are certainly possible as a result of fast growth and short reproductive cycles. Even though annual investments will be higher, maintenance of these tissue culture-raised forests through fertilizing, weeding, pruning, etc., could markedly increase the value of the forest crop, thus yielding higher returns. This could also result in the employment of a large number of people. The era of intensive forest management is sure to come.

By the year 2000, micropropagation, in combination with high-investment tending of farm forests and a breeding strategy, will have an important role to play in forest tree improvement. The other biotechnologies, such as somaclonal variation and genetic engineering, will also supplement these methods for improving the structures of future forests. In this direction, somatic embryogenesis in cell suspension culture and production of artificial seeds offer a great potential in large-scale plantations at low cost (3). Recently, somatic embryogenesis has been achieved from juvenile tissues of *Eucalyptus gunnii* (M. Boulay, pers. comm.) and also in our laboratory from mature seeds of *E. citriodora* (48). Studies using the procedure described earlier (23) are underway to develop embryogenic cell suspension cultures and artificial seed production. Genetic transformation has been reported in poplar (57), loblolly pine (61), and Douglas fir (15).

Recently in our laboratory, tumors have been induced on the plantlets of teak and *E. tereticornis* grown in vivo. Tumors were induced using *Agrobacterium tumefaciens* strain K12 x 167. [The strain was supplied by Calgene, Inc. (Davis, California).] This strain consists of derivatives of wild-type Ti-plasmid pTiA6 with a chimeric foreign gene encoding resistance to the antibiotic kanamycin. All tumors synthesized octopine and were grown on phytohormone-free medium with carbenicillin. Work is in progress to carry out the neomycin phosphotransferase (NPTII) assay and also Southern blot analysis.

ACKNOWLEDGEMENTS

We would like to acknowledge the help and cooperation extended by Mr. I.M. Quereshi and Mr. M.Y. Sowani for collection of bud wood of teak, and by Mr. S. Kondas and Mr. R.J. Tilani for plant material of Eucalyptus and Salvadora persica, respectively. Thanks are also due to Mr. Y.R. Pakkala and Dr. V.G. Chapke for analysis of the soil samples, to Dr. Srilata Gupta for suggestions in the preparation of the manuscript, and to Mr. G. Balu for typing the manuscript. We would also like to acknowledge the financial support of the National Bank for Agriculture and Research Development, Bombay, for carrying out studies on Eucalyptus, Salvadora, and bamboo.

REFERENCES

1. Abo-el-Nil, M.M. (1982) Method for asexual reproduction of coniferous trees. U.S. Patent No. 4.353, 184.
2. Aitken-Christie, J., and C. Jones (1987) Towards automation: Radiata pine shoots hedges in vitro. Plant Cell Tissue Organ Culture 8:185-196.
3. Ammirato, P.V., and D.J. Styer (1985) Strategies for large-scale manipulation of somatic embryos in suspension culture. In Biotechnology in Plant Science: Relevance to Agriculture in the Eighties, M. Zaitlin, P. Day, and A. Hollaender, eds. Academic Press, Inc., Orlando, Florida, pp. 161-178.
4. Bonga, J.M. (1984) Adventitious shoots and root formation in tissue culture of mature Larix decidua. In International Symposium on Recent Advances in Forestry Biotechnology, J. Hanover, D. Karnowsky, and D. Keathley, eds. Traverse City, Michigan, pp. 64-68.
5. Bonga, J.M. (1986) Clonal propagation of mature trees: Problems and possible solutions. In Cell and Tissue Culture in Forestry, Vol. 1, J.M. Bonga and D.J. Durzan, eds. Martinus Nijhoff Publishers, Boston, pp. 249-271.
6. Boulay, M. (1978) Multiplication rapide du Sequoia sempervirens en culture in vitro. Ann. AFOCEL 1977, pp. 36-65.
7. Boulay, M. (1983) Micropropagation of frost resistant Eucalyptus. General Technical Report PSV-69, Pacific Southwest Forest and Range Experiment Station, Forest Service, U.S. Department of Agriculture, Berkeley, California, pp. 102-107.
8. Boulay, M., and A. Franclet (1983) Micropropagation de clones ages d'Eucalyptus selectiones pour leur resistance au froid. In Colloquium International sur Les Eucalyptus Resistants au Froid, IUFRO CSIRO-AFOCEL, France, pp. 587-601.
9. Brando, L.G. (1984) The New Eucalyptus Forest (The Marcus Wallenberg Foundation Symposia Proceedings), Falun, Sweden, pp. 3-15.
10. Campinhos, E., and Yara Kiemi Ikemori (1983) Production of vegetative propagules of Eucalyptus spp. by rooting of cuttings. In Proceedings of the Second Symposium on Plantation Forests in the Neotropics--Its Role as Source of Energy, IUFRO Group, SI.07.09, Vienna, pp. 1-8.

11. Chadha, Y.R., ed. (date undetermined) Wealth of India, Vol. IX, Publications and Information Directorate, CSIR, New Delhi, pp. 193-195.
12. Chaturvedi, A.N., and K.G. Venkataraman (1973) Volume and weight table for Eucalyptus hybrid. Ind. Forester 99:599-608.
13. Corley, R.H.V., C.H. Lee, L.H. Law, and C.Y. Wong (1986) Abnormal flower development in oil palm clones. Planter (Kuala Lumpur) 62:233-240.
14. Cresswell, R.J., and C. Nitch (1975) Organ culture in Eucalyptus grandis L. Planta 125:87-90.
15. Dandekar, A.M., P.K. Gupta, D.J. Durzan, and V. Knauf (1987) Genetic transformation and foreign gene expression in micropropagated Douglas-fir. Bio/Technology (in press).
16. De la Goublaye de Nantois, H. (1980) Reneunissement chez le Douglas (Pseudotsuga menziesii) en vue de la propagation végétative étude sur la flagiotropic des parties aeriewnes et rainaires DEA. Universite de Paris, Vl.
17. Doorebos, J. (1953) Rejuvenation of Hedera helix in graft combinations. Preb 115, Wageningen, November 28.
18. Durzan, D.J. (1984) Special problems: Adult vs juvenile explants. In Handbook of Plant Cell Culture, Vol. 2, W.R. Sharp et al., eds. Macmillan Company, New York, pp. 471-492.
19. Eldridge, K.G. (1978) Genetic improvement of eucalypts. Silvae Genet. 27:205-209.
20. Franclet, A., and M. Boulay (1982) Micropropagation of frost resistant Eucalyptus clones. Aust. For. Res. 13:8389.
21. Franclet, A., M. Boulay, F. Bekkaoui, Y. Fourest, B. Verschoore-Martouzet, and N. Walker (1987) Rejuvenation. In Cell and Tissue Culture in Forestry, Vol. 1, J.M. Bonga and D.J. Durzan, eds. Martinus Nijhoff Publishers, Boston, pp. 232-247.
22. Ghosh, R.C. (1977) Handbook on Afforestation Techniques, F.R.I. Press, Dehra Dun, India.
23. Gupta, P.K., and D.J. Durzan (1987) Somatic polyembryogenesis and plantlet regeneration of loblolly pine. Bio/Technology 5:146-149.
24. Gupta, P.K., and A.F. Mascarenhas (1983) Essential oil production in relation to organogenesis in tissue cultures of Eucalyptus citriodora Hook. In Plant Cell Culture in Crop Improvement, S.K Sen and K.L Giles, eds. Plenum Press, New York, pp. 299-308.
25. Gupta, P.K., and A.F. Mascarenhas (1987) Eucalyptus. In Cell and Tissue Culture in Forestry, Vol. 3, J.M. Bonga and D.J. Durzan, eds. Martinus Nijhoff Publishers, Dordrecht, The Netherlands, pp. 385-399.
26. Gupta, P.K., A.F. Mascarenhas, and V. Jagannathan (1981) Tissue culture of forest trees: Clonal propagation of mature trees of Eucalyptus citriodora Hook by tissue culture. Plant Sci. Lett. 20:195-201.
27. Gupta, P.K., U. Mehta, and A.F. Mascarenhas (1983) A tissue culture method for rapid multiplication of mature trees of Eucalyptus torelliana and E. camaldulensis. Plant Cell Reports 2:296-299.
28. Gupta, P K., A.L Nadgir, A.F. Mascarenhas, and V. Jagannathan (1980) Clonal multiplication of Tectona grandis Linn. (Teak) by tissue culture. Plant Sci. Lett. 17:259-268.
29. Hanover, J.W. (1984) Screening and breeding trees for high value chemicals. In Proceedings of the International Symposium on Recent Advances in Forest Biotechnology, J. Hanover, D. Karnowsky, and D. Keathley, eds. Traverse City, Michigan, pp. 92-103.

30. Hasnain, S., and W. Cheliak (1986) Tissue culture in forestry: Economic and genetic potential. For. Chron. 62:219-225.
31. Hasnain, S., R. Pigeon, and R.P. Overend (1986) Economic analysis of the use of tissue culture for rapid forest improvement. For. Chron. 62:240-245.
32. Hedegart, T., E.B. Lauridsen, and H. Keiding (1975) Teak. In Seed Orchards, R. Faulkner, ed. Forestry Commission Bulletin No. 54, Her Majesty's Stationery Office, London, pp. 139-142.
33. Hodgson, L.M. (1976) Some aspects of flowering and reproductive behaviour in Eucalyptus grandis (Hill). Maiden at J.D.M. Keest Forest Research Station, South Africa. For. J. 99:53-58.
34. Jackson, M.L. (1967) Soil Chemical Analysis, Prentice-Hall of India Pvt. Ltd., New Delhi, India, p. 498.
35. Kendurkar, S.V., and A.F. Mascarenhas (1987) Micropropagation of forest trees through tissue culture. In Proceedings National Symposium on Recent Advances in Plant Culture of Economically Important Plants, Osmania University, India (in press).
36. Khuspe, S.S., P.K. Gupta, D.K. Kulkarni, U. Mehta, and A.F. Mascarenhas (1987) Increased biomass production by tissue culture of Eucalyptus. Can. J. Forestry (in press).
37. Krugman, S.L. (1986) The ethical question. J. Forestry pp.40-41.
38. Kulkarni, V.M., P.K. Gupta, U. Mehta, and A.F. Mascarenhas (1981) Tissue culture of woody trees: Clonal propagation of Tamarindus indica Linn. by tissue culture. In Proceedings VI All India Plant Tissue Culture Conference, Poona University, Poona, India, pp. 3-4.
39. Larsen, C.S. (1966) Genetics in teak (Tectona grandis L.). Royal Veterinary Agricultural College Yearbook for 1966, Copenhagen, pp. 234-235.
40. Magalewski, R.L., and W.P. Hackett (1979) Cutting propagation of Eucalyptus ficifolia using cytokinin-induced basal trunk sprouts. Proc. Int. Plant Prop. Soc. 29:118-124.
41. Mascarenhas, A.F. (1987) Studies on tissue culture raised trees. In Proceedings of the Seminar on Tissue Culture of Forest Species, Forest Research Institute, Malaysia (in press).
42. Mascarenhas, A.F., S. Hazra, U. Potdar, D.K. Kulkarni, and P.K. Gupta (1982) Rapid clonal multiplication of mature forest trees through tissue culture. In Proceedings of the Fifth International Congress of Plant Tissue Culture, A. Fugiwara, ed. Tokyo, Japan, pp. 719-720.
43. Mascarenhas, A.F., S.V. Kendurkar, P.K. Gupta, S.S. Khuspe, and D.C. Agrawal (1987) Teak. In Cell and Tissue Culture in Forestry, Vol. 3, J.M. Bonga and D.J. Durzan, eds. Martinus Nijhoff Publishers, Dordrecht, The Netherlands, pp. 300-315.
44. Mascarenhas, A.F., S. Nair, V.M. Kulkarni, D.C. Agrawal, S.S. Khuspe, and U.J. Mehta (1987) Tamarind. In Cell and Tissue Culture in Forestry, Vol. 3, J.M. Bonga and D.J. Durzan, eds. Martinus Nijhoff Publishers, Dordrecht, The Netherlands, pp. 316-325.
45. McComb, J.A., and I.J. Benett (1986) Eucalypts (Eucalyptus spp.). In Biotechnology in Agriculture and Forestry, Vol. 1, Y.P.S. Bajaj, ed. Springer-Verlag, Berlin, pp. 340-363.
46. McKeand, S.E., and L.J. Frampton (1984) Performance of tissue culture plantlets of loblolly pine in vitro. In Proceedings of the International Symposium on Recent Advances in Forest Biotechnology, J. Hanover, D. Karnowsky, and D. Keathley, eds. Traverse City, Michigan, pp. 82-103.
47. Minier, R. (1985) Cypress de Duprez: Ses Aptitudes, sa resistance vis a vis du Chancre. C.R. Acad. Sci. Agric. Fr. 71:147-151.

48. Muralidharan, E.M., and A.E. Mascarenhas (1987) In vitro plantlet formation by organogenesis in Eucalyptus camaldulensis and somatic embryogenesis in E. citriodora. Plant Cell Reports (in press).
49. Murashige, T., and F. Skoog (1962) A revised medium for rapid growth and bioassays with tobacco tissue cultures. Physiol. Plant. 15:473-497.
50. Muzik, T.J., and H.J. Crucado (1958) Transmission of juvenile rooting ability from seedlings to adults of Heva braziliensis. Nature 101:1288-1290.
51. Nadgir, A.L., C.H. Phadke, P.K. Gupta, V.A. Parasharami, S. Nair, and A.F. Mascarenhas (1984) Rapid multiplication of bamboo by tissue culture. Silvae Genet. 33:221-223.
52. Namkoong, G. (1966) Breeding effects on estimation of genetic additive variance. For. Sci. 12:8-13.
53. Namkoong, G. (1972) Foundations of Quantitative Forest Genetics: A Text for the IUFRO Short Course on Applications of Quantitative Genetics in Forestry, Department of Genetics, North Carolina Agricultural Experimental Station, Raleigh, North Carolina, Paper No. 3863, p. 85.
54. Namkoong, G., R.D. Barnes, and J. Burley (1980) A philosophy of breeding strategy for tropical forest trees. Tropical Forestry Papers No. 16, pp. 1-67.
55. Oka, A.F. (1976) Teak nursery practices in Maharashtra State. Research and Education Circle, Nagpur, India.
56. Olson, S.R., G.V. Cole, F.W. Watanabe, and L.A. Dean (1954) Estimation of Available Phosphorus in Soils by Extraction with Sodium Bicarbonate, U.S. Department of Agriculture Circular 939.
57. Parson, T.S., V.P. Sinkar, R.F. Stettler, E.W. Nester, and M.P. Gordon (1986) Transformation of poplar by Agrobacterium tumefaciens. Bio/Technology 4:535-536.
58. Perur, N.G., C.K. Subramanium, G.R. Muhr, and H.E. Ray (1973) Soil Fertility Evaluation to Serve Indian Farmers, Mysore Department of Agriculture, Mysore University of Agricultural Life Sciences, U.S. Agriculture for International Development, Bangalore, India, p. 124.
59. Rao, S.M. (1987) In vitro studies on Salvadora persica Linn. M.Sc. Thesis, University of Poona, Poona, India.
60. Read, P. (1985) Novel plant growth regulator delivery systems for in vitro culture of horticultural plants. In Symposium on In Vitro Problems Related to Mass Propagation of Horticultural Plants, Abstr. 5, Gembloux, Belgium.
61. Sederoff, R., A.-M. Stomp, W.S. Chilton, and L.V. Moore (1986) Gene transfer into loblolly pine by Agrobacterium tumefaciens. Bio/-Technology 4:647-649.
62. Smith, D.M. (1955) A comparison of two methods for determining the specific gravity of small samples of secondary growth of Douglas-fir. U.S. Forest Products Laboratory Report No. 2033, pp. 13-20.
63. Smith, D.R. (1986) Forest and nut trees. 1. Radiata pine, Pinus radiata. In Biotechnology of Tree Improvement for Rapid Propagation and Biomass Energy Production, Springer-Verlag, Berlin, pp. 274-291.
64. Snedecor, G.W., and W.G. Cochran (1968) Statistical Methods, Oxford and IBM Publishing Company, New Delhi.
65. Timmis, R. (1984) Factors influencing the use of clonal material in commercial forestry. In International Union Forest Research Organisation Conference on Crop Physiology of Forest Trees, Helsinki, Finland, pp. 259-272.

66. Timmis, R., M.M. Abo-el-Nil, and R.W. Stonecypher (1987) Potential genetic gain through tissue culture. In Cell and Tissue Culture in Forestry, Vol. 1, J.M. Bonga and D.J. Durzan, eds. Martinus Nijhoff Publishers, Boston, pp. 198-215.
67. Walkely, A., and T.A. Black (1934) Determination of organic carbon by rapid titration method. Soil Sci. 37:29.
68. Zimmerman, R.H. (1985) Application of tissue culture propagation to woody plants. In Tissue Culture in Forestry and Agriculture, R.R. Henke, K.W. Hughes, M.J. Constantin, and A. Hollaender, eds. Plenum Press, New York, pp. 165-177.

MULTIPLICATION OF MERISTEMATIC TISSUE:

A NEW TISSUE CULTURE SYSTEM FOR RADIATA PINE

Jenny Aitken-Christie, Adya P. Singh,
and Helen Davies

Ministry of Forestry
Forest Research Institute
Rotorua, New Zealand

ABSTRACT

A subculturable meristematic tissue system capable of plantlet regeneration has been developed for *Pinus radiata*. Multiplication was achieved by the continuous production of meristematic tissue on a modified Lepoivre medium containing 5 mg/l BAP (LP5). This tissue has been maintained for 2.5 years to date. Meristematic nodules multiplied for one year consisted of three zones: the outer meristematic layer, a bulky layer of vacuolated cells, and friable cells containing tannins and degrading cell walls. Cavities (hollows) were often present near the center of meristematic nodules. Natural separation of meristematic nodules contributed to the multiplication process. Factors affecting the success of the system were studied because there was a large variation in response. Embryos formed shoots and multipliable meristematic tissue in similar percentages. Seedlot was not a major factor in multiplication of meristematic tissue. Half-strength LP5 medium proved best for the first 12 weeks in culture, whereas LP5 medium proved best after six months in culture. Both gelrite and liquid induced vitrification, and habituation on LP5 medium had not occurred after 20 months in culture. One of the best embryos produced 5,480 pieces of meristematic tissue in 13.5 months. In a separate experiment with the same clone, an average of 68.4 shoots elongated from each piece of tissue. It was estimated that 260,000 trees could be produced from a single, good-reacting seed in 2.5 years. Genetic stability and automation, two prerequisites for commercial use, are also discussed as well as the potential for gene transfer.

INTRODUCTION

A long-standing objective in conifer tissue culture has been to develop a callus culture system that could be subcultured indefinitely and that would produce an unlimited number of propagules. The first conifer callus was produced from cambial explants of *Pinus pinaster* and *Abies pectinata* from trees 15 to 50 cm in diameter by Gautheret (11). Since then, callus has been produced from needle, cotyledon, hypocotyl, seedling shoot, axillary bud, and root tissue (7). Generally this callus has been friable and unorganized, and attempts to induce shoot formation have been unsuccessful (5,34) or the callus has not lasted more than two or three subcultures (7,17). Very little attention has been paid to growing meristematic cells as a subculturable callus, although it has been suggested by Mott (22,23) and attempts with Pinus radiata were reported by Reilly and Brown (25).

Most studies with conifers have shown that shoots are formed from meristematic tissue, meristemoids, meristematic centers, or domes, and it has been suggested that Pinus radiata shoots can only be induced from this type of tissue (4). With radiata pine, formation of meristematic tissue is a prerequisite for first-generation adventitious shoots (1). Meristematic tissue, which forms on whole embryos, seedling shoot tips, and excised cotyledons after three weeks on a cytokinin medium, subsequently forms shoots on a cytokinin-free medium (1,13,25,26).

The anatomy and histochemistry of shoot formation in radiata pine has been studied extensively (4,24,32,33,35). These investigations showed that subepidermal layers of cells are involved in the formation of meristemoids, and finally shoots, and that distinct changes in cellular metabolism occur.

The aim of this investigation was to multiply meristematic tissue, elongate shoots from it, and then regenerate plants. The developmental sequence for induction and multiplication of meristematic tissue and the anatomy of meristematic tissue during multiplication were studied. A series of experiments to improve the system and to estimate the shoot regeneration potential of meristematic tissue were carried out.

MATERIALS AND METHODS

Culture Conditions

All cultures received a 16-hr photoperiod (irradiance of 80 $\mu Em^{-2}s^{-1}$) and were maintained at 26 ± 1°C/22 ± 1°C (day/night). Fluorescent tubes were 80 W cool white. Meristematic tissue cultures could also be cold-stored for up to four months at 5 ± 1°C under a 16-hr photoperiod (irradiance of 8 $\mu Em^{-2} \cdot s^{-1}$).

For the initiation of meristematic tissue, control-pollinated seeds of Pinus radiata D. Don seedlot number 9/0/80/012, except for experiment 2, were sterilized, stratified, and resterilized as described by Reilly and Washer (26). Embryos were dissected from the seeds and inverted in a modified Lepoivre medium (Tab. 1) containing 3% commercial sucrose, 0.8% Difco Bacto agar, and 5 mg/l benzylaminopurine (BAP) in 90 x 15 mm petri dishes (LP5 medium). Whole embryos were subcultured every three weeks onto LP5 medium, and by week 12, pieces of meristematic tissue became detached from the original embryo. Thereafter, all embryos and their pieces were transferred to LP5 medium every four weeks. To reduce labor during transfers, a small domestic fork was used to collect and transfer the pieces of meristematic tissue. Approximately 60 pieces of meristematic tissue varying in size from approximately 1 to 15 mm^2 were transferred to each dish. In initiation experiments, development of meristematic tissue was assessed visually using a score of 1 to 3, with 3 being the best and 1 the worst, i.e., 3 = good embryos (meristematic tissue on all cotyledons),

Tab. 1. Modified Lepoivre medium used for radiata pine (mg/l).

Component	mg/l
Major	
KNO_3	1,800
$Ca(NO_3)_2 \cdot 4H_2O$	1,200
NH_4NO_3	400
$MgSO_4 \cdot 7H_2O$	360
KH_2PO_4	270
Iron	
$FeSO_4 \cdot 7H_2O$	30
Na_2EDTA	40
Minor	
$ZnSO_4 \cdot 7H_2O$	8.6
H_3BO_3	6.2
$MnSO_4 \cdot 4H_2O$	20.0*
$CuSO_4 \cdot 5H_2O$	0.25*
KI	0.08
$Na_2MoO_4 \cdot 2H_2O$	0.25
$CoCl_2 \cdot 6H_2O$	0.025
Vitamins	
Thiamine HCl	0.4
Inositol	1,000

*Modifications by K. Horgan and J. Aitken-Christie on July 17, 1985, based on analysis of tissue-cultured shoots and normal levels expected in radiata foliage from the nursery and the field.

2 = poor embryos (meristematic tissue on only a few cotyledons or greening of the embryo with no meristematic tissue), and 1 = dead or nonreacting embryos which remain white.

Shoots elongated from meristematic tissue on Lepoivre medium containing 3% commercial sucrose and 0.8% Difco Bacto agar (LPO medium). For the first four-week transfer LPO was used, and for the second four-week transfer LPO containing 0.5% Merck activated charcoal proved best. Subsequently, LPO without charcoal was used every four weeks until shoots reached a length of 20 to 30 mm. During transfers, clumps of shoots were cut into smaller clumps and the cut surface placed in contact with fresh medium. Numbers of shoots formed per piece of meristematic tissue were counted.

For root formation, shoots were placed on water-agar with 1.0 mg/l indolebutyric acid (IBA) and 0.5 mg/l naphthaleneacetic acid (NAA) for five days and then planted in a peat/pumice/perlite mix (50/25/25) in a high humidity box under a 16-hr photoperiod (irradiance of 200 $\mu Em^{-2} \cdot s^{-1}$) and temperature of 22 ± 1°C/17 ± 1°C (day/night).

Microscopy

Hand sections of fresh meristematic tissue were examined under a Wild stereomicroscope. For light and electron microscopy, meristematic tissue, including nodules, was cut into small pieces and fixed in 3% glutaraldehyde (in 0.05 M sodium cacodylate buffer) for 1 hr at room temperature and for the next 15 hr in the refrigerator. Tissue was rinsed in buffer, post-fixed in 2% osmium tetroxide prepared in the same buffer as above, and then rinsed again in buffer. Tissue was dehydrated in acetone and embedded in Spurr's epoxy resin. Sections 1 to 2 μm in thickness were cut with glass knives and stained with toluidine blue for light microscopy. Ultra-thin sections were obtained on an LKB III ultramicrotome with a diamond knife and stained in uranyl acetate and lead citrate prior to being examined in a Philips 300 transmission electron microscope.

Computer Image Analysis

The image processing system utilized was the Vax Image Processing System (VIPS). The image capture system comprised an MIP-512M imaging board and a WV-CD50 National Panasonic image processing camera. VIPS software was developed by Dr. D.G. Bailey (University of Canterbury, New Zealand).

For measurement of meristematic tissue, a silhouette of the tissue in a petri dish was obtained using transmitted light by placing the petri dishes on a light box and capturing an image of the pieces with a camera positioned above. Analysis of this image enabled the total meristematic tissue area in the petri dish to be calculated. The average area per piece of meristematic tissue was then calculated from the total meristematic tissue area in the petri dish divided by the number of pieces of meristematic tissue in the dish.

MULTIPLICATION OF MERISTEMATIC TISSUE

The development and anatomy of meristematic tissue was examined to address three questions:

(1) How does meristematic tissue multiply?

(2) How does a piece (group of nodules) or a single nodule of meristematic tissue become detached from its neighbor?

(3) Is this tissue similar to tissue formed on cotyledons and needle sections after three weeks on cytokinin medium?

Developmental Sequence of Initiation and Multiplication

The developmental sequence of initiation and multiplication of meristematic tissue is shown in Fig. 1. At day 0, excised embryos were placed in an inverted position with their cotyledons submerged (Fig. 1A). By week 3, yellow nodular, compact meristematic tissue formed on cotyledons in contact with the medium and sometimes on the hypocotyl (Fig. 1B). Development to this stage is the same as that for shoot formation. A continuous supply of BAP enabled further proliferation of meristematic tissue on the embryo after six weeks (Fig. 1C). By week 9, meristematic tissue became more nodular and small pieces separated easily from the original embryo (Fig. 1D). Meristematic tissue did not always proliferate well on the surface of the embryo not in contact with the medium, and needle primordia grew out instead (Fig. 1C and D). However, needle primordia could be induced to form meristematic tissue if pieces were inverted during transfer and placed in contact with LP5 medium. Contact of tissue with

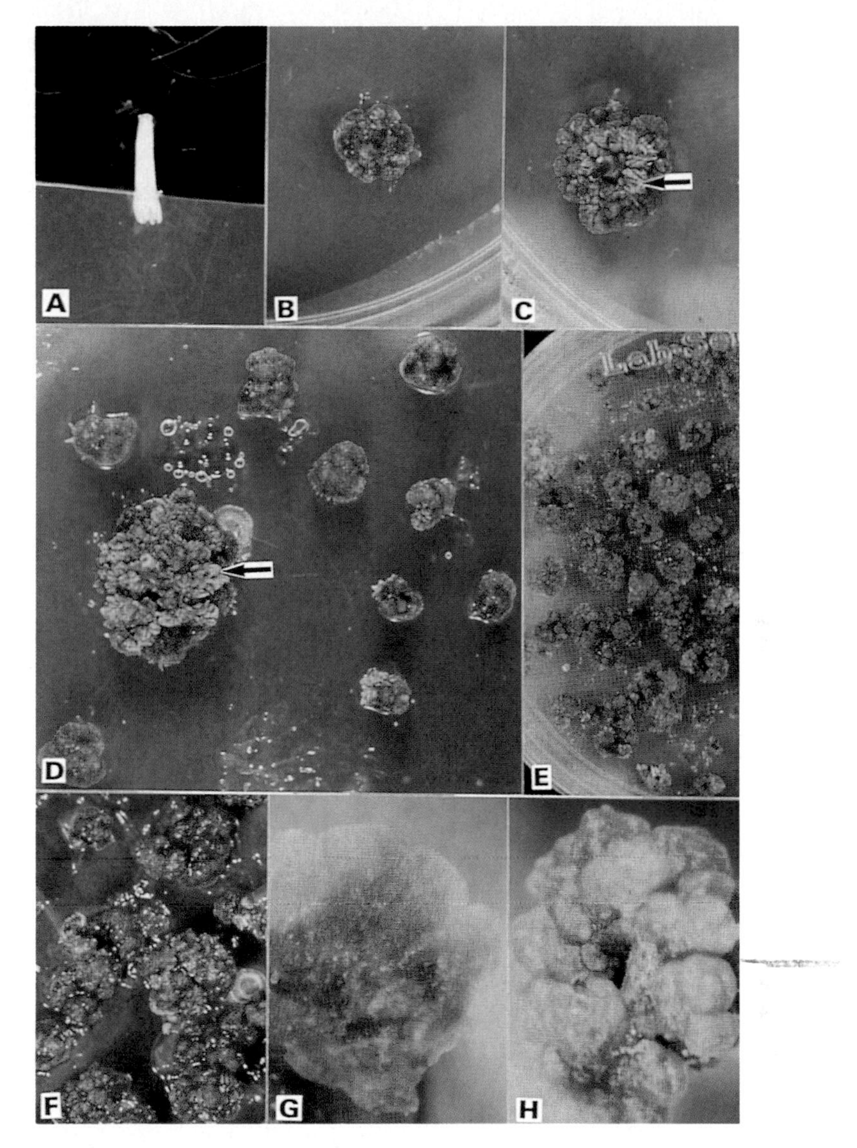

Fig. 1. Developmental sequence of initiation and multiplication of meristematic tissue. (A) Excised embryo inverted in LP5 agar medium. (B) Meristematic tissue formed on cotyledons and hypocotyl after three weeks. (C) Continued growth of meristematic tissue on embryo after six weeks. Needle primordia on upper surface not in contact with medium. (D) Multiplication of meristematic tissue from one embryo after nine weeks. Ten small pieces of meristematic tissue have separated from the original embryo. (E) Multiplied meristematic tissue after one year in culture. (F) Dark-green meristematic tissue which gives rise to vitrified shoots. (G) Hand section through meristematic nodule with hollow center containing friable callus cells and necrotic cells and meristematic cells on external surface. (H) Healthy meristematic tissue surrounding hollow dead nodule. Note: (B) through (F) are viewed from above.

BAP was considered very important for continued multiplication of meristematic tissue. It was also important not to cut meristematic tissue during transfers to avoid necrosis. The pieces either separated spontaneously or were separated by gently prying them apart using forceps. Separated meristematic tissue multiplied to form larger pieces containing many nodules, and the process was then repeated. Figure 1E shows some of the meristematic tissue from one embryo multiplied for one year on LP5 medium. Two clones of meristematic tissue have been multiplied for 2.5 years on LP5 medium. This was only 5% of the embryos initiated. However, 75% of embryos initiated have formed multipliable meristematic tissue after six months using half-strength LP5 medium.

Dark-Green Meristematic Tissue

Of interest was the formation of a dark-green line of meristematic tissue among normal yellow meristematic tissue (Fig. 1F). It was observed that this tissue formed when transferred to fresh LP5 medium that was less than two or three days old. Dark-green meristematic tissue formed predominantly translucent or vitrified shoots when transferred to LPO medium. When the dark-green tissue was transferred to LP5 medium that was two weeks old, some new nodules of the normal yellow type formed. However, most tissue became hollow and died. These preliminary results indicate that vitrification appears to be determined at an early stage of shoot formation and that some tissues can grow out of it. To avoid this problem, media aged for 14 days rather than freshly prepared media should be used.

Necrotic Areas

Small necrotic areas were present on some pieces of meristematic tissue (Fig. 1E), but these were not detrimental to overall multiplication. Closer examination of meristematic tissue revealed that nodules were hollow with friable cells in the center (Fig. 1G). Some of these friable cells were dead and contained tannins (see Fig. 4). This necrosis may contribute in part to the eventual death of a nodule as shown in Fig. 1H, and it may be a normal part of the growth and multiplication process of forming new meristematic tissue. Nodules may also have died because they were initiated under the agar and were vitrified.

Development and Anatomy of Meristematic Tissue

The following describes a preliminary investigation of meristematic tissue multiplied for one year on LP5 medium. The diagram in Fig. 2A gives an overall view of the cellular organization of a meristematic nodule which is also shown by Fig. 1G and 3. The part of the nodule that was in contact with the LP5 medium developed a layer of small meristematic cells comprised of an epidermal cell layer and subepidermal tissue that was one to several cells deep. In small-sized nodules the entire face of the nodule in contact with the medium became meristematic (Fig. 2A). As the nodules grew, large meristematic areas usually became more discrete. Whereas cells of the epidermal layer divided predominantly in two planes, anticlinal and periclinal, the majority of subepidermal cells divided periclinally (Fig. 2B). Cell divisions in the epidermal layer produced cells which became part of the subepidermal meristem and also contributed to the surface growth of the epidermal layer which was necessary in order for this layer to keep pace with the increase in the size of the nodule. While some cells underwent repeated divisions, adding to the bulk of the

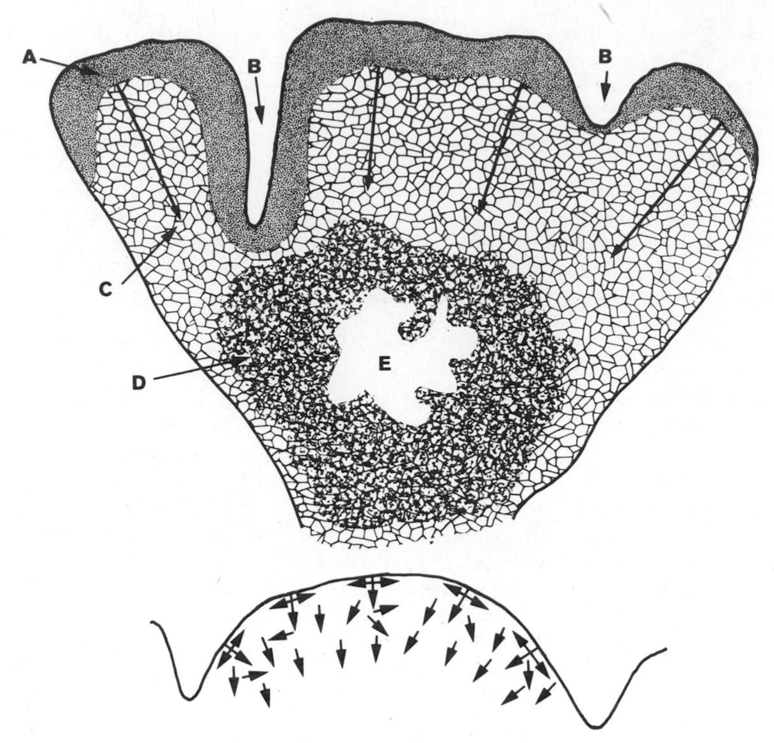

Fig. 2. (Upper) A diagram illustrating the development of a Pinus radiata meristematic nodule in culture. (A) Regions where surface and subsurface cells are meristematic. (B) Grooves between meristematic areas. (C) Region where cells are vacuolating and enlarging; arrows indicate the direction in which cell vacuolation increases. (D) Region showing browning on account of phenolic materials in cells. (E) Cavity. (Lower) A diagram showing directions (arrows) in which surface and subsurface cells divide in meristematic areas of the nodule.

meristem, others differentiated in a direction away from the epidermis (basipetally), enlarging in size and becoming vacuolate (Fig. 2 and 3). A sequential and continuous gradient existed to the extent that cells farthest from the epidermis were the most differentiated. The bulk of the nodule consisted of vacuolated cells. This description is similar to meristematic tissue formed on Pinus radiata needle sections and cotyledons after three weeks on cytokinin medium (25,35).

Two other distinctive events occurred while the nodule increased in size, one involving indentation of the surface of the nodule and the other involving cavity formation in regions where cells were most differentiated. These two events together with actively dividing meristem centers pushing nodules together led to eventual detachment of lobed parts of the nodule, thus enabling the nodules to multiply in number. The indentations appeared initially as small grooves on the surface of the nodule, and then progressively deepened (Fig. 2 and 3). Concurrently, cavities initiated primarily from the disintegration of differentiated cells (see the following section), many of which were filled with tannins and enlarged (Fig. 2, 3,

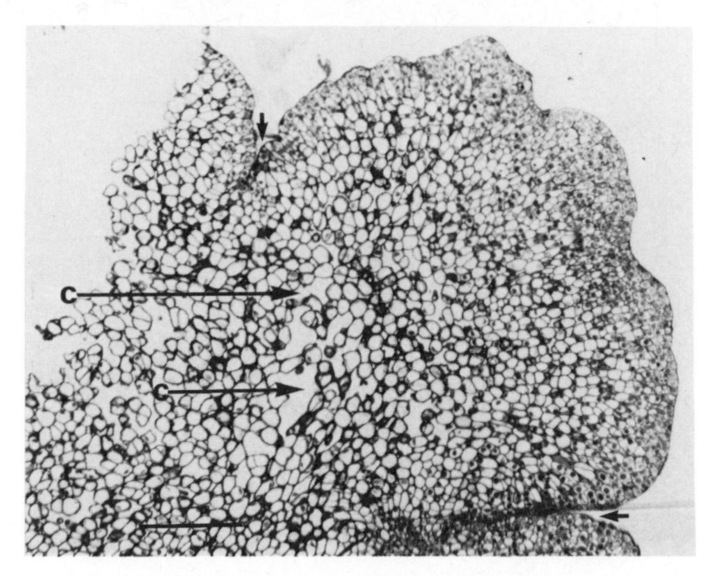

Fig. 3. Part of a nodule showing tissue organization, meristematic cells in outer layer, and developing cavities and indentations (arrows). LM scale = 200 μm.

and 4). Eventually, the tip of the furrow and the margin of the cavity made contact with each other or came near one another, being separated by only one or two vacuolated cells, and a gentle touch detached the lobe from the parent nodule.

Ultrastructure of the Meristematic Nodule

Meristematic cells had thin walls and a large, slightly lobed and centrally located nucleus (Fig. 5). Intercellular spaces were not normally present. The nucleus contained prominent nucleoli and chromatin that showed even distribution and little or no condensation. The remaining part of the meristematic cell was filled with a granular cytoplasm which contained plastids with few internal membranes, cisternae of endoplasmic reticulum, dictyosomes, and abundant mitochondria. Plastids, which were distinguished from mitochondria because of their larger size and somewhat denser stroma, were of different forms, circular to oval being the most common. Meristematic cells also contained a few small vacuoles (Fig. 5).

Cells that did not divide further underwent marked changes during differentiation (Fig. 6). Increase in the size of vacuoles was one of the first signs that these cells were differentiating (Fig. 6). A range in the extent of vacuolation was observed depending upon the location of differentiating cells within the nodule relative to the meristem. Early stages of differentiation also included some other changes: plastids became more oblong and accumulated starch, and their internal membranes began to overlap to form incipient grana (Fig. 6). Cell walls became thickened.

Fully differentiated cells were located in areas farthest from the surface of the nodule where a cavity was developing. The cells had a large, central vacuole and thin peripheral cytoplasm. Cell walls were

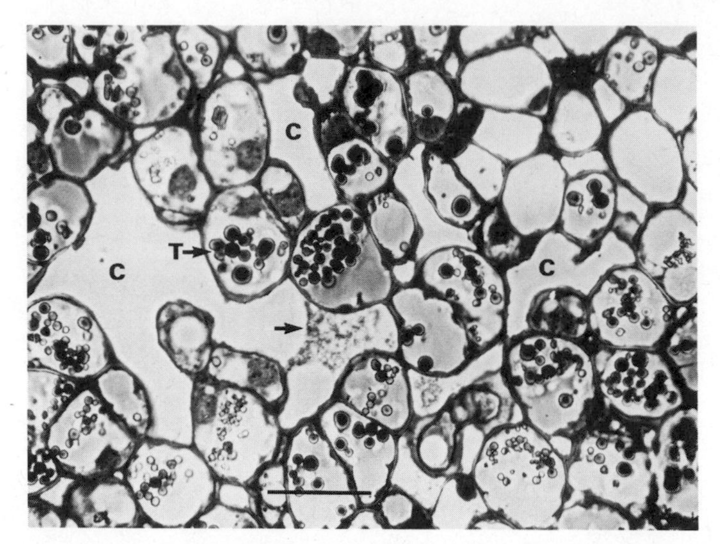

Fig. 4. Part of a nodule where cavities (C) are developing. Cells in this region are highly vacuolated and many contain tannins (T). The arrow points to cytoplasmic residues. LM scale = 40 μm.

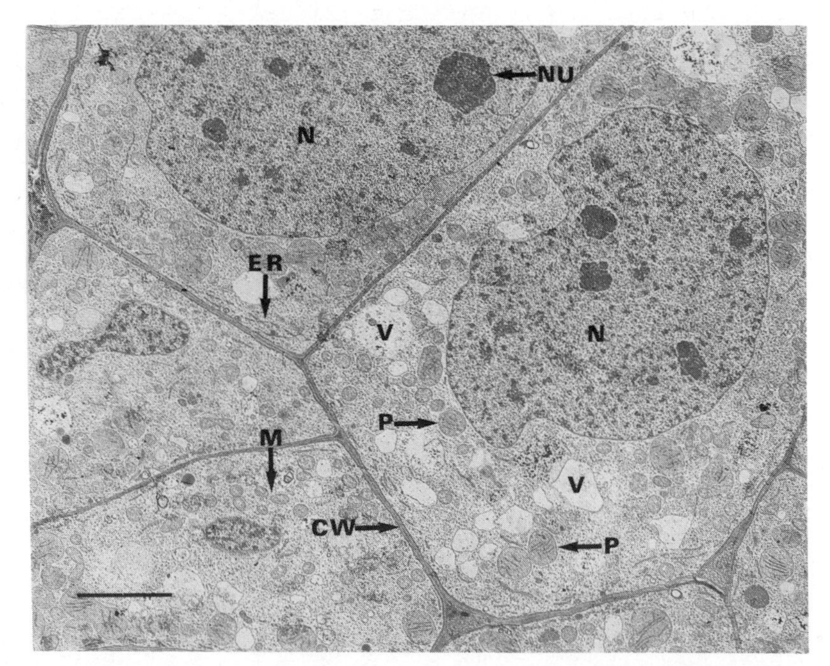

Fig. 5. A group of subepidermal meristematic cells containing thin cell walls (CW), a large centrally located nucleus (N) with prominent nucleoli (NU) and diffuse chromatin, several small vacuoles (V), a high population of mitochondria (M), plastids (P) containing a few internal membranes, and cisternae of endoplasmic reticulum (ER). TEM scale = 5 μm.

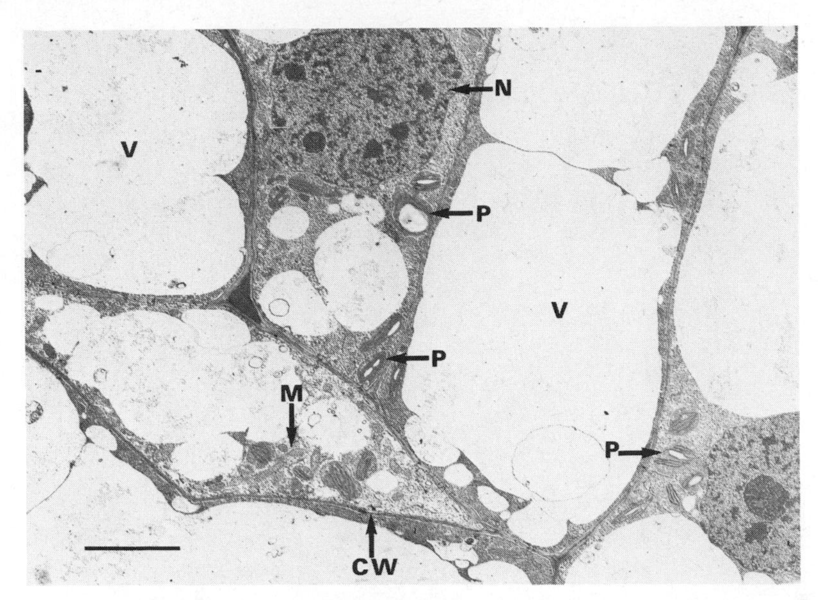

Fig. 6. Differentiating subepidermal cells containing large vacuoles (V), a nucleus (N) which is smaller than the nuclei of meristematic cells and contains several smaller nucleoli and condensing chromatin, mitochondria (M), and plastids (P) containing starch. CW, cell wall. TEM scale = 5 μm.

considerably thicker than in meristematic cells and appeared electron-lucent. The nucleus was smaller than that of meristematic cells and contained more compact chromatin (Fig. 7). Plastids were lenticular and contained moderately developed grana (Fig. 7). Cavities formed in areas with vacuolated cells, many of which contained tannins (Fig. 1G, 2A, and 4). Prior to cavity formation, the middle lamella darkened first in cell corners, and later elsewhere, eventually becoming granular. Figure 7 shows a view that is suggestive of sequential stages leading to formation of cavities in the middle lamella region which may also be accompanied by some mechanical disruptions, as evidenced by the presence of a thin residual wall in distorted form (Fig. 7).

EXPERIMENTS TO IMPROVE THE SYSTEM

It is important to have good representation of clones for clonal forestry, and therefore multiplication of meristematic tissue from at least 60 to 70% of embryos in a seedlot is desirable. A method where only 10 to 20% of embryos respond to treatment may be of limited use for a commercial operation. Experiments were carried out to study factors affecting the ability of embryos to form multipliable meristematic tissue.

Shoot Formation and Meristematic Tissue Multiplication from the Same Seedlot

Thirty-five embryos were treated for shoot formation (three weeks on LP5 medium and then transfers once every three weeks on LP0 medium)

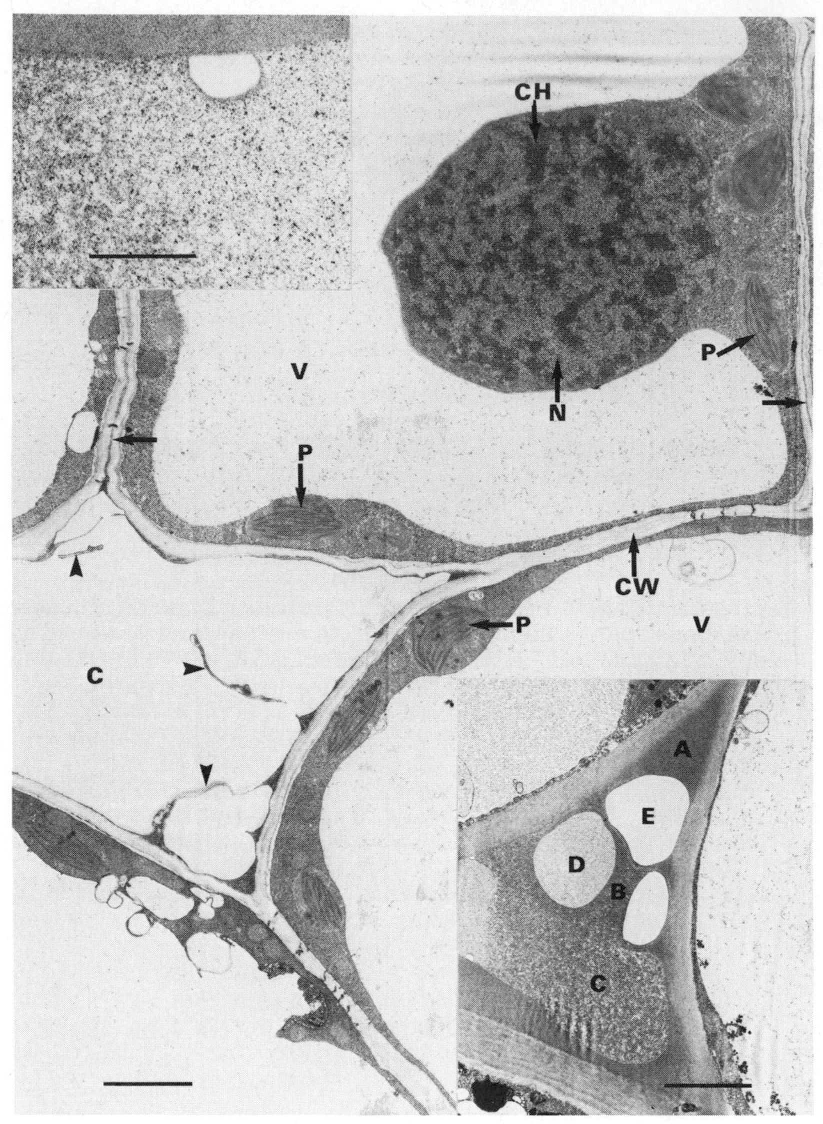

Fig. 7. Vacuolated cells surrounding a developing cavity (C). The nucleus (N) contains moderately condensed chromatin (CH), and plastids (P) contain moderately developed grana and osmiophilic bodies. Cell walls (CW) are thicker and electron-lucent. Middle lamella in some parts of wall is seen as darker, slender structure (arrows). Remains of middle lamella visible along walls lining the cavity and as a distorted and fragmented structure (arrowheads). V, vacuole. Lower inset shows stages in the dissolution of middle lamella in a cell corner: A, dense and compact middle lamella; B, area where granules are densely packed; C, area where granules are less densely packed than in B; D, area where granules are considerably smaller than those in B and C and are sparse; and E, small cavity. Upper inset: higher magnification of area corresponding to C in lower inset. TEM scale = 2 µm. Lower inset scale = 2 µm. Upper inset scale = 250 µm.

and 35 embryos were treated for multiplication of meristematic tissue to see if there was any difference in the number of genotypes responding after 12 weeks. The difference between the two groups was tested by a non-paired t-test.

There was no significant difference between percentages of embryos that were nonreacting or dead, poor, or good for either pathway with this particular seedlot (Tab. 2). The average number of shoots per embryo for poor embryos was 21.5 whereas for good embryos it was 72.5. The average number of pieces of meristematic tissue per embryo for poor embryos was 4.0 whereas for good embryos it was 18.7. It was concluded that embryos had a similar capacity to form good numbers of shoots and good meristematic tissue after 12 weeks in culture. After three weeks on LP5 medium, poor embryos should be induced to form shoots and multiplied by topping to produce more shoots rather than treated for multiplication of meristematic tissue.

Multiplication of Meristematic Tissue from Five Seedlots

In the previous experiment only 20% of embryos formed good meristematic tissue, although a further 54.3% formed some meristematic tissue. The objective of this experiment was to determine if this was a trend among other seedlots. A completely randomized design with five seedlots, 12 replicates (petri dishes) with seven embryos per dish was used. Results of least significant difference (LSD) tests after analysis of variance (ANOVA) are given in Tab. 3.

Seedlot 4 produced a significantly lower percentage of good embryos compared with the other four seedlots after 12 weeks in culture (Tab. 3). One of the parents from seedlot 4 (850.121) was infected with Dothistroma at the time of control pollination and was in poor health. Seeds from seedlot 4 were also noticeably smaller than the rest. These factors may have affected the performance in culture. The percentage of good embryos was not above 31% for any seedlot, suggesting that there was further room for improvement.

Multiplication of Meristematic Tissue on Four Media

To improve the number of embryos forming good meristematic tissue, four different media were tested: LP5 control; LP5 half-strength (nutrients only) (half LP5); LP5 containing 250 mg/l glutamine and arginine (LP5

Tab. 2. Comparison of shoot and meristematic tissue formation from embryos after 12 weeks.

	Percent nonreacting or dead embryos	Percent poor embryos	Percent good embryos
Shoot formation	25.7	48.6	25.7
Meristematic tissue multiplication	25.7	54.3	20.0

Note: No significant difference was found between the two groups.

Tab. 3. Comparison of meristematic tissue multiplication from five seedlots.

Seedlot*	Percent nonreacting or dead embryos	Percent poor embryos	Percent good embryos
1	35.7	36.9	27.4
2	47.6	21.4	31.0
3	34.5	36.9	28.6
4	64.3	19.0	16.7
5	28.6	40.5	31.0
LSD (0.05)	9.82	9.69	11.90

*1-Seed orchard seed (1979). Seedlot R79/A2/2.
2-Seed orchard seed (1984). Seedlot 2/3/83/02/2.
3-Control pollinated seed (1976). Cross 850.55 x 850.100.
4-Control pollinated seed (1979). Cross 850.121 x 850.191.
5-Mixed control pollinated seed (1980). 9/0/80/012 C "875" series.

G/A); and Schenk and Hildebrandt with minor elements modified by Horgan and Aitken (13) containing 5 mg/l BAP (SH5). A completely randomized design was chosen using 12 replicates (petri dishes) with seven embryos per dish for LP5, LP5 G/A, and SH5 media and ten replicates with seven embryos per dish for half LP5 medium. Following ANOVA, an LSD test was used to compare the difference between means. The results are shown in Tab. 4.

Half-strength LP5 medium was a significantly better medium than LP5 or the other two tested for multiplication of meristematic tissue after 12 weeks (Tab. 4). This medium was earlier found to be better than LP5 for initiation of shoots on embryos or excised cotyledons (D.R. Smith, pers. comm.). There was no significant difference among the percentage of good embryos on LP5, LP5 G/A, and SH5 media.

Tab. 4. Comparison of meristematic tissue multiplication on four media.

Medium	Percent nonreacting or dead embryos	Percent poor embryos	Percent good embryos
LP5	33.9	23.8	42.3
Half LP5	20.6	10.0	69.4
LP5 G/A*	44.2	22.8	32.9
SH5	38.0	34.9	37.1
LSD (0.05)	2.02	2.02	2.02

*Glutamine/Arginine.

Since the first experiment, there has been an indication that the percentage of embryos forming good meristematic tissue on LP5 medium increased from 20% (Tab. 2) to 31% (Tab. 3) to 42.3% (Tab. 4) for the same seedlot. We have no firm explanation for this. It is possible that our techniques were improving with time, that the seed was improving with age (which is doubtful as older radiata pine seed usually has a lower germination rate than young seed), or that the time of year influenced the results.

Effect of Gelrite and Liquid

The objective of this experiment was to test gelrite or liquid for the multiplication of meristematic tissue. Four clones of meristematic tissue were divided evenly among the following treatments: LP5 agar control; LP5 gelled with 0.25% gelrite; LP5 gelled with 0.5% Difco Bacto agar and 0.125% gelrite (50/50); LP5 shaken liquid (80 rpm); and LP5 unshaken liquid. Twenty-five ml of the various gelled media was placed in 90 x 15 mm petri dishes and 15 ml of liquid media was placed in 250-ml Erlenmeyer flasks. Liquid medium was replaced three times per week based on the replacement interval used by Vasil and Vasil (31).

After four weeks, meristematic tissue in liquid had become swollen, partially friable, and vitrified (dark green). Tissue in unshaken liquid was healthier than that in shaken liquid. Some dead pieces examined were spongy and had liquid inside.

Observations after eight weeks showed that meristematic tissue on LP5 agar survived better and did not vitrify compared to tissue on gelrite or in liquid (Tab. 5). Once vitrification occurred, death was imminent. Liquid media, which would be ideal for automation, was not suitable for multiplication of meristematic tissue in this experiment. Further experiments are necessary.

Effect of Benzylaminopurine Concentration

The objective of this experiment was to investigate whether meristematic tissue was becoming habituated on LP5 medium. Exactly 1,417 pieces of 20-month-old meristematic tissue from two clones were placed on LP5 medium as a control, and 2,199 pieces were placed on LP medium containing 1 mg/l BAP (LP1). The number of pieces forming needle primordia was assessed after four weeks. The difference between the two treatments was tested by a nonpaired t-test. There were significantly more ($P<0.01$) pieces with needle primordia on LP1 medium than on LP5 medium (Tab. 6). There was no significant difference in the number of pieces per dish for each treatment or between clones.

Tab. 5. Effect of gelrite and liquid on multiplication of meristematic tissue after eight weeks.

	Agar	Gelrite	Agar/Gelrite	Liquid shaken	Liquid unshaken
Percent survival	100	30	70	0	0
Vitrification	-	+	+	+	+

Tab. 6. Effect of benzylaminopurine concentration on needle primordia formation.

Treatment	Percent of pieces with needle primordia
LP5	13.9
LP1	42.7

Habituation to cytokinins was not a problem in the two clones of meristematic tissue proliferated for 20 months as indicated by the requirement for 5 mg/l BAP to keep needle primordia outgrowth in check. Thus no alteration in endogenous growth regulator activity appeared to have taken place. Callus cultures of several "model-system" species, including Nicotiana tabacum and Citrus sinensis, have become habituated on auxin and cytokinin media after long periods of subculturing (29).

Effect of Nutrient Media Strength on Long-Term Multiplication

Though half LP5 was the best medium for multiplication of meristematic tissue from initiation to 12 weeks (Tab. 4), preliminary observations suggested that it was not as good as LP5 for longer-term multiplication. The aim of this experiment was to find the best medium for ten clones of meristematic tissue which had been multiplied for more than six months. Eleven petri dishes (replicates) with an average of 34 pieces of meristematic tissue per dish were placed on half LP5, three-quarters LP5, and LP5 for two four-week transfers. Growth of meristematic tissue was measured by the difference in area of tissue per petri dish between transfers using computer image analysis (Fig. 8). Health was assessed on a 1-4 scale per dish at the end of the experiment (second transfer): 1 = dead, 2 = 50% healthy, 3 = more than 75% healthy, and 4 = more than 90% healthy.

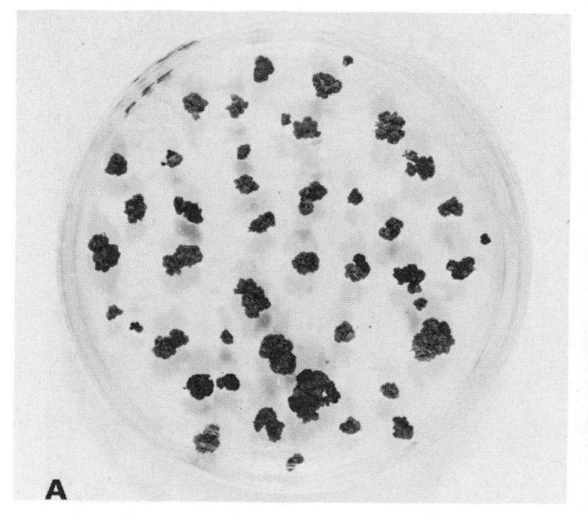

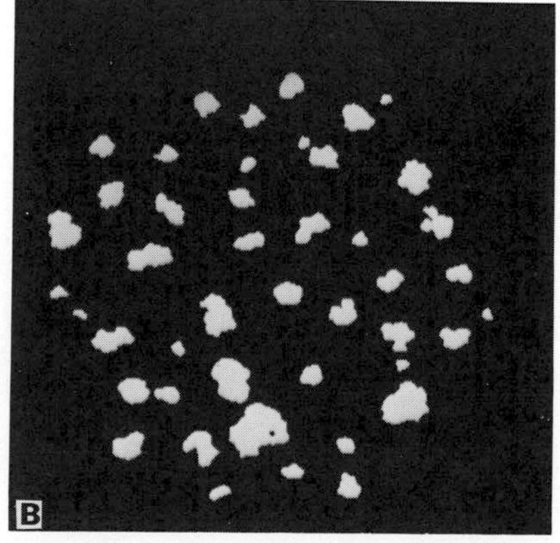

Fig. 8. (A) Petri dish of meristematic tissue before image processing. (B) Processed image of the petri dish of meristematic tissue.

Meristematic tissue grew significantly better ($P<0.01$) on LP5 than on half LP5 or three-quarters LP5 during the first and second transfers (Tab. 7). The area increment was more for the first than the second transfer. Some unhealthy tissue was discarded from all treatments at the end of the first transfer. There was no significant difference among the amount of tissue discarded from LP5, half LP5, and three-quarters LP5. However, this loss may account for the loss of growth during the second transfer. Health of tissue on LP5 and half LP5 was better than that on three-quarters LP5.

REGENERATION POTENTIAL

Multiplication Curve of Meristematic Tissue

Meristematic tissue has been multiplied on LP5 medium for 2.5 years so far. It may be able to be multiplied indefinitely. The multiplication curve for growth of one of the best clones was monitored for 13.5 months (Fig. 9). Growth was exponential and 5,480 pieces of meristematic tissue were produced; however, growth rates for other clones could be different.

Shoot Elongation

Meristematic tissue was transferred to LPO medium to induce shoot formation. After three weeks the tissue was transferred to a medium with charcoal added to enhance growth (compare Fig. 10A and B). However, because charcoal induced vitrification of small shoots with continuous use (I. Steele and J. Aitken-Christie, unpubl. data), shoots were returned to LPO medium for further elongation.

A sample of 366 pieces of meristematic tissue from two of the best clones proliferated on LP5 medium for eight months were elongated to form shoots. After five months, 25,034 shoots had elongated from the meristematic tissue. The average number of shoots formed per piece of meristematic tissue was 68.4. Other clones may produce different numbers of shoots.

Rooting

Because of the high labor input required for transfer and separation of shoots and because established methods for rooting juvenile radiata pine

Tab. 7. Effect of medium strength on area increment and health of meristematic tissue.

Medium	Initial area per piece (mm^2)	Area increment per piece (mm^2) First transfer	Second transfer	Mean health score
Half LP5	14.0	11.3	5.8	3.6
Three-quarters LP5	14.2	12.4	6.4	2.9
LP5	13.1	19.3	12.1	3.6
LSD (0.05)	*	3.99	3.22	0.60

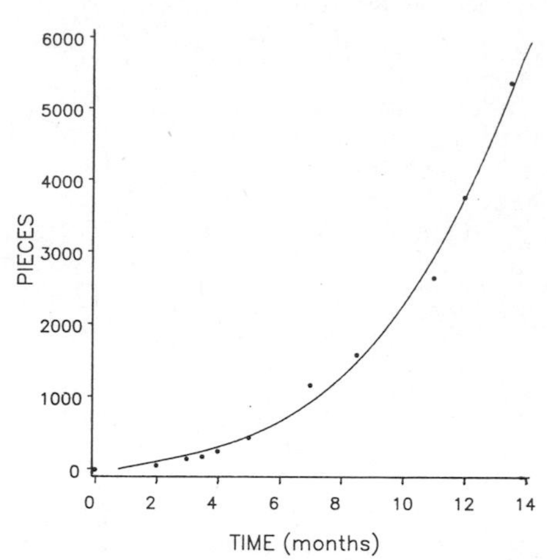

Fig. 9. Multiplication curve of meristematic tissue from one embryo.

shoots (2,13) were being used, only a sample of meristematic tissue-derived shoots were rooted. A total of 443 shoots from ten clones rooted in three different experiments gave 71.8% rooting. Further, more extensive experiments are necessary to determine the long-term effects of BAP on rooting, maturation state, and stability.

High Regeneration Potential

Large numbers of radiata pine trees can now be produced from one seed. For instance, using the example above, if all 5,480 pieces from one embryo formed shoots and 70% of the shoots rooted, approximately 260,000 trees could be produced in 2.5 years to be ready for planting out in the forest (Fig. 11).

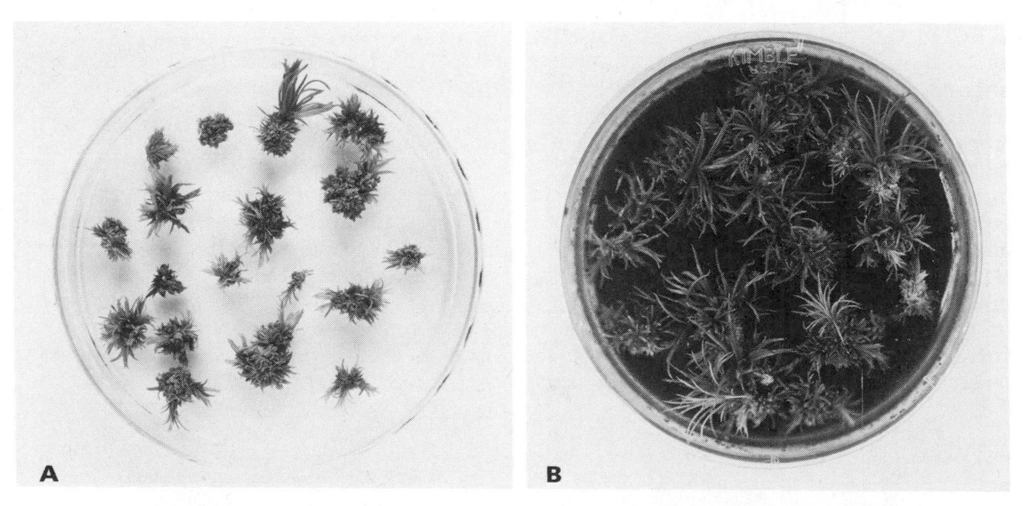

Fig. 10. (A) Shoot elongation on LP medium. (B) Shoot elongation on LP medium containing 0.5% charcoal.

One seed ⟶ 5,480 pieces of meristematic tissue ⟶ 374,830 shoots ⟶ 260,000 trees

Fig. 11. Number of trees that could be produced from one seed in 2.5 years.

Higher multiplication rates would be possible if meristematic tissue was multiplied for longer than 13.5 months. The "cut-off point," i.e., when to stop proliferating meristematic tissue and start elongating shoots from it, for a commercial laboratory would depend on the number of trees required per clone.

DISCUSSION

This is the first report of a subculturable meristematic tissue system capable of plantlet regeneration for a conifer. A similar system with *Populus* nodules has been described by McCown (McCown et al., this Volume). Our morphological and histological description of *Pinus radiata* meristematic tissue appears similar to descriptions of calluses for lily, *Nicotiana rustica*, *Freesia*, *Chrysanthemum*, tomato, daylily, and pea. Most of these calluses have been shown to consist of vacuolated cells covered with superficial meristems (16). Visually they have been described as being self-replicating meristematic centers, continuously proliferating shoot meristems, granular callus fragments, "nubbins," nodular callus, or nodules in a review by George and Sherrington (12). Callus that forms in culture at the base of *Rhododendron* shoots (19) and on roots of different *Prunus* species (9) is also meristematic and nodular.

An important aspect of any tissue culture multiplication method is the genetic stability of the plants produced. The organized callus systems already mentioned have proved to be genetically stable so far (16). Several other authors have stated that meristematic cells are stable (20,31). Because of the similarities in structure between the stable calluses reported and our radiata pine meristematic tissue, it is possible that the radiata tissue will be genetically stable. Hussey (14,15) has proposed that the stability of these calluses may depend on their structure and organization as well as on the genotype concerned. In stable calluses the regenerated shoots appear to arise exclusively from the superficial meristems which may inhibit adventitious regeneration from the inner vacuolated cells, which are more likely to be susceptible to somaclonal changes. In the more typical unstable calluses meristematic regions are small and are dispersed throughout the callus mass. These may have arisen from and continue to arise from differentiated cells with changed genotypes (16). For the meristematic tissue method to be of commercial use, it must be shown that the cultures are genetically stable and that the trees produced are faithful copies of the required clone. There is no simple way of identifying specific gene changes other than regenerating trees and observing their performance to maturity. Field trials of radiata pine plantlets produced from meristematic tissue should be carried out.

The high regeneration potential of meristematic tissue has been demonstrated. However, many labor-intensive transfers are necessary to achieve this. The organization of meristematic tissue lends itself to mechanical division and could provide suitable material for aseptic automated processing. Levin (20) has developed a semiautomated homogenization procedure for meristematic tissue based on the earlier work of Cooke (6). Large numbers of high quality plants of ferns, Lilium, and Philodendron have been produced. The patent states "that the homogenization procedure could be applicable for most species which, during the multiplication stage, could be induced to form a culture consisting of dense clumps of meristematic tissue capable of developing into shoots." Radiata pine meristematic tissue is similar in description to those species already tested, therefore it may have excellent potential for adaptation to this semiautomated procedure. The semiautomated homogenization process separates clumps of meristematic tissue without cutting them and transfers them to a growth or multiplication medium under sterile conditions. This process significantly lowers production costs for tissue culture, and it was estimated that a 75% cost reduction in the recycling multiplication stage was possible (20).

By bulking-up tissue and automating the multiplication stage (i.e., meristematic tissue), elongation could be reduced to a one-stage process since the high numbers have already been obtained. Further reductions in cost, better shoot growth, and more waxy shoots could result if an automated nutrient replenishment system, as suggested by Aitken-Christie and Jones (3), was introduced at the shoot elongation stage. Both Levin's method (20) and the nutrient replenishment method (4) are short-term exposure to liquid only, in order to avoid vitrification problems. It remains to be determined whether automated culture of meristematic tissue combined with automated nutrient replenishment will produce the cheapest micropropagated trees for forestry.

Cell cultures and protoplasts have traditionally been investigated for transferring genes into plants using bacterial, viral, and pollen vectors (8,18,21). The regeneration of plants from protoplast-derived cells has been a major problem for woody species. More recently, transformation and regeneration have been reported for shoot cultures of Populus (Ref. 10; Michel et al., this Volume), Pseudotsuga menziesii (Gupta and Durzan, this Volume), Betula papyrifera, and Alnus glutinosa (MacKay et al., this Volume), and transportation of bacterial genes into seedlings of Pinus taeda has been demonstrated (27). Recovery of transformed shoots has also been increased (28). Thus, tissues other than protoplasts are suitable for genetic transformation. Pinus radiata meristematic tissue, which reliably produces large numbers of shoots and later plantlets, may also be suitable for gene transfer studies.

ACKNOWLEDGEMENTS

We would like to thank Jae Hun Kim and Irene Steele for technical assistance, Michael Hong for statistical analysis and graphs, Paul Miller for image analysis, and Kathy Horgan for helpful discussions and for reviewing this chapter. Dan Cohen and Trevor Thorpe are also thanked for reviewing the manuscript.

REFERENCES

1. Aitken, J., K.J. Horgan, and T.A. Thorpe (1981) Influence of explant selection on the shoot-forming capacity of juvenile tissue of Pinus radiata. Can. J. For. Res. 11:112-117.
2. Aitken-Christie, J., and T.A. Thorpe (1984) Clonal propagation. Gymnosperms. In Cell Culture and Somatic Cell Genetics of Plants,

Vol. 1, I.K. Vasil, ed. Academic Press, Inc., Orlando, Florida, pp. 82-95.
3. Aitken-Christie, J., and C. Jones (1987) Towards automation: Radiata pine shoot hedges in vitro. Plant Cell Tissue Organ Culture 8:185-196.
4. Aitken-Christie, J., A.P. Singh, K.J. Horgan, and T.A. Thorpe (1985) Explant developmental state and shoot formation in Pinus radiata cotyledons. Bot. Gaz. 146(2):196-203.
5. Bornman, C.J. (1987) Picea abies. In Cell and Tissue Culture in Forestry, Vol. 3, J.M. Bonga and D.J. Durzan, eds. Martinus Nijhoff Publishers, Dordrecht, The Netherlands, pp. 2-29.
6. Cooke, R.C. (1979) Homogenization as an aid to tissue culture propagation of Platycerium and Davallia. HortScience 14:21-22.
7. David, A. (1982) In vitro propagation of gymnosperms. In Tissue Culture in Forestry, J.M. Bonga and D.J. Durzan, eds. Martinus Nijhoff/Dr. W. Junk Publishers, The Hague, pp. 72-108.
8. David, A. (1987) Conifer protoplasts. In Cell and Tissue Culture in Forestry, Vol. 2, J.M. Bonga and D.J. Durzan, eds. Martinus Nijhoff Publishers, Dordrecht, The Netherlands, pp. 2-15.
9. Druart, P. (1980) Plantlet regeneration from root callus of different Prunus species. Scienta Horticulturae 12:339-342.
10. Fillatti, J., J. Sellmer, B.H. McCown, B. Haissig, and L. Comai (1987) Agrobacterium mediated transformation and regeneration of Populus. Mol. Gen. Genet. 206:192-199.
11. Gautheret, R.J. (1934) Culture du tissue cambial. C.R. Acad. Sci. Paris 198:2195-2196.
12. George, E.F., and P.D. Sherrington, eds. (1984) Problems of genetic variability. In Plant Propagation by Tissue Culture, Exegetics Ltd., Eversley, England, pp. 73-87.
13. Horgan (nee Reilly), K.J., and J. Aitken (1981) Reliable plantlet formation from embryos and seedling shoot tips of radiata pine. Physiol. Plant. 53:170-175.
14. Hussey, G. (1983) In vitro propagation of horticultural and agricultural crops. In Plant Biotechnology, S.H. Mantell and H. Smith, eds. Cambridge University Press, Cambridge, England, pp. 111-138.
15. Hussey, G. (1985) Vegetative propagation of plants by tissue culture. In Plant Cell Culture Technology, M.M. Yeoman, ed. Blackwell Scientific Publications, London, pp. 29-66.
16. Hussey, G. (1986) Problems and prospects in the in vitro propagation of herbaceous plants. In Plant Tissue Culture and Its Agricultural Applications, L.A. Withers and P.G. Alderson, eds. Butterworths, London, pp. 69-84.
17. Kaul, K., and T.S. Kochhar (1985) Growth and differentiation of callus cultures of Pinus. Plant Cell Reports 4:180-183.
18. Kirby, E.G. (1982) The use of in vitro techniques for genetic modification of forest trees. In Tissue Culture in Forestry, J.M. Bonga and D.J. Durzan, eds. Martinus Nijhoff/Dr. W. Junk Publishers, The Hague, pp. 369-386.
19. Kyte, L., and B. Briggs (1979) A simplified entry into tissue culture production of Rhododendrons. Proc. Int. Plant Prop. Soc. 29:90-95.
20. Levin, R. (1985) Process for plant tissue culture propagation. EP Patent No. 0132414A2.
21. McCown, B.H., and J.A. Russell (1987) Protoplast culture of hardwoods. In Cell and Tissue Culture in Forestry, Vol. 2, J.M. Bonga and D.J. Durzan, eds. Martinus Nijhoff Publishers, Dordrecht, The Netherlands, pp. 16-30.

22. Mott, R.L. (1978) Tissue culture propagation of conifers. In Propagation of Higher Plants through Tissue Culture, K.W. Hughes, R. Henke, and M. Constantin, eds. U.S. Department of Energy, Springfield, Virginia, pp. 125-129.
23. Mott, R.L., and H.V. Amerson (1984) Role of tissue culture in loblolly pine improvement. In Proceedings of Symposium on Recent Advances in Forest Biotechnology, J. Hanover, D. Karnosky, and D. Keathley, eds. Traverse City, Michigan, pp. 24-36.
24. Patel, K.R., and T.A. Thorpe (1984) Histochemical examination of shoot initiation in cultured cotyledon explants of radiata pine. Bot. Gaz. 145(3):312-322.
25. Reilly, K.J., and C.L. Brown (1976) In vitro studies of bud and shoot formation in Pinus radiata and Pseudotsuga menziesii. Georgia For. Res. Paper No. 86.
26. Reilly, K.J., and J. Washer (1977) Vegetative propagation of radiata pine by tissue culture: Plantlet formation from embryonic tissue. New Zealand J. For. Sci. 7:199-206.
27. Sederoff, R., A.M. Stomp, W.S. Chitton, and L.W. Moore (1986) Gene transfer into loblolly pine by Agrobacterium tumefaciens. Bio/-Technology 4:647-649.
28. Sellmer, J.C., and B.H. McCown (1986) A selection system for increased recovery of transformed shoots regenerated from Populus leaf tissue. Abstract of the VI IAPTC Congress, D.A. Somers, B.G. Genebach, D.D. Biesboor, W.P. Hackett, and C.E. Green, eds. University of Minnesota, p. 154.
29. Thorpe, T.A. (1982) Callus organisation and de novo formation of shoots, roots and embryos in vitro. In Application of Plant Cell and Tissue Culture to Agriculture and Industry, D.T. Tomes, B.E. Ellis, P.M. Harney, K.J. Kasha, and R.L. Peterson, eds. University of Guelph, Guelph, Canada, pp. 115-138.
30. Vasil, I.K. (1985) Somatic embryogenesis and its consequences in the Graminae. In Tissue Culture in Forestry and Agriculture, R.R. Henke, K.W. Hughes, M.J. Constantin, and A. Hollaender. Plenum Press, New York, pp. 31-47.
31. Vasil, V., and I.K. Vasil (1984) Isolation and maintenance of embryogenic cell suspension cultures of Graminae. In Cell Culture and Somatic Cell Genetics of Plants, Vol. 1, I.K. Vasil, ed. Academic Press, Inc., New York, pp. 36-42.
32. Villalobos, V.M., E.C. Yeung, and T.A. Thorpe (1985) Origin of adventitious shoots in excised radiata pine cotyledons cultured in vitro. Can. J. Bot. 63:2172-2176.
33. Villalobos, V.M., M.J. Oliver, E.C. Yeung, and T.A. Thorpe (1984) Cytokinin-induced switch in development in excised cotyledons of radiata pine cultured in vitro. Physiol. Plant. 61:483-489.
34. Washer, J., K.J. Reilly, and J.R. Barnett (1977) Differentiation in Pinus radiata callus culture: The effect of nutrients. New Zealand J. For. Sci. 7(3):321-328.
35. Yeung, E.C., J. Aitken, S. Biondi, and T.A. Thorpe (1981) Shoot histogenesis in cotyledon explants of radiata pine. Bot. Gaz. 142(2):494-501.

INTEGRATING BIOTECHNOLOGY INTO WOODY PLANT BREEDING PROGRAMS

D.E. Riemenschneider,[1] B.E. Haissig,[1] and E.T. Bingham[2]

[1]Forestry Sciences Laboratory
North Central Forest Experiment Station
U.S. Department of Agriculture Forest Service
Rhinelander, Wisconsin 54501

[2]Department of Agronomy
University of Wisconsin
Madison, Wisconsin 53706

ABSTRACT

Breeding and biotechnology are being integrated in plant genetic improvement programs. Such integration may be more important in programs for woody than herbaceous species because of the former's often long sexual generations. Main focal points for integration are the two major bases of genetic improvement: selection and genetic recombination. In discussing these foci, we have considered and illustrated three broad genetic conditions: desired gene(s) present in the target species, desired gene(s) absent in the target species but present in a sexually compatible species, and desired gene(s) absent in the target species and also in sexually compatible species. Examples illustrate that an integrated approach to genetic improvement uses one technology (breeding or biotechnology) to foster the other, to the overall long-term benefit of both--and, most importantly, to the benefit of plant genetic improvement.

INTRODUCTION

Breeding has been systematically applied to crop improvement since the firm of Vilmorin in France began selecting for sugar content in beets in 1727. The Darwinian theory of evolution and the rediscovery of Mendel's experiments added much scientific strength to breeding. Historically, plant breeding has adopted newly acquired disciplines and technologies. Thus, plant breeding now uses genetics, biometrics, physiology, and biochemistry, and is rapidly incorporating biotechnology. For example, the cellular and molecular methods of biotechnology foster plant breeding by supplying new basic knowledge about genes and genomes.

There is an especially compelling case for integrating biotechnology into breeding of woody species. In particular, the long time between sexual generations in some woody species makes them well suited for biotechnology, which may shorten the genetic improvement process. In this chapter, we discuss the merits of integrating biotechnology and woody plant breeding, citing supporting evidence. Although forest trees are used as an example, our discussion applies equally to many other woody species. We have also limited discussion to genetic modification of a target higher plant, such as somaclonal selection and recombinant DNA. We have not included, for example, modification of microbes that may benefit woody plant growth, or micropropagation, except where micropropagation limits genetic manipulation.

CHARACTERISTICS OF BREEDING

Breeding is a powerful tool for genetically manipulating plant populations, and is based on the statistical extension of Mendelian genetics. Mendelian (qualitative) traits are distributed in distinct, discontinuous classes; genetic control is exercised by a single gene (Fig. 1), and genetic variation results from the occurrence of two or more gene forms (alleles). However, most economically important plant characteristics are not qualitative. Rather, they are continuously distributed and, therefore, are not describable with any single-gene model. Such continuously distributed (quantitative) traits are controlled by many genes (Fig. 1), whose effects are described statistically and by using statistical terminology. Thus, quantitative genetic control is expressed as the amount of observed variance in a trait that is attributable to genetic effects. Heritability is

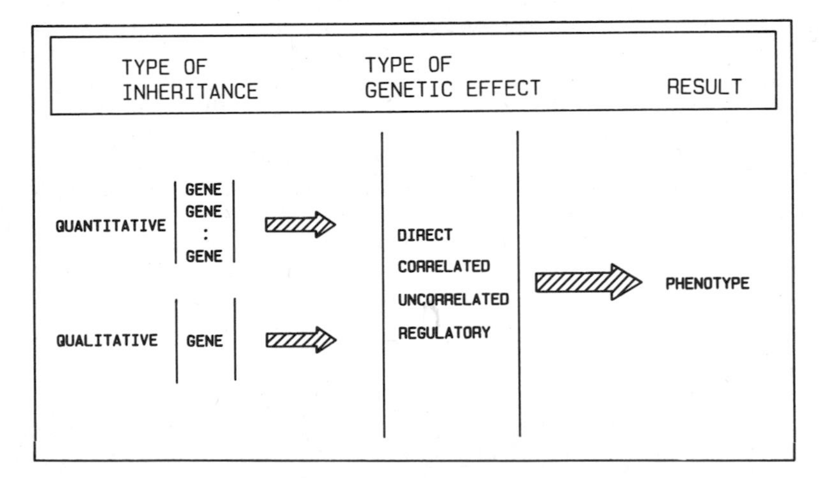

Fig. 1. Diagrammatic representation of genetic inheritances and effects. Qualitative traits are controlled by one or a few genes whose effects result in clearly defined phenotypic classes. Quantitative traits are controlled by many genes whose effects result in continuously distributed phenotypes. Genes may be expressed by their direct correlated, uncorrelated, or regulatory effects. Knowledge of inheritance is useful but not essential in designing breeding programs. (Adapted from Ref. 28.)

a commonly used estimate of the ratio of quantitative genetic variance to total phenotypic variance. Therefore, heritability estimates the importance of genetic effects relative to environmental effects.

The plant breeder's goal is to increase the frequency of desirable alleles, termed genetic value, in the breeding population. Populations thus improved may be used directly as open-pollinated varieties, a use which is common in forest tree improvement; or certain individuals (clones) may be selected from such populations for further hybridizations or for vegetative propagation and use as clonal cultivars.

The frequency of desirable alleles in a population is increased through selection. Successful selection requires only some correlation between the phenotypic value of individuals and the frequency of desirable alleles. For example, no distinction need be made between genes with direct, indirect, correlated, or regulatory effects on the desired phenotype (Fig. 1). Selection-driven increases in the frequency of desirable alleles are usually small per generation and, therefore, several generations may be required to achieve breeding goals.

The principal advantages of breeding are its broadly demonstrated applicability and proven potential for progress. Breeding operates on the phenotype and can be applied to any plant characteristic that is even partially genetically controlled. Progress can be made without a complete understanding of genetic-biochemical relations; however, progress will be accelerated by increased fundamental knowledge. For example, more efficient mating and selection will result as knowledge increases about the number of controlling genes and alleles, and modes of gene action.

Breeding has long been applied to tree improvement (Ref. 16 and 63 and references therein). For example, Dorman (16) has listed ten southern pine species for which growth, quality, or pest resistance traits are genetically determined. Wright (63) lists species from all genera of Pinaceae for which there is evidence of genetic effects on important traits. Genetic effects on growth rate have also been found in cottonwoods and other hardwoods (63). Multigeneration breeding programs have been implemented for conifers such as loblolly pine in the southern United States (61).

In addition to improving selection traits of forest trees, breeding can promote genetic diversity by taking advantage of natural selection pressures. This concept has been implemented with multiple breeding populations (31,40,51). Such breeding populations are located in different environments, where they are subjected to different, random, natural selection pressures. In this scheme, artificial selection improves certain specific traits; natural selection in the separate environments results in genetic diversity in other traits.

As previously stated, breeding results in small increases in the frequency of desirable alleles in each generation. Thus, time is a problem in tree breeding. For example, the selection criterion is often a mature tree characteristic. Evaluation of progenies takes much time unless there are high correlations between juvenile and mature tree traits (23). However, such juvenile-mature correlations are not commonly known. Also, breeding is often delayed because trees may not flower during their first 15 to 25 years of growth. Tree breeding is also sometimes hampered by interspecific and other sexual incompatibilities (58,62).

In summary, plant breeding has several basic characteristics: (a) it manipulates individuals; (b) it genetically alters populations by cumulatively changing allelic frequency; (c) it evaluates population and individual phenotypes; and (d) it manages genetic diversity.

CHARACTERISTICS OF BIOTECHNOLOGY

Biotechnology is difficult to define because of the heterogeneity of its components. In general, biotechnology is a collection of strategies, not including breeding, that can be used to research and manipulate heritable variation. Molecular methods used in biotechnology include DNA restriction enzymes and restriction fragment-length polymorphisms (RFLPs), DNA cloning and sequencing, transfer of recombinant DNA through biological or other gene vectors, and monoclonal and polyclonal antibodies. Cell fusion, somatic cell selection, DNA injection, organelle insertion, and other non-classical genetic methods, conducted at the cellular or subcellular level, are also often included in biotechnology.

Biotechnology is presently proven in some higher plant species for genetic modification of traits controlled by one or perhaps a few defined genes, although neither the genetic traits nor the subject plants may have had commercial value. Such genetic modifications have been achieved, for example, by direct gene insertion (e.g., *Agrobacterium* transformation, electroporation-electroinjection, and microinjection), somaclonal selection, and somatic cellular hybridization. Of these, only *Agrobacterium* transformation and somaclonal selection have been successfully used with forest trees.

Direct gene insertion requires knowledge of gene structure, transcription, translation, and expression, because the genetic changes are precisely targeted. In forest trees, transformations using *Agrobacterium tumefaciens* have been reported for the tumor-inducing portion of the Ti-plasmid (T-DNA), which causes the crown-gall disease, and for small genes that confer antibiotic or herbicide tolerance (14,20,21,46,54). However, regeneration of transformed plants was achieved in only one of these studies (14,20).

Selection of somaclonal variants has been another commonly used biotechnology. Somaclonal variation is Mendelian and non-Mendelian genetic variation sometimes found in plants regenerated from in vitro cultures (11,12,17,18,19,36,37). Somaclonal selection has mostly been used with easily regenerated herbaceous species, because frequency of in vitro regeneration is a major determinant of probable success (26).

It does not presently seem that biotechnology can be used to manage genetic diversity, even though it has great potential for measuring, describing, and tracing genetic diversity. For example, evolutionary relations can be precisely resolved based on homology between nuclear, chloroplast, or mitochondrial DNA segments. But research is required to determine how biotechnology might be used to solve or prevent problems of insufficient genetic diversity in plant populations.

Regardless of the biotechnology, regeneration of plants is required after genetic modification, yet regeneration has been difficult to achieve with forest trees. Biotechnologies cannot now be applied to many forest tree species, especially conifers and determinate-growth angiosperms, because of a lack of high-frequency regeneration in vitro (7,29). For example, some poplars can be regenerated from organ explants, callus, cells, and protoplasts (26,27,53), but conifers have only been regenerated at high frequency from juvenile tissues such as hypocotyls, cotyledons, or embryos (29). In addition, nonbiological gene insertion by microinjection, electroinjection, and electroporation may require regeneration of plants from protoplasts or cells. Cellular-level regeneration of conifers and angiosperms is improving but still deficient (24,25,29).

In summary, biotechnology of all kinds has five basic characteristics: (a) it manipulates molecules, cells, tissues, or organs; (b) it genetically alters individuals; (c) it evaluates individual phenotypes; (d) it evaluates but may not manage genetic diversity; and (e) it is constrained by availability of background knowledge. Biotechnologies for direct gene insertion have two additional basic characteristics, which are requirements for: (a) precise knowledge of genetics and biochemistry for the trait(s) involved; and (b) a nonsexual mechanism to insert ("foreign") genes into the target species.

COMPARISON OF BREEDING AND BIOTECHNOLOGY

Based on the foregoing, there are pronounced general similarities between breeding and biotechnology. They have the identical strategic goals of genetic improvement of plants, and learning which genes are involved. They are also operationally similar (Fig. 2) because they: (a) assess progress toward an improvement goal(s) by rating the phenotype; (b) use sexual reproduction to confirm stable integration of genetic characteristic(s); (c) use hybridization (for example, sexual genetic recombination in breeding and nonsexual genetic recombination in biotechnology); and (d) are constrained by time.

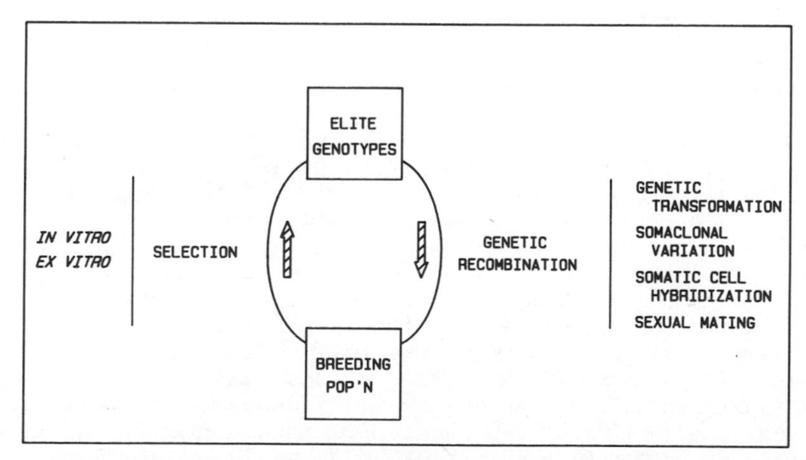

Fig. 2. Scheme showing major similarities between use of breeding and biotechnology for plant genetic improvement. Biotechnologies such as direct gene insertion, somaclonal variation, and somatic cell hybridization, which achieve genetic recombination, are nonsexual analogs of breeding (mating). Selection always follows genetic recombination and may be in vitro (usually in biotechnology) or ex vitro (usually in breeding).

The time constraints on the application of breeding and biotechnology to forest tree improvement have similar or identical biological bases, as follows: (a) lack of knowledge about sexual regeneration, which precludes short sexual generations; (b) lack of knowledge of vegetative regeneration in vitro, which limits general and effective application of every biotechnology that requires plan regeneration in vitro; (c) lack of juvenile-mature correlations, which impairs the rates of progress in tree breeding and biotechnology; and (d) lack of knowledge about, for example, economically important higher plant genes, which substantially impairs the application of biotechnology, and somewhat impairs breeding.

There are also major differences between breeding and biotechnology. By definition, breeding is "classical," biotechnology is "nonclassical." Of more importance, breeding and biotechnology are operationally different. For example, breeding can focus on improvement by manipulation of single genes or on population-based changes in allelic frequency in a large number of genes. The primary breeding unit is the population, which is a repository of desirable alleles, reconstituted in each generation. Individual genotypes are of little long-term importance (excluding vegetatively propagated cultivars) because they are lost to genetic recombination in each breeding cycle. In comparison, biotechnology often focuses only on very specific genetic modification of selected individual genotypes. Nevertheless, the individual genotype links the two technologies--it is an important transient in breeding and central to biotechnology (Fig. 3).

All the foregoing differences may not be so clear in the future. The difference of classical vs nonclassical will fade with time. The different bases of anatomical organization may have somewhat more permanence, although even that is moot. For example, improvements in in vitro flowering might reduce the level of biological organization at which some breeding is applied. Increased fundamental knowledge of biology will most probably overcome problems of vegetative and sexual regeneration, and the present poor understanding of juvenile-mature correlations.

EXAMPLES OF INTEGRATING BREEDING AND BIOTECHNOLOGY

As previously stated, our thesis is that biotechnology and breeding can and will be fully integrated in plant genetic improvement programs, and that such integration may be more important in programs for woody species than for herbaceous species. Obvious focal points for integration are the two major bases for genetic improvement: selection and genetic recombination. Using these bases, we have considered and illustrated three broad genetic conditions: desired gene(s) present in the target species, desired gene(s) absent in the target species but present in a sexually compatible species, and desired gene(s) absent in the target species and also in sexually compatible species. As will be seen from our examples, an integrated approach uses one technology to foster the other, to the long-term benefit of both--and, most importantly, to the benefit of plant genetic improvement.

Selection

Breeding assumes the presence of the desirable gene forms in the target species or in a sexually compatible species. But the major objective of breeding--increasing the frequency of desirable alleles in a population--may be difficult to achieve. The ease of increasing allelic frequency

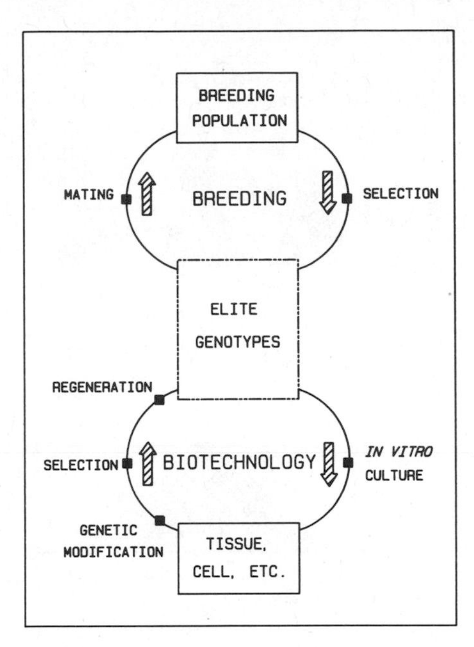

Fig. 3. Idealized relations between cycles of breeding and biotechnology for genetic improvement. The primary focus of breeding is the population and the management of its genes. Individual elite genotypes are important but transient in breeding, because they are lost through genetic recombination in each breeding cycle. Biotechnology makes specific genetic modifications to individual elite genotypes, but, unlike breeding, it does not effectively address the broader aspects of plant genetic improvement such as genetic diversity.

depends on the correlation between phenotype and genotype (which for quantitative traits is the heritability, as previously defined). The pronounced influence of the amount of correlation between phenotype and genotype on increasing allelic frequency is demonstrated by a one-gene, two-allele model (Fig. 4). If genetic effects are distinct (Fig. 4A), the desirable allele (A) may be fixed in a single generation by selecting individuals with the highest phenotypic value (i.e., highest yielding, tallest, etc.). Selection is less efficient when genetic effects are indistinct (Fig. 4B), because desired homozygous individuals (AA) cannot be distinguished from undesirables, which are those individuals heterozygous (Aa) or homozygous (aa) for the unwanted allele.

Unlike the previous example, most economically important tree characteristics are probably quantitative. However, their improvement by breeding is still accurately described by the consequence of having indistinct genetic effects in a Mendelian model (Fig. 4B): gradual increases in the frequency of desirable alleles, with gene fixation requiring many generations. The correlation between phenotype and genotype of trees is often very low because criteria for selection and response may be widely separated in tree development. For example, juvenile tree size has been used

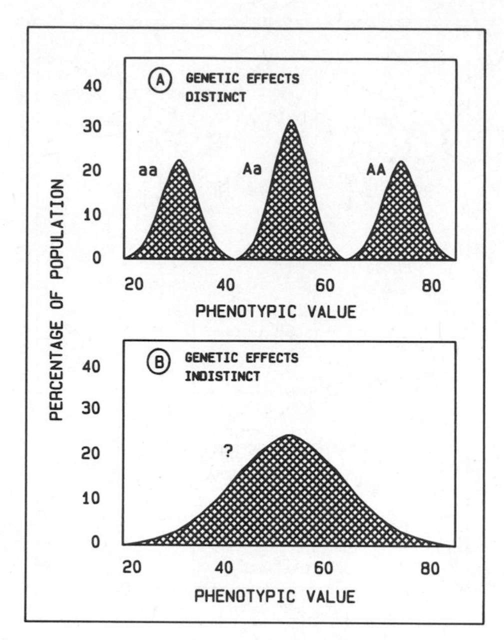

Fig. 4. Single-gene, two-allele model indicating hypothetical distributions of phenotypic values when genetic effects are distinct or indistinct relative to environmental effects. For quantitative traits, analogs of distinct and indistinct genetic effects are high and low heritability, respectively. When genetic effects are distinct (A), a desirable allele may be fixed in one generation of selection, because individuals homozygous for the desirable allele are identifiable without error. When genetic effects are indistinct (B), individuals homozygous for the desirable allele may not be clearly distinguished from less desirable genotypes, and gene fixation may take many generations. (Adapted from Ref. 4.)

as a selection criterion for mature tree size (22,23,32,34,56,60). However, in such instances the correlation between phenotype and genotype is reduced if age-age correlations are imperfect, as is usually true (22,23,34).

A method for increasing the correlation between phenotype and genotype would aid selection. This method could be, for example, a technique whereby desirable alleles were detectable at the molecular level and could be directly associated with the phenotypic characteristic that they encode. Under such circumstances, heritability of a quantitative trait might be increased to unity (i.e., phenotype perfectly expressing genotype). This may be possible through RFLP mapping (Fig. 5). In RFLP analysis, single- or low-copy genomic or cDNA probes (35) are used to identify molecular markers that closely flank important genes (6,10,59). Subsequent selection can be based on the presence of the markers that, in the absence of crossing over, unambiguously identify the desired genes (R.W. Michelmore, pers. comm.).

Breeding is required to make RFLP mapping possible. For example, parents and at least one generation of full-sib progenies are required for linkage analyses. Breeding of annual crops for RFLP analyses can be

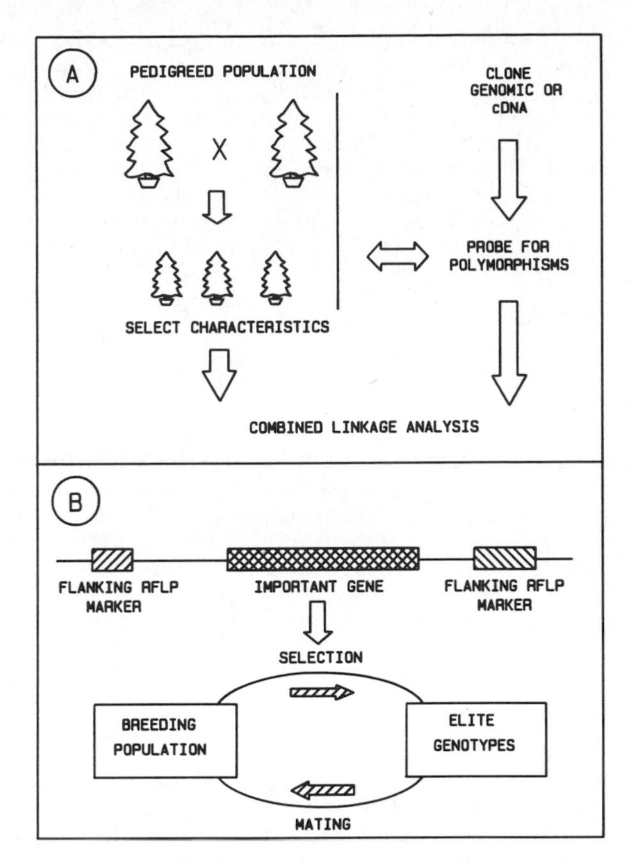

Fig. 5. Diagrammatic representation of using restriction fragment-length polymorphisms (RFLPs) to improve the efficiency of genetic selection. Genomic or cDNA libraries are used to probe DNA polymorphisms. Combined linkage analysis of RFLP and morphological characteristics (A) can provide markers that are closely linked to desirable target genes (B). Closely linked RFLP markers can increase heritability to 1.0., its highest value, thus increasing the rate of genetic improvement.

quickly completed, but more time is required to produce tree populations with suitable family structure. This limits immediate application of this method to trees. Nevertheless, it is an excellent goal for tree breeding.

Recombining Genes

Sexual gene transfer. In some instances, breeding goals cannot be achieved because desired genes are not present in the target breeding population. But if desired genes exist in a sexually compatible population, they can be transferred to the breeding population by hybridization and backcrossing. Backcrossing is needed because linked, undesirable genes from the donor population always accompany desired genes. Backcrossing restores the distorted recipient (target) genotype while selection retains the new, desirable genes. However, fixation of transferred genes through backcross breeding is uncertain even with agronomic crops (48).

Biotechnology may aid sexual gene transfer for the same reason that it may generally aid selection, as described above, by increasing the correlation between genotype and phenotype (i.e., heritability). For example, use of molecular genetic probes to unambiguously identify genes would almost certainly result in gene fixation in a recipient population. In one recent study, use of molecular genetic probes aided transfer of three insect resistance genes from wild tomato to a domesticated variety (3,45). In that research the domesticated variety was hybridized with wild tomato plants containing the insect resistance genes. Variation in insect resistance in 74 F_2 plants was associated with specific RFLP loci. Combined RFLP probes and a biochemical assay were used to identify genes that coded for insect resistance (45), and to select for (maintain) them through two generations of backcrossing (J. Nienhuis, pers. comm.). In comparison, backcrossing without the use of biotechnology would have been much less efficient (48).

The foregoing example may have a particular future application to forest trees. Backcrossing of forest trees is almost always very inefficient if the desired trait is a mature characteristic and the sexual generation is long. But association of important genetic traits with specific RFLP loci would eliminate evaluation of mature trees. Thus, backcrossing would at least be operationally feasible, especially for tree species that naturally flower early in life, or that flower after chemical or cultural treatment(s).

Nonsexual gene transfer. Sometimes breeding goals cannot be achieved because desired genes are not present in the target breeding population, in any sexually compatible species, or in any other higher plant. Biotechnology may overcome such barriers. For example, natural resistance to the herbicide glyphosate has not been found in higher plant species. Glyphosate inhibits the enzyme 5-enolpyruvylshikimate-3-phosphate (EPSP) synthase (EC 2.5.1.19). However, higher plants with increased glyphosate tolerance have been produced (13,15,20,21,55). One method was to isolate a chimeric *aroA* gene from *Salmonella typhimurium* that codes for a chimeric EPSP synthase that tolerates glyphosate better than the wild-type enzyme (13,57). This 1,331-base pair (bp) gene, originally isolated on a 1.3-kilobase (kb) DNA segment, has a single amino acid substitution (57). The gene has been successfully introduced, using *Agrobacterium*-mediated transformation, into tobacco, tomato, and hybrid poplar (20,21,27,52). Successful genetic transformation in poplar was confirmed by NPTII' enzyme activity, Southern blot analysis, and immunological detection of bacterial EPSP synthase by western blotting (20,21). However, further genetic confirmation by breeding is needed.

Breeding has been used as an adjunct to biotechnology to determine whether introduced phenotypic traits were genetically stable, although this has only been done with some herbaceous species (8,12,17,18). In poplars, stability of the introduced phenotypic trait of herbicide tolerance has been examined only over vegetative cycles (27,52) because of the long sexual generation of poplars. However, epigenic effects may diminish over vegetative generations, which makes this technique useful, even though it is not definitive.

Genetic transformation of higher plants by using higher plant genes is under investigation (55), and has two primary hypothetical advantages compared to breeding, especially for forest trees. First, as stated above, genetic transformation would allow genes to be moved between higher plant species that are not sexually compatible. Second, it would avoid lengthy backcrossing because only a single target gene is transferred, which eliminates the problem of linkage with undesirable genes. Thus, nonsexual transfer may prove to be the future method of choice, even when donor and recipient populations are sexually compatible. Recent research (55), including advances with RFLP analyses and transposable elements, suggests that important higher plant genes may soon be used in genetic transformation of trees.

Random genetic variation. Random genetic variation can be induced during the tissue culture process, producing plants that are termed "somaclonal variants." Somaclonal variants apparently arise in vitro because of cytogenetic changes such as polyploidy, translocation, deletion, inversion, mitotic crossing over, or nuclear or cytoplasmic gene mutation (12,18). Somaclonal variation may be especially useful for plant genetic improvement under either of two circumstances. First, as previously stated, breeding requires some correlation between phenotype and genotype. Genes that are not expressed (masked genes) have no correlated phenotypic effect and, therefore, cannot be manipulated by breeding. Second, as in the previous example of herbicide tolerance, required genes may not be present in the target species, a sexually compatible species, or any higher plant. In either case, somaclonal selection may achieve the desired phenotype by causing expression of existing genes or mutations leading to their presence, or both. Somaclonal selection does not require specific genetic or biochemical knowledge (even though it would be helpful), and presumably could be applied to any phenotypic trait, regardless of its genetic complexity.

Somaclonal selection may achieve genetic modification by intentionally applying a selection pressure, such as a normally lethal chemical or pathogen, to target plant cells or tissues in vitro (Fig. 6). Complete plants, putatively tolerant of the selection agent, are regenerated after an appropriate period of exposure. Putative tolerance is confirmed based on the target's ex vitro tolerance of the same selection pressure. As an example, putative tolerance to some herbicides has been achieved in this manner with hybrid poplars (Ref. 44; Michler and Bauer, poster abstract, this Volume). However, somaclonal variants have also resulted from in vitro culture when a selection pressure was not intentionally applied. For example, morphological variants of poplars have resulted in this way (38).

The problem of linked, undesirable genes may occur in somaclonal selection, as it does in backcross breeding, where undesirable genes introduced by the initial hybridization must be removed by subsequent breeding. Such linkage with undesirable genes would not occur in somaclonal selection if induced genetic variation were restricted to the trait being selected, but present knowledge may not be sufficient enough to allow that. For example, somaclonally selected plants that varied in more than one character have been found in wheat (37) and alfalfa (8). The elimination of concurrent genetic variation in nonselected traits, and also epigenic variation, may mandate breeding in somaclonal selection programs, thus increasing the time required to produce a new cultivar (Fig. 6).

Regeneration Ability

As previously stated, application of biotechnology to many tree species, especially conifers and determinate-growth angiosperms, is limited by their inability to regenerate shoots in vitro (7,29). Loblolly pine, for example, constitutes over half of the two billion seedlings planted yearly in

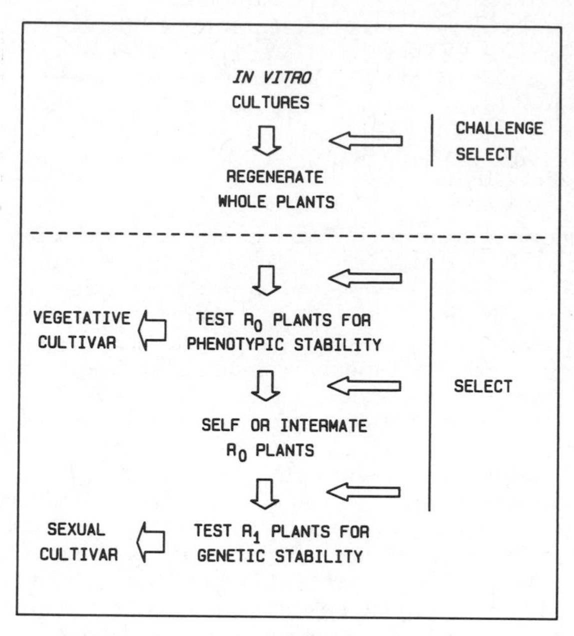

Fig. 6. Scheme for somaclonal selection of genetically based phenotypic traits. Somaclonal selection integrates nonclassical (biotechnology) and classical (breeding) genetic technologies. Challenging in vitro cultures and selecting desirable phenotypes at the cellular, tissue, or organ level involves biotechnology (above dashed line); testing selected phenotypes for genetic stability, and eliminating concurrent, undesirable genetic characteristics mostly involve breeding (below dashed line).

the United States, which indicates its importance as a tree crop (29). However, loblolly pine is very difficult to regenerate in vitro if tissues are taken from mature or maturing (genetically proven) trees. There are numerous similar examples (29).

Inadequate regeneration in vitro is common to many plant species. Failures to obtain high-frequency regeneration with certain herbaceous crop plants have led to recent, pioneering studies concerning the genetic control of regeneration in in vitro cultures. Evidence indicates that conventional breeding can improve the frequency of shoot regeneration in vitro. For example, recurrent selection of tetraploid alfalfa was used to produce varieties with markedly enhanced regeneration frequency from in vitro callus cultures (9). Subsequent research established that bud-forming ability in diploid alfalfa callus cultures is under genetic control (50). The foregoing and related research with alfalfa has been summarized elsewhere (43).

Similarly, research in vitro with red clover callus tissue indicated that the propensity for rooting and somatic embryogenesis was inherited (33). The authors suggested that selection and breeding would enhance the regeneration frequency of plants from callus (33). Differences in rooting ability in vitro between Lycopersicon spp. suggested that necessary regeneration genes might be transferred into tomato from wild-types (39). Marked genotypic differences have also been found, for example, in bud- and embryo-forming abilities of cultures of wheat (41,42), sunflower (47), and barleys (30). Observations with Norway spruce indicated genetic control of adventitious bud initiation (5). Genotypic influences on the in vitro regeneration ability of poplars have also been noted (Ref. 1 and 2; C. Michler and D. Riemenschneider, unpubl. results). Genotypically based potentials for in vitro regeneration ability, such as described above, have been positively and negatively modified by the previous environment to which the parent plant was exposed (41) and by the nature of the culture medium (49). Studies of genetic effects on regeneration in vitro have recently been reviewed (28).

The above studies indicated that breeding can increase adventitious shoot regeneration in vitro and thus overcome a major limitation in the application of biotechnology to tree species. In addition, improvement in shoot regeneration in vitro may be accelerated by biotechnological aids to selection and gene transfer, and thereby become an example of synergism between breeding and biotechnology.

CONCLUSIONS

Genetic improvement of plants, especially trees, requires much time. Breeding will remain important for improvement of trees and other plants. Biotechnology is expected to accelerate genetic improvement directly and indirectly. Some direct uses, such as genetic transformation and somaclonal selection, already have been accomplished, and broader application is expected. Similarly, somatic cell selection holds promise. Combined use of breeding and biotechnology, where possible, should result in the most efficient genetic improvement, because each technology has special applications in selection and genetic recombination (Fig. 2).

Poor regeneration in vitro, which markedly limits direct application of biotechnology to important tree species and many other crops, is a major problem to be solved. Previous research has suggested that regeneration in vitro can be improved and understood by using genetics and breeding. Understanding the genetic control of in vitro regeneration would provide two benefits. First, individuals with high regeneration ability would be available for biotechnology. Second, genetic correlations between regeneration ability and other selection criteria could be estimated. Negative genetic correlations between traditional selection criteria, such as yield and regeneration ability, would indicate that special selection procedures may be required to maintain in vitro regeneration.

Regarding indirect applications, the molecular methods of biotechnology already are being used in plant breeding as tools for rapid plant diagnostics. These include virus detection with RNA-DNA hybridizations and antibodies; plant line identification and analysis of genetic purity with isozymes and molecular probes; and identification of genetic diversity and characterization of male-sterile cytoplasms, also with isozymes and probes.

In addition, gene mapping is being elegantly done by in situ hybridization of DNA and RNA probes to specific chromosomes, and by linkage of economically important genes to specific RFLPs. Thus, biotechnology is increasing our basic understanding of both Mendelian and quantitative traits. This basic knowledge provides a foundation for biotechnology-based selection and genetic recombination, and in the meantime helps chart new strategies for breeding.

The greatest indirect benefit of using biotechnology may simply be the resulting general advances in many types of basic knowledge. Genetic transformation, for example, will be used to study the regulation and function of plant genes regardless of their importance to plant improvement. All breeding approaches and related areas of biology are directly or indirectly benefiting from new basic knowledge about the structure, function, and regulation of plant genomes. New knowledge is likely to result in efficient solutions to old biological problems, such as how to control and shorten the juvenile phase in woody species, thereby promoting early-life flowering, and how to identify and manipulate productive, stable genotypes at early stages of growth. Sexual and somatic approaches to improvement will thereby benefit. Moreover, our rapidly expanding knowledge about heritable variation may help solve important genetic improvement problems that are not yet apparent.

ACKNOWLEDGEMENT

We thank Drs. D. Karnosky and R. Stettler for critically reviewing the manuscript.

REFERENCES

1. Ahuja, M.R. (1983) Somatic cell differentiation and rapid clonal propagation of aspen. Silvae Genet. 32:3-4.
2. Ahuja, M.R., and H. Muhs (1982) Isolation, culture, and fusion of protoplasts: Problems and prospects. Silvae Genet. 31:66-77.
3. Anonymous (1986) Genetic Technology News, Technical Insights, Inc., Englewood, New Jersey, Vol. 6, No. 6, p. 12.
4. Allard, R.W. (1960) Principles of Plant Breeding, John Wiley and Sons, Inc., New York, 485 pp.
5. Arnold, S. von (1984) Importance of genotype on the potential for in vitro adventitious bud production of Picea abies. For. Sci. 30:314-318.
6. Beckmann, J.S., and M. Soller (1983) Restriction fragment length polymorphisms in genetic improvement: Methodologies, mapping, and costs. Theor. Appl. Genet. 67:35-43.
7. Berlyn, G.P., R.C. Beck, and M.H. Renfroe (1986) Tissue culture and the propagation and genetic improvement of conifers: Problems and possibilities. Tree Physiol. 1:227-240.
8. Bingham, E.T., and T.J. McCoy (1986) Somaclonal variation in alfalfa. Plant Breed. Rev. 4:123-152.
9. Bingham, E.T., L.V. Hurley, D.M. Kaatz, and J.W. Saunders (1975) Breeding alfalfa which regenerates from callus tissue in culture. Crop Sci. 15:719-721.
10. Botstein, D., R.L. White, M. Skolnick, and R.W. Davis (1980) Construction of a genetic linkage map in man using restriction fragment length polymorphisms. Am. J. Hum. Genet. 32:314-331.
11. Chaleff, R.S., and M.F. Parsons (1978) Direct selection in vitro for herbicide-resistant mutants of Nicotiana tabacum. Proc. Natl. Acad. Sci., USA 75:5104-5107.

12. Chaleff, R.S., and T.B. Ray (1984) Herbicide-resistant mutants from tobacco cell cultures. Science 223:1148-1151.
13. Comai, L., D. Facciotti, W.R. Hiatt, G. Thompson, R.E. Rose, and D.M. Stalker (1985) Expression in plants of a mutant aroA gene from Salmonella typhimurium confers tolerance to glyphosate. Nature 317:741-744.
14. Dandekar, A.M., P.K. Gupta, D.J. Durzan, and V. Knauf (1987) Transformation and foreign gene expression in micropropagated Douglas-fir (Pseudotsuga menziesii). Bio/Technology 5:587-590.
15. Della-Cioppa, G., S.C. Bauer, M.L. Taylor, D.E. Rochester, B.K. Klein, D.M. Shah, R.T. Fraley, and G.M. Kishore (1987) Targeting a herbicide-resistant enzyme from Esherichia coli to chloroplasts of higher plants. Bio/Technology 5:579-584.
16. Dorman, K.W. (1976) The Genetics and Breeding of Southern Pines. U.S. Department of Agriculture Forest Service, Agricultural Handbook No. 471, Washington, D.C.
17. Evans, D.A., and W.R. Sharp (1983) Single gene mutations in tomato plants regenerated from tissue culture. Science 221:949-951.
18. Evans, D.A., and W.R. Sharp (1986) Applications of somaclonal variation. Bio/Technology 4:528-532.
19. Evans, D.A., W.R. Sharp, and H.P. Medina-Filho (1984) Somaclonal and gametoclonal variation. Am. J. Bot. 71:759-774.
20. Fillatti, J.J., B.H. McCown, J. Sellmer, and B. Haissig (1986) The introduction and expression of a gene conferring tolerance to the herbicide glyphosate in Populus NC 5339. In Proceedings of the Technical Association of the Pulp Paper Industry--1986 Research and Development Conference, Raleigh, North Carolina, TAPPI Press, Atlanta, Georgia, pp. 83-84.
21. Fillatti, J.J., J. Sellmer, B. McCown, B. Haissig, and L. Comai (1987) Agrobacterium mediated transformation and regeneration of Populus. Mol. Gen. Genet. 206:192-199.
22. Foster, G.S. (1986) Trends in genetic parameters with stand development and their influence on early selection for volume growth in loblolly pine. For. Sci. 32:944-959.
23. Franklin, E.C. (1979) Model relating levels of genetic variance to stand development of four North American conifers. Silvae Genet. 28:207-212.
24. Gupta, P.K., and D.J. Durzan (1985) Shoot multiplication from mature trees of Douglas-fir (Pseudotsuga menziesii) and sugar pine (Pinus lambertiana). Plant Cell Reports 4:177-179.
25. Gupta, P.K,. and D.J. Durzan (1986) Isolation and cell regeneration of protoplasts from sugar pine (Pinus lambertiana). Plant Cell Reports 5:346-348.
26. Haissig. B.E. (1987) Tissue culture-based biotechnology for Populus clones. In Energy from Biomass and Wastes X, D.L. Klass, ed. Institute of Gas Technology, Chicago, Illinois, and Elsevier Applied Science Publishers, London, pp. 155-175.
27. Haissig, B.E., and D.E. Riemenschneider (1987) Genetic engineering of hybrid poplars for herbicide tolerance. In Proceedings of the Forest Products Research Conference 1986--Matching Utilization Research with the Needs of Timber Managers, U.S. Department of Agriculture Forest Service, Forest Products Laboratory, Madison, Wisconsin, pp. 37-46.
28. Haissig, B.E., and D.E. Riemenschneider (1988) Genetic effects on adventitious rooting. In Adventitious Rooting by Cuttings, T.D. Davis, B.E. Haissig, and N. Sankhla, eds. Dioscorides Press, Portland, Oregon (in press).

29. Haissig, B.E., N.D. Nelson, and G.H. Kidd (1987) Trends in the use of tissue culture in forest improvement. Bio/Technology 5:52-59.
30. J rgensen, R.B., C.J. Jensen, B. Andersen, and R. von Bothmer (1986) High capacity of plant regeneration from callus of interspecific hybrids with cultivated barley (Hordeum vulgare L.). Plant Cell Tissue Organ Culture 6:199-207.
31. Kang, H. (1980) Designing a tree breeding system. In Proceedings of the 17th Meeting of the Canadian Tree Improvement Association, Gander, Newfoundland, Canadian Forest Service, pp. 51-63.
32. Kang, H. (1985) Juvenile selection in tree breeding: Some mathematical models. Silvae Genet. 34:75-84.
33. Keyes, G.B., G.B. Collins, and N.L. Taylor (1980) Genetic variation in tissue cultures of red clover. Theor. Appl. Genet. 58:265-271.
34. Lambeth, C.C. (1980) Juvenile-mature correlations in Pinaceae and implications for early selection. For. Sci. 26:571-580.
35. Landry, B.S., and R.W. Michelmore (1985) Selection of probes for restriction fragment length analysis from plant genomic clones. Plant Mol. Biol. Reporter 3:174-179.
36. Larkin, P.J., and W.R. Scowcroft (1981) Somaclonal variation--A novel source of variability from cell cultures for plant improvement. Theor. Appl. Genet. 60:197-214.
37. Larkin, P.J., S.A. Ryan, R.I.S. Brettell, and W.R. Scowcroft (1984) Heritable somaclonal variation in wheat. Theor. Appl. Genet. 67:443-445.
38. Lester, D.T., and J.G. Berbee (1977) Within-clone variation among black poplar trees derived from callus culture. For. Sci. 23:122-131.
39. Locy. R.D. (1983) Callus formation and organogenesis by explants of six Lycopersicon species. Can. J. Bot. 61:1072-1079.
40. Lowe, W.J., and J.P. van Buijtenen (1981) Tree improvement philosophy and strategy for the western gulf forest tree improvement program. In Research Needs in Tree Breeding--Proceedings of the 15th North American Quantitative Forest Genetics Group Workshop, Coeur d'Alene, Idaho, School of Forest Resources. North Carolina State University, Raleigh, North Carolina, pp. 43-50.
41. Maddock, S.F., V.A. Lancaster, R. Risiott, and J. Franklin (1983) Plant regeneration from cultured immature embryos and inflorescences of 25 cultivars of wheat (Triticum aestivum). J. Exp. Bot. 34:915-926.
42. Mathias, R.J., and E.S. Simpson (1986) The interaction of genotype and culture medium on the tissue culture responses of wheat (Triticum aestivum L.) callus. Plant Cell Tissue Organ Culture 7:31-37.
43. McCoy, T., and K. Walker (1983) Alfalfa. In Handbook of Plant Cell Culture, Vol. 3, P.V. Ammirato, D.A. Evans, W.R. Sharp, and Y. Yamada, eds. Macmillan Publishing Company, New York, pp. 171-193.
44. Michler, C., and B.E. Haissig (1987) Somaclonal selection of hybrid poplars that tolerate herbicides. In Proceedings of the IUFRO Working Party on Somatic Cell Genetics, Grosshansdorf, Federal Republic of Germany (in press).
45. Nienhuis, J., T. Helentjaris, M. Slocum, B. Ruggero, and A. Schaefer (1987) Restriction fragment length polymorphism analysis of loci associated with insect resistance in tomato. Crop Sci. 27:797-803.
46. Parsons, T.J., V.P. Sinkar, R.F. Stettler, E.W. Nester, and M.P. Gordon (1986) Transformation of poplar by Agrobacterium tumefaciens. Bio/Technology 4:533-536.

47. Paterson, D.E., and N.P. Everett (1985) Regeneration of Helianthus annuus inbred plants from callus. Plant Sci. 42:125-132.
48. Reddy, B.V.S., and R.E. Comstock (1976) Simulation of the backcross breeding method. I. Effects of heritability and gene number on fixation of desired alleles. Crop. Sci. 16:825-830.
49. Reddy, V.S., S. Leelavathi, and S.K. Sen (1985) Influence of genotype and culture medium on microspore callus induction and green plant regeneration in anthers of Oryza sativa. Physiol. Plant. 63:309-314.
50. Reisch, B., and E.T. Bingham (1980) The genetic control of bud formation from callus cultures of diploid alfalfa. Plant Sci. Lett. 20:71-77.
51. Riemenschneider, D.E. (1985) Heritability and intertrait correlations in breeding subpopulations of jack pine. In Proceedings of the Fourth North Central Tree Improvement Conference, East Lansing, Michigan, Department of Forestry, University of Wisconsin, Madison, Wisconsin, pp. 12-22.
52. Riemenschneider, D.E., and B.E. Haissig (1987) Transfer of a bacterial herbicide tolerance gene to hybrid poplar. In Proceedings of the IUFRO Working Party on Somatic Cell Genetics, Grosshansdorf, Federal Republic of Germany (in press).
53. Russell, J.A., and B.H. McCown (1986) Techniques for enhanced release of leaf protoplasts in Populus. Plant Cell Reports 5:284-287.
54. Sederoff, R., A. Stomp, W.S. Chilton, and L.W. Moore (1986) Gene transfer into loblolly pine by Agrobacterium tumefaciens. Bio/Technology 4:647-649.
55. Shah, D.M., R.B. Horsch, H.J. Klee, G.M. Kishore, J.A. Winter, N.E. Tumer, C.M. Hironaka, P.R. Sanders, C.S. Gasser, S. Aykent, N.R. Siegel, S.G. Rogers, and R.T. Fraley (1986) Engineering herbicide tolerance in transgenic plants. Science 233:478-481.
56. Squillace, A.E., and C.R. Gansel (1974) Juvenile-mature correlations in slash pine. For. Sci. 20:225-229.
57. Stalker, D.M., W.R. Hiatt, and L. Comai (1985) A single amino acid substitution in the enzyme 5-enolpyruvylshikimate-3-phosphate synthase confers resistance to the herbicide glyphosate. J. Biol. Chem. 260:4724-4728.
58. Stettler, R.F., R. Koster, and V. Steenackers (1980) Interspecific crossability studies in poplars. Theor. Appl. Genet. 58:273-282.
59. Suiter, K.A., J.F. Wendel, and J.S. Case (1983) Linkage-1: A Pascal computer program for the detection and analysis of genetic linkage. J. Hered. 74:203-204.
60. Wakely, P.C. (1971) Relation of thirtieth-year to earlier dimensions of southern pines. For. Sci. 17:200-209.
61. Weir, R.J. (1981) North Carolina State University-industry cooperative tree improvement program. In Research Needs in Tree Breeding--Proceedings of the Fifth North American Quantitative Forest Genetics Workshop, Coeur d'Alene, Idaho, School of Forest Resources, North Carolina State University, Raleigh, North Carolina, pp. 57-70.
62. Wright, J.W. (1955) Species crossability in spruce in relation to distribution and taxonomy. For. Sci. 1:319-349.
63. Wright, J.W. (1976) Introduction to Forest Genetics, Academic Press, Inc., New York, 463 pp.

CONTENTS: POSTER ABSTRACTS

POSTER ABSTRACTS

MORPHOGENESIS IN CALLUS CULTURE OF PISTACHIO (PISTACIA VERA L.) AS A TOOL FOR GENETIC IMPROVEMENT

M. Barghchi

Plant Physiology Division
Department of Scientific and Industrial Research
Palmerston, New Zealand

Pistachio is a dioecious plant and it is not possible to simply combine the best characteristics of two cultivars, both of which are female, by hybridization. Micropropagation methods for Pistacia were previously reported by the author. The potential of morphogenesis in callus culture for genetic manipulation and improvement of Pistachio was investigated. Shoot growth and rhizogenesis was achieved.

THE POTENTIAL OF CALLUS, CELL SUSPENSION, AND SOMATIC REGENERATION IN THE GENETIC MANIPULATION AND IMPROVEMENT OF ROBINIA PSEUDOACACIA L.

M. Barghchi

Plant Physiology Division
Department of Scientific and Industrial Research
Palmerston North, New Zealand

Following the establishment of a micropropagation method for Robinia pseudoacacia, somatic regeneration methods for in vitro genetic manipulation and improvement were investigated. Callus and cell suspension cultures were established and maintained. Embryogenesis from callus and adventitious shoot growth from leaf explants were obtained. The potential of somaclonal variation for genetic manipulation and improvement is discussed.

COMPLETE PLANT REGENERATION FROM NORWAY SPRUCE SOMATIC EMBRYOS: MATURATION, "GERMINATION," AND INITIATION OF SECONDARY GROWTH FROM RESTING TERMINAL BUDS

M.R. Becwar, S.R. Wann, and M.A. Johnson

The Institute of Paper Chemistry
Forest Biology Division
Appleton, Wisconsin 54912

Quantitative data are presented on the frequency at which Norway spruce somatic embryos reached three stages of development: mature cotyledonary embryos, primary root growth ("germination"), and continued growth in soil. Maturation frequencies ranged from 1 to 15% among embryogenic callus lines derived from individual immature embryo explants using the developmental protocol previously described [Becwar et al. (1987) Plant Cell Reports 6:35]. Addition of 10 μM buthionine sulfoximine doubled the maturation frequency of one embryogenic line. By avoiding immersion of the radicle of somatic embryos in agar medium, germination frequencies as high as 82% (mean of 56%) have been obtained. Nine of 31 (29%) somatic embryo plantlets survived transfer to the greenhouse, set a resting terminal bud, overwintered (to as low as -5°C), and initiated new vegetative growth synchronously with control zygotic seedlings.

INITIATION OF EMBRYOGENIC CALLUS IN *PINUS STROBUS* (EASTERN WHITE PINE) FROM IMMATURE EMBRYO EXPLANTS

M.R. Becwar,[1] S.R. Wann,[1] and H.B. Kriebel[2]

[1]Forest Biology Division
The Institute of Paper Chemistry
Appleton, Wisconsin 54912

[2]School of Natural Resources
Division of Forestry
The Ohio State University
Wooster, Ohio 44691

A white to translucent, mucilaginous embryogenic callus was initiated at low frequencies (less than 8% of the explants per treatment responded) from immature embryos on defined media supplemented with 2,4-dichlorophenoxyacetic acid (2 to 3 mg/l) and benzyladenine (0.5 to 1 mg/l). Four of the five embryogenic callus lines initiated and maintained were derived from precotyledonary embryos, suggesting that the optimum stage of embryo-explant development in white pine is earlier than in Norway and white spruce where the optimum embryo stage is postcotyledonary. The density of somatic embryos ranged from 200 to 1,000 per gram of callus among five lines derived from different embryo explants, suggesting genotypic differences in embryogenic capacity.

SOMATIC EMBRYOGENESIS IN AESCULUS

B.A. Bergmann, W.P. Hackett, and H. Pellett

Department of Horticultural Science
and Landscape Architecture
University of Minnesota
St. Paul, Minnesota 55108

Somatic embryogenesis was observed in Aesculus with: (a) four-week-old, in vitro-germinated, excised embryo roots and shoots (A. x arnoldiana); (b) three- and 14-day forced shoots on three-year-old seedlings (A. glabra); and (c) three- and seven-day forced shoots on a 30-day-old tree (A. 'Autumn Splendor'). Embryogenesis was greatest with juvenile tissue and least with mature tissue. Explants (root, nodal, internodal) were cultured under a 16-hour photoperiod at 20°C on agar-solidified woody plant medium, pH 5.6, supplemented with 2% sucrose. Embryogenic callus developed after three months only on senescing tissues which had been exposed to cytokinin (benzyladenine at 5 μM). Embryogenic callus emerged from proximal cut ends of the explants. After another three to four months, embryos with distinct cotyledonary structures and root/-shoot axes were observed. Occasionally, plumules developed and/or radicles elongated, but no intact plantlets were retrieved. Liquid medium did not enhance development of embryogenic callus. Dark culture of tissue segments with "dormant" adventitious buds for eight weeks at 5°C increased frequency, uniformity, and rapidity of somatic embryo formation upon returning cultures to original environmental conditions.

IN VITRO SHOOT ORGANOGENESIS ON LEAF TISSUE FROM INTACT AND MICROPROPAGATED MATURE LIQUIDAMBAR STYRACIFLUA

Mark H. Brand and R. Daniel Lineberger

Department of Horticulture
Ohio State University
Columbus, Ohio 43210

Leaf and petiole explants were taken from shoot-proliferating cultures and intact plants of Liquidambar styraciflua 'Variegata' and 'Moraine.' Woody plant medium supplemented with benzyladenine (2.5 mg/l) supported optimum shoot organogenesis. Naphthaleneacetic acid (0.1 mg/l) altered the pattern of shoot organogenesis. Shoots formed on major vasculature of the lamina and on the petiole stump in three to six weeks. Wounding of leaf explants increased the total number of buds formed and the proportion of the total buds formed on the lamina. Young, expanding leaves or half-expanded leaves were more organogenic than folded, emerging leaves or fully expanded leaves. Increasing developmental age increased the proportion of total buds formed on the leaf blade. Green leaves from 'Variegata' and white leaves from a nearly albino 'Variegata' selection were cultured in two intensities of white, blue, and red light, and in darkness. Light quality and intensity altered explant organogenic patterns. Regenerated plants are being evaluated for variation.

ISOLATION OF FUNCTIONAL RNA FROM CULTURED COTYLEDONS OF DOUGLAS FIR

M.A. Campbell, J.J. Gaynor, and E.G. Kirby

Department of Biological Sciences
Rutgers University
Newark, New Jersey 07102

We have established a system for the examination of early molecular events associated with shoot formation in cotyledon cultures of Douglas fir (Pseudotsuga menziesii). Total RNA was isolated from cultured cotyledons after 0, 4, 12, 16, and 24 hr using a guanidinium thiocyanate extraction followed by cesium chloride density gradient ultracentrifugation. Poly(A^+)-selected RNA was translated in vitro in a wheat germ system and analyzed using two-dimensional NEPHGE. Results indicate discrete pattern of incorporation of [^{35}S]methionine as visualized by fluorography of two-dimensional gels. The isolation of functional RNA associated with the shoot-forming response enables us to examine early events associated with phytohormone-induced development.

A POSSIBLE NOVEL ANTIBIOTIC FROM BACILLUS SUBTILIS AGAINST PATHOGENIC FUNGI TO TREE SPECIES

Ming Tu Chang and Steven M. Eshita

U.S. Department of Agriculture Forest Service
Delaware, Ohio 43015

An antibiotic-producing strain of Bacillus subtilis from the xylem of an American elm has been studied. Crude antibiotic preparation is inhibitory to the fungal pathogens for Dutch elm disease, Ceratocystis ulmi; the fungal pathogen for verticillium wilt, Verticillium dahliae; and the fungal pathogen for oak wilt, Ceratocystis fagacearum. The inhibitory activity is heat-stable in culture broth concentrate, but becomes unstable after solvent extraction. Sephadex LH-20 column and C-18 reverse phase high pressure liquid chromatography were used to purify antibiotics to high purity ($\geq$99%). Absorbance spectrum analysis using the in-line scan capacity of Waters M490 detector indicates two peak areas after solvent baseline subtraction. The major peak is at 204 to 207 nm, and a second peak is at 285 to 286 nm. Pico-tag amino acid analysis at 10 pmol level of sensitivity indicates that this antibiotic contains no amino acid or primary/secondary amine residue. Antimicrobial activity of this antibiotic differs from that known to B. subtilis suggesting the possibility of a novel antibiotic.

SOMATIC EMBRYOGENESIS AND PLANT REGENERATION FROM TISSUE AND CELL SUSPENSION CULTURES OF SELECTED MATURE HIMALAYAN POPLAR

G.S. Cheema

Tata Energy Research Institute
New Delhi 110 003, India

Cultures of Himalayan poplar (Populus ciliata) were established from leaves of mature trees in the year 1979 on a defined medium (MS + 2 mg/l BAP) in total darkness (Mehra and Cheema, 1980). These cultures have continued to proliferate and have shown high frequency regeneration for almost eight years now. The mode of regeneration on this medium is by organogenesis. The regenerated shoots, on being transferred to fm medium, form well-developed plantlets which have been transferred to soil outdoors from time to time. Callus was induced from "leaf" mass on MS + 0.2-0.5 mg/l 2,4-D. It was transferred to liquid medium of the same composition to raise cell suspensions. High levels of 2,4-D (0.5 mg/l) yielded fine suspensions of single cells and small cell clusters, whereas lower concentrations of 2,4-D (0.2 mg/l) showed a tendency to form larger cell clumps in four to six weeks. The organization of somatic embryos at different stages of growth was observed by third subcultures in MS + 0.2 mg/l 2,4-D. The different stages of embryogeneic organization ranging from minute to globular to serially advanced stages leading to fully developed embryo and multiple embryo formation through secondary embryogenesis were observed to occur in the same culture. The total removal of auxin (2,4-D) from the medium did not help further growth and germination of embryos. The plating of such embryos on fm medium showed somatic embryos developing through a variety of developmental patterns such as precocious germination, arrested shoot development, arrested root development, and good growing seedling development. A few of these could be successfully hardened and transferred to field soil in polypots and grown in natural outdoor conditions.

ADVENTITIOUS BUD FORMATION ON EMBRYO CULTURES OF JACK PINE AND WHITE PINE

E. Chesick, B.A. Bergmann, W.P. Hackett, and C.A. Mohn

Department of Forest Resources
University of Minnesota
St. Paul, Minnesota 55108

Jack pine (JP) (Pinus banksiana) and white pine (WP) (P. strobus) embryos were cultured on Schenck and Hildebrandt, Murashige and Skoog (MS), Gresshoff and Doy (GD), and LePoivre (LP) agar-solidified media supplemented with 3% sucrose and benzyladenine (BA) at 0, 1, 5, 10, 20, and 50 μM. After four months, JP embryo survival and adventitious bud formation were greatest on GD. Survival of WP embryos was equal on LP, MS, and GD, but bud formation was best on LP. The BA increased JP embryo survival and was required for bud formation on both JP and WP.

Concentrations of BA between 5 and 20 μM were equally promotive for survival and bud formation with JP. The WP bud formation was enhanced equally using BA between 1 and 50 μM. A maximum of 234 and 56 adventitious buds was obtained per WP embryo and JP embryo, respectively, with optimal treatments. Embryos survived only if adventitious buds were present. Survival and bud formation were greater with WP than JP.

INFLUENCE OF SUBCULTURING PERIOD AND DIFFERENT CULTURE MEDIUM ON COLD STORAGE OF *POPULUS ALBA* X *POPULUS GRANDIDENTATA* PLANTLETS

Young Woo Chun and Richard B. Hall

Department of Forestry
Iowa State University
Ames, Iowa 50011

In vitro cultured hybrid poplar, *Populus alba* x *Populus grandidentata*, could be stored at 4°C air temperature in darkness for 24 months without adversely affecting its potential for rapid multiplication. Nevertheless, the subculturing period preceding cold storage, plantlet condition, and culturing medium had a significant influence on survival at 4°C storage in darkness. A one-month subculturing period before cold storage was better than a 0-month or two-month subculturing period for cold storage. Shoot proliferation medium (MS + BA 0.3 mg/l) was better than rooting medium (MS + IBA 0.2 mg/l) for long-term cold storage. After 24 months storage, the survival percentage of plantlets with 4-6 axillary branching shoots that were subcultured on shoot proliferation medium for one month preceding cold storage was 70%. Continuous subculture is labor intensive and requires extensive culture space. Cold storage of in vitro cultured *Populus* plantlets could serve to alleviate these problems as well as provide methods to facilitate germplasm conservation.

MORPHOGENETIC POTENTIAL OF LEAF, INTERNODE, AND ROOT EXPLANTS FROM *POPULUS ALBA* X *POPULUS GRANDIDENTATA* PLANTLETS

Y.W. Chun, N.B. Klopfenstein, and R.B. Hall

Department of Forestry
Iowa State University
Ames, Iowa 50011

Morphogenetic responses of explants of *Populus alba* x *Populus grandidentata* plantlet depend significantly upon the explant source and upon the combination of exogenously applied plant growth regulators. Two clones of this hybrid poplar, Crandon and Hansen, exhibited interclonal variation in their morphogenetic responses in leaf, internode, and root explant cultures at various benzylaminopurine (BA) and naphthaleneacetic acid (NAA) combinations. Among the three explant sources, leaf explants were the most responsive to the BA and NAA combinations tested, and root explants were the least responsive. Abaxial side culture of entire leaf explants was best suited for inducing adventitious shoot buds from two clones of this hybrid poplar. Other aspects associated with the biotechnology applications of in vitro cultured plantlets are also discussed.

INDUCTION OF TUMORS BY VARIOUS STRAINS OF AGROBACTERIUM TUMEFACIENS ON ABIES NORDMANNIANA AND PICEA ABIES

D.H. Clapham and I. Ekberg

Department of Forest Genetics
Swedish University of Agricultural Sciences
Uppsala, Sweden

Abies nordmanniana and Picea abies seedlings were inoculated with a wild-type strain (C58) or an attenuated strain (rooter or shooter mutants) of Agrobacterium tumefaciens. Large tumors were formed on Abies nordmanniana in response to the wild-type and rooter strains, 60% of the seedlings being susceptible. Smaller tumors were formed on Picea abies in response to the wild-type strain, at least 12% of the seedlings being susceptible. Tumors from both species induced by the wild-type strain have grown in vitro on medium without added phytohormones for eight months to date, and produce a compound with the electrophoretic mobility of nopaline.

THE 25S AND 18S RIBOSOMAL DNAs FROM PINUS RADIATA

C.A. Cullis,[1] G.P. Creissen,[1] S.W. Gorman,[1] and R.D. Teasdale[2]

[1]Department of Biology
Case Western Reserve University
Cleveland, Ohio 44106

[2]Calgene-Pacific
Victoria, Australia

The genes for the 25S and 18S ribosomal RNAs of Pinus radiata have been isolated from a genomic library constructed in EMBL4. Flax rDNA was used as the heterologous probe. Six clones are being mapped with restriction sites and the RNA coding regions to obtain the organization of the complete repeating unit in P. radiata. The organization and quantitative variation of the rDNAs in a series of individual trees and a cell culture line are also being determined. These individual trees represent five separate native populations. The flax rDNA has also been used for in situ hybridization studies to localize the pine rDNA. There appear to be at least ten major sites of rDNA on the P. radiata chromosomes.

IN VITRO PROPAGATION OF MATURE ROBINIA PSEUDOACACIA L. USING BUD EXPLANTS OBTAINED DURING WINTER DORMANCY

J.M. Davis and D.E. Keathley

Department of Forestry
Michigan State University
East Lansing, Michigan 48824

Lateral buds excised from stems of a 21-year-old dormant black locust tree were placed on MS basal medium containing 0.01 μM or 0.1 μM kinetin, or 0.01 or 0.1 μM 6-benzylaminopurine (BAP) (20 buds per treatment). The most rapid elongation of shoots occurred on medium with 0.1 μM BAP. Shoot cultures were successfully initiated from 15 of the 20 buds in this treatment. Only four shoot cultures were initiated from treatments which contained kinetin. Shoots longer than 1.0 cm produced roots after being placed on 1/10 MS basal medium with 0.1 μM IBA and no sucrose. The use of bud culture appears to be a viable means for propagating mature black locust trees.

IN VITRO NODULATION OF MICROPROPAGATED PLANTS OF LEUCAENA LEUCOCEPHALA WITH RHIZOBIUM

V. Dhawan[1] and S.S. Bhojwani[2]

[1]Tata Energy Research Institute
New Delhi 110 003, India

[2]Department of Botany
Delhi University
Delhi 110 007, India

The nitrogen fixation efficiencies of nine strains of Rhizobium, seven isolated from locally grown Leucaena leucocephala and NGR-8 and CG81 obtained from Papua, New Guinea, were compared. For in vitro nodulation of leucaena seedlings, on the basis of number of nodules per plant, nitrogenase activity of the nodules and the growth of nodulated plants in nitrogen-free medium, NGR-8 proved to be the best followed by the local strain Lcn-8. Micropropagated plants [culture initiation and shoot multiplication on MS with 4% sucrose, 4×10^{-6} M glutamine and with 3×10^{-6} M BAP; rooting on MS (inorganics reduced to half strength) with IAA 5×10^{-6} M] were transferred to sterilized quartz sand in feeding bottles for hardening and were irrigated with quarter strength MBH medium with Rhizobium.

The micropropagated plants took almost five weeks to form visible nodules as compared to two weeks taken by the seedling plants perhaps due to their photosynthetic inefficiency. Initially, NGR was used but later strains TAL600 and TAL1145 obtained from Niftal, Hawaii, were also successfully applied to nodulate micropropagated plants.

AGROBACTERIUM-MEDIATED GENE TRANSFER IN EUROPEAN LARCH

A.M. Diner and D.F. Karnosky

BioSource Institute
Michigan Technological University
Houghton, Michigan 49916

Several strains of Agrobacterium rhizogenes and Agrobacterium tumefaciens bearing selectable plasmid markers, such as that for neomycin phosphotransferase synthesis, were shown virulent to young Larix decidua seedlings. Calli generated either from hairy roots (A. rhizogenes pARC8) or hypertrophic inoculated hypocotyls (A. tumefaciens pWB101) survived and grew on media supplemented with kanamycin to levels of 30 and 300 μg/ml, respectively. These levels of tolerance to kanamycin are consistent with those reported for agronomic species transformed using these vectors. Work is now underway in our laboratory both to incorporate economically useful genetic loci to plasmids of these virulent strains, and to develop procedures for regeneration of plantlets from transformed tissues.

IMMUNOASSAY OF CYTOKININS IN DOUGLAS-FIR COTYLEDONS CULTURED IN VITRO

P. Doumas, B. Goldfarb, A. Bataille,
and J.B. Zaerr

Department of Forest Science
Oregon State University
Corvallis, Oregon 97331

Immunological techniques were used to purify and assay N^6-benzylaminopurine (BAP), BAP-riboside, and endogenous cytokinins in cultured Douglas-fir cotyledons exposed to media containing 0, 0.5, and 5.0 μM BAP. Monoclonal antibodies raised against conjugated N^6 (Δ_2-isopentenyl) adenosine demonstrated excellent cross-reactivity with BAP, BAP-riboside, and other BAP derivatives. These antibodies, and ones raised against zeatin-riboside, were used in immunoaffinity columns to purify extracts from cultured cotyledons. Purified cytokinin-like compounds were separated by high pressure liquid chromatography and quantitated by radioimmunoassay and enzyme-linked immunosorbent assay. After 27 hr of culture, the BAP concentration in the cotyledons increased in proportion to the BAP concentration in the media. Synthesis of naturally occurring cytokinins was not induced by BAP treatment.

These techniques are now being used to investigate BAP uptake and metabolism in morphogenetic and nonmorphogenetic tissue cultures of Douglas fir.

SOMATIC POLYEMBRYOGENESIS FROM PROTOPLASTS OF LOBLOLLY PINE (PINUS TAEDA)

Don J. Durzan and Pramod K. Gupta

Department of Pomology
University of California
Davis, California 95616

Somatic embryos were regenerated from protoplasts of loblolly pine derived from embryonal suspensor masses (ESMs). Ninety-five percent viable and totipotent protoplasts were recovered from three- to four-day-old cell suspension cultures of proliferating ESMs. Protoplasts from embryonal cells contained a dense neocytoplasm with strong acetocarmine staining. Microcolonies of cells were obtained after six to seven weeks in a thin layer of agarose in a specially formulated medium using myoinositol as an osmoticum. Cells of microcolonies were composed mainly of embryonal cells, which were confirmed by double-staining with acetocarmine and Evans blue. New ESMs were developed from microcolonies on agar-solidified medium. All stages of embryo development were identical to earlier reported stages of somatic and zygotic polyembryogenesis. Within 16 to 17 weeks, over 100 somatic embryos were regenerated by sequential transfer of different media.

RESTRICTION FRAGMENT-LENGTH POLYMORPHISM ANALYSIS OF THE CHLOROPLAST GENOMES OF ROCKY MOUNTAIN JUNIPER AND EASTERN RED CEDAR

S. Ernst and D.F. Van Haverbeke

University of Nebraska
Lincoln, Nebraska 66583

Rocky Mountain juniper (*Juniperus scopulorum* Sarg.) and eastern red cedar (*Juniperus virginiana* L.) accounted for 84% of all conifer seedlings, and 37% of all conifer and hardwood seedlings (combined) distributed through the Clarke-McNary Tree Distribution Program in 1982 and 1983. However, while eastern red cedar and Rocky Mountain juniper are the most extensively planted tree species in the Great Plains, there is very little genetic information available for either species. Because so many seedlings of both species are planted each year in the Great Plains region, it is important to better understand the genetic control of desirable traits in a variety of environments. To accomplish this, the two "pure species" genomes must first be characterized. This poster describes some of our preliminary work using heterologous chloroplast DNA probes to detect restriction fragment-length polymorphisms in the chloroplast DNA of sympatric and allopatric seed sources of Rocky Mountain juniper and eastern red cedar.

COMPARISON OF EMBRYOGENIC AND NONEMBRYOGENIC CONIFER CALLI: CHLOROPLAST ULTRASTRUCTURE AND PROTEIN SYNTHESIS

R. Feirer, S.R. Wann, M.R. Becwar,
R. Nagmani, and J. Carlson

Forest Biology Division
The Institute of Paper Chemistry
Appleton, Wisconsin 54912

Embryogenic conifer calli are characteristically white or translucent in appearance even when grown in the light. Ultrastructural examination of embryogenic Norway spruce callus has revealed that somatic embryos contain plastids resembling the proplastids found in zygotic embryos. Mature chloroplasts, containing grana, are found in nonembryogenic callus, which is green when grown under identical conditions. Proplastids are also found in embryogenic calli of other conifer species. The morphology of the plastids changes as the somatic embryos develop. Soluble proteins extracted from both types of callus have been compared by SDS-PAGE. Differences in the proteins extracted from the two tissue types have been observed, some of which may relate to chloroplast development/ultrastructure. These findings provide additional evidence that the development of somatic embryos corresponds to that of zygotic embryos.

SOMATIC EMBRYOGENESIS IN EASTERN REDBUD (*CERCIS CANADENSIS* L.)

Robert L. Geneve and Sharon T. Kester

Department of Horticulture
University of Kentucky
Agricultural Science Center North
Lexington, Kentucky 40546-0091

Somatic embryos were initiated from zygotic embryo explants of *Cercis canadensis*. There was a significant effect of explant age on the ability of the explants to form somatic embryos and adventitious roots on a medium supplemented with 2,4-D. Somatic embryos could only be induced from zygotic embryos collected 100 to 115 days post anthesis. The number of adventitious roots per explant increased with the age of the embryo explant. Both somatic embryos and adventitious root initials were produced on the hypocotyl of the explant, but their cellular origin was distinctly different. Somatic embryos arose from epidermal cells while adventitious roots were produced from cortical parenchyma cells adjacent to the vascular system.

PINUS ELDARICA DE NOVO SHOOT ORGANOGENESIS REFINED AND APPLIED TO OTHER CONIFERS

H.J. Gladfelter and G.C. Phillips

Department of Agronomy and Horticulture
Plant Genetic Engineering Lab
New Mexico State University
Las Cruces, New Mexico 88003-0003

De novo shoot organogenesis in long-term callus cultures of *Pinus eldarica* involves the discrete developmental steps of bud induction, maturation, apical organization, and elongation, each step requiring different cultural manipulations. Growth regulator requirements were refined for each developmental step using inflexible transfer schedules. Competence was improved by manipulating tissues through each developmental step using a flexible transfer schedule, holding the media sequence constant. *Pinus eldarica* cell suspension cultures were finely dispersed and callus colonies were recovered readily. Bud induction was observed in callus recovered from cell suspensions. This regeneration approach was evaluated with other conifer species. Bud induction was observed in six- to eight-month-old callus cultures of *Pseudotsuga menziesii* and *Abies concolor*, indicating potential for this regeneration approach to provide a useful model for in vitro conifer cultures.

THE 5S RNA GENES FROM PINUS RADIATA

S.W. Gorman,[1] R.D. Teasdale,[2] and C.A. Cullis[1]

[1]Department of Biology
Case Western Reserve University
Cleveland, Ohio 44106

[2]Calgene-Pacific
Victoria, Australia

Genomic DNA from *Pinus radiata*, enriched for the ribosomal DNA on a cesium chloride/actinomycin D gradient, was used to clone into the *Bam*HI site of pUC19. The clones containing the 5S sequences were identified using an heterologous probe (the cloned 5S genes from flax). Partial restriction digests indicate that the 5S DNA is organized in tandem arrays. Sequencing shows little variation in the coding region compared to nine angiosperms. Preliminary results reveal few restriction fragment-length polymorphisms for the 5S genes in individuals taken across the *P. radiata* gene pool.

FOREIGN GENE TRANSFER AND EXPRESSION IN DOUGLAS FIR (*PSEUDOTSUGA MENZIESII*)

Pramod K. Gupta, Don J. Durzan,
and Abhaya M. Dandekar

Department of Pomology
University of California
Davis, California 95616

Tumors have been induced on micropropagated shoots from mature trees and on seedlings of Douglas fir grown in vitro. Tumors were induced by two strains of *Agrobacterium tumefaciens* K12x562E and K12x167. These strains contain derivatives of the wild-type Ti-plasmid pTiA6 with a chimeric foreign gene encoding resistance to the antibiotic kanamycin. These derivatives were constructed by co-integrating plasmid pCGN562 or pCGN167, respectively, into pTiA6 by homologous recombination. Tumors were excised and grown on DCR medium without phytohormones in the presence of carbenicillin. Tumor growth was inhibited after eight to ten weeks in culture on phytohormone-free medium. All tumors synthesized octopine. Presence of a foreign DNA sequence and expression of a chimeric, bacterial kanamycin-resistant gene were demonstrated by Southern blot analysis and aminoglycoside phosphotransferase [APH(3')II] activity. This is the first report of transfer and expression of a foreign gene in Douglas fir, one of the most important conifer species in North America.

THE IN VITRO ACTIVITY OF RIBOSOMES IN PINE BUDS DURING WINTER AND ACTIVE GROWING PERIODS

H. Häggman[1] and S. Kupila-Ahvenniemi[2]

[1]Finnish Forest Research Institute
Kolari Research Station
SF-95900 Kolari, Finland

[2]Biocenter and Department of Botany
University of Oulu
SF-90570 Oulu, Finland

Ribosome assemblies were isolated from vegetative pine buds (*Pinus sylvestris* L.) collected from a natural stand in Oulu, Finland (25° 30' E; 65° N), in January and in May. The long-term mean daily temperatures in Oulu are 10.4°C in January and +7.1°C in May. The in vitro translation capacity of the ribosomes was tested using [^{3}H]leucine and a cell-free protein synthesizing system prepared from wheat germ. The translated proteins were analyzed using fluorography and a liquid scintillation spectrometer.

Acid-insoluble radioactivity could be detected both in May and in January. This means that the ribosomes of the bud tissues are able to maintain the in vitro activity during the winter, too, although at a reduced level. The results from protein analyses are presented.

USING LEAVES FROM TISSUE CULTURE PROPAGATED PLANTS FOR PROTOPLAST ISOLATION

Thomas D. Hillson and Richard C. Schultz

Department of Forestry
Iowa State University
Ames, Iowa 50011

Shoot tip propagation methods were used to establish several woody species in vitro (Acer, Populus, Quercus, and Fraxinus). These plants provided a source of sterile leaves for use in protoplast isolation, eliminating the need to sterilize the leaves and reducing the trauma to the cells. The intact leaves were placed in an enzyme solution, as these leaves have little or no cuticle. Therefore, no cutting or peeling is necessary to expose the cells to the enzymes which, again, reduces the trauma to the cells. These methods have been used to isolate protoplasts from Acer saccharinium, Fraxinus pennsylvanica, Populus alba L. x Populus grandidendata Michx., and Quercus robur L. We are using these as a base to study methods for isolation and culture of protoplasts from a wide variety of woody species.

TISSUE AND SUSPENSION CULTURES FROM MATURE SCOTS PINE TREES

Anja Hohtola and Sirkka Kupila-Ahvenniemi

Biocenter and Department of Botany
University of Oulu
SF-90570 Oulu, Finland

The aim of the project is to develop a method to propagate Scots pine (Pinus sylvestris L.) from buds of mature trees. Buds are collected once a week at Oulu, Finland (23° 30' E; 65° N), using ten- to 40-year-old trees. The best callus proliferation occurs on a modified MS medium. The amount of cytokinins and auxins required for cell proliferation depends on the season; in March and May, explants need more auxins than cytokinins, while during the summer and in November, the opposite hormone balance is beneficial. Best proliferating calli are obtained from buds taken from the youngest part of a tree. Minute apex initiation has been observed. However, differentiation occurs more frequently on calli started from seed embryos.

Suspension cultures have been established from four- to six-week-old calli. They reach the logarithmic phase of growth in nine days. When suspension cells are subcultured on a solid medium, calli become covered with elongated cells and branched formations.

Browning of the tissues is a major problem in cultures. Electron-microscopic studies indicated that the number of organelles in the cells decreases simultaneously with browning of the tissues.

DNA ANALYSIS OF AN AGROBACTERIUM VECTOR VIRULENT IN LARCH

Y. Huang, A.M. Diner, T.S. Snyder,
and D.F. Karnosky

BioSource Institute
Michigan Technological University
Houghton, Michigan 49931

Characterizing *Agrobacterium* plasmid DNA is essential for transferring selected genetic loci. We have developed methods for characterizing the DNA of *Agrobacterium tumefaciens* plasmid pWB101, a vector for NPT expressed in transformed *Larix*. Bacteria grown in suspension to late log phase were treated with chloramphenicol to amplify the plasmid. Cells were then lysed with proteinase K (50 μg/ml) and SDS (1% w/v). Chromosomal DNA was removed by alkaline (pH 12.4) denaturation and precipitation with NaCl. The resultant crude plasmid DNA was precipitated in 50% (w/v) polyethylene glycol and subsequently purified on CsCl/EtBr gradients. After extraction of EtBr with n-butanol and dialysis to remove CsCl, the DNA was precipitated then stored in TE (Tris/EDTA, 10/1 mM) buffer. Plasmid DNA was analyzed by restriction endonuclease and the enzyme digests were separated by horizontal agarose gel electrophoresis. The average yield of plasmid DNA from a 100 ml of bacterial suspension was 160 μg. This procedure permitted the selective isolation of plasmid DNA appropriate for nick translation, transformation, and DNA cloning experiments.

MOLECULAR ANALYSIS OF GENE OF EXPRESSION DURING THE DEVELOPMENT AND MATURATION OF LARCH

Keith W. Hutchison,[1] Patricia B. Singer,[1] and
Michael S. Greenwood[2]

[1]Department of Biochemistry
[2]Department of Forest Biology
University of Maine
Orono, Maine 04469

We have initiated a study to determine the underlying genetic mechanisms governing gene expression during growth and maturation of conifers. Complementary DNA libraries have been constructed using RNA meristemic tissue and foliage from juvenile and mature larch (*Larix laricina*). Currently, we have identified clones representing: (a) the small subunit of Rubisco, (b) the chlorophyll a/b-binding protein, and (c) several clones which appear to be differentially expressed between juvenile and mature trees. The small subunit of Rubisco is strongly conserved relative to published angiosperm sequences and is expressed in the foliage but not in the roots. Our current data on differentially expressed clones are presented.

MICROPROPAGATION AS AN AID TO CLONAL PROPAGATION OF GUAVA CULTIVAR

V.S. Jaiswal and M.N. Amin

Department of Botany
Banaras Hindu University
Varanasi, India

Development of the micropropagation method for one of the important guava (Psidium guajava L. cv. 'Chittidar') was established from nodal explants of field-grown adult trees. Agitation of explants in 0.5% PVPP and two to three changes of medium for the initial ten days were essential for the establishment of cultures. On MS-revised medium containing BA (1 mg/l), axillary buds grew out within three to four weeks. On transfer to fresh medium of the same composition, these shoots attained 3-5 cm length, having four to six nodes after four weeks of further culture. Nodal segments (three to four propagules/shoot) taken from in vitro proliferated shoots gave rise to two to four shoots by precocious axillary branching without an initial lag period. By repeated subculture, a large number of shoots were built up with a multiplication rate of three- to four-fold per subculture. Shoots obtained from the proliferation stage were rooted on one-half salt MS medium having a combination of IBA and NAA with about 60% rooting after the second and 85% rooting after the fifth and subsequent subcultures. Regenerants were successfully established in soil under field conditions.

PRODUCTION OF TRANSGENIC APPLE PLANTS USING ENGINEERED PLASMIDS IN AGROBACTERIUM SPP.

D.J. James and A.J. Passey

Institute of Horticultural Research
East Malling
Maidstone, Kent, ME19 6BJ, United Kingdom

Under optimal conditions, shoot regeneration from apple leaf tissues approaches 100% with several shoots per disc. Using these conditions, transformed apple callus and shoots have been produced after infection with disarmed strains of Agrobacterium tumefaciens that carry the engineered binary vector, Bin6. Transformation has been shown by the presence of nopaline in leaf tissues and the ability of the plants to root in the presence of kanamycin. Further molecular evidence is being sought.

Wild-type and engineered vectors of Agrobacterium rhizogenes have been used in attempts to regenerate both genetic chimeras and wholly transformed plants of several fruit tree subjects.

FACTORS INFLUENCING IN VITRO MICROPROPAGATION OF WHITE PINE

K. Kaul

CRS Plant and Soil Science Research
Kentucky State University
Frankfort, Kentucky 40601

A procedure for in vitro plant regeneration from cotyledon-hypocotyl (C/H) explants of *Pinus strobus* L. was established. Plant regeneration was achieved in four discrete steps: shoot induction, shoot growth, root induction, and root growth. The influence of certain factors on the above-mentioned steps was studied. There was a great deal of genotypic variation in the response of individual C/H explants. The number of plants regenerated was influenced by the age of the seedling from which the explant was taken. Best results were obtained if explants were taken within a week of the emergence of cotyledons from the seed coat. A combination of auxin and cytokinin was required for shoot induction. The medium of Litvay et al. (LM) without any growth hormones gave good shoot growth. Inclusion of activated charcoal in LM drastically inhibited the number of shoots regenerated. Inhibition of shoot growth also occurred if a hormone-free Murashige and Skoog medium was used instead of LM. Root induction seemed to be affected by auxin concentration in the medium.

PARTIAL NUCLEOTIDE SEQUENCE OF AN ENTIRE PINE ACTIN GENE

J. Kenny

Department of Forestry
University of Alberta
Alberta, Canada

As deduced from Southern analysis, lodgepole pine contains a multigene family. The complete nucleotide sequence of the carboxy-terminal portion of one of the actin genes, PAc1-A, has previously been reported. A second clone carrying an entire actin coding region, PAc2, has now been isolated from a bacteriophage library. The partial nucleotide sequence of this gene, corresponding to amino acids 151 to 376 of the SAc3 soybean actin gene, has been determined. The sequence was compared with PAc1-A and actin genes from soybean (SAc3 and SAc1), maize (MAc1), chicken, and yeast. The highest homology values are observed between the two pine actins. Both of the pine actin amino acid sequences are also very homologous to the SAc3 sequence.

SOMATIC EMBRYOGENESIS AND PLANT REGENERATION IN CALLUS FROM HYPOCOTYL SEGMENTS OF *TILIA AMURENSIS*

Jae-Hun Kim, H. Moon, J. Park, and B. Lee

Institute of Forest Genetics
Suwon, Korea

The appearance of yellow and brown friable callus was assured on MS media containing 2,4-D alone or in combination with kinetin using hypocotyl segments of Tilia amurensis after six to eight weeks culture. The best results for embryoid induction were obtained after transferring the proembryonic-stage calli cluster cultured on MS with 2.0 mg/l 2,4-D to one-half MS containing 0.01 or 0.05 mg/l NAA under diffuse light conditions. The embryoids thus obtained developed normal plantlets on one-half MS control or on one-half MS with added casein hydrolysate or NAA. They were transferred to pots containing vermiculite and perlite. Two of them have survived and showed normal growth.

ISOLATION OF A PINE ADH cDNA CLONE

C.S. Kinlaw, D.E. Harry, D. Sleeter, and R. Sederoff

U.S. Forest Service
PSW Forest and Range Experiment Station
Berkeley, California 94701

A number of angiosperms, including maize, respond to brief periods of anaerobic stress by increasing their synthesis of a specific set of 20 anaerobic response proteins (ANPs). All of the ANPs which have been identified are enzymes involved in anaerobic energy metabolism, such as alcohol dehydrogenase (ADH). We have evidence that a gymnosperm, Monterey pine, also induces the synthesis of ADH in response to anaerobic stress. Thus, gymnosperms may share with angiosperms the ability to respond to anaerobiosis by inducing the expression of a specific set of genes.

In order to compare the molecular mechanisms responsible for the regulation of ADH expression in gymnosperms and angiosperms, we have isolated a recombinant cDNA clone for a Monterey pine ADH gene. The pine ADH mRNA shares about 70% sequence homology with maize and Arabidopsis ADH sequences.

TOWARD TRANSFORMATION OF POPULUS SPECIES BY AGROBACTERIUM BINARY VECTOR SYSTEMS

N.B. Klopfenstein, Y.W. Chun, H.S. McNabb, Jr.,
R.B. Hall, E.R. Hart, R.C. Schultz,
and R.W. Thornburg

Departments of Forestry, Plant Pathology, Entomology,
and Biochemistry and Biophysics
Iowa State University
Ames, Iowa 50011

The host range of wild-type Agrobacterium strains on various Populus clones, and the effective concentration of kanamycin for selection of transgenic tissue were evaluated in preliminary studies to facilitate transformation of Populus by Agrobacterium binary vector systems. To determine the kanamycin concentration for effective selection of transgenic tissue expressing a neomycin phosphotransferase gene, nine different kanamycin concentrations (0, 10, 20, 30, 40, 50, 100, 150, and 200 mg/l) were incorporated into MS regeneration medium. Leaf explants (entire leaf, terminal half, and basal half) of P. x rouleauiana were placed in abaxial and adaxial culture and monitored for shoot, root, and callus formation. In the host range study, several Populus species and hybrids were inoculated with A. tumefaciens wild-type strains A281 and A348 and monitored for crown gall formation. Using Agrobacterium binary vector systems (pGA472, pRT45, and pRT50 as binary vector; pTiBo542 and pAL4404 as helper Ti-plasmid), we are now conducting transformation studies on Populus.

BamHI-REPEAT FAMILY IN PROSOPIS FULIFLORA

M. Lakshmikuraran, V. Gupta, A. Agnihotri, S. Ranade, and V. Jagannathan

Tata Energy Research Institute
New Delhi 110003, India

The nuclear genome of higher plants and animals contains different families of repeated DNA sequences. Since a high proportion of plant genomes consists of repetitive DNA, it is important to understand their organization and function. In order to study the repeat families in Prosopis, the nuclear DNA was extracted from leaves. The DNA was subjected to restriction enzyme analysis for the identification of repeat families. Southern blot hybridization analyses have revealed the presence of distinct repeat families in BamHI digests. These families have been cloned into pUC plasmid vectors. These cloned DNAs are also being characterized by nucleotide sequencing. The sequence organization of these repeat families are discussed.

DUPLICATION OF THE psbA GENE BUT LACK OF rDNA CONTAINING INVERTED REPEATS ON THE CHLOROPLAST GENOME OF LODGEPOLE PINE (PINUS CONTORTA)

Jonas Lidholm,[1] Alfred Szmidt,[2] and Petter Gustafsson[1]

[1]Institute of Cell and Molecular Biology
University of Umeå
S-901, 87 Umeå, Sweden

[2]Department of Forest Genetics and Plant Physiology
Faculty of Forestry
Swedish University of Agricultural Sciences
S-901 83 Umeå, Sweden

The structure of the chloroplast (cp) genomes from a large number of angiosperms and algae is well described. In contrast, very little is known about gymnosperms. We have studied the cp genome of lodgepole pine, Pinus contorta, by restriction analysis and heterologous hybridization. From the sum of restriction fragment sizes, we estimate the genome size to be 115-120 kb. In Southern blot analysis, a spinach psbA probe hybridizes equally strong to two cpDNA fragments per digest. This implies a duplication of the psbA gene which encodes the herbicide-binding D1 protein of photosystem II. Normally, duplicated genes are located within the rDNA containing inverted repeats of the cp genome. However, neither a

23S nor a 16S rDNA probe hybridizes to the same cpDNA fragments as does the psbA probe; however, surprisingly, they hybridize to only a single band each. This strongly indicates that there is no rDNA containing inverted repeat in this cp genome. From a cpDNA library we have isolated both copies of psbA and obtained partial DNA sequences. These are identical for the two genes and upon comparison to the corresponding spinach sequence, 87% homology was revealed. The two psbA clones show identical restriction maps over their coding and 5' regions.

PROTOPLASTS AND RIBOSOME ASSEMBLIES FROM SCOTS PINE TISSUES

Aija Lindfors, Anja Hohtola,
and Sirkka Kupila-Ahvenniemi

Biocenter and Department of Botany
University of Oulu
SF-90570 Oulu, Finland

Protoplasts and ribosome assemblies were isolated from callus and suspension cultures of pine (Pinus sylvestris L.) buds. Buds were collected from a natural stand at Oulu, Finland (23°30'; 65°N), using ten- to 40-year-old trees.

Protoplasts were isolated by 6-hr enzyme treatment (1% cellulysin, 0.5% hemicellulase, 0.5% macerozyme) and purified on the sucrose-Percoll gradient. Protoplast yield was abundant from calli and suspensions which were in logarithmic growth phase. Protoplast isolation was not successful from slowly growing and browning tissues. Most of the protoplasts were capable of resynthesizing new cell walls.

The amount of total ribosome assemblies and the in vitro translation capacity were determined from both slowly and actively dividing cells. We observed marked differences in the quantity of ribosome assemblies. When calculated on a ribosome unit basis, the in vitro translation activity was fairly constant.

PROMOTION OF ADVENTITIOUS BUD FORMATION ON PICEA GLAUCA AND PICEA MARIANA WITH 4-PHENYL UREA-30

K.A. Louis, W.P. Hackett, and C. Mohn

University of Minnesota
St. Paul, Minnesota 55108

Several Picea glauca (Pg) and Picea mariana (Pm) explants were cultured: (a) 14-day-old shoot tips; (b) isolated cotyledons; (c) "four-year-old" shoot tips; (d) three developmental stages of isolated needles from (c); and (e) excised quiescent buds from 30+-year-old trees. Schenk and Hildebrandt basal medium, pH 5.0, with 2% sucrose was supplemented with 5.0 µM BA plus 5.0 µM 2ip or 4-phenyl urea-30 (4PU-30) at 1, 10, or 100 µM. Bud formation on cotyledons cultured with 4PU-30 at 100 µM (Pg = 86%; Pm = 61%) was at least equal to that in cytokinin cultures (Pg = 82%; Pm = 65%). Bud formation on the least developed needles cultured on 4PU-30 at 100 µM (Pg = 40%; Pm = 37%) was less than that in cytokinin supplemented cultures (Pg = 45%; Pm = 58%). Adventitious bud formation on seedling and greenhouse shoot tips was also achieved, but not on fully developed needles or mature explants. Urea derivatives (i.e., 4PU-30 and thidiazuron) have been shown to have cytokinin-like activity with tobacco callus; however, this is the first known report of 4PU-30 on conifer species.

SUBMERSION INCREASED ADVENTITIOUS BUD DEVELOPMENT OF *PICEA GLAUCA* AND *PICEA MARIANA*

K.A. Louis, W.P. Hackett, and C. Mohn

University of Minnesota
St. Paul, Minnesota 55108

Shoot tips of greenhouse-forced, two- to 0-year-old *Picea glauca* (white spruce) and *Picea mariana* (black spruce) were cultured on solidified Schenk and Hildebrandt (SH) medium supplemented with 5 μM benzyladenine plus 5 μM isopentenyladenine and 2% sucrose at pH 5.0 and submerged under 0.0, 1.0, 2.0, or 4.0 ml liquid medium of similar composition. After submersion for 0, one, two, or three weeks, explants were transferred to solidified SH medium containing 0.2% charcoal without cytokinin and submersion. After six weeks, 67% of the *P. glauca* shoot tips that had been submerged under 1.0 or 2.0 ml for three weeks had greater than 20 buds per explant, and 38% of the *P. mariana* explants that had been submerged under 1.0 ml for three weeks had greater than 20 buds per explant. Buds had formed at the sterigma (point of needle attachment) and were usually evenly distributed from base to apex. All other treatments ranged from 0.29 to 6.1 buds for white spruce and 0.95 to 4.17 buds for black spruce, the buds usually being located at the base of the explant.

GENETIC TRANSFORMATION OF *BETULA PAPYRIFERA* AND *ALNUS GLUTINOSA* WITH WILD-TYPE STRAINS OF *AGROBACTERIUM TUMEFACIENS*

J. MacKay, A. Seguin, L. Simon, and M. Lalonde

C.R.B.F. Faculté de Foresterie et de Geodesie
Université Laval, Ste-Foy, Quebec, Canada G1K 7P4

Shoots of *Betula papyrifera* and *Alnus glutinosa*, obtained through in vitro micropropagation, were inoculated with *Agrobacterium tumefaciens* strains C58 and Ach5. Thirty percent to 80% of shoots formed tumors. The excised tumors grew in absence of phytohormones, even though better growth was obtained by addition of 2,4-D. Octopine or nopaline, according to the bacterial strain, was detected in most tumorous calli, indicating that transformation had occurred. Southern blot hybridization using the *tmr* gene as a probe confirmed T-DNA integration into the plant genome.

With *B. papyrifera*, we obtained three organogenic tumors producing up to 20 shoots. Shoots are now being propagated for further analysis.

OVERCOMING IN VITRO VITRIFICATION THROUGH MEDIA MANIPULATION FOR LARCH

J. McLaughlin and D.F. Karnosky

School of Forestry
Michigan Technological University
Houghton, Michigan 49931

While our micropropagation system for Larix decidua yields an average of about 70 adventitious buds per seedling, nearly 70% of these buds generally prove vitreous. Attempts were made to eliminate vitreousness by both restricting initiation of these aberrant buds, and by reverting vitreous buds to a normal state. Reducing vitrification frequency to 30% was realized using a 90% reduction of the initiation medium cytokinin concentration. However, this was accompanied by a proportional decrease in bud production. Established vitreous buds were reverted by 80% following three weeks on Gresshoff and Doy (GD) medium with one-half strength N, as compared to 69% for GD with full-strength N and 22% for GD with full-strength N and 1.46 g/l L-glutamine.

DEVELOPMENT OF A MICROPROPAGATION SYSTEM FOR RED PINE

R. Meilan and R.C. Schultz

Department of Forestry
Iowa State University
Ames, Iowa 50011

Foliar applications of 200 mg/l 6-benzylaminopurine (BAP) + 0.075% Tween 20 led to the proliferation of adventitious buds at the apex of red pine (Pinus resinosa Ait.) seedlings. This cluster of buds was surface-sterilized with 15% commercial bleach for 15 min, followed by one rinse with 70% ethyl alcohol and three rinses with sterile distilled water. When plated on Murashige and Skoog (MS) media containing 1.0 μM BAP + 1.0 μM 1-naphthaleneacetic acid (NAA), the buds grew out.

Callus derived from mature embryos, extracted from surface-sterilized red pine seed, was induced to form adventitious shoots when plated on MS media with 1.0 μM BAP + 1.0 μM NAA. Shoot development proceeded after callus was subcultured on basal media. Callus from plates containing 10.0 μM BAP + 0.1 μM NAA formed numerous adventitious shoots upon transfer to basal media.

A rooting medium is being sought for both of these approaches to micropropagation. Once this is accomplished, an attempt will be made to transform the resulting seedlings with Agrobacterium tumefaciens.

MODIFIED EMBRYOGENESIS IN MALUS USING IRRADIATED POLLEN

E.C. Menhinick and D.J. James

Institute of Horticultural Research
East Malling, Maidstone
Kent ME19 6BJ, United Kingdom

Gamma-irradiated marker pollen carrying a dominant homozygous gene for anthocyanin pigmentation has been used to obtain matromorphic shoot cultures of two apple cultivars, Spartan and Idared. All red-colored hybrid plants, which may or may not be aneuploid, are obtained in crosses where pollen is irradiated in the range 0.5 to 40 krads. Green matromorphic plants, assessed by absence of the marker gene and isoenzyme evidence, are only obtained above 50 krads pollen irradiation. At 70 krads, haploid embryos may form and these can be successfully micropropagated, although they do not root in vitro.

TRANSFORMATION OF POPLARS BY AGROBACTERIUM TUMEFACIENS

M.F. Michel,[1] F. Delmotte,[2] and C. Depierreux[2]

[1]Station d'Amelioration des Arbres Forestiers
I.N.R.A. Orleans
Ardons F-45160 Olivet, France

[2]Centre de Biophysique Moleculaire
C.N.R.S.
Université d'Orleans
F-45071 Orleans, Cedex 2, France

Galls were induced by Agrobacterium tumefaciens wild strains in poplar hybrids belonging to the Leuce Section. Shoot regeneration was observed on gall tissues grown on hormone-free medium. The transformed nature of these shoots was confirmed by the presence of opines.

SELECTION FOR HERBICIDE TOLERANCE IN POPULUS TO SULFOMETURON METHYL IN VITRO

C.H. Michler and E. Bauer

North Central Forest Experiment Station
U.S. Department of Agriculture Forest Service
Rhinelander, Wisconsin 54501

Weed competition limits plantation establishment of poplars which may be killed by environmentally safe, broad-spectrum herbicides, such as sulfometuron methyl (Oust). To allow weed control without killing poplars, we have selected sulfometuron methyl-tolerant poplar plantlets in vitro. Leaf pieces of Populus maximowiczii x Populus trichocarpa (NC11390) were explanted on Murashige and Skoog (MS) media supplemented with 1 mg/l benzyladenine (BA), 0.1 mg/l naphthaleneacetic acid, 20 g/l sucrose, and various concentrations of sulfometuron methyl (0, 10, 25, 50, 75, and 100 ppb). After 30 days in darkness, surviving leaf pieces were subcultured onto the identical herbicide-containing media for three consecutive 30-day periods in the light. Adventitious shoots that survived three subcultures at toxic levels (>5 ppb) of sulfometuron methyl were subcultured on MS media with 0.1 mg/l BA for three consecutive 30-day periods. Surviving shoots were rooted for further testing in the greenhouse.

INDUCTION OF CALLUS AND ADVENTITIOUS BUDS IN *PINUS SYLVESTRIS*

S.M. Jain, S. Mohan, R.J. Newton, and E.J. Soltes

Department of Forest Science
Texas A&M University
Texas Agricultural Experiment Station
College Station, Texas 77843-2135

Pinus sylvestris L. seeds were sterilized, soaked on wet filter paper, and kept in the dark for 48 hr. Zygotic embryos were isolated and cultured on Murashige and Skoog (MS) and B_5 media for callus formation and on SH medium for the induction of adventitious buds, under 2,000 lux light intensity at 25°C. The callus formation was poor on B_5 medium and was better on MS medium having 0.5 to 1.0 mg/l 2,4-dichlorophenoxyacetic acid (2,4-D). Half- and three-quarter-strength MS salts had no significant effect on callus growth as compared to the full-strength salts. For shoot induction, calli were transferred to MS medium supplemented with 1 to 12 mg/l benzylaminopurine (BAP) and 0.5 mg/l 2,4-D. Twenty percent of the calli turned dark green and did not form shoots. The adventitious buds were induced on SH medium containing 0 to 5 mg/l BAP. At 3 mg/l BAP concentration, 45% of the zygotic embryos formed adventitious buds. Well-developed shoots were obtained by continuous subculture of the adventitious buds after every seven to ten days. Fifty percent root induction resulted on DCR and MS media supplemented with 0.1 mg/l naphthaleneacetic acid.

PEACH FRUIT mRNA POPULATIONS VARY DURING FRUIT DEVELOPMENT AND BETWEEN CULTIVARS

P.H. Morgens, A. Callahan, E. Walton, R. Scorza, and J.M. Cordts

Appalachian Fruit Research Station
Agricultural Research Service
U.S. Department of Agriculture
Kearneysville, West Virginia 25430

Cold hardiness and high fruit quality are considered complex multigenic traits that are difficult to combine in peach through sexual hybridization. We plan to circumvent this problem by transferring genes associated with high quality fruit into cold-hardy cultivars using gene isolation, plant transformation, and plant regeneration techniques. As a preliminary step in the search for the appropriate genes, we have extracted RNA from the fruit of both cold-hardy cultivars which bear poor quality fruit and cold-sensitive commercial cultivars which bear high quality fruit. We have translated these RNAs in vitro and show that there are some major differences between cultivars and between developmental stages within cultivars. We are using the in vitro translation patterns to guide a cloning strategy for putative genes associated with high fruit quality.

REGENERATION OF PEACH FROM SOMATIC EMBRYOS/AGROBACTERIUM TUMEFACIENS-MEDIATED TRANSFORMATION OF PEACH TISSUE

R. Scorza, J.M. Cordts, A.M. Callahan, and P.H. Morgens

Appalachian Fruit Research Station
Agricultural Research Service
U.S. Department of Agriculture
Kearneysville, West Virginia 25430

Peach (*Prunus persica* L. Batsch) plants have been regenerated in vitro from immature (50- to 60-day-old) embryos. Regenerative calli have continued to produce somatic embryos for over two years, although shoot regeneration rates are low in long-term cultures. Peach stem tissue has been transformed with an armed, engineered *Agrobacterium tumefaciens* strain containing the neomycin phosphotransferase (NPT) gene, which confers resistance to kanamycin. Transformed calli are growth regulator-independent and tolerate up to 200 mg/l kanamycin sulfate. Transformation of embryogenic calli and the development of a leaf disc transformation/regeneration system for peach are in progress.

TRANSFORMATION OF PINACEOUS GYMNOSPERMS BY AGROBACTERIUM

John W. Morris, Linda A. Castle, and Roy O. Morris

Department of Agricultural Chemistry
Oregon State University
Corvallis, Oregon 97331

Agrobacterium tumefaciens is known to effect the transformation of a variety of dicotyledonous hosts, but is not generally regarded as a pathogen on gymnosperms. We have assessed the ability of a series of over 60 *Agrobacterium* strains to induce galls on four pinaceous gymnosperms: *Abies procera*, *Pinus ponderosa*, *Pseudotsuga menziesii*, and *Tsuga heterophylla*. Galls were induced upon infection of the apical regions of shoots from very young seedlings. Induction occurred on all four species, but at very different frequencies among the strains. A few strains induced galls at high frequencies on all four hosts. A high proportion (>70%) of the *A. rhizogenes* strains induced galls on all four hosts, *Tsuga* and *Abies* having the highest frequencies. Overall, the frequency of gall induction was greatest on *Abies* and least on *Pinus*. Transformation was confirmed by identification of opines. (Research support provided by U.S. Department of Agriculture grant 85-FSTY-9-0146.)

GROWTH AND SOLUTE CHANGES IN LOBLOLLY PINE CALLUS IN RESPONSE TO WATER STRESS

R.J. Newton, S. Sen, and J.D. Puryear

Department of Forest Science
Texas A&M University
Texas Agricultural Experiment Station
College Station, Texas 77843-2135

Solute levels in plant cells have an important role in adaptation to drought. We measured four solute fractions (potassium, sugars, organic acids, amino acids) relative to growth and osmotic potential in callus tissue of *Pinus taeda* L. over an eight-week period. Water stress was induced by adding polyethylene glycol (PEG) to the nutrient media. Callus fresh weight reductions due to decreasing media water potential were associated with reductions in the following: (a) cell volume, (b) callus osmotic potential, (c) solute levels (μg/mg dry weight), and (d) callus water content. Less than 5% of the measured change in callus osmotic potentials was attributed to PEG and 40 to 60% was attributed to the sum of the four solute fractions. Reduced volume rather than solute accumulation appears to be a primary response of pine callus cells to prolonged, severe water stress.

SOMACLONAL VARIATION FOR DISEASE RESISTANCE IN FOREST TREES

Michael E. Ostry and Darroll D. Skilling

North Central Forest Experiment Station
U.S. Department of Agriculture Forest Service
St. Paul, Minnesota 55108

Tissue culture techniques are being used to obtain somaclonal variation in the resistance of *Populus* sp. to the foliar pathogen *Septoria musiva* and resistance of *Larix* sp. to scleroderris canker caused by *Gremmeniella abietina*. Poplars with putative resistance to *Septoria* leaf spot have been found among tissue culture-derived plants from a clone previously susceptible to the disease. Variant plants were selected from the regenerants using a bioassay that rapidly distinguishes plants resistant to the fungus. Nearly 30% of the regenerated plants were clearly more resistant than the original clone. These trees have been planted in the field and are being evaluated for field resistance. Similarly, variation in resistance to scleroderris has been found among plants regenerated from cotyledon cultures and inoculated with the fungus in vitro. Tissue culture and somatic variation have the potential to significantly reduce the time required for selecting and improving forest trees.

AN ULTRASTRUCTURAL STUDY OF FERTILIZATION AND CYTOPLASMIC INHERITANCE IN DOUGLAS FIR

J.N. Owens and S.J. Simpson

Department of Biology
University of Victoria
Victoria, British Columbia, Canada V8W 2Y2

The pollen germinates and elongates along the essentially dry micropylar canal to the nucellus. Secretory cells at the tip of the nucellus release their contents and degenerate. This liquid appears to stimulate pollen tube development. Pollen tubes penetrate the nucellus by separating nucellar cells, many of which collapse adjacent to the pollen tube. The tube nucleus followed by the body cell migrates down the pollen tube. Pollen tubes reach the archegonia four to five days after pollen tube formation. The body cell divides, forming two male gametes, each surrounded by dense cytoplasm containing distinct mitochondria and less distinct plastids but no cell wall. One male gamete with a small portion of cytoplasm migrates to the egg nucleus. The egg nucleus is surrounded by a perinuclear zone rich in maternal mitochondria which are morphologically distinct from paternal mitochondria. Further studies will concentrate on the fate of maternal and paternal organelles during fertilization and proembryo development.

ISOLATION AND CULTURE OF MESOPHYLL PROTOPLASTS FROM *POPULUS* SPECIES

Y.G. Park,[1] S.H. Son,[1] and K.-H. Han[2]

[1]College of Agriculture
Kyungpook National University
Daegu 635, Korea

[2]Department of Forestry
Michigan State University
East Lansing, Michigan 48824

Factors affecting the isolation and culture of protoplasts from mesophyll were investigated in *Populus alba* x *Populus glandulosa*, *P. glandulosa*, and *Populus davidiana*. The best yields of mesophyll protoplasts were obtained using leaves in vitro with 0.2% cellulase P-10, 0.8% macerozyme R-10, 1.2% hemicellulase, 2.0% driselase, 0.05% pectolyase Y-23, and 0.6% M mannitol, in DTT and MES buffer adjusted to pH 5.6. When protoplasts were cultured in MS basal medium (minus NH_4NO_3) enriched with 0.5 mg/l benzylaminopurine and 2.0 mg/l 2,4-D, cell division was highly stimulated by liquid plating methods. Among four protoplast culture methods, the semisolid agar method produced good results for making colonies of cells investigated after 14 days in culture, while liquid plating methods showed faster cell division than other methods after seven days in culture.

DEVELOPMENT OF CELL CULTURE SYSTEMS FOR SELECTED MATURE CLONES OF EUCALYPTUS GRANDIS

R.M. Penchel,[1,2] Y.K. Ikemori,[1] and E.G. Kirby[1]

[1]Department of Biological Sciences
Rutgers University
Newark, New Jersey 07102

[2]Departmento de Biotecnologia Florestal
Aracruz Florestal S.A.
Aracruz E.S., Brazil

Exploitation of tissue culture systems for genetic improvement of Eucalyptus grandis requires the establishment of cell and protoplast culture systems for selected superior clones. A method has been developed for production of multiple buds by culture of nodal explants of epicormic shoots of selected clones on a modified Quoirin and Lepoivre (QL) medium containing 1 μM benzylaminopurine (BAP) and 50 nM naphthaleneacetic acid (NAA). Induction of friable callus lines has been achieved by culture of internodes and leaves produced by multiple buds on similar medium containing 10 μM 2,4-dichlorophenoxyacetic acid (2,4-D), 2 μM NAA, and 0.1 μM BAP. Fine cell suspension cultures were initiated from selected callus lines and grown on a modified QL medium containing 2 μM kinetin, 2 μM NAA, 2 μM 2,4-D, 15 mM glutamine, 2% sucrose, without ammonium, and were grown on a gyrotory shaker (50 rpm) at 19°C in darkness. Development of the cell culture system will be discussed in light of applications to the genetic improvement of Eucalyptus spp.

CALLUS AND SUSPENSION CULTURES OF PINUS BANKSIANA, PICEA GLAUCA, AND PINUS MARIANA

P. Périnet and F.M. Tremblay

Petawawa National Forestry Institute
Canadian Forestry Service
Chalk River, Ontario, Canada

Eight salt formulations and four light treatments were compared for callus initiation from seedling material of Pinus banksiana, Picea glauca, and Picea mariana. The media always had a stronger effect on callus growth than the light quality (Gro-Lux, Vita-Lite, and cool-white). Jack pine calli grew better on Litvay's medium in darkness or in dim light conditions. Spruce calli grew well on Nagata and Takebe's (NT), Litvay's, von Arnold and Eriksson's, and Murashige and Skoog's media under dim light. Cell suspension cultures were initiated from friable and fast-growing calli of jack pine and white spruce using Litvay's and NT media. Suspension cultures have been maintained for four to ten months by weekly subcultures.

PLANTLETS FROM LEAF DISCS OF POPULUS DELTOIDES

C.S. Prakash and B.A. Thielges

Department of Forestry
University of Kentucky
Lexington, Kentucky 40546

Calli were initiated by incubating leaf discs of *Populus deltoides* cv. K417 on Murashige and Skoog (MS) medium with 2,4-dichlorophenoxyacetic acid (2,4-D) (1 μM) and benzylaminopurine (BAP) (0.1 μM) in the dark, and were then subcultured on MS medium with naphthaleneacetic acid (NAA) (1 μM) and kinetin (0.1 μM) under a 16-hr photoperiod. Multiple shoot regeneration occurred when compact, green calli were transferred on MS or woody plant (WP) medium supplemented with 1 μM BAP (under a 12-hr photoperiod). Rooting was achieved by pulsing shoots with 1 to 2 μM BAP (in WP medium) for two to four days. The technique is being further refined to obtain somaclonal variants for resistance to leaf rust disease (*Melampsora medusae*) and also for *Agrobacterium*-mediated gene transfer.

CALLUS REGENERATION FROM PROTOPLASTS OF A NONSEEDLING GYMNOSPERM TREE

Julie A. Russell and Brent H. McCown

Department of Horticulture
University of Wisconsin
Madison, Wisconsin 53706

Previous success with gymnosperm tree protoplasts has relied mainly on seedling tissue as a protoplast source. We have circumvented this limitation by applying techniques developed for the protoplast culture of non-seedling *Populus* (*Plant Sci.* 46:133-142) to a mature gymnosperm tree and have achieved sustained protoplast division. Shoot cultures of *Thuja occidentalis* 'Woodwardii' which were established from mature plants served as the protoplast source tissue. Protoplasts were isolated and cultured according to the *Populus* techniques. Sustained division was achieved when 2% (v/v) coconut water and 0.025% (w/v) casein hydrolysate were included in the culture medium. Cell colonies divided rampantly as the medium osmolarity was reduced and subculture to fresh medium was necessary in order to minimize the inhibitory effect of the resultant high cell populations. The protoplast-derived cell colonies could be transferred to a medium solidified with 0.2% (w/v) Gelrite and development continued into large calli.

RIBOSOMAL DNA POLYMORPHISM IN PINUS SYLVESTRIS

Folke Sitbon and Petter Gustafsson

Unit for Applied Cell and Molecular Biology
University of Umeå
Umeå, Sweden

To study *Pinus sylvestris* genome organization and restriction fragment-length polymorphism (RFLP) among different individuals, a screen for ribosomal DNA (rDNA) RFLP was performed using the Southern blot technique. The DNA was extracted from needles from ten different trees and digested with six restriction enzymes. The DNA was then hybridized with probes for 25S and 18S rDNA genes. The 25S hybdridization pattern for the four-cutter enzyme RsaI clearly divided the material into two groups, with one group lacking a 1,250-bp fragment. The relative band strength of certain bands also varied, indicating a variation in the number of fragments. The 18S pattern divided the material in the same way. This result could be explained by placing the DNA sequence alteration in the non-transcribed spacer region. In a similar screen of Pinus contorta using five trees and three enzymes, no RFLP was found.

CLONING OF A cDNA ENCODING THE CHLOROPHYLL a/b-BINDING PROTEIN FROM SCOTS PINE (PINUS SYLVESTRIS L.)

Stefan Jansson and Petter Gustafsson

Department of Plant Physiology
Institute of Cell and Molecular Biology
University of Umeå
Umeå, Sweden

A method for preparing mRNA from Scots pine (Pinus sylvestris L.) was developed. The method is a modification of the standard guanidinium isothiocyanate/cesium chloride protocol. The yield of total RNA from seedlings grown in the dark and exposed to light for three days is about 0.4 mg/g fresh weight. In northern hybridization of poly(A)-enriched RNA to a pea cDNA clone encoding the chlorophyll a/b-binding protein (cab), a single RNA of 1,100 nucleotides is detected. Messenger RNA encoding the cab protein was also present in seedlings grown in absolute darkness. From a cDNA library constructed in λgt10, the cab gene from Scots pine was cloned using the pea cab gene as probe.

EVIDENCE FOR PATERNAL INHERITANCE OF PLASTIDS IN INTERSPECIFIC HYBRIDS OF PICEA

M. Stine and D.E. Keathley

Department of Forestry
Michigan State University
East Lansing, Michigan 48824

Procedures were developed for the purification of chloroplast DNA (cpDNA) from Picea pungens and Picea glauca. Interspecific restriction fragment-length polymorphisms (RFLPs) were used to follow the inheritance of plastids in hybrid crosses. In the seven crosses examined, all progeny exhibited the cpDNA restriction pattern of the pollen parent. This pattern was observed in progeny from crosses in both directions (P. pungens x P. glauca or P. glauca x P. pungens). Intraspecific RFLPs have also been identified which will allow the mode of plastid inheritance in intraspecific crosses to be analyzed.

CHLOROPLAST RESTRICTION SITE AND GENE MAPS OF DOUGLAS FIR AND RADIATA PINE

S.H. Strauss,[1] J.D. Palmer,[2] G. Howe,[1] and A. Doerksen[1]

[1]Department of Forest Science
Oregon State University
Corvallis, Oregon 97331-5704

[2]Division of Biological Sciences
University of Michigan
Ann Arbor, Michigan 48109

Chloroplast genome structures of Douglas fir and radiata pine were determined by Southern blot analyses. The DNA fragments were generated from restriction enzyme digests with *PvuII*, *SmaI*, *KpnI*, and *SacI*, and from double digests with each enzyme and *PvuII*. Blots of these fragments were hybridized with *PvuII* fragments from Douglas-fir chloroplast DNA that were recovered from agarose gels, cloned chloroplast fragments from petunia and mung bean, and small gene fragments that code for known proteins in spinach, pea, and tobacco. Both Douglas fir and radiata pine lack the large inverted repeats which characterize most higher plants, accounting for their relatively small genome size (approximately 121 kb). They appear to differ from each other by a single major inversion and both possess inversions and translocations not present in *Gingko* and *Petunia*. These results support the hypothesis that lack of the inverted repeats potentiates "rapid" evolutionary changes in chloroplast structure.

GENETIC POLYMORPHISM FOR RIBOSOMAL GENE COPY NUMBER IN DOUGLAS FIR

S.H. Strauss and C.-H. Tsai

Department of Forest Science
Oregon State University
Corvallis, Oregon 97331-5704

Genetic variability for the number of copies of ribosomal genes was studied in DNA extracted from foliage of 54 trees that were derived from seed collected nearly throughout the natural range of Douglas fir. Slot blots, filters on which carefully measured quantities of DNA are fixed in precise, narrow bands, were hybridized with a radiolabeled, heterologous ribosomal gene probe, exposed to film, and the relative copy number estimated from densitometer scans of the resulting autoradiographs. Genetic polymorphism for ribosomal gene copy number was substantial and resided almost entirely within, rather than between, populations; among-tree variance accounted for nearly half of the densitometric variance observed. Clinal patterns were evident, but weak; copy number increased with latitude, elevation, and longitude. The tendency for copy number to increase northward concurs with clines found for genome size in Douglas fir, in other conifers, and among wild and cultivated herbaceous plant species.

TISSUE CULTURE STUDIES ON MAHOGANY (SWEITENIA SPECIES)

F. Sultanbawa, S. Venketeswaran,
M.A.D.L. Dias, and Ursula V. Weyers

Department of Biology
University of Houston
Houston, Texas 77004

Callus tissues were established from aseptic tissue segments of cotyledons and leaves of mahogany (Sweitenia mahogany and Sweitenia macrophylla) on a modified B_5 or Murashige and Skoog (MS) medium supplemented with 2 mg/l 2,4-dichlorophenoxyacetic acid (2,4-D) and 1 mg/l α-naphthaleneacetic acid (NAA). In two to three weeks, the explants proliferated extensively on the agar medium in a growth chamber at 25 ± 1°C. The cotyledonary callus was very soft and white in color and further subcultures resulted in a loose "snowflake" appearance. Callus obtained from leaves were almost transparent and colorless. Experiments on in vitro organogenesis and development of plantlets by micropropagation methods are be presented. Mahogany, which grows as natural populations in tropical America and parts of Asia and Africa, is a highly valued timber tree known for its redwood color, strength, durability, water resistance, and aesthetic appeal. (This research supported in part by the University of Houston Coastal Center funds and NASA grant No. NAG 9-214.)

RESTRICTION ANALYSIS OF CHLOROPLAST DNA DIVERSITY IN PINES

Alfred E. Szmidt and Jan-Erik Hallgren

Department of Forest Genetics and Plant Physiology
Swedish University of Agricultural Sciences
S-901 83 Umea, Sweden

Chloroplast DNA (cpDNA) diversity was studied in nine species in the genus Pinus by restriction analysis. The cpDNA was extracted from fresh needles, digested with six different restriction endonucleases (BamHI, BclI, HindIII, SacI, EcoRI, and KpnI), and separated by agarose gel electrophoresis. The cpDNA restriction patterns of the twelve species investigated were different for all endonucleases used. Chloroplast genome size for each species was estimated to be about 121 kbp. Estimates of genetic divergence between species were obtained. Using these estimates, a dendrogram showing genetic relatedness between chloroplast genomes was constructed. Phylogenetic implications of the observed diversity are discussed.

CULTURE OF PINUS RADIATA CELLS: STUDY OF A CELL VIABILITY FACTOR

R.D. Teasdale and D.K. Richards

Centre for Forest Biotechnology
GIAE
Churchill, Victoria, Australia

Suspension-cultured *Pinus radiata* cells lose their viability when diluted below a critical inoculum density. The threshold can be extended by adding cell-free supernatant from separately grown cultures. Fresh medium is conditioned rapidly, indicating production of a potent cell viability factor (CVF). This effect is not mimicked by auxins, cytokinins, polyamines, or vitamins. The factor has moderate temperature (65°C) and storage stability, is adsorbed onto charcoal, and is not peptidyl. Limited concentration-response curves increase monotonically, indicating that still greater effect may be obtained if the factor can be separated from other medium components. Ultrafiltration studies indicate a molecular weight of less than 1,000 daltons.

NUTRIENT REQUIREMENTS OF PINUS RADIATA CELLS: INTERACTION BETWEEN BORON, CALCIUM, AND MAGNESIUM

R.D. Teasdale and D.K. Richards

Centre for Forest Biotechnology
GIAE
Churchill, Victoria, Australia

Suspension-cultured *Pinus taeda* cells in LM medium exhibited, inter alia, an unexpectedly high requirement (5X MS) for boron. Such a high boron requirement was not found in later studies with *Pinus radiata* cells in an alternative (P6) medium. When P6 medium was modified to the LM levels of calcium (20X decrease) and magnesium (5X increase), the boron requirement increased dramatically. Concentration studies showed marked nutrient interaction between calcium, magnesium, and boron. This is consistent with a simple model where calcium and boron are both required to reversibly bind to separate sites on an acceptor to activate it, with magnesium competitively displacing calcium. Moreover, moderate levels of mannitol (which binds boron) exacerbate boron deficiency, whereas, surprisingly, high levels (0.132 M) alleviate the stress, possibly through osmotic support, since boron-deficient cells are typically enlarged or ruptured. The metabolic function of boron in plants remains elusive, but this clear link with calcium may indicate commonality of primary roles.

SOMATIC EMBRYOGENESIS FROM MATURE ZYGOTIC EMBRYOS ISOLATED FROM STORED SEEDS OF PICEA GLAUCA

F.M. Tremblay

Petawawa National Forestry Institute
Canadian Forestry Service
Chalk River, Ontario, Canada K0J 1J0

Zygotic embryos were isolated from white spruce (Picea glauca) seeds stored at 4°C for three years. The seeds were cultured on half-strength Litvay's medium (LM) supplemented with 10 μM 2,4-dichlorophenoxyacetic acid, 5 μM benzylaminopurine, 1 g/l casein hydrolysate, 500 mg/l glutamine, and 1% sucrose. After three to five weeks, a high percentage of the embryos produced embryogenic calli with typical suspensors and proembryo structures. Fully developed somatic embryos were recovered from these calli.

PLANT TISSUE CULTURE OF SCOTS PINE (PINUS SYLVESTRIS L.)

H.S. Tsai and F.H. Huang

Department of Horticulture and Forestry
University of Arkansas
Fayetteville, Arkansas 72701

Plantlet regeneration of Scots pine (Pinus sylvestris L.) through cultured tissues has been attained; mature embryos and seedling cotyledon whorls were used as explants and cultured on one-third-strength Murashige and Skoog (MS) medium, shoot multiplication medium B supplemented with kinetin (1 mg/l), or Durzan's medium 2. The adventitious buds or adventitious shoots initiated on the embryonic cotyledon or on the basal area of the cotyledon whorl were encouraged into elongation by subculturing onto one-half-strength Durzan's medium 2 with the addition of 0.1 mM spermidine which improved the survival rate to 100%. Elongated adventitious shoots were rooted on one-half-strength Durzan's medium 2 free of growth hormones in one month after 24 hr naphthaleneacetic acid (NAA) pulse treatment (125 mg/l). The NAA pulse treatment not only improved the rooting percentage to 33% but also increased the number of roots in individual propagule. The regenerated plantlets were given the hardening-off treatment by transferring to one-half-strength Durzan's medium 2 without any carbon sources and with high concentration of agar (10 g/l) for three months before being transferred to soil mix. Embryogenesis was attempted with several tissues on 12 media.

TRANSFORMATION OF SALIX CLONES BY AGROBACTERIUM TUMEFACIENS

Tiina Vahala, Priska Stabel, and Tage Eriksson

Department of Physiological Botany
University of Uppsala
Uppsala, Sweden

Different genotypes of fast-growing willows (*Salix dasyclados*, *Salix scwerinii*, and *Salix viminalis*) have been transformed using *Agrobacterium tumefaciens* as a vector. In explant transformation experiments, transformed callus was obtained when axenic as well as greenhouse-grown leaf and stem material was used. Both kanamycin resistance and hormone-independent growth could be used for selection of transformants. C58 was oncogenic, whereas inoculation with octopine wild-type strains did not result in tumor formation. Tumors were also obtained with "shooter" and "rooter" mutant strains, but no organogenesis was observed. The frequency of tumor formation varied from 0 to 88% between genotypes.

INDUCTION OF STRESS PROTEINS IN PINUS ELLIOTTII

J.V. Valluri, E.J. Soltes,
R.J. Newton, and J. Castillon

Department of Forest Science
Texas A&M University
Texas Agricultural Experiment Station
College Station, Texas 77843-2135

A 55-kDa stress protein was induced in tissue-culture-grown slash pine seedlings infected with conidial suspensions of pitch canker fungus. Induction was evident in both hypocotyl and root sections of the seedlings after a lag phase of 16 hr. Seedlings challenged with nonvirulent extract also responded with induction of new proteins. The stress protein is best extracted from stressed tissues with slightly alkaline Tris-HCl (pH 8.65) buffer.

We have also investigated protein synthesis in water-stressed slash pine seedlings and calli. Mannitol-induced water stress inhibited overall protein synthesis, with induction of unique stress proteins. At -18 and -25 bars, stress proteins were detected by SDS-PAGE and fluorography. A direct relationship appears between protein synthesis and water potential. As water potential decreased, incorporation of L-[^{35}S]methionine into protein also decreased.

HIGH EFFICIENCY REGENERATION OF SHOOTS FROM *POPULUS* TISSUE CULTURES

Peter Viss and Steven E. Ruzin

Plant Genetics, Inc.
1930 Fifth Street
Davis, California 95616

We have regenerated shoots from callus and suspension cultures of 25 clones of *Populus* originating from ten species and varieties. Callus cultured on BA (5 μM) combined with kinetin (1 μM) in a DKW salt base on solid medium (regeneration medium) was found to yield shoots with at least one internode (versus nonelongated shoot buds) in 96% of the calli after four weeks in culture. The addition of either 3% maltose or glucose as the sole carbohydrate source increased the efficiency of regeneration from 20% to 64% in *Populus* clones that had been previously low or recalcitrant regenerators. Callus derived from suspension cultures grown in Schenk and Hildebrandt medium containing picloram (2.5 μM) yielded rapidly growing friable callus and dense callus clumps that regenerated shoots at a high frequency after transfer to solid regeneration medium. The highest frequency of dense callus clumps and of shoot regeneration occurred when suspension cultures were taken off picloram medium for one pass and then returned to a picloram-containing medium.

HAPLOID EMBRYOGENESIS OF *LARIX DECIDUA*

Patrick von Aderkas and Jan Bonga

Fredericton, E3B 5P7
New Brunswick, Canada

Induction of haploid embryoids from isolated and cultured megagametophytes is dependent on the timing of explant removal and the use of 2,4-D in the medium. Embryoids arise from two types of pathways: (a) reiterative alternation of long and short cell types, and (b) following a free-nuclear state. Both pathways are unlike natural embryogenesis. The various stages of embryoid development will be illustrated and described.

ASSOCIATIONS OF CHLOROPLAST DNA AND CONE MORPHOLOGY IN THE SYMPATRIC REGION OF *PINUS BANKSIANA* AND *PINUS CONTORTA*

D.B. Wagner and D.R. Govindaraju

Department of Forestry
University of Kentucky
Lexington, Kentucky 40546

We have screened 786 individuals from two populations of the *Pinus banksiana* x *Pinus contorta* sympatric region for cone morphology and for a chloroplast (cp) DNA marker. Five percent of the individuals had cpDNA atypical of the allopatric ranges of the parental species. The atypical variants in the sympatric region might be due to: (a) establishment of cpDNA mutations or intramolecular recombinants, enhanced by interspecific hybridization; (b) maternal leakage of chloroplasts, leading to cpDNA heteroplasmy and/or intermolecular recombination (cpDNA is apparently paternally inherited in P. contorta x P. banksiana hybridization events); or (c) interactions among selection, mutation, and genetic drift in marginal populations of each species. Three hundred forty-five individuals had cpDNA typical of one species, but had cone morphology typical of the other species or hybrids. This observation indicates that cpDNA of one species is functional in genetic backgrounds of the other, if cone morphology is an accurate indicator of taxonomic species in this sympatric region.

BIOCHEMICAL DIFFERENCES BETWEEN EMBRYOGENIC AND NONEMBRYOGENIC CALLUS OF CONIFERS

S.R. Wann, M.A. Johnson, R.P. Feirer,
M.R. Becwar, and R. Nagmani

Forest Biology Division
The Institute of Paper Chemistry
Appleton, Wisconsin 54912

Embryogenic and nonembryogenic calli of loblolly pine (*Pinus taeda* E.), white pine (*Pinus strobus*), pond pine (*Pinus rigida* var. *serotina*), white spruce (*Picea glauca*), and European larch (*Larix decidua*) were analyzed for biochemicals previously shown to be indicative of an embryogenic state in Norway spruce [Wann et al. (1987) *Plant Cell Reports* 6:39-42]. Concentrations of glutathione, total reductants as well as ethylene evolution, and protein synthesis rates in the two callus types were consistent with previously measured values for Norway spruce. The use of these parameters as markers for embryogenesis is indicated by the observation that embryogenic potential in loblolly pine and pond pine was predicted by biochemical analysis in advance of the appearance of somatic embryos.

DIFFERENCE IN THE METHYLATION OF DNA EXTRACTED FROM TREES OF NORWAY SPRUCE (*PICEA ABIES*) OF DIFFERENT AGES

Roger Westcott

Unilever Research
Colworth House
Sharnbrook, Bedford, England

The methylation of cytosine nucleotides in extracted DNA has been measured directly after high performance liquid chromatography and indirectly by comparing restriction enzyme digests of two isoschizomers *HpaII* and *MspI*. Methylation of cytosine was greatest in older (18- to 35-year old) than in younger (0-, two-, and six-year old) trees. The results are discussed in relation to micropropagation of mature trees.

EXPRESSION OF rbcS AND cab GENES IN BOTH DARK AND LIGHT CONDITIONS IN PINE (PINUS THUNBERGII) SEEDLINGS

N. Yamamoto,[1] Y. Mukai,[1] M. Matsuoka,[2] Y. Ohashi,[2] Y. Kano-Murakami,[3] Y. Tanaka,[4] and Y. Ozeki[5]

[1]Forest and Forest Product Research Institute
[2]National Institute of Agrobiological Resources
[3]Fruit Tree Research Station
[4]National Institute of Agro-Environment
Tsukuba Science City
[5]University of Tokyo
Tokyo, Japan

In the Angiospermae, light has been shown to induce the expression of a so-called "photogene" such as rbcS and cab encoding a small subunit of ribulose bisphosphate carboxylase and light-harvesting chlorophyll a/b binding protein, respectively. We have examined the expression of rbcS and cab genes in pine seedlings grown in the dark and in light conditions. Two cDNA clones encoding the two major proteins were isolated and used to estimate the transcript of each gene in both seedlings. Northern blotting analysis shows that, in pine seedlings, two genes are expressed in the dark and in the light.

ROSTER OF SPEAKERS, PARTICIPANTS, AND PERSONNEL

Ahuja, Raj	Institute of Forest Genetics, Groshansdorf, WEST GERMANY
Aitken-Christie, Jenny	Forest Research Institute, Rotorua, NEW ZEALAND
Allen, Robert	Clemson University, Clemson, SC 29634-0375
Amerson, Henry D.	North Carolina State University, Raleigh, NC 27695-7612
Back, Arie	P.B. Industries, Mobil Post Ashrat, ISRAEL
Ballester, A.	CSIC, 15080 Santiago de Compostela, SPAIN
Bana, O.P.S.	University of Minnesota, Minneapolis, MN 55455
Barrows-Broaddus, Jane	U.S. Department of Agriculture Forest Services, Athens, GA 30602
Becwar, Michael	Institute of Paper Company, Appleton, WI 54912
Bekkaoui, Faouzi	Saskatoon, Saskatchewan, CANADA S7 N 0J4
Bercetche, Joelle	AFOCEL, Domaine de l'Etancon, 77370 Nangis, FRANCE
Bergmann, Ben	University of Minnesota, St. Paul, MN 55108
Berlin, Jochen	Gesellschaft für Biotechnologische Forschung, Braunschweig, WEST GERMANY
Bernard-Dagan, Colette	Université de Bordeaux, Talence, Cedex, FRANCE
Bhardwaj, Shrwan	Utah State University, Logan, UT 84322
Bhat, M.L.	University of Minnesota, Minneapolis, MN 55455
Bloese, Paul	Michigan State University, East Lansing, MI 48824
Bonga, Jan M.	Maritimes Forest Research Center, Fredericton, N.B., CANADA
Bousquet, Jean	Université Laval, Quebec, CANADA
Boxus, Philippe	Agricultural Research Center, Claussee de Charleroi, Gembloux, BELGIUM
Brand, Andrew	University of Connecticut, Storrs, CT 06268
Brand, Mark	Ohio State University, Columbus, OH 43210
Bullard, Ray	University of Tennessee, Chickamauga, GA 30707
Campbell, Charles	Michigan State University, East Lansing, MI 48824
Campbell, Michael	Rutgers University, Newark, NJ 07102
Canavera, David	Westvaco, Summerville, SC 29484
Chang, Jye	Michigan State University, East Lansing, MI 48824

Chang, Ming Tu	U.S. Department of Agriculture Forest Service
Chaparro, Jose	North Carolina State University, Raleigh, NC 27695
Chavan, Shivaji	Ohio State University, Columbus, OH 43210-1085
Chesick, Emily	University of Minnesota, St. Paul, MN 55108
Christianson, Michael	Zoecon Corporation, Palo Alto, CA 94304-0859
Chun, Young Woo	Iowa State University, Ames, IA 50011
Clapham, David	SLV, 575007 Uppsala, SWEDEN
Coke, Jay Eric	University of California, Riverside, CA 92507
Coleman, Gary	University of Nebraska, Lincoln, NE 68583
Cordts, John	Appalachian Fruit Research Station, Kearneysville, WV 25430
Coston, D.C.	Clemson University, Clemson, SC 29634-0375
Coumans, Marc	Allelix, Inc., Mississauga, Ontario, CANADA L4V 1P1
Cullis, Chris	Case Western Reserve University, Cleveland, OH 44105
David, John	Michigan State University, East Lansing, MI 48824
Davidson, Campbell	Agriculture Canada Research Branch, Morden, Manitoba, CANADA
De Groot, B.	Research Institute Ital, 6705 AG Wageningen, THE NETHERLANDS
Dhillon, Sukhraj S.	North Carolina State University, Raleigh, NC 27695-7612
Diner, A.M.	Michigan Technological University, Houghton, MI 49331
Dunstan, David	Plant Biotechnology Institute, Saskatoon, Saskatchewan, CANADA S7N0W9
Durzan, Don	University of California, Davis, CA 95616
Eriksson, Tage	Uppsala University, Uppsala, SWEDEN
Ernst, Stephen	University of Nebraska, Lincoln, NE 68583
Evers, Peter	Dorschkamp Research Institute, Holland, THE NETHERLANDS
Fillatti, JoAnne	Calgene, Inc., Davis, CA 95616
Froberg, Cal	Plant Reproduction International, Humble, TX 77338
Furnier, Glenn	University of California, Riverside, CA 92521
Gale, Wanda	Hilltop Trees, The Nursery Corp., Hartford, MI 49064
Gasque, C. Edward	University of Wisconsin, Stevens Point, WI 54481
Geneve, Robert	University of Kentucky, Lexington, KY 40546
Gingas, Vicki	Ohio State University, Columbus, OH 43210
Gladfelter, Heather	New Mexico State University, University Park, NM 87131
Goldfarb, Barry	Oregon State University, Corvallis, OR 97331

Goldie, Ron	North Carolina State University, Raleigh, NC 27695-7009
Gorman, S.	Case Western Reserve University, Cleveland, OH 44105
Green, Thomas	Auburn University, Auburn, AL 36849
Greenwood, Michael	University of Maine, Orono, ME 04469
Gruber, Karl	Michigan State University, East Lansing, MI 48824
Gudin, Serge	Selection Meilland, 06600 Antibes, FRANCE
Gustafsson, Petter	Umeå University, S-901 87 Umeå, SWEDEN
Hackett, Wesley	University of Minnesota, St. Paul, MN 55108
Häggman, Hely	The Finnish Forest Research Institute, SF-5900 Kolari, FINLAND
Häggman, Juhani	The Finnish Forest Research Institute, SF-5900 Kolari, FINLAND
Haissig, Bruce	U.S. Department of Agriculture Forest Service, Rhinelander, WI 54501
Hakman, Inger	Uppsala University, S-75221 Uppsala, SWEDEN
Handley, Lee	Westvaco Corporation, Summerville, SC 29483
Hanover, James W.	Michigan State University, East Lansing, MI 48824
Harry, David E.	University of California, Berkeley, CA 94720
Heidekamp, Freek	Research Institute Ital, 6700AA Wageningen, THE NETHERLANDS
Hillsan, Thomas	Iowa State University, Ames, IA 50011
Hohtola, Anja	University of Oulu, SF-90570, Oulu, FINLAND
Howe, Glenn	Oregon State University, Corvallis, OR 97331
Huang, Feng Hou	University of Arkansas, Fayetteville, AR 72702
Huang, Yinghua	Michigan Technological University, Houghton, MI 49331
Huhtinen, Ossi	Enso-Gutzeit, SF-55800 Imatra, FINLAND
Hutchison, Keith	University of Maine, Orono, ME 04469
Iezzoni, Amy	Michigan State University, East Lansing, MI 48824
Jain, Shri Mohan	Texas A&M University, College Station, TX 77843
Jambulingam, R.	North Carolina State University, Raleigh, NC 27695-8002
James, David	Institute of Horticulture Research, Maidstone, Kent, ENGLAND
Jelaska, Sibila	University of Zagreb, 41001 Zagreb, YUGOSLAVIA
Jones, Myrtle	Michigan State University, East Lansing, MI 48824
Karnosky, David	Michigan Technological University, Houghton, MI 49931
Kaul, Karan	Kentucky State University, Frankfort, KY 40601

Keathley, Daniel	Michigan State University, East Lansing, MI 48824
Kenny, Joyce	University of Alberta, Edmonton, Alberta, CANADA T6S5W8
Khajuria, H.N.	School of Forest Resources, Raleigh, NC 27695
Kim, Jae-Hun	Institute of Forest Genetics, Suwon, Kyong-gido 170, SOUTH KOREA
Kinlaw, Claire S.	U.S. Forest Services, Berkeley, CA 94704
Kirby, Edward G.	Rutgers University, Newark, NJ 07102
Klimaszewaska, Krystyna	Petawawa National Forestry Service, Chalk River Ontario, CANADA KOJ 1JO
Klopfenstein, Ned	Iowa State University, Ames, IA 50011
Kohn, Hubertus	Western Washington University, Bellingham, WA 98225
Kumar, B. Mohan	Utah State University, Logan, UT 84322
Lakshmikumaran, Malathi	Tata Energy Research Institute, New Delhi, INDIA 110003
Laliberte, Jean-Francois	Institut Armand-Frappier, Quebec, CANADA H7N 4Z3
Lalonde, Maurice	Université Laval, Quebec, CANADA
Laurer, Christian	Unilever Research Laboratory Colworth, Sharnbrook, MK44 ILQ, ENGLAND
Layton, Patricia	Oak Ridge National Laboratory, Oak Ridge, TN 37831-6352
Lazarte, Jaime	Plant Reproduction International, Humble, TX 77338
Leong, Merlin	University of Victoria, Victoria, B.C., CANADA V8W 2Y2
Lesney, Mark S.	University of Florida, Gainesville, FL 32611
Lindberger, Daniel	Ohio State University, Columbus, OH 43210
Lindholm, Jonas	University of Umeå, S-901 87 Umeå, SWEDEN
Lindors, Aija	University of Oulu, Oulu, FINLAND
Loo-Dinkins, Judy	University of British Columbia, Vancouver, B.C., CANADA VGT 1W4
Loopstra, Carol	U.S. Department of Agriculture Forest Service, Berkeley, CA 94704
Louis, Kathryn	University of Minnesota, St. Paul, MN 55108
MacKay, John	Université Laval, Quebec, CANADA
Malac, Barry	Union Camp Corporation, Savannah, GA 31402
Marquard, Bob	Texas A&M University, El Paso, TX 79927
Martin, Melinda	Rutgers University, Newark, NJ 07102
Mascarenhas, A.F.	National Chemical Laboratory, Pune, INDIA
Maynard, Charles	State University of New York, Syracuse, NY 13210
McCown, Brent	University of Wisconsin, Madison, WI 53706
McLaughlin, James	Michigan Technological University, Houghton, MI 49331
McRae, John	International Forest Seed Co., Odenville, AL 35120
Meagher, Laura	North Carolina State University, Raleigh, NC 27695
Mebrahtu, Tesfai	Michigan State University, East Lansing, MI 48824

Meilan, Richard	Iowa State University, Ames, IA 50011
Mergener, Richard	Michigan Department of Natural Resources, Howell, MI 48843
Michel, Marie	INRA, Orleans, FRANCE 45000
Michels, Charles	U.S. Department of Agriculture Forest Service, Rhinelander, WI 54501
Miller, Raymond	Michigan State University, East Lansing, MI 48824
Minocha, Rakesh	University of New Hampshire, Dover, NH 03820
Minocha, Subhash	University of New Hampshire, Durham, NH 03824
Morgens, Peter	Appalachian Fruit Research Station, Kearneysville, WV 25430
Morris, John	Oregon State University, Corvallis, OR 97331
Mott, Ralph	North Carolina State University, Raleigh, NC 27695-2727
Mukewar, Anand	Mississippi State University, MS
Nance, Warren	Micro Computer Consultants, Gulfport, MS 39503
Neale, David	U.S. Department of Agriculture Forest Service, Berkeley, CA 94701
Nester, Eugene W.	University of Washington, Seattle, WA 98195
Neuman, Mark	Southern Illinois University, Carbondale, IL 62901
Newell, Christine	Monsanto Company, St. Louis, MO 63017
Newell, Nanette	Newell Associates, San Francisco, CA 94131
Ostry, Michael	U.S. Department of Agriculture Forest Service, North Central, St. Paul, MN 55108
Owens, John	University of Victoria, Victoria, CANADA V8W2Y2
Park, Young Goo	Kyungpook National University, KOREA
Patil, C.S.P.	UAS, Dharwad, INDIA
Penchel, Ricardo	Rutgers University, Newark, NJ 07102
Perinet, Pierre	Canadian Forestry Service, Chalk River, Ontario, CANADA KOJ 1JO
Phillips, Gregory	New Mexico State University, Las Cruces, NM 88003
Pijut, Paula	Ohio State University, Columbus, OH 43210
Pogany, Mary	Ohio State University, Columbus, OH 43210
Post, Boyd	U.S. Department of Agriculture, Washington, DC 22201
Prakash, C.S.	University of Kentucky, Lexington, KY 40546-0073
Prentice, Roy	Michigan State University, East Lansing, MI 48824
Pullman, Jerry	Weyerhaeuser Company, Tacoma, WA 98477
Rang, Allah	Michigan State University, East Lansing, MI 48824-1222
Rangaswamy, Nagmani	Institute of Paper Chemistry, Appleton, WI 54912
Read, Paul	University of Minnesota, Minneapolis, MN 55455

Rhyne, Charles	Jackson State University, Jackson, MS 39217
Riemenschneider, Don	U.S. Department of Agriculture Forest Service, Rhinelander, WI 54501
Rier, John	Howard University, Washington, DC 20059
Ruzin, Steven	Plant Genetics, Inc., Davis, CA 95616
Schlarbaum, Scott	University of Tennessee, Knoxville, TN 37901-1071
Schwarz, Otto	University of Tennessee, Knoxville, TN 37996-1100
Scorza, Ralph	Appalachian Fruit Research Station, Kearneysville, WV 25430
Sederoff, Ronald R.	U.S. Department of Agriculture Forest Service, Berkeley, CA 94701
Seguin, Armand	Université Laval, Quebec, CANADA
Sehgal, Ravinder	Michigan State University, East Lansing, MI 48824-1222
Sikora, Len	Goodyear Tire & Rubber Company, Akron, OH 44305
Simon, Luc	Université Laval, Quebec, CANADA
Singer, Patty	University of Maine, Orono, ME 04469
Sitbon, Folke	Swedish University, 90187 Umeå, SWEDEN
Skilling, Darrol	U.S. Department of Agriculture Forest Service, North Central, St. Paul, MN 55108
Smith, Andrew	Michigan State University, East Lansing, MI 48824
Smith, Harriet	U.S. Department of Agriculture, CRGO, Washington, D.C. 20250
Staba, E. John	University of Minnesota, Minneapolis, MN 55455
Stabel, Priska	Uppsala University, 75121 Uppsala, SWEDEN
Stiff, Carol	University of Idaho, Moscow, ID 83843
Stine, Michael	Michigan State University, East Lansing, MI 48824
Stomp, Anne-Marie	IFG, Placerville, CA 95667
Strauss, Steven	Oregon State University, Corvallis, OR 97331
Stricklen, Mariam	Michigan State University, East Lansing, MI 48824
Tauer, Charles	Oklahoma State University, Stillwater, OK 74078
Teasdale, Bob	Calgene Pacific, Ivanhoe, AUSTRALIA
Tewari, Salil	Ohio State University, Columbus, OH 43210-1085
Thompson, David	Kemira Oy, FINLAND
Thorpe, Trevor A.	University of Calgary, Alberta, CANADA T2N 1N4
Timmis, Roger	Weyerhaeuser Company, Tacoma, WA 98477
Torrey, John	Harvard Forest, Petersham, MA 01366
Tremblay, Francine	Canadian Forestry Service, Chalk River, Ontario, CANADA KOJ 1JO
Trotter, Patrick	Weyerhaeuser Company, Tacoma, WA 98477
Tulecke, Walt	Antioch College, Yellow Springs, OH 45387
Vahala, Tiina	Uppsala University, Uppsala, SWEDEN

Valluri, Jagan	Texas A&M University, College Station, TX 77841
Venketeswaran, S.	University of Houston, Houston, TX 77004
Von Aderkas, Patrick	Canadian Forestry Service, Fredericton, N.B., CANADA E3B3P7
von Arnold, Sara	University of Uppsala, Uppsala, SWEDEN
Wagner, David	University of Kentucky, Lexington, KY 40546-0073
Wann, Steven	Institute of Paper Chemistry, Appleton, WI 54912
Welander, Margareta	Swedish University of Agricultural Science, Alnarp, SWEDEN
Westcott, Roger	Unilever Research, Bedford, ENGLAND
Wilson, Claire M.	Council for Research Planning, Washington, DC 20036-2077
Yamamoto, Naoki	Forestry & Forest Products Research Institute, Ibaraki, 305, JAPAN
Yamamoto, Yoshikazu	Nippon Paint Company, Ltd., Osaka, JAPAN
Zuzek, Kathy	University of Minnesota, Chanhassen, MN 55317

INDEX